Sea cucumbers

A global review of fisheries and trade

FAO
FISHERIES AND
AQUACULTURE
TECHNICAL
PAPER

516

Edited by
Verónica Toral-Granda
FAO Consultant
Puerto Ayora, Santa Cruz
Galapagos Islands, Ecuador

Alessandro Lovatelli
Fishery Resources Officer (Aquaculture)
Aquaculture Management and Conservation Service
FAO Fisheries and Aquaculture Department
Rome, Italy

and

Marcelo Vasconcellos
Fishery Resources Officer
Fisheries Management and Conservation Service
FAO Fisheries and Aquaculture Department
Rome, Italy

FOOD AND AGRICULTURE ORGANIZATION OF THE UNITED NATIONS
Rome, 2008

ISBN 978-92-5-106079-7

Preparation of this document

Prompted by concerns about the status of sea cucumbers stocks worldwide, because of the demand in international markets for bêche-de-mer, different initiatives have been implemented in recent years aimed at improving the understanding of these resources and fisheries, as well as to provide technical guidance for their conservation and sustainable exploitation. Two international meetings were held to review the situation of fisheries and to discuss management measures. The FAO Technical Workshop on Advances in Sea Cucumber Aquaculture and Management (ASCAM) was held in Dalian, People's Republic of China, in 2003. The CITES Technical Workshop on Conservation of Sea Cucumbers in the Families Holothuridae and Stichopodidae was held in Kuala Lumpur, Malaysia, in 2004. Building on the results of these meetings, both FAO Members and CITES Parties concurred on the urgent need to improve capacity of countries to manage sea cucumber fisheries through the provision of scientific information and management tools.

With this in mind, FAO has been implementing a Japanese-funded project on "CITES and commercially-exploited species, including the evaluation of listing proposals" which aims, among other things, to collate and disseminate information on the global status of commercially exploited sea cucumber stocks and to assist fishing nations in the conservation and sustainable exploitation of these benthic marine organisms. The main goal of the project is to develop technical guidelines to assist fisheries managers in deciding regulations and processes for the better management, conservation and sustainable exploitation of their sea cucumber resources. In support of the development of Technical Guidelines, regional reviews and hotspot analyses were commissioned to leading experts in sea cucumber fisheries and used as background documents in an FAO Technical Workshop on "Sustainable use and management of sea cucumber fisheries" held in Puerto Ayora, Galapagos Islands, Ecuador, from 19 to 23 November 2007.

This publication collects all the regional reviews and hotspot analysis prepared for the project and presented at the workshop. Together they provide a comprehensive and up-to-date evaluation of the global status of sea cucumber populations, fisheries, trade and management, constituting an important information source for researchers, managers, policy-makers and regional/international organizations interested on sea cucumber conservation and exploitation.

To facilitate the reading of this document and to accurately distinguish China from China, Hong Kong Special Administrative Region, the comma in the official name of the China, Hong Kong Special Administrative Region has been intentionally omitted.

Abstract

The present document reviews the population status, fishery and trade of sea cucumbers worldwide through the collation and analysis of the available information from five regions, covering known sea cucumber fishing grounds: temperate areas of the Northern Hemisphere; Latin America and the Caribbean; Africa and Indian Ocean; Asia; and Western Central Pacific. In each region a case study of a "hotspot" country or fishery was conducted to highlight critical problems and opportunities for the sustainable management of sea cucumber fisheries. The hotspots are Papua New Guinea, Philippines, Seychelles, Galapagos Islands and the fishery for *Cucumaria frondosa* of Newfoundland in Canada.

Across the five regions, the scale of catches and the number of exploited species varies widely, the Asian and Pacific regions being those with the highest catches and species diversity. Most fisheries are multispecific, or have evolved from single-species to multispecies fisheries as the more valuable species became overexploited. There are many typologies of sea cucumber fisheries, ranging from artisanal, to semi-industrial and industrial. The bulk of the catches are exported to supply the Asian bêche-de-mer market, with China Hong Kong Special Administrative Region (SAR) the main export destination for the totality of countries reviewed. With the exception of some stocks in the temperate areas of the northern hemisphere, sea cucumber stocks are under intense fishing throughout the world. In Latin America and the Caribbean it appears that high valued commercial species have been depleted. In the majority of the countries reviewed in the Africa and Indian Ocean region stocks are overfished. Likewise in the Asian Pacific region the most sought-after species are largely depleted.

Despite the fact that sea cucumber fishing is not a traditional activity, a large number of coastal communities have developed a strong dependency on it as alternative source of income. Reconciling the need for conservation with the socio-economic importance that these fisheries have acquired will require effective management efforts, which are currently lacking in many places. The hotspot case studies show for instance that, despite the adoption of management plans in some countries, the lack of enforcement capacity poses considerable constraints on the effectiveness of adopted management measures, besides exacerbating illegal, unreported and unregulated fishing and trade.

The papers also discuss some of the factors behind the unsustainable use of sea cucumbers and the role and potential benefits of alternative management measures, such as the Convention on International Trade in Endangered Species of Wild Fauna and Flora (CITES). The lack of capacity to gather the basic information needed for management plans, weak enforcement, the high demand from international markets and the pressure exerted from resource-dependent communities figure high as important factors responsible for the critical status of sea cucumber fisheries worldwide. Authors concur on the need for immediate actions to stop the trend of sequential depletion of species if we are to conserve stocks biodiversity and sustain the ecological, social and economic benefits of these resources.

Toral-Granda, V.; Lovatelli, A.; Vasconcellos, M. (eds).
Sea cucumbers. A global review of fisheries and trade.
FAO Fisheries and Aquaculture Technical Paper. No. 516. Rome, FAO. 2008. 317p.

Contents

Preparation of this document iii
Abstract iv
Acknowledgements vii
Contributors viii
Abbreviations and acronyms ix

Executive summary **1**

Population status, fisheries and trade of sea cucumbers in the Western Central Pacific **7**
Jeff Kinch, Steven Purcell, Sven Uthicke and Kim Friedman

Papua New Guinea: a hotspot of sea cucumber fisheries in the Western Central Pacific **57**
Jeff Kinch, Steve Purcell, Sven Uthicke and Kim Friedman

Population status, fisheries and trade of sea cucumbers in Asia **81**
Poh-Sze Choo

The Philippines: a hotspot of sea cucumber fisheries in Asia **119**
Poh-Sze Choo

Population status, fisheries and trade of sea cucumbers in Africa and the Indian Ocean **143**
Chantal Conand

Seychelles: a hotspot of sea cucumber fisheries in Africa and the Indian Ocean **195**
Riaz Aumeeruddy and Chantal Conand

Population status, fisheries and trade of sea cucumbers in Latin America and the Caribbean **213**
Verónica Toral-Granda

Galapagos Islands: a hotspot of sea cucumber fisheries in Latin America and the Caribbean **231**
Verónica Toral-Granda

Population status, fisheries and trade of sea cucumbers in temperate areas of the Northern Hemisphere **257**
Jean-François Hamel and Annie Mercier

Precautionary management of *Cucumaria frondosa* in Newfoundland and Labrador, Canada **293**
Jean-François Hamel and Annie Mercier

ANNEXES

1. Workshop agenda 307
2. List of participants 309
3. Participant profiles 311
4. Experts group photograph 317

Acknowledgements

Numerous individuals contributed to the successful organization and implementation of the sea cucumber workshop in the Galapagos Islands, which resulted in the preparation of draft technical guidelines for managing sea cucumber fisheries and the compilation of the present global review. All of them are thanked for their efforts and contributions during the preparatory phases and at the workshop itself.

Special thanks are due to the Charles Darwin Foundation, Galapagos, for hosting the workshop and for providing excellent logistical support. Much appreciation goes to Verónica Toral-Granda and Tom Poulsom for their hospitality. The preparation of the workshop programme, identification of the various experts and scientific and editorial support throughout this activity was possible thanks to the immense work of all members of the Scientific Committee established almost one year before the workshop took place. The dedication of Chantal Conand, Steven Purcell, Sven Uthicke, Jean-François Hamel, Annie Mercier and Verónica Toral-Granda were invaluable to the success of the workshop and to the quality of the present document.

The organization of the workshop and the preparation of this document were possible thanks to funds provided to FAO by the government of Japan through the Trust Fund Project on "CITES and commercially-exploited aquatic species, including the evaluation of listing proposals". The workshop organizers also wish to thank the institutions that have permitted their experts to prepare for and attend the workshop. The Secretariat of the Pacific Community (SPC) is thanked for supporting the participation of Kim Friedman, and the University of Nagoya, Japan, for that of Jun Akamine.

Finally, the opportunity is taken to thank all the FAO staff members in Rome and Quito (Ecuador), who have contributed in one way or another in the organization of the workshop. The maps of the regional reviews and hotspot areas were prepared by Fabio Carocci. The layout creation was by José Luis Castilla Civit.

Contributors

Jun AKAMINE
School of Humanities and Social Sciences
Nagoya City University
Nagoya City, Aichi, Japan

Poh Sze CHOO
WorldFish Center
Penang, Malaysia

Chantal CONAND
Laboratoire ecologie marine
Université de la Réunion
Saint-Denis, France

Eduardo ESPINOZA
Galapagos National Park Service
Puerto Ayora, Santa Cruz
Galapagos, Ecuador

Kim FRIEDMAN
Secretariat of the Pacific Community
Nouméa, New Caledonia

Ruth GAMBOA
Department of Biology
University of the Philippines Mindanao
Davao City, Philippines

Jean-François HAMEL
Society for the Exploration and Valuing of the Environment (SEVE)
Newfoundland, Canada

Alex HEARN
Charles Darwin Foundation
Puerto Ayora, Santa Cruz
Galapagos Ecuador

María Dinorah HERRERO-PÉREZRUL
La Paz, Baja California Sur
Mexico

Jeff KINCH
Coastal Management Advisor
Secretariat for the Pacific Regional Environment Program
Apia, Samoa

Priscilla C. MARTÍNEZ
World Wildlife Fund
Puerto Ayora, Santa Cruz
Galapagos, Ecuador

Annie MERCIER
Ocean Sciences Centre
Memorial University
St. John's Newfoundland & Labrador
Canada

Steven PURCELL
WorldFish Center
Nouméa, New Caledonia

Verónica TORAL-GRANDA
Puerto Ayora, Santa Cruz
Galapagos, Ecuador

Sven UTHICKE
Australian Institute of Marine Science
Queensland, Australia

Matthias WOLFF
Charles Darwin Foundation
Puerto Ayora, Santa Cruz
Galapagos, Ecuador

Abbreviations and acronyms

ACIAR	Australian Centre for International Agricultural Research
AFLP	amplified fragment length polymorphism
ASCAM	Advances in Sea Cucumber Aquaculture and Management
BACI	before–after, control–impact
BFAR	Bureau of Fisheries and Aquatic Resources
CAFID	Canada/Newfoundland Cooperation Agreement for Fishing Industry Development
CCC	Coral Cay Conservation
CCFI	Canadian Center for Fisheries Innovation
CDF	Charles Darwin Foundation
CFMDP	Coastal Fisheries Management and Development Programme
CICIMAR	Centro Interdisciplinario de Ciencias Marinas (Mexico)
CITES	Convention on the International Trade in Endangered Species of Wild Fauna and Flora
CN-MAD	National Committee in Madagascar
CNMI	Commonwealth of the Northern Mariana Islands
CPUE	catch per unit effort
CRED	Coral Reef Ecosystem Division
CSIRO	Australia Commonwealth Scientific and Research Organization
CUD	Belgian University Corporation for Development
DA-BFAR	Department of Agriculture-Bureau of Fisheries and Aquatic Resources
DFA	Department of Fisheries and Aquaculture (Canada)
DFMR	Department of Fisheries and Marine Resources
DFO	Department of Fisheries and Oceans (Canada)
DOST	Department of Science and Technology
DPRK	Democratic Peoples' Republic Korea
EEZ	exclusive economic zone
EIO	Eastern Indian Ocean
EU	European Union
FAO	Food and Agriculture Organization of the United Nations
FFAW	Fish, Food and Allied Workers
FSM	Federated States of Micronesia
FUNZEL	Fundación Zoológica de El Salvador (Honduras)
GATT	General Agreement on Tariffs and Trade
GBR	Great Barrier Reef
GEM-USAID	Growth with Equity in Mindanao
GMR	Galapagos Marine Reserve
GNPS	Galapagos National Park Service
ICAR	Indian Council of Agricultural Research
ICNAF	International Convention for the Northwest Atlantic Fisheries
IMA	Inter-Institutional Management Authority
INVEMAR	Instituto Investigaciones Marinas y Costeras (Colombia)
IPN	Instituto Politécnico Nacional (Mexico)
IQF	individually quick frozen
ITQ	individual transferable quotas
IUU	illegal, unregulated and unreported

IWP	International Waters Project
JPA	Joint Project Agreement
KNA	Kenya National Archives
LRFFT	Live Reef Fish Food Trade
MCS	monitoring, control and surveillance
MI	Marine Institute
MOA	Memorandum of Agreement
MOF	Ministry of Fisheries
MOU	Memorandum of Understanding
MPAs	marine protected areas
MSI-UP	Marine Science Institute, University of the Philippines
MSY	maximum sustainable yield
NAFO	Northwest Atlantic Fisheries Organization
NB	New Brunswick
NFA	National Fisheries Authority
NFC	National Fisheries Corporation
NFRDI	National Fisheries Research and Development Institute (Philippines)
NGO	Non-governmental Organization
NIPAS	National Integrated Protected Areas System
NL	Newfoundland and Labrador
NOAA	National Oceanic and Atmospheric Administration (USA)
NS	Nova Scotia (Canada)
NTZ	No-Take Zone
ONETH	The National Association of Sea Cucumber Producers
OSC	Ocean Sciences Centre (Memorial University, Canada)
PAMB	Protected Area Management Board
PAMS	Participatory and Adaptive Management Scheme
PCAMRD	Philippine Council for Aquatic and Marine Research and Development
PICT	Pacific Island Countries and Territories
PMB	Participatory Management Board
PNG	Papua New Guinea
PROCFish/C	Pacific Regional Oceanic and Coastal Development Project Coastal Component
RDA	Resource Development Associates
RZS	Rotational Zoning Scheme
SAR	Special Administrative Region
SCUBA	Self Contained Underwater Breathing Apparatus
SEAFDEC	Aquaculture Department of the Southeast Asia Fisheries Development Center
SEVE	Society for the Exploration and Valuing of the Environment
SFA	Seychelles Fishing Authority
SFAC	Sea-area Fishery Adjustment Commission
SFCA	Hokkaido Semposhi Fisheries Cooperative Association
SITC	Standard International Trade Classification
SOM	Size of Maturity
SPC	Secretariat of the Pacific Community (formerly the South Pacific Commission)
TAC	total allowable catch
TINRO	Pacific Fishery and Oceanography Research Institute (Russian Federation)
TL	total length

TNC	The Nature Conservancy
TOP	Technical Operational Procedure
UAE	United Arab Emirates
UICN	The World Conservation Union
UMAS	Unidad de Manejo para la Vida Silvestre
UPCH	Universidad Peruana Cayetano Heredia
USD	United States dollar
VMS	vessel monitoring system
WIO	Western Indian Ocean
WIOMSA	Western Indian Ocean Marine Science Association
WMCIP	Western Mindanao Community Initiatives Project
WTO	World Trade Organization
WWF	World Wildlife Fund

Executive summary

Sea cucumbers (Echinodermata: Holothuroidea), or their dried form (bêche-de-mer), have been a dietary delicacy and medicinal cure for Asians over many centuries. The collection of sea cucumbers to supply the market has seen a depletion of this resource in the traditional fishing grounds close to Asia and more recently the expansion of this activity to new and more distant fishing grounds. Currently, there are fisheries harvesting sea cucumbers across most of the resource range, including remote parts of the Pacific, the Galapagos Islands, Chile and the Russian Federation. This global review shows that sea cucumber stocks are under intense fishing pressure in many parts of the world and require effective conservation measures. It also shows that sea cucumbers provide an important contribution to economies and livelihoods of coastal communities, being the most economically important fishery and non-finfish export in many countries. Reconciling the need for conservation with the socio-economic importance of sea cucumber fisheries is shown to be a challenging endeavour, particularly for the countries with limited management capacity. Furthermore, no single management measure will work optimally due to the many idiosyncrasies of these fisheries, which are outlined in this document through a comprehensive review of their biological and human dimensions.

The present document reviews the population status, fishery, trade, management and socio-economic importance of sea cucumbers worldwide. It includes regional reviews and hotspot case studies prepared by leading experts on sea cucumber fisheries and their management. These documents were made available prior to the FAO Workshop on Sustainable Use and Management of Sea Cucumber Fisheries, held in Puerto Ayora, Galapagos Islands, Ecuador, from 19 to 23 November 2007 (workshop agenda, list of participants and their profiles are appended).

Reviews were prepared for five regions: temperate areas of the Northern Hemisphere (including Canada, Iceland, Russian Federation and the United States of America); Latin America and the Caribbean; Africa and Indian Ocean; Asia; and Western Central Pacific (including Australia). In each region, specialists conducted a case study of a "hotspot" country or fishery to highlight topical or critical problems and opportunities for the sustainable management of sea cucumber fisheries. The five hotspots are: Papua New Guinea (Western Central Pacific); Philippines (Asia); Seychelles (Africa and Indian Ocean), Galapagos Islands (Latin America and the Caribbean); and the *Cucumaria frondosa* fishery of Newfoundland in Canada (Temperate areas of Northern Hemisphere).

A multitude of sea cucumber species are being exploited worldwide, with new species being placed on the market whilst valuable species become scarcer and more difficult to find. Across the five regions, the number of commercially exploited species varies widely, with the highest number of species exploited in the Asia (52 species) and Pacific (36 species) regions partially due to the higher natural diversity in these areas. Still, little is known of the ecology, biology and population status of most commercial species, and in many cases, species are being commercialised without a clear taxonomic identification (e.g. the "pentard" in the Seychelles, *Actinopyga* sp. in Yap). Information on catches is also scarce, as these fisheries operate over large scales in often remote locations. In view of the importance of international trade, export and import statistics of bêche-de-mer are in many cases the only information available to quantify the magnitude of fisheries catches. Based on the most recent available catch and trade data, Asia and the Pacific are the top producing regions despite the long history of exploitation. Depending on the conversion factor used for the dry:wet weight of sea cucumbers, it is possible to infer that the combined catches for the Asia and Pacific regions are in the order of 20 000 to 40 000 tonnes/year. The temperate areas of the Northern Hemisphere are also responsible for a substantial share of the world catches (in the order of 9 000 tonnes/year); catches being sustained almost exclusively by one species (*Cucumaria frondosa*). Sea cucumber catches are relatively less important in Africa and in the Indian Ocean (2 000–2 500 tonnes/year) region and, particularly low in Latin America and the Caribbean region (<1 000/year).

Sea cucumber capture fisheries generally target a large variety of species, which adds complexity to management and trade reporting; and those that commenced as single-species fisheries have now progressed to include "new" species in their catches (i.e. fisheries in Peru and Chile). Four of the five hotspots reviewed in this document present multi-species fisheries, and in all cases they have moved from low quantity-high value to high quantity-low value ventures, as the more valuable species become fully-exploited or overexploited. Some regions have seen a dramatic increase in the number of species under commercial exploitation (e.g. Galapagos Islands, Philippines, Papua New Guinea, Solomon Islands), yielding more species to fishery pressure, as well as masking overexploitation and species-specific decreases in catches.

The majority of sea cucumbers are exported for the bêche-de-mer market and few species for the live trade (aquarium) market, which is currently underreported. There is also an emerging market for the use of sea cucumbers in the pharmaceutical, nutriceutical and cosmetic industries. The type of fishery varies by region and the species under exploitation. Examples of different fishery types, ranging from artisanal (Papua New Guinea and Philippines), to semi-industrial (Galapagos Islands) and industrial (*Cucumaria frondosa* fishery in northern Canada), are described in the hotspot case studies. Hand collecting, gleaning, lead bombs, SCUBA diving, hookah and dredging are examples of fishing methods used.

Effective management plans for sea cucumber fisheries are uncommon. For example, the Seychelles and Papua New Guinea have adopted management plans for their fisheries, which came into place following concerns about declines in catches caused by unregulated harvesting. The Galapagos Islands sea cucumber fishery also started as an open-access activity, but after a long ban it re-opened in 1999 with an adaptive and participatory management scheme. The Philippines do not have a management plan in place despite some species now becoming locally endangered through overfishing. The *C. frondosa* fishery in Newfoundland (Canada) is still under an exploratory stage but aims at being managed through ecosystem-based guidelines. This fishery, unlike other sea cucumber fisheries, has the advantage of using a precautionary approach at the start of commercialisation, and has some of the most complete set of biological and population information on which to base a management plan. In spite of the lack of management plans in most locations described in the regional reviews, some management measures have been adopted to regulate fishing pressure, including closed seasons, minimum sizes, total allowable catches, gear restrictions, spatial and temporal closures and the establishment of marine protected areas. However, the lack of enforcement capacity has posed considerable constraints on the effectiveness of such management measures. The lack of enforcement and compliance is a common denominator for the majority of the fisheries reviewed in this document, and has exacerbated illegal, unreported and unregulated fishing and trade.

Sea cucumber populations are in dire straits in many parts of the world. This unfortunately includes high profile conservation sites such as the Great Barrier Reef Marine Park and the Galapagos Islands National Park (both World Heritage listed). These locations have well documented cases of sea cucumber population collapse, and subsequent population surveys showed no recovery of overfished stocks. The temperate areas of the Northern Hemisphere offer perhaps the few exceptional cases of abundant stocks still moderately exploited and with some potential for expansion.

Despite the limited information available on sea cucumber fisheries in Latin America and the Caribbean, it appears that high valued commercial species have been depleted and the risk of fishery collapses is high due to the small size of stocks, the strength of market forces and the unregulated nature of these fisheries. In Africa and the Indian Ocean region, at least 12 out of 17 countries, where sea cucumber fisheries have been documented, indicate evidence of overfishing of sea cucumber stocks. Species are under heavy fishing pressure throughout the Asian Pacific region, whilst the most sought-after species in the western Pacific are largely depleted.

Recognizing the importance of international trade as a threat to the conservation of sea cucumber species, consideration has been given to the possible role of the Convention on International Trade in Endangered Species of Wild Fauna and Flora (CITES) as a complementary measure for regulating the sustainable use of sea cucumber fisheries. One species is currently listed in a CITES Appendix (*Isostichopus fuscus* in Appendix III by Ecuador) and the Galapagos Islands hotspot case study (see report in this document) lists advantages and disadvantages based on this experience. The analysis of the situation in the Seychelles, Papua New Guinea and the Philippines asserts the possible benefits of CITES to sea cucumber populations, but indicate that a listing could lead to socio-economic impacts as well as an increased administrative burden where institutional capacity is limited.

In many countries, particularly in the Western Pacific region, some sea cucumbers, and their organs, are considered as delicacies and a protein component to traditional diets. However, in the majority of countries reviewed in this document, sea cucumbers are harvested to supply the Oriental market of bêche-de-mer. Indonesia is the major exporter of sea cucumber from capture fisheries. Of all importing countries and territories, China Hong Kong Special Administration Region (SAR), is the most important, with products arriving from most countries worldwide; whilst some countries, such as the United Arab Emirates in the Indian Ocean, have become "intermediate" markets.

Although in most current fishing grounds, sea cucumber fishing is not a traditional activity the majority of rural coastal communities are dependent on it, as alternative opportunities for income are often limited. High prices and the increasing consumer demand have seen the expansion of the range covered by marine products agents, the development of capacity in these fisheries and an ongoing search for new species. In many regions, the socio-economic dependency on bêche-de-mer is so vast that fishers continue collecting sea cucumbers despite scant catches, further affecting the stocks capacity to reproduce and repopulate the fishing grounds. Generally, when one commercial species is depleted, or "economically extinct", traders will encourage fishers to search for new species, or fish deeper or further a field, in order to continue their business.

Aquaculture, sea ranching and restocking have been evaluated as possible solutions to wild sea cucumber overexploitation, and some countries have started such ventures (e.g. Australia, China, Kiribati, Philippines, Viet Nam and Madagascar). Restocking has been considered an expensive remedy to overfishing. Presently, China is successfully producing an estimated 10 000 tonnes, dry weight, of *Apostichopus japonicus* from aquaculture, mainly to supply local demand. This value, when converted into wet weight, is in the same order of magnitude of the total world wild catches. A feasibility study is presently being carried out in Chile to evaluate the possible introduction of *A. japonicus,* as an alternative to capture fisheries of two wild species of sea cucumbers. In the Asia Pacific region aquaculture is still in the early development stages, with one species of sea cucumber (*Holothuria scabra*) in trials to ascertain the commercial viability of culture and farming options.

Many additional threats have been identified for sea cucumber populations worldwide, including global warming, habitat destruction, unsustainable fishing practices (e.g. blasting), the development of fisheries with little or no information on the species, and lack of natural recovery after overexploitation. Illegal, Unregulated and Unreported (IUU) fisheries are widespread in all regions, representing an indirect threat as it fuels unsustainable practices and socio-economic demand.

The critical status of sea cucumber fisheries worldwide is compounded by different factors including i) the lack of financial and technical capacity to gather basic scientific information to support management plans, ii) weak surveillance and enforcement capacity, and iii) lack of political will and socio-economic pressure exerted by the communities that rely on this fishery as an important source of income. The fast pace of development of sea cucumber fisheries to supply the growing international demand for bêche-de-mer is placing most fisheries and many sea cucumber species at risk. The pervasive trend of overfishing, and mounting examples of local economic extinctions, urges immediate action for conserving stocks biodiversity and ecosystem functioning and resilience from other stressors than overfishing (e.g. global warming and ocean acidification), and therefore sustaining the ecological, social and economic benefits of these natural resources.

Population status, fisheries and trade of sea cucumbers in the Western Central Pacific

Jeff Kinch, Steve Purcell, Sven Uthicke and Kim Friedman

Papua New Guinea: a hotspot of sea cucumber fisheries in the Western Central Pacific

Jeff Kinch, Steve Purcell, Sven Uthicke and Kim Friedman

Population status, fisheries and trade of sea cucumbers in the Western Central Pacific

Jeff Kinch
Secretariat for the Pacific Regional Environment Program
Apia, Samoa
E-mail: jeffk@sprep.org

Steven Purcell
The WorldFish Center
Noumea Cedex, New Caledonia
E-mail: SteveP@spc.int

Sven Uthicke
Australian Institute of Marine Science
Queensland, Australia
E-mail: s.uthicke@aims.gov.au

Kim Friedman
Secretariat of the Pacific Community
Noumea Cedex, New Caledonia
E-mail: KimF@spc.int

Kinch, J.; Purcell, S.; Uthicke, S.; Friedman, K. 2008. Population status, fisheries and trade of sea cucumbers in the Western Central Pacific. In V. Toral-Granda, A. Lovatelli and M. Vasconcellos. Sea cucumbers. A global review of fisheries and trade. *FAO Fisheries and Aquaculture Technical Paper*. No. 516. Rome, FAO. pp. 7–55.

SUMMARY

In the Western Central Pacific region, most sea cucumber fisheries have exhibited boom-and-bust cycles since the late nineteenth century. Since the 1980s, elevated export prices and demand from Asian markets have been the catalysts for increased fishing. At many localities, high-value species have been depleted and previously unfished species are now exploited. The sustainability of these fisheries is of widespread concern.

Australia and Melanesian countries are the largest exporters of bêche-de-mer in the region. While annual exports from Melanesian countries have not declined markedly over the last two decades, those from Polynesia and Micronesia have. The declining exports appear to be attributed to unsustainable fishing pressure and naturally low abundances of many commercial species in remote Pacific islands and atolls.

Currently, 35 sea cucumber species in the families Holothuriidae and Stichopodidae are thought to be harvested. Greater endemicity occurs in Melanesian countries with sea cucumber species richness generally declining eastward of Papua New Guinea (PNG). On average, 13 species are harvested per country.

The vast majority of sea cucumbers are exported as dried bêche-de-mer; relatively small amounts are exported frozen or salted. A few species are exported as ornamentals and this component of trade is commonly under-reported. Many reports showed that some form part of subsistence diets, particularly for Polynesians. In some of these cases, just the gonads and/or intestine are taken and the animal is released to regenerate these organs for re-harvesting.

Especially in the Pacific Island Countries and Territories (PICTs), sea cucumbers are collected by hand in coral reefs and shallow lagoons. The exploitation often involves a high number of artisanal fishers, accessing stocks from shore or using small boats. Values of catch-per-unit-effort varied greatly among the published studies, and generally declined over time. Rural poverty in Papua New Guinea is causing some fishers to continue to collect sea cucumbers even when returns fall below 1 specimen per 10 hours of diving.

The multispecies nature of these fisheries adds difficulty for management and trade reporting. Export data are sometimes inaccurate, amalgamated across species groups, or missing, which adds to the difficulty of monitoring catches. Comparisons of past and recent trade data show an alarming trend of increasing proportions of low-value species in exports and a greater range of species in exports. This is particularly evident in Papua New Guinea and the Solomon Islands, where biodiversity is high.

The authors compare data from past and recent field surveys, and present a case study of *Holothuria whitmaei* densities among fished and unfished locations. Populations of most higher-priced species in the Western Central Pacific are, apparently, grossly depleted compared to virgin densities.

For some coastal villages, sea cucumber fishing is the primary source of income to residents. Financial benefits are generally distributed widely, at the village level, although processing by exporting companies is an increasing trend. In most fisheries, the depletion of sea cucumber stocks is already impacting the potential incomes of coastal and island communities and national revenues. In some cases, overfishing is affecting the sustainability of these fisheries for the long term.

The development of sustainable management in the Western Central Pacific region has been difficult. Management tools like size limits, gear restrictions, spatial and temporal closures, quotas and marine reserves have not curbed overfishing. Much of their ineffectiveness can be attributed to a lack the necessary funds and technical capacity for adequate awarness raising in most PICTs. Commonly, there are also conflicts of interest within differing levels and agencies of government, politicians and influential business people. Fishing moratoria have been declared in some countries, including Solomon Islands, Fiji, Tonga and Vanuatu. Although breeding populations at some localities have recovered, empirical studies show that populations for other species have failed to rebuild after years of respite. International support is needed to evaluate CITES listing for the conservation of rare and threatened species.

Restocking using hatchery-produced juveniles is technically feasible, but will be an expensive remedy to overfishing. International translocation of stocks for restocking or sea ranching is discouraged. Recent research has focussed on underwater population surveys, to assess population densities, and socio-economic surveys. In particular, the SPC PROCFish/C programme has trained fisheries officers in these survey methods and is providing comparative analyses of stock status in PICTs. Effort must now turn to aiding PICTs to develop practical management frameworks that allow breeding populations to recover to productive levels with a limited institutional capacity for compliance and enforcement of regulations.

CONVENTIONS

To be consistent with the terminology for this fishery, "holothurians" or "sea cucumbers" are used throughout this report when referring to live animals and "bêche-de-mer" is used when referring to the dead animal when processed for commercial purposes.

Where older taxonomic classifications have been used in referenced texts or in information provided by colleagues, these have been changed to their new taxonomic determinations. For example, in this report, the authors adhere to the results of a recent morphological and genetic study (Uthicke *et al.*, 2004) that suggests that all black teatfish in the area covered in this review are *H. whitmaei*; with the presumption that *H. nobilis* does not exist in the Western Central Pacific region. All white teatfish are referred to as *H. fuscogilva*, though there is a possibility that this species may also have its taxonomic designation revised in the future.

Similarly, where imperial or colonial names have been used for countries, districts or islands in the past, their modern names, post-independence or associations have now been used instead.

1. THE WESTERN CENTRAL PACIFIC REGION

The Western Central Pacific region (for purposes of this report) is an area that encompasses the 22 Pacific Island Countries and Territories (PICTs) that are affiliated to the Secretariat of the Pacific Community (SPC) (Figure 1), Australia and New Zealand; but excludes, Hawaii (United States of America) and Easter Island (Chile).

Information on sea cucumber fisheries is available for 23 of the 24 countries and territories under this review.

The Western Central Pacific region can be broken into the following geographical and cultural areas:

- Micronesia: Palau, Guam, the Commonwealth of the Northern Mariana Islands (CNMI), the Federated States of Micronesia (FSM), the Marshall Islands, Nauru and Kiribati;
- Polynesia: Tuvalu, Wallis and Futuna, Tokelau, Samoa and American Samoa, Tonga, Pitcairn Islands, Niue, the Cook Islands, and French Polynesia;

FIGURE 1

Region under study and countries covered in this review (highlighted in yellow). The dashed line denotes the boundary of the SPC Pacific Island Countries and Territories. The region encompasses mainly the FAO Statistical Areas 71 (Western Central Pacific), 77 (Eastern Central Pacific) and 81 (Pacific Southwest)

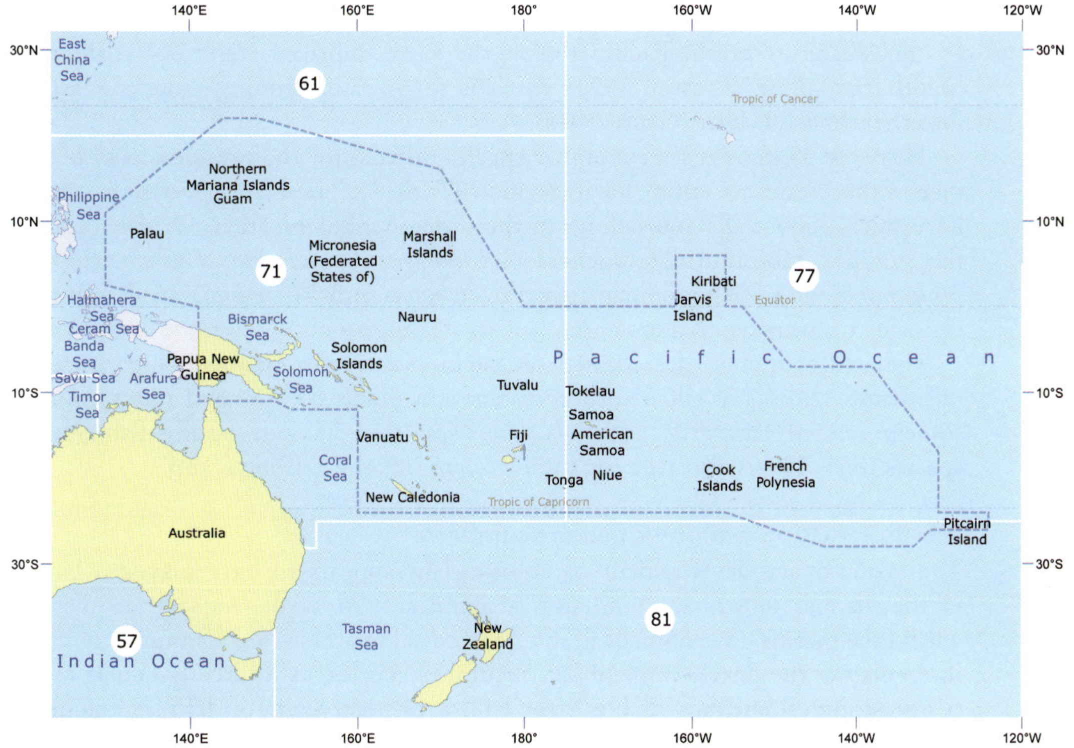

FIGURE 2
Varied habitats of the tropical Western Central Pacific region. Left: *H. scabra* in a muddy inshore seagrass bed. Right: *A. mauritiana*, in complex habitat on the reef crest

COURTESY OF S. PURCELL

- Melanesia: Papua New Guinea (PNG), the Solomon Islands, Vanuatu, New Caledonia and Fiji; and
- Australia and New Zealand (the latter also culturally and geographically part of Polynesia).

Most of the sea cucumber fisheries in this region exist in shallow tropical waters. Whereas the waters in Southeast Asia and the Indian Ocean are nutrient rich, many localities in the Western Central Pacific region are isolated and nutrient poor, and do not support a high biomass of sea cucumbers. However, the warm tropical waters of the Western Central Pacific region have afforded a high level of speciation. Habitats for sea cucumbers in the Western Central Pacific Region are predominantly coastal seagrass beds near mangroves and the soft and hard substrata of coral reefs (Figure 2). It is both the warm temperatures and shallow habitats that permit easy harvesting by coastal fishers and make these fisheries vulnerable to degradation from impacts associated with global climate change.

2. BIOLOGICAL AND POPULATION STATUS

2.1 Key taxonomic groups

In the Western Central Pacific region there are approximately 300 shallow-water holothurian species (Preston, 1993), with species diversity tending to decline eastwardly into the tropical Pacific (Clark, 1946).

There are 35 known sea cucumber species utilized for the production of bêche-de-mer in the Western Central Pacific region (Table 1). These are species that generally have thick body walls and belong to the order Aspidochirotida. Of this order, only the families Holothuriidae (genera *Actinopyga, Bohadschia, Pearsonothuria* and *Holothuria*) and Stichopodidae (genera *Stichopus* and *Thelenota*) are harvested and exported to overseas markets.

The actual number of exploited sea cucumber species maybe as high as 57 when misidentifications are taken into consideration (Table 1). Conand (1998) notes that some processed *Actinopyga* and *Bohadschia* spp. are not commercially distinguished to species level. This is also true for some *Holothuria* and *Stichopus* spp.

2.2 Biology and ecology of sea cucumbers

The evolutionary development of holothurians appears to have played a large role in the manner in which population assemblages (or taxocoenoses) are distributed within the marine environment (Table 2). Coastal processes are also important factors in regulating the distribution of sea cucumber species, as hydro-dynamics influence sediment granulometry, which is a key habitat characteristic for defining the niches of holothurians, and larval dispersion (Massin and Doumen, 1986).

TABLE 1
Holothurians used for the production of bêche-de-mer in the Western Central Pacific region. The table includes sea cucumber species known to be utilized (shaded) and other possible species that are misidentified once processed

Holothuriidae		Stichopodidae
Actinopyga albonigra	*Holothuria arenicola*	*Stichopus chloronotus*
A. caerulea	*H. atra*	*S. godeffroyi*
A. echinites	*H. cinerascens*	*S. herrmanni*
A. lecanora	*H. coluber*	*S. horrens**
A. mauritiana	*H. difficilis*	*S. monotuberculatus**
A. miliaris	*H. edulis*	*S. naso**
A. palauensis	*H. flavomaculata*	*S. pseudohorrens*
A. spinea	*H. fuscocinerea*	*S. vastus*
Bohadschia anaes	*H. fuscogilva*	*S. ocellatus*
B. argus	*H. fuscopunctata*	*Australostichopus mollis*
B. bivitatta	*H. grises*	*Thelenota ananas*
B. geoffreyi	*H. guamensis*	*T. anax*
B. maculisparsa	*H. hilla*	*T. rubralineata*
B. marmorata	*H. impatiens*	
B. similis	*H. leucospilota*	
B. subrubra	*H. maculata*	
B. tenuissima	*H. pardalis*	
B. vitiensis	*H. pervicax*	
	H. scabra	
	H. scabra var. *versicolor**	
	H. verrucosa	
	H. whitmaei	
	Holothuria sp. (Hongpai – Solomon Islands)	
	Holothuria sp. (Tulele – Solomon Islands)	
	P. graeffei	

* taxonomic review in progress.

Tropical coral reefs offer a broad range of habitats with high floral and faunal diversity, often resulting in highly speciose sea cucumber populations. This high species richness in the Western Central Pacific region is probably attributed largely to these diverse micro-habitats.

Many sea cucumber species in the Western Central Pacific region have been reported to exhibit episodic spawning behaviour throughout the year, with a period of enhanced activity from October–February (Conand, 1981; Lokani, 1990; Ramofafia, Gervis and Bell, 1995; Ramofafia, Battaglene and Bryne, 2001; Ramofafia, Byrne and Battaglene, 2001, 2003; Battaglene and Bell, 2004; Kinch, 2004a) (Table 3).

The mechanisms and triggers for the settlement of holothurians are still somewhat unknown because the larvae are, as yet, difficult to identify by species and the juveniles are generally cryptic. Conand (1993) and Uthicke (1994) suggest settlement of *S. herrmanni* and *S. chloronotus* in reef flat zones and subsequent migration towards other areas. *H. scabra* was demonstrated to settle in shallow seagrass beds (Mercier, Battaglene and Hamel, 2000a) and some eventually happen to make it to deeper waters (Mercier, Battaglene and Hamel, 2000b; Hamel *et al.*, 2001). Recruitment in sea cucumber populations is thought to be highly irregular and variable.

Several common holothurians in the Western Central Pacific region also reproduce asexually through transverse fission, whereby the body is split into an anterior and posterior section (Conand, 1996; Uthicke, 1997; Purwati, 2001; Purwati and Thinh Luong Van, 2003), and this may be linked to anthropogenic or ecological disturbances in some species and areas (Doty, 1977; Ebert, 1978; Harriott, 1982; Conand, 1989, 1996; Uthicke, 1997). Fission in the Western Central Pacific region has been observed in

TABLE 2
Characteristics of selected commercial holothurian species in the Western Central Pacific region

Species	Average length (cm)	Average wet weight (kg)	Body wall thickness (cm)	Habitat preference	Depth range (m)
Actinopyga echinites	20	0.3	0.7	Reef flats of fringing and lagoon-islet reefs, rubble reefs and compact flats	0–12
A. lecanora	25	0.4	0.6	Hard substrates (nocturnal)	0–20
A. mauritiana	20	0.3	0.6	Outer reef flats and fringing reefs, mostly in the surf zone	0–20
A. miliaris	25	0.4	0.6	Reef flats of fringing and lagoon-islet reefs, never found on barrier reefs	0–10
Bohadschia argus	36	1.8	1.0	Barrier reef flats and slopes, or outer lagoons on white sand	0–30
B. similis	18	0.3	0.4	Coastal lagoons and inner reef flats, often burrowed in sandy-muddy bottoms	0–3
B. viticnsis	32	1.2	0.7	Coastal lagoons and inner reef flats, often burrowed in sandy-muddy bottoms	0–20
Holothuria atra	20	0.2	0.4	Inner and outer reef flats and back reefs or shallow coastal lagoons	0–20
H. coluber	40	0.3	0.4	Inner and outer reef flats and back reefs or shallow coastal lagoons	0–15
H. edulis	20	0.2	0.3	Inner reef flats of fringing and lagoon-islets reefs, and shallow coastal lagoons	0–30
H. fuscogilva	42	2.4	1.2	Outer barrier reefs and passes, also on shallow seagrass beds	0–40
H. fuscopunctata	36	1.5	1.0	Reef slopes and shallow seagrass beds	0–25
H. leucospilota	20	0.2	0.2	Rubble, consolidated rubble and boulder	0–5
H. scabra	22	0.3	0.6	Inner reef flats of fringing reefs, lagoon-islets	0–15
H. whitmaei	37	1.7	1.2	Reef flats, slopes and shallow seagrass beds	0–20
Pearsonothuria graeffei	35	0.7	0.4	Reef slopes, close to the coast	0–25
Stichopus chloronotus	18	0.1	0.2	Reef flats and upper slopes, mostly on hard substrates	0–15
S. herrmanni	35	1.0	0.8	Seagrass beds, rubble and sandy-muddy bottoms	0–25
S. horrens	20	0.2	0.2	Reef flats and upper slopes, mostly on hard substrates	0–15
Thelenota ananas	45	2.5	1.5	Reef slopes and near passes, hard bottoms with large rubble and coral patches	0–25
T. anax	55	3.5	1.5	Reef slopes, outer lagoon and near passes, large rubble and sand patches	10–30

Source: SPC, 2003.

H. atra (Doty, 1977; Harriott, 1982; Conand, 1993, 1996; Seeto, 1994; Uthicke, 1997), *H. edulis* (Harriott, 1985; Uthicke, 1997), *S. chloronotus* (Franklin, 1980; Uthicke, 1997), *H. coluber* (Conand, Morel and Mussard, 1997) and *T. ananas* (Reichenbach, Nishar and Saeed, 1996).

Information on growth rates of holothurians has been difficult to ascertain because conventional methods to measure growth of marine organisms are difficult to apply. Methods used to date have included marking the calcareous (epipharyngeal) rings (Ebert, 1978), chemical marking of spicules (Purcell and Simutoga, 2008), external tagging (Shelley, 1981; Conand, 1989), internal tagging (Lokani, 1992), by following the mean weight of a population over time (Chao, Chen and Alexander, 1994), and Modal Progression Analysis (Franklin, 1980; Shelley, 1985; Conand, 1988; Uthicke, 1994). From these studies, growth rates of holothurians in the Western Central Pacific region have been determined to range between 3 and 30 g mo^{-1}. Genetic tagging and recapture studies confirmed growth rates in that range for *H. whitmaei*, but also indicated that larger individuals can shrink over time (Uthicke and Benzie, 2002; Uthicke, Welch and Benzie, 2004).

Mortality rates for sea cucumbers in the Western Central Pacific region have been estimated for *S. chloronotus* with a life span of about five years, whilst *T. ananas*,

TABLE 3
Peak spawning periods (shaded) for selected commercial holothurian species in the Western Central Pacific region

Species	Location	J	F	M	A	M	J	J	A	S	O	N	D	Reference
Actinopyga echinites	New Caledonia													Conand, 1988, 1989, 1993
	PNG													Shelley, 1981
A. mauritiana	Guam													Richmond, 1996a
	New Caledonia													Conand, 1989
	Solomon Islands													Ramofafia, Byrne and Battaglene, 2001
A. miliaris	New Caledonia													Conand, 1988, 1989, 1993
Bohadcshia vitiensis	PNG													Shelley, 1981
	Palau													Hendler and Meyer, 1982
B. argus	Australia (GBR)													Uthicke, 1994
	New Caledonia													Conand, 1988, 1989, 1993
Holothuria atra	Australia (GBR)													Harriott, 1980, 1982, 1985
	Fiji													Seeto, 1994
H. edulis	Australia (GBR)													Harriott, 1985
H. fuscogilva	New Caledonia													Conand, 1981, 1988, 1989, 1993
	Solomon Islands													Ramofafia *et al.*, 2000
H. fuscopunctata	Australia (GBR)													Uthicke, 1994
	New Caledonia													Conand, 1988, 1989, 1993a
H. leucospilota	Australia (GBR)													Franklin, 1980
	Australia (NT)													Purwati, 2001
	Cook Islands													Drumm, 2004
	Cook Islands													McCormack, 1984
H. scabra	Australia (GBR)													Harriott, 1980
	Australia (GBR)													Morgan, 2000
	Australia (NT)													DEH, 2004
	New Caledonia													Conand, 1988, 1989, 1993
	PNG													Shelley, 1981
	Solomon Islands													Ramofafia, Byrne and Battaglene, 2003
H. whitmaei	Australia (GBR)													Shiell and Uthicke, 2005
	Australia (WA)													Shiell and Uthicke, 2005
	Guam													Richmond, 1996a
	New Caledonia													Conand, 1981, 1988, 1989, 1993
Pearsonothuria graeffei	Australia (GBR)													Uthicke, 1994
Stichopus chloronotus	Australia (GBR)													Uthicke, 1994
S. herrmanni	Australia (GBR)													Uthicke, 1994
	New Caledonia													Conand, 1988, 1989, 1993
Thelenota ananas	Guam													Richmond, 1996a
	New Caledonia													Conand, 1981, 1988, 1989, 1993

Note: Australia – **GBR** = Great Barrier Reef; **NT** = Northern Territory; **WA** = Western Australia.

A. echinites, A. mauritiana have life spans in excess of 12 years (Shelley, 1981; Conand, 1990). Natural mortalities ranged from 16–60 percent y^{-1} for the latter species. Ebert (1978) estimated a natural survival rate of 40 percent annually for *H. atra* at Enewetak Atoll in the Marshall Islands.

Population genetics studies have shown that some populations are highly connected. For example, studies conducted by Uthicke and Benzie (2001, 2003) determined that contiguous range expansion for *H. whitmaei* in the Australian region probably began in the late Pleistocene, with the existing population genetic structure probably formed prior to the last ice age. Uthicke and Benzie (2001, 2003) also found *H. whitmaei* populations could not be distinguished genetically from each other within the Great Barrier Reef, but did exhibit some restrictions in gene-flow with populations in West Australia and the Coral Sea. This suggests that on evolutionary time scales, sea cucumber stocks could be replenished from a large variety of sources, but are not highly relevant on the ecological time scales required for fisheries management.

Genetics studies on *H. scabra* have shown that populations from Australia (the Northern Territory, the Torres Strait, the Solomon Islands, the Great Barrier Reef), the Solomon Islands, and New Caledonia are distinct populations with little gene-flow between these populations (Uthicke and Benzie 1999, 2001; Uthicke and Purcell, 2004). The fragmented meta-population is a likely result of shorter larval time when compared to *H. whitmaei*, but also probably due to the hydro-dynamic retainment in the coastal areas and bays where *H. scabra* occurs. Gene-flow estimations for *H. atra* and *S. chloronotus* from the Great Barrier Reef, the Torres Strait and Réunion, have been hampered by high rates of asexual reproduction. However, it appears that gene-flow is high for these species, though there are some limitations in population connectivity between inshore and mid-shelf areas of the Great Barrier Reef (Uthicke, Benzie and Ballment, 1998, 1999; Uthicke, Conand and Benzie, 2001).

Apart from harvesting, sea cucumbers in the Western Central Pacific region generally suffer low predation owing to their chemical and physical defense mechanisms (Bakus, 1968; 1973). These include the release of cuvierian tubules (collagenous fibres that are extremely sticky), and the ability to eviscerate parts of their internal organs or body to evade predators (Mercier and Hamel, 2000). One common holothurian, *H. atra,* possesses strong toxins (mainly saponins), generally referred to as "holothurine", which is thought to interfere with the action of the fish branchiae (Bakus, 1973; FAO, 1990). The calcareous ossicles (or "spicules") in the outer body wall of holothurians also provides structural defence to their body wall; for example, *H. scabra* spicules are more densely packed in the dermis of juveniles than adults (Purcell, Blockmans and Nash, 2006). Predators that will take sea cucumbers in the Western Central Pacific region include seagulls, sharks, gastropods (in particular *Tonna* spp.), fish (notably Balistidae, Labridae, Lethrinidae and Nemipteridae), sea stars and crustaceans (Kropp, 1982; Francour, 1997; Mercier and Hamel, 2000; Dance, Lane and Bell, 2003), and loggerhead turtles (Cannon and Silver, 1987).

2.3 Exploitation of sea cucumbers for subsistence purposes

In many countries in the Western Central Pacific region, *A. miliaris, A. echinites, A. mauritiana, H. atra, H. scabra, H. leucospilota, H. verrucosa, H. fuscopunctata, B. argus, B. similis, B. vitiensis, T. ananas, S. horrens* and *S. herrmanni* are consumed (or their intestines and/or gonads) as delicacies or as a protein component to traditional diets (Conand, 1990; Dalzell, Adams and Polunin, 1996; Lambeth, 1999, 2000; Mathews, 1995; Mathews and Oiterong, 1991, 1995; Smith, 1992). Local consumption is particularly important in times of hardship and following cyclones (Adams, 1992). The top three species consumed across the countries in which subsistence use was reported are *A. mauritiana, H. atra* and *S. horrens* (Table 4).

Micronesia

Mathews and Oiterong (1991, 1995) noted the consumption of gonads and/or intestines of *S. vastus, A. echinites, H. scabra* and *H. verrucosa* in Palau. The leathery body wall of an unidentified *Actinopyga* sp. is also consumed. The ejected cuvierian tubules of certain *Bohadschia* spp. are also used by the youth of Palau to coat the soles of their feet to protect them while walking on the reef (Adams *et al.*, 1994). Smith (1986) reported *S. horrens* and *H. atra* consumed in Guam.

Due to economic hardship, Nauruans have started to exploit whatever marine resources are edible, and subsequently target *A. mauritiana* and *H. atra* for subsistence purposes (Vunisea, A., Secretariat of the Pacific Community, personal communication).

Polynesia

Conand (1990) reported that sea cucumbers were consumed in Wallis and Futuna. Conversely, sea cucumbers and their body parts are not reported in the traditional diet of Tuvalu islanders (Belhadjali, 1997).

TABLE 4
Subsistence use of holothurian species in the Western Central Pacific Region

Species	Micronesia			Polynesia						Melanesia		Total No. of countries
	Palau	Guam	Nauru	Wallis-Futuna	Samoa	Cook Islands	Tonga	Nuie	French Polynesia	Papua New Guinea	Fiji	
Actinoypa echinites	✓											1
A. mauritiana	✓		✓	✓	✓	✓			✓	✓		7
A. miliaris	✓											1
Bohadschia argus				✓	✓							2
B. vitiensis					✓						✓	2
Holothuria atra		✓	✓		✓	✓		✓	✓			6
H. cinerascens						✓						1
H. fuscogilva	✓			✓					✓			3
H. hilla					✓	✓						2
H. leucospilota					✓	✓	✓					3
H. scabra	✓			✓						✓	✓	4
H. scabra var. *versicolor*										✓	✓	2
H. verrucosa	✓											1
H. whitmaei	✓								✓			2
Stichopus chloronotus				✓					✓			2
S. herrmanni					✓	✓						2
S. horrens		✓		✓	✓	✓	✓					5
S. vastus	✓											1
Thelenota ananas	✓			✓					✓			3
Total	9	2	3	7	8	7	2	1	6	3	3	–

FIGURE 3
Roadside stall for locally consumed sea cucumber product (mainly the body wall of *B. vitiensis* and some viscera of *S. horrens*) in Samoa

COURTESY OF H. ERIKSSON

In the southern Cook Islands, *H. leucospilota* and *H. cinerascens* are harvested for their gonads (Zoutendyk, 1989a). *H. leucospilota* is heavily exploited by women and children throughout the year, with harvesting intensifying during the summer months from October–March (Baquie, 1977; Drumm and Lonergan, 2005; Drumm, 2004; Dzeroski and Drumm, 2003; Hoffman, 2001; Tiraa-Passfield, 1997). The removal of gonads appears to have no impact on the survival, with the body-wall of gutted animals healing within 7–14 days, while the gonads regenerate after 41 days. It is thought, however, that their body weight, general sheltering, feeding behaviours and spawning is affected (Drumm and Loneragan, 2005).

During recent surveys in Samoa, the subsistence sea cucumber fishery was considered to be an important element to fishers' livelihoods and their families (Figure 3), with 29 percent of all fishers surveyed selling either all or part of their sea cucumber catch at local markets (Eriksson, 2006; Friedman *et al.*, 2006). Assessments of markets by the Samoan Fisheries Division has shown that there has been a decline in the sale of *S. horrens*, with the less sought after *B. vitiensis* increasing in sales between 2000 and 2004. This is possibly an indication that the availability of *S. horrens* is declining, or possibly not satisfying local consumer demands (Eriksson, 2006; Friedman *et al.*, 2006). Both *S. horrens* and *H. leucospilota* are also exploited for local consumption and sale in Tonga.

Sea cucumbers are not a common sea food in French Polynesia, with consumption limited to some of the southern islands of the Austral Archipelago, especially Rurutu Island (Gibert, A., Tahiti Eco Clams Project, personal communication), where *A. mauritiana, H. atra, S. chloronotus* and *T. ananas* are commonly consumed (Stein, A., Polynésie française, Service de la pêche, personal communication).

Melanesia

Apparently, there is no subsistence use of sea cucumbers in the Solomon Islands (Adams *et al.*, 1992), though it has been reported that bêche-de-mer (most probably *H. scabra*) is baked with cabbage and coconut cream in North Malaita (Foale, S., James Cook University, personal communication). In Papua New Guinea, some *Actinopyga* spp. are consumed in the West New Britain Province, and some areas in Manus use the toxins of *H. atra* to fish for octopus (Lokani, 1990). *H. scabra* is regularly consumed grilled in the Western Central Province and in the Trobriand Islands in the Milne Bay Province (Kinch, 2002; 2004b; Kinch *et al.*, 2007).

Qalovaki (2006) reports *H. scabra* and *H. scabra* var. *versicolor* used as food in Fiji, whilst Adams (1992) includes *B. vitiensis*. Dalzell, Adams and Polunin (1996) estimated that approximately seven tonnes of sea cucumbers were sold for food in markets and produce stores in Fiji in the mid-1990s.

There is also probably considerable domestic consumption and probably small levels of harvesting by Asian residents and workers in countries throughout the Western Central Pacific region.

2.4 Background of the commercial sea cucumber fishery

In the Western Central Pacific region, sea cucumber fisheries have existed for several centuries. These initially centred on the north-west peninsular of present day West Papua Province of Indonesia (Souter, 1963) and in the Northern Territory of Australia (McKnight, 1976; Cannon and Silver, 1987). Sea cucumber harvesting in the Northern Territory by Macassans from Indonesia continued right up until 1907 when the South Australian government which then controlled the Northern Territory ceased issuing fishing licenses (McKnight, 1976). Oral history of Torres Strait Islanders also indicates that Chinese and Malays regularly visited the islands to harvest sea cucumbers long before European explorers arrived (Laade, 1966).

Increased exploitation of sea cucumber stocks in the Western Central Pacific region began in the latter years of the eighteenth century with European and Japanese colonial expansion. Bêche-de-mer exports became important components of early English, Dutch, German, Spanish, French and Japanese colonial administrations (Ward, 1972; Moore, 2003).

All throughout the Western Central Pacific region, the bêche-de-mer trade followed a similar pattern: the sea cucumber fisheries peaked in the early nineteenth century and then slowly declined. This was partly due to over-harvesting in some areas, and partly because bêche-de-mer was replaced by other goods, such as copra (Ward, 1972; Moore, 2003). Production of bêche-de-mer in the Western Central Pacific region also remained at low levels for much of the early part of the twentieth century (Conand, 1990). Outside of Micronesia, the Sino-Japanese War and World War II also caused a decline in exports (Richmond, 1996b), and it was only in the early 1960s that sea cucumber fisheries regained some importance throughout the region.

In the 1980s, price increases and an enhanced demand in China and in Chinese communities around the world led to increasing effort in the re-development of sea cucumber fisheries in many PICTs (Preston, 1993). Increased trade was also facilitated by the removal of trade barriers to China and its concurrent increasing affluence and concomitant drops in supplies from traditional source countries closer to Asia, such as Sri Lanka. This greater demand and higher prices paid to fishers also provided a strong incentive for many coastal and island people in the region to shift from other "traditional" fisheries, to neglect agricultural cash-cropping and to relax religious, both "traditional" and Christian taboos on the harvesting of sea cucumbers (Kinch *et al.*, 2006, 2007). An example of the latter is the harvesting of sea cucumbers for cash by Seventh Day Adventists who are prohibited under the Bible scripture, Leviticus 11: 9–12; to touch or eat marine animals that do not have fins or scales (Kinch *et al.*, 2006).

Even with this re-emergence of exploitation, only a subset of species was mainly targeted. These initial species were *H. scabra*, *H. whitmaei*, *H. fuscogilva*, *A. miliaris* and *T. ananas* (Sachithananthan, 1971a, b; Crean, 1977; Shelley, 1981; Kinch, 1999).

As noted in Section 2.1, today there are approximately 35 sea cucumber species in trade, with possibly another 22 also utilized to varying degrees in countries of the Western Central Pacific region (Table 5).

Micronesia

The 1830s was the main period of exploitation in Micronesia with sea cucumber fisheries starting in the CNMI (Morrell, 1832), before starting in Yap (Cheyne, 1852) and in the Marshall Islands (Coulter, 1847). Exploitation continued through the Japanese mandate era (prior to World War II), with an estimated 5 124 tonnes exported during this period (Richmond, 1996b). Chuuk Atoll, in the Federated States of Micronesia, is reported to have exported nearly 454 tonnes y^{-1} during the early part of the twentieth century (SPC, 1979). Smith (1947) identifies the chief centres of production during this period as Chuuk Atoll (producing 61% of exports), Palau (18%), Pohnpei (11%), Saipan (6%) and Yap (4%). Sea cucumber populations were also heavily impacted during World War II, when large numbers of Japanese soldiers were left without food, and subsequently harvested significant amounts of sea cucumbers to supplement their diets.

Palau does not have an active export trade for bêche-de-mer at present due to a moratorium that was implemented a decade ago, though several species are used for subsistence purposes.

Guam does not have an active commercial fishery, although there is an increase in local consumption due to recent influx of other Micronesian Islanders (Kerr, A., University of Guam, personal communication).

The sea cucumber fishery re-started in the CNMI in 1995. Because *A. mauritiana* and *S. chloronotus* were available in relative abundance, they were the main species targeted on the island of Rota, and fishing continued there through to 1996, when operations moved to Saipan due to a drop in harvest rates because of seasonal weather conditions and over-exploitation of accessible areas (Trianni, 2001, 2002). The Saipan sea cucumber fishery also targeted *A. mauritiana* (making up 99% of all exports) and some *H. whitmaei* but stopped in 1997 due to declining CPUE (Trianni, 2002). Resource surveys, in the period shortly after harvesting had ceased, estimated that the remaining population of *A. mauritiana* was between 10 and 22 percent of its initial population size (Trianni, 2002). A 10-year moratorium on the harvest of all sea cucumbers was put in place across the CNMI in 1998.

In the FSM, only *A. mauritiana* and *S. chloronotus* were thought abundant enough to support commercial exploitation (Kerr, 1994; Edward, 1997). These two species have been commercially harvested since 1996, with *A. mauritiana* being targeted principally (Lindsay, 2001a). Between 2005 and late 2007, *A. miliaris* and another local *Actinopyga* sp. has been commercially harvested from Yap. Due to uncontrolled expansion in the fishery a moratorium was institued in late 2007.

Based on the information collected by SPC-Pacific Regional Oceanic and Coastal Development Project Coastal Component (PROCFish/C), the only species with

FIGURE 4
A fisher processing his catch at Wallis Island

COURTESY OF K. FRIEDMAN

potential for commercial harvests in Nauru appears to be *A. mauritiana* (Kim Friedman, unpublished data).

In Kiribati, small sea cucumber fisheries operate on Christmas (Kiritimati) Island and within the Gilbert Island Group. The fishery boomed from 2000–2002, but is now considered depleted. Commercial sea cucumber species targeted include *H. fuscogilva* and *T. ananas,* with smaller quantities of *H. whitmaei, S. chloronotus, B. argus, B. vitiensis* and *A. mauritiana* also exploited and exported (Kronen, M., Secretariat of the Pacific Community, personal communication). *H. atra* was not collected previously, but due to the scarcity of other higher-value species, it is now also being harvested ('Tekanene, M., Kiribati Ministry of Fisheries and Marine Resources Development, personal communication).

Polynesia

Fishing in Polynesia began early with the first records of sea cucumber exports around 1810 from French Polynesia (Ward, 1972), and 1825 from Hawaii (Morrell, 1832) expanding elsewhere throughout the region soon after.

With increasing prices paid for bêche-de-mer and subsequent increasing interest from fishers, many countries began to take advantage of the new economic opportunities and began exploiting their sea cucumber stocks for export.

On Wallis, there is a small but growing sea cucumber fishery, which exports bêche-de-mer to New Caledonia for on-routing to Asian markets (Figure 4) (Kronen, M., Secretariat of the Pacific Community, unpublished data). Futuna does not have an active fishery due to transport and marketing problems. Species taken at Wallis include *H. scabra, S. chloronotus, S. herrmanni, T. ananas, H. whitmaei, H. fuscogilva*, and *A. mauritiana*.

In Tuvalu, the Fisheries Department received funding from the United Nations Development Program in 1978 to redevelop its sea cucumber fishery. Shortly after, a total of 1.8 tonnes of bêche-de-mer was bought from fishers in Nukufetau and exported to Fiji (Belhadjali, 1997). Production then fell, because fishers lacked interest, until 1993 when the fishery was re-established by Fiji-based Asian companies. From 1993–1995, the main species targeted for export were *H. fuscogilva* (64%), *T. ananas* (18%), *H. fuscopunctata* (8%) and *H. whitmaei* (3%) (Belhadjali, 1997). Since April 2007 fishing has recommenced in at least 3 of the islands including the main island of Funafuti

The bêche-de-mer trade in Samoa recommenced in the 1960s, with ad-hoc exports up until the mid-1990s (Mulipola, 1994). Exports in 1993–1994 consisted of *B. vitiensis* (44%), *A. mauritiana* (30%), *B. argus* (19%), *S. chloronotus* (3%), *H. whitmaei* (2%), *H. atra* (2%), and very small amounts of *H. fuscogilva* and *T. ananas* (Mulipola, 1994). After 1994, the commercial fishery stalled and bêche-de-mer exports have been closed since then, though some sea cucumbers have been sold for the aquarium trade (Mulipola 2002). Currently, harvesting is concentrated on supplying the domestic market, targeting mostly *S. horrens, B. vitiensis,* and *H. atra* (Eriksson, 2006). In 2004, 1.7 tonnes of *S. horrens* and *B. vitiensis* were sold at local markets (Friedman *et al.*, 2006).

American Samoa does not have an export trade for bêche-de-mer although there is a small subsistence fishery (Fenner, D., American Samoa Department of Marine and Wildlife Resources, personal communication).

The early-1980s saw the re-commencement of the sea cucumber fishery in Tonga (as an extension of the Fiji fishery) using SCUBA and hookah. Sea cucumber harvests were thought to have peaked in 1994 (Tonga Ministry of Fisheries, 1995; Kailola, Petelo and Gillett, 1995). A moratorium on harvesting and exports, declared in 1999 after a recognized depletion in stocks, persists in Tonga, but is due to be lifted in 2008.

A fisheries survey was conducted in 1994 in the Pitcairn Islands, whereby 640 kg gutted wet weight of *H. whitmaei* was collected as part of an assessment of the potential of the sea cucumber fishery and taken to New Zealand (Sharples, 1994). The industry has not developed further (Dunn, E., Office of the Commissioner of Pitcairn Islands, personal communication).

Following a marine resources assessment of Niue in 1990, Dalzell, Lindsay and Patiale (1993) concluded that sea cucumber stocks offered little commercial potential, unless the fishery was to target the low-valued *H. atra*. Recent assessments provided the same conclusion (Kronen, M., Secretariat of the Pacific Community, personal communication).

The Cook Island fishery recommenced in the mid-1980s, with the exploitation and export of *T. ananas, S. chloronotus, A. mauritiana, H. atra, H. whitmaei* and *B. argus* (Zouthendyk, 1989b).

Currently, French Polynesia does not have an active bêche-de-mer trade (Gibert, A., Tahiti Eco Clams Project, personal communication). However, there is a small trade from the Tuamotu Islands supplying the Chinese community in Tahiti (Stein, A., Polynésie française, Service de la pêche, personal communication). In Moorea, commercial sea cucumber species observed in recent underwater surveys include *B. argus, H. fuscogilva, H. whitmaei* and *T. ananas* (SPC, 2006a).

Melanesia

The sea cucumber fishery in PNG was first described in 1873 but it was most likely exploited earlier than that (Russell, 1970; Shelley, 1981; Kinch, 2004b). During the period 1878-1900, bêche-de-mer exports were reported at around 37 tonnes y^{-1}, but this is likely a gross under-estimation as most shipments went unreported (Russel, 1970). Fort (1886) highlights in official dispatches during the 1880s, that approximately 500 tonnes y^{-1} of bêche-de-mer was being produced from the Louisiade Archipelago in the Milne Bay Province alone. In the early-to-mid-1930s, the Territories of Papua and New Guinea were exporting an estimated 158 tonnes y^{-1} (Shelley, 1981). In the early 1980s, Shelley (1981) reported seven sea cucumber species in trade in Papua New Guinea consisting of *A. echinites, A. mauritiana, A. miliaris, H. scabra, H. whitmaei, H. fuscogilva* and *T. ananas.* In 1989, the high-valued *H. scabra* accounted for 70 percent of the total bêche-de-mer exports from Papua New Guinea (Lokani, 1990). More recently, catches have shifted to mostly low-value species, particularly *B. vitiensis* and *H. atra* (Kinch, 2004b).

The sea cucumber fishery was active in the Solomon Islands in 1844 (Cheyne, 1852; Bennett, 1987; Ward, 1972), with exports to Australia during the 1870–1880s averaging around 90 tonnes y^{-1} (Bennett, 1987). In 1966, the sea cucumber fishery re-commenced in the Solomon Islands, and a processing plant was established in the early 1970s, but was unsuccessful (Sachithananthan, 1971a, 1971b). In 1977, exports consisted of *H. fuscogilva, H. whitmaei, A. miliaris, T. ananas* and *A. echinites.* Fifteen species were being harvested in 1988, which increased to 18 in 1991 (Adams *et al.*, 1992; Holland, 1994a, 1994b). In 2004, Kinch (2004c) identified 28 sea cucumber species as having commercial value in the Solomon Islands, including the rare *T. rubralineata* (Lane, 1999; Kinch, 2005). The sea cucumber fishery in the Solomon Islands peaked in 1992, when 715 tonnes were exported (Kinch, 2004c). In 1999, the high-valued *H. fuscogilva* contributed 50 percent of exports from the Solomon Islands and dropped to 2 percent in 2002; in comparison, the low-valued *H. atra* made up 22 percent of exports in 2000, increasing to 60 percent in 2003 (Ramofafia, 2004). In 2004, several exporters did not renew their trading licenses due to dwindling catches (Ramofafia,

FIGURE 5
A sea cucumber purchasing board of an agent in Vanuatu

COURTESY OF D. SCHULTZ

2004), and a moratorium was placed by the Solomon Island government on the harvesting and export in December 2005 (Nash and Ramofafia, 2006). This moratorium was lifted for humanitarian reasons following an earthquake and tsunami in April 2007, and an interim management plan has been devised, which will be revised in December 2007 (Nash, W., WorldFish Center, personal communication).

There is a small sea cucumber fishery in Vanuatu, though stocks are generally regarded as depleted around the more populated areas (Figure 5). Areas under Customary Marine Tenure have managed to protect some high-value populations. Recently, an Australian-based company has invested in trial sea ranching of *H. scabra* in Vanuatu using juveniles produced in their Australian-based hatchery.

The fishery in New Caledonia began in the 1840s (Cheyne, 1852; Conand, 1990). Catches during the 1920s ranged from 100–150 tonnes y^{-1} (Conand 1990). Most previous and current harvesting in New Caledonia has centred around the country's main island of La Grande Terre. In the Loyalty Islands Province, *H. whitmaei* has been fished on Ouvéa, with light harvesting on Lifou and Maré due to customary restrictions. However, *H. whitmaei* is now being fished regularly in Maré (Purcell, Gossuin and Agudo, in press). In New Caledonia, the numerically dominant species in catches are *H. scabra*, *H. whitmaei*, *A. miliaris*, *A. palauensis*, *A. spinea* and *T. ananas*. Species caught in moderate quantities include *A. echinites*, *A. mauritiana*, *H. scabra* var. *versicolor*, *H. fuscogilva*, and *S. herrmanni*. Since 2003, export statistics show declines in catches of *H. scabra* but concomitant increases in catches of reef-dwelling species. The fishery now comprises a network of village fishers gleaning reef flats or by skin divers utilizing small boats, and semi-industrial fishing companies using larger boats of 10–20 m in length (Purcell, Gossuin and Agudo, in press).

In 1813, the sea cucumber fishery started in Fiji and from 1827–1835, approximately 600 tonnes of bêche-de-mer were exported (Ward, 1972). By 1834, sea cucumber populations on reefs of the Western Central and northern Vanua Levu and south-east Viti Levu were considered depleted (Ward, 1972). In the early 1980s, fishing recommenced in Fiji, peaking in 1988 when 717 tonnes of bêche-de-mer were exported. The actual export figure is thought to be closer to 1 000 tonnes, as some product was labeled as "miscellaneous mollusks" (Adams, 1992). Prior to 1988, *A. miliaris*, *H. fuscogilva*, *H. whitmaei* and *H. scabra* were the most important commercial species followed by *A. echinites*, *A. lecanora* and *H. atra*. Other species, such as *B. argus*, *B. vitiensis*, *H. edulis*, *H. fuscopunctata*, *Pearsonothuria graeffei*, *S. chloronotus*, *T. ananas* and *T. anax* were considered to have no or low commercial value (Preston, 1988). The establishment of centralised processing facilities in the mid-1980s enabled mass-processing of *A. miliaris* which made up approximately 95 percent of all exports in 1988 (Preston, 1990). With subsequent decline in *A. miliaris* stocks, the exploitation of *S. chloronotus*, *A. mauritiana*, *H. fuscogilva*, *H. whitmaei*, *H. scabra*, *S. herrmanni*, *H. atra* and *B. vitiensis* increased in importance (Adams, 1992). Currently, the sea cucumber fishery in Fiji is controlled by licensed companies who prefer to buy wet products from local fishers but also have their own teams of 15–30 divers who use SCUBA (Qalovaki, 2006).

Australia and New Zealand

The first reports of fishing on the Great Barrier Reef date to 1804, with further developments occurring in the 1840–1850s (Sumner, 1981). In 1846, the fishery was established in the Torres Strait (Beckett, 1977), and by 1870, the fishery was considered to be over-harvested. From 1896 to 1928, between 16 and 558 tonnes of bêche-de-mer were exported from Thursday Island in the Torres Strait annually (Harriott, 1984). The sea cucumber fishery in Torres Strait is almost exclusively fished by indigenous inhabitants. The major species harvested in the Torres Straits are *H. scabra* and *A. mauritiana,* with some smaller landings of *H. whitmaei, H. fuscogilva, H. atra, A. echinites* and *H. fuscopunctata* (AFMA, 2004, 2005).

TABLE 5
Past and present commercially exploited holothurian species in the Western Central Pacific region

	Micronesia							Polynesia							Melanesia					Australia and New Zealand							
Species	**Palau**	**Guam**	**CNMI**	**Federated States of Micronesia**	**Marshall Islands**	**Nauru**	**Kiribati**	**Tuvalu**	**Wallis and Futuna**	**Samoa**	**Tonga**	**Niue**	**Cook Islands**	**French Polymrsia**	**Papua New Guinea**	**Solomon Islands**	**Vanuatu**	**New Caledonia**	**Fiji**	**Australia (Coral Sea)**	**Australia (Torres Strait)**	**Australia (Great Barrier Reef)**	**Australia (Morteon Bay)**	**Australia (Northern Territory)**	**Australia (Western Australia)**	**New Zealand**	**Total**
Actinopyga caerulea															✓												**1**
A. echinites	✓	✓	✓	✓							✓				✓	✓		✓	✓		✓	✓		✓	✓		**13**
A. lecanora	✓			✓							✓				✓	✓	✓	✓	✓		✓						**9**
A. mauritiana	✓	✓	✓	✓	✓	✓	✓	✓	✓	✓	✓	✓	✓	✓	✓	✓	✓	✓	✓	✓	✓	✓					**22**
A. miliaris	✓		✓	✓			✓	✓	✓	✓	✓	✓			✓	✓	✓	✓	✓	✓	✓	✓					**17**
A. palauensis	✓			✓							✓	✓				✓		✓									**6**
A. spinea																		✓				✓					**2**
Australostichopus mollis																										✓	**1**
Bohadschia anaes																			✓								**1**
B. argus	✓	✓	✓	✓			✓	✓	✓	✓	✓	✓	✓	✓	✓	✓	✓	✓	✓								**17**
B. similis	✓			✓							✓			✓	✓	✓	✓	✓	✓								**9**
B. vitiensis	✓	✓	✓	✓			✓	✓	✓	✓	✓		✓	✓	✓	✓	✓	✓	✓		✓						**17**
Holothuria atra	✓	✓	✓	✓	✓	✓	✓	✓	✓	✓	✓	✓	✓	✓	✓	✓	✓	✓	✓	✓		✓		✓	✓		**23**
H. cinerascens		✓																									**1**
H. coluber	✓			✓					✓		✓				✓	✓	✓	✓	✓								**9**
H. edulis	✓	✓	✓	✓			✓				✓				✓	✓	✓	✓	✓		✓						**12**
H. flavomaculata	✓			✓												✓	✓	✓									**5**
H. fuscogilva	✓			✓			✓	✓	✓	✓	✓			✓	✓	✓	✓	✓	✓	✓	✓	✓		✓	✓		**18**
H. fuscopunctata	✓	✓	✓	✓			✓	✓	✓	✓	✓			✓	✓	✓	✓	✓	✓		✓	✓					**17**
H. grisea																✓											**1**
H. leucospilota		✓		✓	✓		✓			✓	✓		✓		✓	✓			✓								**10**
H. scabra				✓			✓		✓						✓	✓	✓	✓	✓		✓	✓	✓	✓	✓		**13**
H. scabra var. *versicolor*											✓				✓	✓	✓	✓	✓		✓	✓					**8**
Holothuria sp. (Hongpai – the Solomon Islands)																✓											**1**
Holothuria sp. (Tulele – the Solomon Islands)																✓											**1**
H. whitmaei	✓	✓	✓	✓		✓	✓	✓	✓	✓	✓	✓	✓	✓	✓	✓	✓	✓	✓	✓	✓	✓		✓	✓		**23**
Pearsonothuria graeffei		✓		✓			✓								✓	✓		✓	✓								**7**
S. chloronotus	✓	✓	✓	✓	✓		✓	✓	✓	✓	✓		✓	✓	✓	✓	✓	✓	✓	✓	✓	✓					**20**
S. herrmanni	✓			✓			✓		✓	✓	✓			✓	✓	✓	✓	✓	✓		✓	✓					**14**
S. horrens	✓	✓		✓			✓		✓	✓	✓		✓	✓	✓	✓	✓	✓	✓								**14**
S. ocellatus																				✓	✓	✓		✓			**4**
S. pseudohorrens															✓												**1**
S. vastus	✓			✓											✓	✓	✓				✓	✓					**7**
Thelenota ananas	✓	✓		✓		✓	✓	✓	✓	✓	✓	✓	✓	✓	✓	✓	✓	✓	✓	✓	✓	✓		✓	✓		**22**
T. anax	✓	✓		✓			✓	✓	✓	✓	✓			✓	✓	✓	✓	✓	✓	✓	✓						**16**
T. rubralineata															✓	✓											**2**
Total	**21**	**15**	**10**	**24**	**4**	**4**	**17**	**11**	**4**	**14**	**21**	**7**	**9**	**13**	**26**	**29**	**21**	**25**	**23**	**8**	**16**	**14**	**1**	**6**	**6**	**1**	**–**

In 1986, the fishery recommenced in Queensland (Beumer, 1992). The targeted species in this fishery are *A. echinites, A. mauritiana, A. miliaris, H. atra, H. fuscogilva, H. fuscopunctata, H. whitmaei, H. scabra, H. scabra* var. *versicolor, S. chloronotus, S. herrmanni* and *T. ananas* (DEH, 2006).

In the Northern Territory, the main target species is *H. scabra* (DEH, 2004). The Northern Territory fishery generally has a hiatus from November-April due to monsoonal weather (DEH, 2004). Fishing operations are vessel-based, with a "mother boat" anchoring in deeper offshore waters, while dinghies supplying compressed air follow divers through the harvesting areas (DEH, 2004).

The Western Central Australian sea cucumber fishery began in 1995, peaking at 382 tonnes in 1997, then declining to a more current average of 80 tonnes y^{-1} (DEH, 2005). Six species are harvested: *H. scabra, H. whitmaei, H. fuscogilva, T. ananas, A. echinites* and *H. atra* (DEH, 2005). Harvesting in Western Central Australia is year round, but confined to neap tides in some areas (DEH, 2005).

Australostichopus mollis is common along many coastlines of New Zealand and southern coasts of Australia (Pawson, 1970). In 1990 *A. mollis* was commercially exploited in New Zealand, when one company was granted a special permit to harvest by SCUBA up to 74 tonnes wet weight of *A. mollis* from the south-eastern fjords to evaluate the potential of this fishery (Morgan and Archer, 1999).

2.5 Species in trade

As mentioned above, the vast majority of sea cucumbers harvested in the Western Central Pacific region are those that have thick body walls and are subsequently processed for bêche-de-mer. Exports of salted sea cucumbers are limited to small shipments from New Caledonia and possibly Kiribati. Frozen sea cucumbers are also exported from Australia and New Zealand, and previously Tonga.

Several species of sea cucumbers are used for the aquarium trade. For example, *T. ananas, H. leucospilota, B. argus* and *H. hilla* are exported as ornamentals from the Solomon Islands (Kinch, 2004a), and some probably also from the Cook Islands, Fiji, French Polynesia, Tonga and Kiribati. Small amounts of colourful sea cucumber species are collected from Moreton Bay for the Queensland Marine Aquarium Fishery (DEH, 2006). There are no current data available on sea cucumber exports for the aquarium trade. Wabnitz *et al.* (2003) and Green (2003) have highlighted problems with trade statistics and the Global Marine Aquarium Database as sea cucumbers are often aggregated simply as invertebrates (which also includes other echinoderms such as starfish, and marine molluscs) or as "tropical fish".

Within the Western Central Pacific region, one company in Vanuatu, Unicorn Pacific, exports TBL-12, an immuno-therapy treatment that is produced from sea cucumbers and other marine organisms (Carroll, 2005).

2.6 Population status

Most surveys of sea cucumber resources have aimed to assess the potential for developing a fishery, or in response to determining status of a declining resource. When comparing density records, one must be aware that the data collection method, position and scale of the study can affect results. Equally, the state of the habitat available or the bio-geographical position can affect the potential of populations and species diversity (e.g. the Cook Islands has a restricted species range due to its easterly position in the Pacific). Higher densities in some countries may also reflect the presence of moratoriums which may have allowed for some regeneration of sea cucumber stocks (e.g. Tonga, Palau, Federated States of Micronesia and Samoa), stronger management controls either at the government or local level (e.g. New Caledonia and Vanuatu) or remoteness from marketing opportunities.

Micronesia

In Palau, Birkeland *et al.* (2000) believed that over-fishing was the most likely reason for the low densities of the eight sea cucumber species recorded on Helen Reef. On the main islands of Palau, *H. whitmaei* was reported as relatively common (Ilek, 1991) and this observation was confirmed in recent SPC surveys in 2007 (>10 ind. ha^{-1} in broad-scale surveys). In a study from Airai (Anon, 2003), high densities of *S. vastus* (1 800 ind. ha^{-1}), an unidentified species of *Actinopyga* (137 ind. ha^{-1}) and *H. scabra* (40 ind. ha^{-1}) were recorded.

Assessments in CNMI in the late-1980s suggested that sea cucumbers have not recovered from the heavy exploitation during the 1920–1940s (Tsuda, 1997; Trianni, 2001, 2002). Trianni (2002) showed changes in densities after the 1997 boom at Saipan had finished.

In the Federated States of Micronesia, starting with Yap, surveys of Ngulu Atoll in 1985 found *T. ananas* in greater abundance then *H. whitmaei* and *H. fuscogilva*, with smaller populations of *B. argus*, *S. herrmanni* and *H. fuscopunctata* (Moore and Marieg, 1986). On Yap itself, 17 species of commercial sea cucumbers were recorded, mainly in low densities for the high value species (SPC, 2006a). In Chuuk, 19 species were recorded, but despite wide distribution across the lagoon, populations were considered depleted (SPC, 2006a). Surveys in Pohnpei in 2000 found populations of the commercially important species, *H. whitmaei*, *H. fuscopunctata*, *B. vitiensis*, *S. herrmanni*, *S. chloronotus* and *T. ananas* in relatively high densities, though Lindsay (2000a, 2001b) notes suitable sea cucumber habitats are not abundant.

In the Marshall Islands, surveys in 1970s found weak potential for a fishery, with the exception of *H. atra* because suitable habitat was limited (McElroy, 1990; see also Ebert, 1978; Lawrence, 1979). Lindsay (2001a) and Lindsay and Abraham (2004) also found low densities of commercially valuable species, except *H. atra*. At Jaluit Atoll, *H. whitmaei*, *H. fuscopunctata*, *B. vitiensis*, *S. herrmanni* and *T. ananas* were also scarce (Bungitaki and Lindsay, 2004).

In Nauru, six commercial species were recorded during recent in-water surveys, with only *A. mauritiana* moderately common in certain areas (Kim Friedman, unpublished data).

In Kiribati, assessments at the populated atoll of Tarawa (Pauly, 2000) found *H. atra* and *B. vitiensis* to be fairly common. Eleven species were recorded in the Gilbert group (Fufudate, 1999). More current SPC PROCFish/C surveys recorded very low densities of sea cucumbers in the Gilbert Group and in the Line Islands (Kim Friedman, unpublished data).

Polynesia

In Tuvalu in 1978, only the atolls of Funafuti and Nukufetau were identified as having commercial densities of *H. fuscogilva*, *B. argus*, *T. ananas*, *H. fuscopunctata*, *A. miliaris*, *A. mauritiana*, *H. whitmaei* and *B. vitiensis* (Belhadjali, 1997). Later surveys in the islands of Funafuti, Nukufetau, Vaitupo and Niutau in 2005 recorded 10 commercial species, with *H. fuscogilva* and *T. ananas* of interest for small-scale commercialization.

In Tokelau, sea cucumbers were generally at low density in surveys, except for *H. atra*, which was noted at 8 000–12 000 ind. ha^{-1}. At Fakaofo Atoll, both *B. argus* and *A. mauritiana* were recorded in moderately high densities in some places by Passfield (1998), but were considered to be at low abundance by Fisk, Axford and Power (2004a). Similar findings were noted for Fakaofo and Atafu (Fisk, Axford and Power, 2004b, c).

From Samoa, there is little sea cucumber stock density data, except for what can be gleaned from marine protected area planning studies (e.g. Fisk, 2002). Recent surveys found 11 commercial species (Friedman *et al.*, 2006). High-value sea cucumber species were found in low densities (no *H. scabra* was recorded) and, apart from *S. chloronotus*, there was a general paucity of medium value species available for exploitation.

In Tonga, surveys in the 1990s showed stocks to be overfished following rapid commercialization of sea cucumber resources (Preston and Lokani, 1990; Lokani, Matoto and Ledau, 1996). Following the institution of a 10-year moratorium on commercial fishing, a survey in 2004 found that there was recovery of *H. fuscogilva* in the nutrient poor, isolated island group of Ha'apai, but *H. whitmaei* was still at depleted levels (Friedman *et al.*, 2004).

A broad fisheries resource survey in Pitcairn Islands visited all the islands in the mid 1990s (Sharples, 1994) and made cursory searches for marketable sea cucumbers. Significant densities were only observed for *H. whitmaei* at Ducie and Oeno Atolls.

During surveys of Niue in the late-1980s, 95 percent of all observed sea cucumbers were the low-valued *H. atra* (Dalzell, Lindsay and Patiale, 1993). Surveys in mid 2005 by SPC PROCFish/C re-iterated the fact that there is a limited number of sea cucumber species available for commercial fishing and the exposed environment (plus effects of cyclone Heta in 2004) possibly limits abundances (see Fisk, 2004, 2005).

In the Cook Islands, most sea cucumber surveys have been conducted at Rarotonga (Drumm, 2004), Aitutaki (Zoutendyk, 1989b) and Palmerston Atoll (Preston *et al.*, 1988) in the southern group. At the latter site, *B. argus, H. fuscogilva* and *H. whitmaei* were sparse, although *A. mauritiana* was relatively abundant. At Rarotonga, the low-valued *H. atra* and *H. leucospilota* were numerous, averaging 9 942 and 8 330 ind. ha^{-1} respectively (Drumm, 2004). Re-survey of all these sites were completed by SPC PROCFish/C in 2007 (SPC, 2007a, b).

In the late-1990s, surveys of sea cucumber population in Tahiti, Rangiroa and Moorea were conducted for the Ministère de la mer de Polynésie française (Anon., 1997, 1998, 1999), whereby a non-conservative annual catch was set at between 20 to 33 percent of the total estimated biomass for each of the sites surveyed. At Tahiti, the estimated catch was 4.1 kg ha^{-1} $year^{-1}$ for *H. whitmaei*, 70.6 kg ha^{-1} $year^{-1}$ for *B. argus* and 67.7 kg ha^{-1} $year^{-1}$ for *H. atra*. At Rangiroa, no *H. whitmaei* were recorded, but potential catches for *B. argus* were listed at 4.1 kg ha^{-1} $year^{-1}$ and 1 210 kg ha^{-1} $year^{-1}$ for *H. atra*. The island of Moorea generally had lower estimates than Rangiroa and Tahiti (Anon, 1998). The overall potential production (wet weights) for Tahiti was suggested to be 2 500 tonnes, 10 792 tonnes for Rangiroa and 142 tonnes for Moorea. Other species noted over the three sites were *T. ananas, T. anax, B. vitiensis* and *A. mauritiana*. In these assessments no deeper-water surveys were completed, although Costa (1995) noted that *H. fuscogilva* was currently fished. More recent assessments in 2004 by SPC PROCFish/C found restricted ranges of species at Tahiti, the Tuamotu, Austral group and Moorea, although moderately high densities for some species were recorded.

Melanesia

Recent surveys in Papua New Guinea show that stocks are depleted. In the Milne Bay Province, Skewes *et al.* (2002a) found low densities of commercial holothurians (average of 21 ind. ha^{-1}). Low survey densities and a comparison of historical and recent catch data indicate that *H. scabra* and *H. whitmaei* populations have been grossly overexploited. Surveys in the Manus Province in 2006 found most shallow water species depleted but *H. fuscogilva* still present in moderate numbers in deeper water, despite active fishing (SPC, 2006b). In New Ireland Province, sparse populations were observed of *H. scabra,* which was targeted at all sizes (SPC, 2006b; NFA, 2007a).

In the Solomon Islands, a survey conducted by several conservation non-governmental organizations and the Division of Fisheries and Marine Resources commonly recorded only two low-valued species: *H. edulis* and *P. graeffei* (Ramohia, 2006). *T. ananas, A. lecanora, S. chloronotus, H. whitmaei* and *S. herrmanni* were seen only in low numbers, while *A. caerulea*, *A. mauritiana*, *H. coluber*, *H. scabra*, *B. similis*, *S. horrens*, *S. pseudohorrens* and *T. rubralineata* were rare (Ramohia, 2006). Surveys conducted under the International Waters Program in the Marovo Lagoon also report sea cucumber densities to be low (Kinch *et al.*, 2006). Some data are also available from a long term resource survey of the Arnavon Marine Conservation Area in the Isabel Province (Lincoln-Smith *et al.*, 2000). The generally bleak picture of sea cucumber resource status was again recorded in 2006, when sparse populations were found at four survey sites in Guadalcanal, Central and Western Central Provinces (SPC, 2006b).

Vanuatu's volcanic islands generally lack large protected lagoons, but reasonable densities of sea cucumbers exist. A past survey found *S. chloronotus* and *H. atra* abundant at Gaua Island (Baker, 1929). Chambers (1989) found relatively dense populations of *A. miliaris* (785 ind. ha^{-1}) and *H. scabra* (43 ind. ha^{-1}) in studies in 1987. Recent surveys found a wide range of species present, but densities were low at the island of Efate, compared with results from the island of Malekula (Friedman, K., unpublished data).

New Caledonia's holothurians have been well documented by Conand's (1989) thesis on the ecology and biology and densities of commercial sea cucumber species. Inshore surveys by the WorldFish Center in New Caledonia during 2003–2005, found variable densities of *H. scabra*, indicating some over-harvesting. Of 35 sites surveyed, 33 of them had mean densities under 30 ind. ha^{-1}, with only two with densities over 100 ind. ha^{-1} (Purcell, S., unpublished data). Likewise, *H. scabra* was found at only one of the five SPC PROCFish/C survey sites (Friedman, K., unpublished data). In a 2004 survey of New Caledonia by conservation NGOs, *H. whitmaei* and *T. ananas* were observed at low densities (Lindsay and McKenna, 2006). Field surveys of 50 lagoon and barrier reef sites in a ZoNéCo project have found generally low densities of commercial species, but occasional dense (>100 ind. ha^{-1}) patches of certain species, like *A. palauensis*, *A. spinea*, *S. chloronotus*, *S. herrmanni*, *S. horrens* and *T. ananas* (Purcell, Gossuin and Agudo, in press).

In Fiji, Stewart (1993) observed *H. scabra* at 625 ind. ha^{-1} and SPC PROCFish/C in 2003 estimated reasonably high densities of 160 ha^{-1} close to Suva (Friedman, K., unpublished data). In the 1980s, *A. miliaris* made up a large proportion of the commercial catch, and occurred at high density at some sites (Preston *et al.*, 1989). Surveys of the Vanua Levu Lagoon sites over a decade later found both a more restricted distribution and lower densities for this species (Friedman, K., unpublished data).

Australia and New Zealand

Recent research has shown that most stocks in three of Australia's sea cucumber fisheries are overexploited. These are the Torres Strait (Skewes, Burridge and Hill, 1998), the Timor Sea MOU74 Box (Skewes *et al.*, 1999), and Queensland (Uthicke and Benzie, 2000a). Uthicke and Benzie (2000a) found an approximately 1:5 ratio for populations of *H. whitmaei* in fished and non-fished areas, respectively, on the Great Barrier Reef.

In 1996, a survey of Warrior Reef in the Torres Strait showed suppressed stocks of *H. scabra*, comprised of small individuals (Long *et al.*, 1996). A follow-up survey in 1998 confirmed these observations, so the management body closed the fishery (Skewes, Burridge and Hill, 1998). Surveys in 2000 (Skewes, Dennis and Burridge, 2000), 2002 (Skewes *et al.*, 2002b) and 2004 (Skewes *et al.*, 2004) demonstrated a slow recovery of *H. scabra* stocks on Warrior Reef (Skewes *et al.*, 2006). After the *H. scabra* closure in 1998, fishers targeted other species, particularly *H. whitmaei*, *H. fuscogilva* and *A. mauritiana*. By 2002, there was evidence of some depletions elsewhere, with population densities for many species <10 ind. ha^{-1} (Skewes *et al.*, 2006). Following a large survey of the east Torres Strait fishery, and catch analysis, fishing was closed for many other species in 2003 (Skewes *et al.*, 2006). Even though Australia is one of the most developed countries in the region and has adopted several management measures, most of the exploitation has resulted in stock depletion of the high-value species.

Surveys of *Australostichopus mollis* in New Zealand conducted in the early-1990s found densities of 10–20 ind. ha^{-1} (Mladenov and Gerring, 1996).

Fishing pressure across the Western Central Pacific region

From the density records reported, and noting the general decline in availability of sea cucumbers, it can be seen that sea cucumbers across the Western Central Pacific region are, or have been, under high fishing pressure. In most cases, the depletion of stocks is negatively impacting the potential incomes of coastal communities and in some cases is affecting the sustainability of fisheries for the long term.

Recognizing weaknesses in past research, sea cucumber surveys are now being approached with a more regional focus, using standardized methods to assess status. Despite this recent regional approach, there is still insufficient data to describe definitively the densities required to sustain sea cucumber populations within an active fishery. What is becoming apparent is that availability and condition of habitats, and their connectivity with nearby reef systems plays a major role in defining the potential of populations. The outlook for understanding questions of "stock health" will rely on on-going monitoring of stocks and catches at relevant scales, to determine changes in populations from different reef systems under various management regimes. Data from populations recovering from periods of fishing, (e.g. in areas under moratorium such as Samoa and Tonga) and unfished populations in marine protected areas (e.g. green zones of the Great Barrier Reef compared with open zones fished on three year rotations) will also be of great value. Finally, fisheries management decisions should also incorporate new information on life histories of commercial species from aquaculture and sea ranching studies.

Fishing pressure* H. whitmaei: *an example

H. whitmaei is one of the most valuable species across sites in the Western Central Pacific region (and Western Australia). This geographical spread represents nearly the entire distributional range of this species, with the exception of populations in Southeast Asia.

Several studies have been conducted on unfished reefs, including, "closed" (green zones) reefs on the Great Barrier Reef (Uthicke, Welch and Benzie, 2004), isolated reefs near Lord Howe Island (Oxley *et al.*, 2004) and Ningaloo Reefs in Western Australia (Shiell, 2004). Without exception, all of these reefs have densities above 12.5 ind. ha^{-1} (Figure 6), and on rare occasions densities as high as 108.6 ind. ha^{-1} at one site in New Caledonia (Purcell, S., unpublished data). Noting the range of densities found at "closed" sites it seems a conservative assumption that densities above 12.5 ind. ha^{-1} represent a "natural" density for this species on suitable habitat.

When compared to reefs open to fishing, it becomes clear that most fished reefs have populations with less than ca. 5 ind. ha^{-1} (Figure 6). Although some of this variation may be due to habitat differences, the fact that most of the "open" reefs hold less than 25 percent of the "natural" *H. whitmaei* populations suggests that these reefs are overfished.

A second important finding from this case study is that "open" reefs that have been "closed" after a period of fishing, have not generally shown signs of strong recovery after a reasonable closure period (e.g. 10 years). This is supported by surveys on the Great Barrier Reef where stocks of *H. whitmaei* have not increased after four years of fishery closure (Uthicke, Welch and Benzie, 2004), and surveys in Ha'apai in Tonga where densities recovered only marginally after seven years of closure (0.3–1.2 ind. ha^{-1}, Friedman *et al.*, 2004).

H. whitmaei, like most holothurians, is a broadcast spawner. Subsequently, fertilization success for this species, as in other sea cucumber species, is density

FIGURE 6

H. whitmaei densities recorded from fished and unfished sites in the Western Central Pacific region

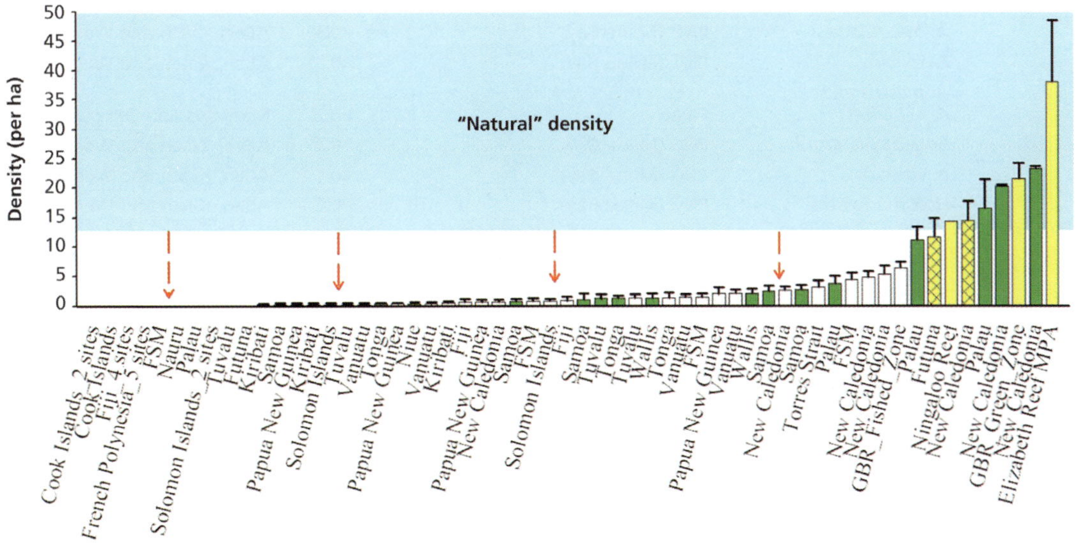

Note: Fished sites across the Western Central Pacific region are represented by white bars (SPC and Timothy Skewes, Commonwealth Scientific and Industrial Research Organization). Unfished sites are split into three categories. Firstly, remote marine protected area sites (yellow bars) such as Elizabeth reef in the Great Barrier Reef and Ningaloo Reef in Western Australia. Secondly SPC PROCFish/C and WorldFish unfished sites in populated areas which are represented by yellow hatched bars. Thirdly, green bars represent areas where fishing has been halted for a decade or more.

dependent, and a reduction to fewer than 5 ind. ha^{-1} could more than double average distance between individuals. It is thus possible that remnant populations on most reefs have "effective" population sizes close to zero, and the reproduction and larval supply is mostly from populations in marine reserves.

Catch per unit effort

Catch rates usually refer to the number of individuals collected per fisher per hour. Monitoring the effort and related catches in a sea cucumber fishery is useful, but care should be taken in comparing values between sites, as fishing conditions, gear used and a range of economic factors (such as community demands, incentives paid by companies, debt cycles, etc.) can have a major impact on the productivity of divers. Fishers are also often involved in multiple livelihoods, or harvesting of sea cucumbers is integrated into other fishing activities, which can mask signals of declining catch rates. It should be noted that unless these complexities of monitoring the fishing effort are noted, and there is some spatial understanding of the effort expended, then CPUE records can be a poor indicator of population status.

A summary of CPUE data for the Western Central Pacific region is given in Table 6.

2.7 Catch

The main areas of production of bêche-de-mer across the Western Central Pacific region are Australia and the countries within Melanesia. In the early 1990s, the Western Central Pacific Region bêche-de-mer trade was thought to represent 7 percent of all trade in marine resources by weight and 15 percent by value (excluding Australia and New Zealand) (Dalzell and Adams, 1994).

There is no commercial-level mariculture or sea ranching at present in the Western Central Pacific region. There have been however, some small research projects by The WorldFish Center, and a couple of experimental projects in Australia and Kiribati. Thus, all bêche-de-mer exports from the Western Central Pacific region are considered to be from capture fisheries.

The main harvesting methods in the Western Central Pacific region are gleaning and free-diving. Bombs (a weighted barb on a string line to harvest sea cucumbers at

TABLE 6
Catch per unit of effort (CPUE) rates for the Western Pacific Central region

Species	Location	No./diver/h	Reference
Actinoypa echinites	New Caledonia	118.0	Conand, 1989
A. mauritiana	PNG (Milne Bay)	<0.1	Allen, Kinch and Werner, 2003
A. miliaris	PNG (Milne Bay)	<0.1	Allen, Kinch and Werner, 2003
A. miliaris and *A. echinites*	Palau	68.2	Mathews and Oiterong, 1991
Bohadschia argus	PNG (Milne Bay)	0.4	Allen, Kinch and Werner, 2003
B. vitiensis	PNG (Milne Bay)	<0.1	Allen, Kinch and Werner, 2003
Holothuria atra	PNG (Milne Bay)	0.3	Allen, Kinch and Werner, 2003
	Solomon Islands (Western)	<519.0	Adams *et al.*, 1992
H. echinites	PNG (Central)	110.0	Shelley, 1981
H. fuscogilva	Fiji	~13.0	Conand, 1989
	Fiji	16.0	Gentle, 1979
	PNG (Central)	4.5	Shelley, 1981
	PNG (Milne Bay)	0.4	Sebatian and Foale, 2007
	PNG (Milne Bay)	0.2	Allen, Kinch and Werner, 2003
H. fuscopunctata	PNG (Milne Bay)	0.2	Allen, Kinch and Werner, 2003
H. scabra	New Caledonia	~13.0	Conand, 1989
	Palau	65.5	Mathews and Oiterong, 1991
	PNG (Central)	103.0	Shelley, 1981
H. whitmaei	Australia (Great Barrier Reef)	2.0–3.0	Uthicke and Benzie, 2000a
	New Caledonia	31.0	Conand, 1989
	PNG (Central)	15.1	Shelley, 1981
	Solomon Islands (Ontong Java)	11.1	Crean, 1977
	Solomon Islands (Ontong Java)	3.5	Bayliss-Smith, 1986
Stichopus chloronotus	PNG (Milne Bay)	0.3	Allen, Kinch and Werner, 2003
S. herrmanni	PNG (Milne Bay)	0.1	Allen, Kinch and Werner, 2003
	Palau	61.7	Mathews and Oiterong, 1991
Australostichopus mollis	New Zealand	52.0	Mladenov and Gerring, 1996
Thelenota ananas	New Caledonia	42.0	Conand, 1989
	PNG (Central)	10.0	Shelley, 1981
	PNG (Milne Bay)	0.2	Allen, Kinch and Werner, 2003
T. anax	PNG (Milne Bay)	0.1	Allen, Kinch and Werner, 2003

between 12–30 m depth), trawling, and the use of SCUBA and hookah, are also used occasionally. Women and children are often involved in gleaning on reef flats, whereas men do most of the diving. The use of trawl gear has been reported in Ontong Java in the Solomon Islands, whereby a small net is dragged along the sea bed behind two small boats (Ramofafia, 2004). Sea cucumbers also comprise some of the bycatch of trawl fisheries in Australia, but these are not allowed to be retained (DEH, 2006).

Usually, all species in accessible marine habitats are harvested. Fishing in recent years has changed in most PICTs to include a greater proportion of low-value species in exports and a greater range of species.

Catch figures are largely unobtainable, except for some limited data from Australia. The production of bêche-de-mer is mostly an export-driven industry. To obtain estimates of wet weight of catch, export figures need to be increased by a ratio of 1:10 (Preston, 1993). The authors present only the export data here (Figure 7; Appendix A). Of note though, is that the export data do not take into consideration such issues of buying "wet" or "first-boiled" sea cucumbers, shrinkage during processing and storage, rejection of undersize or damaged sea cucumbers/bêche-de-mer at point of sale, and sea cucumbers collected for subsistence use. Therefore, even using conversion ratios will under-estimate catches.

The export "production" of bêche-de-mer has varied more than an order of magnitude among some countries. The Melanesian countries and Australia are clearly the larger exporters. These countries have relatively large land mass and coastal lagoon area to support sea cucumber populations compared to Micronesian and Polynesian countries.

Within each country, exports have been markedly different over time. There is some indication of boom-and-bust cycles, for example in Fiji, over time periods of 3–10 years. This may, however, be accentuated by sociological and economic affects as much as the natural variability in resource abundance.

Interestingly, while bêche-de-mer exports from Melanesian countries have not particularly declined over the last two decades, those from Polynesia and Micronesia have. Steady declines in exports from most Polynesian and Micronesian countries are likely due to unsustainable fishing pressure and the fact that the sea cucumber populations are relatively sparse over relatively small areas. Sea cucumber populations in these countries are most likely limited, mostly depleted, and under the greatest risk of local extinctions of some species.

FIGURE 7

Exports of bêche-de-mer from the Western Central Pacific Region

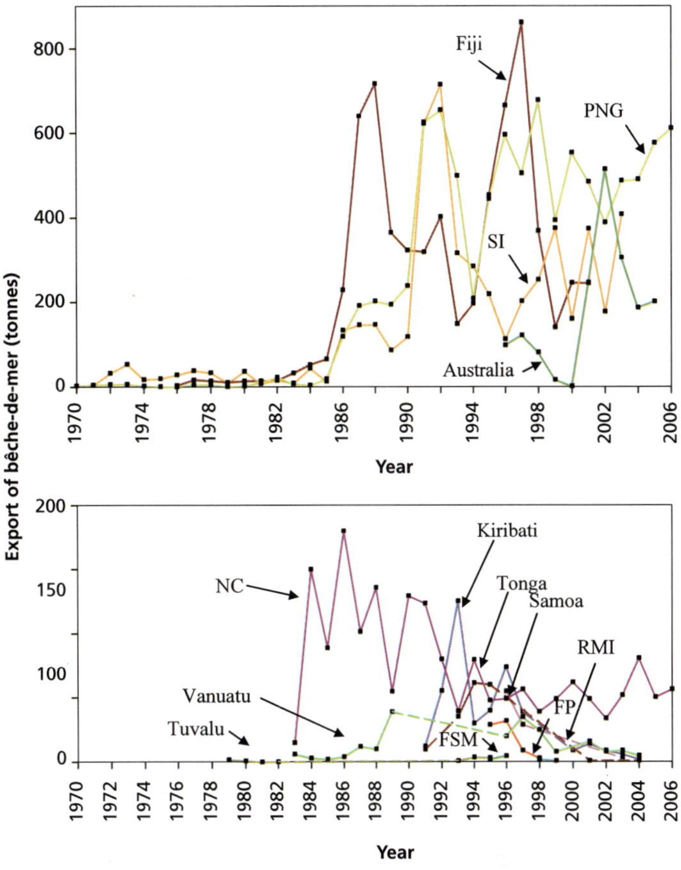

Note: High-volume countries (above) and medium- to low-volume countries (below). FP = French Polynesia, FSM = Federated States of Micronesia, NC = New Caledonia, PNG = Papua New Guinea, RMI = Republic of the Marshall Islands, and SI = Solomon Islands. No data for some years may simply reflect missing data rather than no exports. Each dashed line represents a period between two datum points for which data were unavailable.

2.8 Management measures

Sea cucumber fisheries are multi-species and thus present more complex management problems than a single-species fishery such as trochus (*Trochus niloticus*). Part of the complexity is owing to the diverse range of benthic habitats in which different species occur and differing demographic traits among species. SPC and ACIAR produced underwater identification cards of many of the common, commercially important holothurians, with biological traits and preferred habitats (Figure 8). The development of sustainable management in the Western Central Pacific has not been easy and current management tools like size limits, gear restrictions, spatial and temporal closures, quotas and marine reserves have not been overly effective (Preston, 1993; Dalzell, Adams and Polunin, 1996; Kinch, 2004b). An added problem is that, many countries in the Western Central Pacific region, except Australia and New Zealand, lack the necessary funds and technical capacity for adequate surveillance and enforcement (Adams, 1992; Jimmy, 1995, 1996; Kinch, 2002, 2004b; Kinch *et al.*, 2007; Trianni, 2002). In some countries, there are also conflicts of interest within differing levels and agencies of government, politicians and influential business people.

Gear restrictions and size limits are the most commonly used fisheries management tools in the Western Central Pacific region (Tables 7 and 8), with the harvesting of undersize animals a considerable problem (Kinch, 2002, 2004b, 2004c; Kinch *et al.*, 2007; Qalovaki, 2006). An economic analysis conducted by Callaghan (1996) in Micronesia showed that net revenue would fall as the size of sea cucumbers became smaller. Also,

FIGURE 8

Cover page and example (*H. scabra*) of waterproof identification cards for live and dried sea cucumbers of the Western Pacific Central region

COURTESY OF A. DESURMONT

Source: Prepared and distributed by the SPC.

fishers would actually lose money by collecting smaller sea cucumbers because the unit price remains low, while the cost of collection and processing is constant. Despite this, it is common practice by fishers in the PICTs to measure profit by the amount of cash that goes into their pocket, and subsequently continue to harvest any size of sea cucumber to gain immediate income from sale (Kinch *et al.*, 2007).

Micronesia

In Guam, current regulations permit the harvesting of 100 sea cucumbers $pers^{-1}$ d^{-1} (Kerr, A., University of Guam, personal communication). There are current moratoriums in Palau and the CNMI. In 1997, the government of Kosrae in the FSM placed a moratorium on commercial fishing, but it was later lifted due to lack of scientific data. Subsequently, a brief survey showed limited stocks (Lindsay, 2001b). The moratorium was therefore reinstated in 2000. In Yap a moratorium was instituted after the uncontrolled expansion of the fishery in 2007. The moratorium is due to be lifted in 2008 once a management plan for the fishery has been developed.

Polynesia

In Samoa, a moratorium was placed in 1994 (Mulipola, 1994). King and Fasili (1999) also reported that commercial fishing was banned in 41 percent of Samoan villages; and a number of villages have banned the harvesting of *S. horrens* for subsistence use (Eriksson, 2006). Following resource surveys in 1996, the Tonga Fisheries Authority recommended a moratorium in 1997 when the fishery was in serious decline after heavy fishing in the early 1990s (Preston and Lokani, 1990).

Melanesia

The National Fisheries Authority (NFA) is responsible for the conservation, management, development and sustainable use of PNG's fisheries under the 1998 *Fisheries Management Act.* The NFA has gazetted the 2001 *National Bêche-de-mer Management Plan* for regulation and management of the sea cucumber fishery (Polon, 2004). Its aim is to ensure that Papua New Guinea and its people obtain the maximum economic benefit from the fishery, and that fishing is sustainable and that fishing has a minimal impact on the marine and coastal environment. Following recent research, and the high proportion of undersized bêche-de-mer being purchased by exporting companies (Friedman and Gisawa, 2008), the NFA is now considering a three-year moratorium on the sea cucumber fishery, with consultation currently being conducted amongst stakeholders.

Previously in the Solomon Islands, there were no national fishery regulations, except for the ban on fishing for *H. scabra* in 1998, which was repealed in 2000. Following results of resource surveys by the Department of Fisheries and Marine Resources (DFMR) and other groups, fishing and exporting all sea cucumbers was banned in December 2005 until a management plan was put in place (Nash and Ramofafia, 2006). The ban was later lifted in order to allow communities in the tsunami-stricken Western Central and Choiseul provinces a means of gaining income but has recently been closed again.

In Vanuatu, an annual quota of 35 tonnes was established by legislation in 1991, and finally implemented in 1996 (Jimmy, 1995, 1996). A cooperative management scheme exists in which the Vanuatu Fisheries Division provides scientific information and advice, and coastal villages handle surveillance and local enforcement of the fishery regulations. Today, many villages employ temporal and spatial closures on harvesting sea cucumbers. Due to concernes on unsustainable harvesting of the sea cucumber a moratorium on fishing was imposed in 2008, and will be in place for a period of 5 years.

Prior to 2006, the only regulations for harvesting sea cucumbers in New Caledonia were a ban on the use of SCUBA and hookah, and no collection at night or in marine reserves. The government of the Northern Province has now imposed species-specific size limits for dried product and live animals. In the north, the Arama community suspends sea cucumber fishing during the April-January crab season (Northern Province Fisheries Division, 1993). In the Loyalty Islands Province, healthy stocks of many valuable species still persist, largely owing to district chiefs who forbid commercial fishing in a desire to preserve ecosystem health.

In 1984, Fiji introduced the *Bêche-de-Mer Exploitation Guidelines* (Qalovaki, 2006), and in 1988, imposed a size limit of 7.6 cm for all dried bêche-de-mer (Adams, 1988; Seeto, 1994). This strategy was initially effective, and exports dropped after the law came into force in 1989 (though this is probably linked to a resource decline). However, quantities of undersized product were reported in 1990, so new laws enabled the Fiji Fisheries Division to enforce the size limit at the exporting premises. There is evidence today that undersized product is still being exported from Fiji (Ram, in prep.). Although the number of exporters was also limited at one time, conflicts arose among stakeholders and this measure was rescinded (Adams, 1992). Fiji also prevents export of *H. scabra*, since it is a local food source.

Australia and New Zealand

In Australia, sea cucumbers are considered as regulated native species under national law (DEH, 2004, 2005; Shelly and Puig, 2004). The Torres Strait and Coral Sea fisheries are managed by the Australian Commonwealth through the Australian Fisheries Management Authority, while the remainder are managed by the their respective fisheries agencies in each State.

The Torres Strait fishery regulations include limiting the method of taking sea cucumber to either hand or a hand held non-mechanical implement, a ban on the use of hookah or SCUBA gear, a limit to boat length, limiting fishing of the one non-Islander licensed operator, minimum size limits, and a competitive TAC measured in gutted wet weight of harvested sea cucumbers. Minimal traditional fishing for subsistence is also allowed.

Commercial fishing on the Great Barrier Reef in Queensland is managed principally through a mixed species TAC of 380 tonnes gutted wet weight. In 2004/2005, the TAC allowed no *H. whitmaei*, 127 tonnes of *H. fuscogilva* and 253 tonnes of other species.

TABLE 7
Past and previous management measures adopted in selected countries in the Western Central Pacific region

	Micronesia				Polynesia		Melanesia					Australia and New Zealand							
Species	Palau	Guam	CNMI	FSM	Samoa	Tonga	Papua New Guinea	Solomon Islands	Vanuatu	New Caledonia	Fiji	Coral Sea	Torres Strait	Great Barrier Reef	Moreton Bay	Northern Territory	Western Australia	New Zealand	Total
Area restrictions												✓		✓	✓	✓	✓		5
Bag limits		✓										✓	✓	✓					4
Effort controls												✓	✓	✓	✓	✓	✓		6
Gear restrictions				✓		✓	✓			✓	✓	✓	✓	✓	✓	✓	✓		11
Moratorium	✓		✓	✓	✓	✓		✓											6
No night harvesting							✓			✓									2
Quotas (for companies)									✓			✓		✓				✓	4
Restricted fishery (Nationals only)							✓				✓								2
Rotational harvesting												✓		✓					2
Size limits					✓	✓	✓			✓	✓	✓	✓	✓	✓	✓	✓		11
Temporal closures							✓								✓				2
Total allowable catches							✓					✓	✓	✓	✓				5

TABLE 8
Live size limits (cm – body length) adopted in selected countries in the Western Central Pacific region (dry size limits in parenthesis)

	Polynesia	Melanesia			Australia				
Species	Samoa	PNG	New Caledonia (Province Nord)	Fiji	Torres Strait	Great Barrier Reef	Moreton Bay	Northern Territory	Western Australia
Actinopyga echinites		25 (12)			12	20		12	12
A. lecanora		15 (10)							
A. mauritana		20 (8)	25 (12)		22				
A. miliaris		15 (10)	25 (12)		22				
Bohadschia argus		20 (10)							
B. similis		25 (7)							
B. vitiensis		20 (10)							
Holothuria atra		30 (15)			15	20		15	15
H. edulis		25 (10)							
H. fuscogilva		35 (15)	35 (16)		32	40		32	32
H. fuscopunctata		45 (15)			24				
H. scabra		22 (10)	20 (10)		18	16	17	16	16
H. scabra var. *versicolour*			30 (11)						
H. whitmaei		22 (10)	30 (16)		25	30		26	26
Stichopus chloronotus	(6.3)	20 (10)							
S. herrmanni		25 (10)	35 (15)		27				
S. horrens							17		
Thelenota ananas		25 (10)	45 (20)		30	50		30	30
T. anax		20 (10)							
All species				(7.6)					
All species with exception of *S. chloronotus*	(7.6)								

For other species, catch trigger points have been set. Once a trigger point is reached, the sea cucumber industry must carry out a stock assessment of that particular species to determine a TAC (Lowden, 2005). There are currently 18 licences in the fishery held by two companies. Minimum size limits are also in place for all the major species (Roelofs, 2004, 2005). A Rotational Zoning Scheme (RZS) was implemented in 2004, allowing limited harvesting of a zone once every three years. Recreational harvesting is limited to an in-possession limit of five specimens, excluding *H. whitmaei* which is banned (Roelofs, 2005).

The Moreton Bay fishery, near Brisbane, is managed through permits as a developmental fishery and stringent harvest controls. Only *H. scabra* and *S. horrens* can be fished (McCormack, 2005), though *S. horrens* is currently not taken (Skewes, T., Commonwealth Scientific and Industrial Research Organization, personal communication). Management regulations also include minimum size limits, a seasonal closure, limits to number and size of boats and the number of fishers, fishing zones, gear restrictions (harvesting by hand collection only), an annual TAC, zonal trigger limits and rotational harvesting (McCormack, 2005).

In the Northern Territory, fishing is restricted to waters seaward of the high water mark to three nautical miles. The fishery is divided into two zones. Three licenses are permitted in each zone, now acquired by one company. There is no TAC set for the sea cucumber fishery in the Northern Territory, though a management action is triggered if the annual harvest of sea cucumbers exceeds 640 tonnes. This figure is set at the level that equates to twice the highest reported catches taken since 1992 and allows for continued growth in the fishery, within what is considered to be a sustainable harvest level. There are concerns however that this reference point does not take into account available research on the sustainability of the target stock and is not sufficiently precautionary to ensure the ecological sustainability of the fishery (DEH, 2004).

In Western Central Australia, the fishery management regulations include limited entry, species size limit and limits on the number of divers permitted to harvest under each licence. Currently six licences and one exemption are in operation.

3. TRADE

As noted above, monitoring of the bêche-de-mer trade is difficult and complicated, and the collection of accurate trade data from the Western Central Pacific region has proven difficult. The predominant form of sea cucumber product exported from the Western Central Pacific region is bêche-de-mer with very small amounts exported frozen or salted.

Bêche-de-mer is primarily exported from producing countries in the Western Central Pacific region to central markets in China Hong Kong SAR, and Singapore, and then re-exported to China and to other Chinese consumers worldwide. Apart from species and size, factors that affect price for the dried product are the general appearance (colour, shape), odour, moisture content and spoilage. A pleasing smooth surface, uniform shape and clean (not ragged) cut in the body wall demonstrate proper handling and processing and a pleasant smell reflects proper storage.

As a result of demand and increasing prices, both the volume and total value of bêche-de-mer exports from the region have increased in recent years. Prices on the international market for bêche-de-mer also vary considerably between importing country, though an increasing upward trend is easily observed. Export prices vary between the different producer countries as well. It is very difficult to get Asian import prices for several reasons. Firstly, prices for bêche-de-mer vary, often according to the amount and quality sent. Secondly, prices include profit margin and wider knowledge of this may push the threshold price up, resulting in lower profits. Thirdly, prices vary according to how they are purchased, for example if a foreign financier advances money to a in-country licence holder, the financier is the one who

TABLE 9

Bêche-de-mer imports to China Hong Kong SAR from the Western Central Pacific region (tonnes)

	Micronesia			Polynesia		Melanesia				Australia and New Zealand	
Year	Palau	Marshall Islands	Kiribati	Samoa	Tonga	PNG	Solomon Islands	Vanuatu	Fiji	Australia	New Zealand
1993			99.0			179.0	319.0	6.0	119.0		
1994			130.0			150.0	247.0	40.0	176.0		
1995						236.0	161.0		402.0		
1996			23.2	17.5		351.6	141.7	25.1	550.4	78.9	9.5
1997			6.5	10.7		470.1	134.5	35.1	562.7	76.1	2.8
1998			14.9	4.4		639.8	260.6	25.4	304.3	119.3	
1999			6.5	5.7		350.3	49.7	8.0	168.3	125.3	0.5
2000			9.1			531.9	149.1	28.5	364.4	146.5	11.0
2001			22.8			541.1	259.7	16.6	291.1	186.9	31.2
2002	6.4		8.6			380.6	248.8	8.4	235.5	138.7	15.8
2003	2.6	2.7	5.5		0.3	447.6	222.8	9.0	264.3	136.4	3.5
2004	17.8		1.9		1.1	518.3	153.3	5.3	282.3	128.1	1.7
2005	24.9		3.6		0.1	555.3	155.1	9.9	313.9	144.8	6.9
Total	**51.6**	**2.7**	**331.6**	**38.3**	**1.5**	**5351.6**	**2502.1**	**217.3**	**4034.0**	**1280.9**	**83.0**

Source: TRAFFIC Hong Kong Special Administrative Region; Ferdhouse, 1999, 2004; CITES, 2006; Conand, 2006. Note: Where figures vary from sources, the highest value has been used.

TABLE 10

Singapore bêche-de-mer imports from the Western Central Pacific region (tonnes)

	Melanesia			Australia and New Zealand	
Year	PNG	Solomon Islands	Fiji	Australia	New Zealand
1993	146.0	3.0	3.0	13.0	4.0
1994	80.0	4.0	8.0	47.0	9.0
1995	89.0		17.0	80.0	
1996					
1997				12.0	5.0
1998	2.0			7.0	13.0
1999	1.0		20.0	14.0	12.0
2000	1.0			4.0	19.0
Total	**319.0**	**7.0**	**48.0**	**177.0**	**62.0**

Source: Ferdhouse, 1999, 2004; Conand, 2006. Note: Where figures vary from sources, the highest value has been used.

FIGURE 9

Average prices per kg for bêche-de-mer exported from the Western Central Pacific region

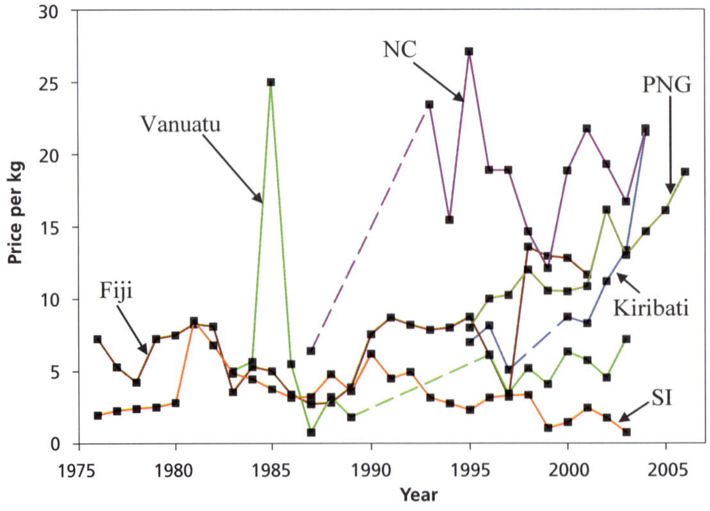

Note: Compiled from export data. NC = New Caledonia, PNG = Papua New Guinea, and SI = Solomon Islands. Data are for years in which both the export quantity and value was recorded in national data; each dashed line represents a period between two data points for which data were unavailable.

takes the risk and subsequently expects to pay less for product purchased. There are also considerable costs along the chain of custody such as direct costs (e.g. processing, packing, freight, insurance, commissions, wastage, interest), and indirect costs or overheads (e.g. power, transport, wages, fuel, travel, communications) before adding a profit margin.

A summary of bêche-de-mer imports from the Western Central Pacific region into China Hong Kong SAR, and Singapore is provided in Tables 9 and 10. Small amounts of bêche-de-mer are known to be imported into the United States of America, Malaysia and China, but no accurate or current data are available. There is also probably a small amount of bêche-de-mer exported by air for personal consumption from countries that have subsistence fisheries in Micronesia and Polynesia, as well as by Asians residing in PICTs.

An analysis of average export prices of bêche-de-mer over time among different countries revealed two key findings (Figure 9). Firstly, some countries have received up to 10 times greater prices than others for the bêche-de-mer they exported. This could be due three reasons: the fishers are receiving a better price for the same product than those in other countries; or countries exported a greater percentage of the species of higher value on the Asian seafood market; or the grade exported bêche-de-mer is better in some countries because animals are larger or better processed. Notably, New Caledonia, Papua New Guinea and Kiribati mostly received relatively good prices for the bêche-de-mer exported, while Vanuatu and Solomon Islands generally received poor prices.

Secondly, the average prices of exported bêche-de-mer have evolved considerably over time. For Fiji and Kiribati, the export prices received for bêche-de-mer have increased markedly in recent years. Again, this could be due to several factors notably that the importers, or foreign buyers, gave increasingly attractive prices and/or that fishers started collecting and exporting higher-value species (although this is less likely). In at least the Solomon Islands, export prices have declined since the early 1980s, most likely due to exports comprising progressively higher proportions of low-value species each year.

4. SOCIO-ECONOMIC IMPORTANCE TO LOCAL FISHING COMMUNITIES

Human populations in the Western Central Pacific region are at various stages of Western-style development. Subsequently, socio-economic conditions vary widely between and within countries. In general, Australians, New Zealanders and peoples of the dependent territories of metropolitan countries have better access to goods and services than those of the independent nations. In general, urban residents live a more consumer merchandised lifestyle than those with relatively low incomes who follow a more subsistence way of life in small isolated islands and remote coastal areas. It is this latter group that the sea cucumber fishery has a direct impact on their sociological and economic well-being by providing income-earning opportunities and direct financial benefits where other earning opportunities are limited (Figure 10) (Kinch, 2002; Kinch *et al.*, 2007).

Details are not generally available on the depth of dependency of coastal communities on the sea cucumber fishery in the Western Central Pacific region. The only sources identified were household census information, country specific studies and the regional SPC PROCFish/C data sets. Only a limited number of household censuses taken in the Pacific contained questions pertaining directly to sea cucumbers. A scan of questionnaires from other countries in the Pacific revealed many questions relating to fisheries, but no particular sections targeting information on sea cucumbers.

Country-specific data collection have been conducted in the Solomon Islands (Ramofafia, 2004; Ramofafia *et al.*, 2007), Papua New Guinea (Kinch *et al.*, 2007; NFA, 2005, 2006, 2007b), and Samoa (Eriksson, 2006), and New Caledonia (Purcell, Gossuin and Agudo, in press). Each of these studies found sea cucumber fisheries to be of great importance to fisher communities, being in some cases, the major source of income.

FIGURE 10

Artisanal sea cucumber fishers and processors in the Western Central Pacific region. Fishers in Fiji taking their catch to be sold to a local buyer. Centre: Fishers in New Caledonia with home-dried bêche-de-mer. Right: Children processing *H. scabra* in the Western Province, Papua New Guinea

COURTESY OF K. FRIEDMAN, S. PURCELL, J. KINCH

During recent surveys in Samoa, the subsistence sea cucumber fishery was considered to be an important element to fishers' livelihoods and their families, with 29 percent of all respondents selling either all or part of their sea cucumber catch at local markets (Eriksson, 2006; Friedman *et al.*, 2006).

During the lead author's (Jeff Kinch) PhD work at Brooker Island in the Milne Bay Province, bêche-de-mer sales contributed 47 percent of all household incomes (Kinch, 1999). A socio-economic survey conducted by Coastal Fisheries Management and Development Program (CFMDP) in the Milne Bay Province showed that 22 percent of households surveyed, harvested sea cucumbers for income (NFA, 2006). A similar survey in the New Ireland Province showed that 12 percent of households harvested sea cucumbers for income (NFA, 2005). In the West New Britain Province, Koczberski *et al.* (2006) also found the sale of bêche-de-mer to be an important source of income.

A recent socio-economic assessment of the sea cucumber fisheries in the Western Central, Central and Manus Provinces of Papua New Guinea by Kinch *et al.* (2007) showed that on average, households that harvested sea cucumbers could make between USD 1 000–3 000 y^{-1} from the sale of bêche-de-mer. In Manus Province, of all households surveyed, 75 percent stated that harvesting sea cucumbers was their most important income stream (Kinch *et al.*, 2007). During this survey, between 24 and 40 percent of all households stated they would move to other forms of fishing and harvesting other marine resources if they could no longer harvest sea cucumbers, and thus possibly starting a spiral of fishing down other valuable species. Of particular concern would be the fishing for sharks for sharkfins, and sea turtles.

In Solomon Islands the 1999 census showed that village wards (a local-level census unit that may consist of part of a village, a village or a group of villages) with a predominantly coastal location had 12 percent of households involved in selling sea cucumbers as bêche-de-mer. Sea cucumber fishery is of more importance in the smaller island groups of Ontong Java, the Reef Islands and Temotu, and especially the Western Province (Figure 11). The number of people fishing for sea cucumbers is believed to have increased greatly subsequent to the census as the ethnic tension closed down national systems for exporting copra and cocoa, leaving many rural communities with no other source of income. In the Rennell and Belona island groups, where cyclones have recently destroyed plantations, bêche-de-mer remains the main source of cash, while at Ontong Java, bêche-de-mer has been the main source of income for decades

FIGURE 11

A depiction of the percentage of households that sell bêche-de-mer in the Solomon Islands. Wards in yellow support the average numbers of sellers whereas those in orange and red have the highest level of household involvement in the trade. Box inserts are Ontong Java in the North East, and the Reef Islands and Temotu in the South East

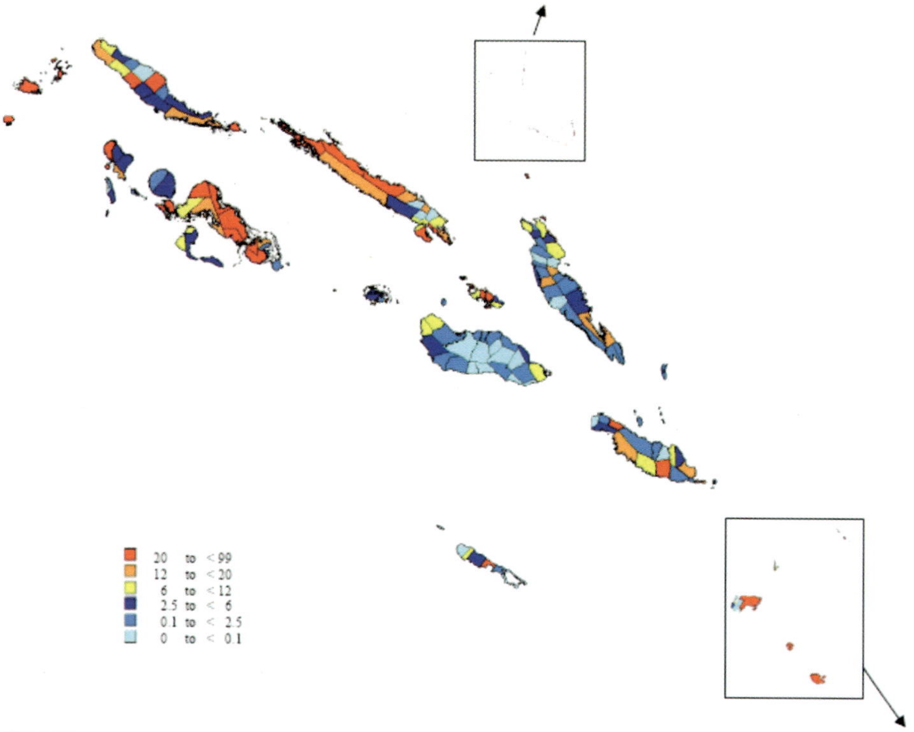

COURTESY OF SPC

(Bayliss-Smith, 1986). In 1999, Donnelly (2001) also conducted a household survey, towards research into the economics of the Live Reef Fish Food Trade (LRFFT). He stated that a quarter of all households in the Marovo Lagoon were obtaining cash from the harvesting sea cucumbers. During the recent International Waters Project (IWP) surveys in 2004, it was determined that 63 percent of households at Mbili Passage stated that bêche-de-mer contributed to household income, whilst 27 percent at Chea stated likewise (Kinch *et al.*, 2006). Kia community had a high reliance on the bêche-de-mer fishery for income (Ramofafia *et al.*, 2007). Despite the fishery showing signs of overfishing, little effort had been made neither to address those warning signs nor to pro-actively investigate alternative means of providing community income.

In New Caledonia, sea cucumbers were the most economically important wild fishery export in 2007, being twice the value of finfish exports (Purcell, Gossuin and Agudo, in press). Sea cucumbers are the most important income source for most of the fishers who collect them, while some others rely more on fishes, molluscs, and crusteans. Fishers remote from urban processing centres tend to process the sea cucumbers into bêche-de-mer themselves and rely on selling this and other non-perishable transformed marine animals more than fishes or bivalves for their income (Purcell, Gossuin and Agudo, in press).

When comparing the benefits of other fisheries such as tuna versus sea cucumbers, it is important to note that the extremely diffuse, village-level nature of the sea cucumber fishery means that the financial benefits are very widely distributed amongst coastal and island villages. Although tuna fisheries and their licensing fees also bring in foreign currency, they contrast with sea cucumber fisheries in being centralized and highly capital-intensive, with benefits delivered at the national scale over a long term, rather than locally and immediately.

5. CURRENT RESEARCH AND PROJECTS

Fisher, buyer and exporter surveys are currently being conducted in 17 PICTs as part of the SPC PROCFish/C programme (funded by the European Union). This includes the collection of species specific catch and value information through fully structured questionnaires conducted with fishers, buyers and exporters.

In the Solomon Islands, a project implemented by the WorldFish Centre (funded by ACIAR) examined the importance of the sea cucumber fishery for the community at Kia, in the Santa Isabel Province, one of the major areas of bêche-de-mer production in that country. This project entitled "Improving Sustainability and Profitability of Village Sea cucumber Fisheries in Solomon Islands" was halted when the Solomon Islands government placed a moratorium on the sea cucumber fishery in December 2005.

Biological and ecological surveys are underway across the Pacific by the SPC-PROCFish/C programme to get a better understanding of current densities of sea cucumbers (Figure 12). In-water surveys are being conducted in 17 PICTs with the purpose of providing governments and communities with information that will assist them in improving management and governance of reef fisheries and coastal environments. In this regional overview, four to six sites are surveyed for each PICT and the same survey techniques are employed throughout the region, allowing comparison between, as well as within, countries. Further analysis will assist in distilling the various survey techniques employed in this study, to develop indicators for monitoring sea cucumber fisheries in the future.

With a similar regional approach, but concentrating more on ecosystems than fisheries, NOAA is surveying US-affiliated Pacific Islands. The Coral Reef Ecosystem Division (CRED) of NOAA uses consistent survey methodology to get data on applied management-relevant issues for coral reef ecosystems, which includes data on sea cucumbers (Brainard, R., National Aeronautical and Oceanographic Agency, personal communication).

A recent project within the ZoNéCo programme evaluated population densities, abundances and sizes of sea cucumbers at 50 sites along New Caledonia's main island, La Grande Terre (WorldFish Center and SPC). Along with landing and sociological surveys, the underwater survey data were used to provide recommendations for better management of sea cucumber resources (Purcell, Gossuin and Agudo, in press).

With continued Asian market demand, mariculture could provide a substantial share of the bêche-de-mer market, and partially replace wild stocks. Baine (2004) suggests that the push for hatchery and restocking programmes is linked to the difficulties of managing sea cucumber fisheries. It should be cautioned here, that whilst these programmes have potential, they are only in their experimental stages at present, with just one (possibly two) species under investigation. Restocking and sea ranching should not be heralded as replacements for sound management. Moreover, inter-regional

FIGURE 12

Surveying for sea cucumbers under the SPC-PROCFish/C Project. Left: A diver with manta board, conducting a population survey. Right: The data sheet and tally counter on the manta board for recording species counts and habitat features

COURTESY OF K. FRIEDMAN

translocation of the main species of interest, sandfish *H. scabra,* is strongly discouraged, as it can lead to introgression of foreign genes into the existing populations, or allow for a non-interbreeding population to out compete local stocks (Uthicke and Purcell, 2004). Unless done carefully, and with local broodstock, translocations can also risk the introduction of new parasites and diseases.

Over the past decade, several hatchery, grow-out and optimal releasing projects have been undertaken by the WorldFish Center (previously ICLARM) in the Solomon Islands and New Caledonia with the aim of assessing the potential for restocking to restore depleted sea cucumber fisheries (Purcell, Gardener and Bell, 2002). Purcell and Simutoga (2008) recently indicated that it takes at least 2–3 years for *H. scabra* released in the wild to grow to a good market size of around 700 g, with roughly 7–20 percent survival if the habitats for releasing juveniles and the size-at-release are well chosen. Early experiments showed that *H. scabra* could be co-cultured with juvenile shrimp (Purcell, Patrois and Fraisse, 2006b), but further trials dismissed commercial culture with larger shrimp (Bell *et al.*, 2007). Recent findings give hope to commercial on-growing of sandfish in monoculture in earthen ponds, with fast growth rates of 1–3 g ind. day^{-1} (Bell *et al.*, 2007).

As a component of ACIAR's "Sustainable Aquaculture Development in the Pacific Islands Region and Northern Australia", production technology was transferred from the WorldFish Center to the Queensland Department of Primary Industries and Fisheries in Cairns, Australia. Broodstock of *H. scabra* from the Moreton Bay area was used in three spawnings. Juveniles from these productions were used for grow-out trials in shrimp farms in Townsville and Cairns, but these also indicated that *H. scabra* juveniles could not be co-cultured with shrimp (Hair, C., Queensland Department of Primary Industries and Fisheries, unpublished data).

A new sea ranching/restocking research project began in 2007 and will have four years to apply the experimental hatchery and release methods developed for sandfish, *H. scabra* (WorldFish Center and partners, funded by ACIAR). This project aims to assess the economic and social feasibility of sea ranching and restocking through commercial-scale trials with communities in the Northern Territory, Australia, and in the Philippines.

In addition, a private company from Queensland, Australia, has also been successful in mass producing *H. scabra* in the hatchery. They have released thousands of juvenile *H. scabra* in designated aquaculture areas over the last few years to try to enhance yields (Ivy and Giraspy, 2006). The company also intends to transfer juvenile *H. scabra* from Australia to communities in the Pacific, in joint fishing arrangements, but this has been met with controversy over the translocation issues discussed above.

The Fisheries Division of Kiribati has maintained a sea cucumber hatchery for many years and has cultured *H. fuscogilva* and released the juveniles in the wild for stock enhancement. Tens of thousands of juvenile *H. fuscogilva* have been placed on the reefs around Tarawa and neighbouring Abaiang Atoll, but survival has been poorly monitored, so there is limited information on optimal habitats and sizes for release (Friedman and Tekanane, 2006).

6. ADDITIONAL THREATS TO SEA CUCUMBER POPULATIONS

The biggest immediate threat to sea cucumber fisheries in the Western Central Pacific region is over-harvesting. As high value species become overexploited the focus shifts to lower-value species. Another trend is fishers exploiting new grounds further a field once they have removed all animals from a location. Until recently, deep water and remoteness have provided refuge for some heavily fished species, because most collection was done by wading or snorkeling. Some studies indicate that overexploited populations of sea cucumbers may require as much as 50 years in the absence of fishing pressure to rebuild (Bruckner, Johnson and Field, 2003; Uthicke, 2004).

Overfishing has been documented in Tonga (Lokani, Matoto and Ledau, 1996), Fiji (Preston, 1988), the Solomon Islands (Nash and Ramofafia, 2006), Papua New Guinea (Skewes *et al.*, 2002a; Kinch, 2002, 2004a; Kinch *et al.*, 2007), New Caledonia (Conand, 1989, 1990), Australia (Shiell, 2004; Skewes *et al.*, 2000, 2002b, 2006; Uthicke and Benzie, 2000a; Uthicke, 2004; Uthicke, Welch and Benzie, 2004). The reefs located in the Timor MOU Box in Western Australian reef are also heavily fished by Indonesian fisherman (Vail and Russel, 1990; Skewes *et al.*, 1999; Smith *et al.*, 2001, 2002; Rees *et al.*, 2003).

Continued over-harvesting of sea cucumbers can lead to reduced densities, and this can lead to disproportionally reduced larval production (Uthicke, 2004), a form of Allee effect (Allee *et al.*, 1949). Because holothurians are broadcast spawners, when adults are at low densities, eggs released by females may not encounter sperm by distant males, thus continuing a spiral of declining local population size.

Sea cucumbers can also be affected by coastal run-off, particularly those species that inhabit areas near riverine environs.

Global climate change is recognized as a major threat to the coral reefs. Significant coral bleaching events have been observed in recent years throughout the Western Central Pacific region (Wilkinson, 2004), thus affecting some sea cucumber species' habitats.

Human-induced increases in of atmospheric carbon dioxide (CO^2) concentrations are expected to cause rapid changes in the earth's climate causing significant effect on coastal ecosystems, especially estuaries and coral reefs through changes in temperature, sea level rise, the availability of water and associated nutrients from precipitation and runoff from land, wind patterns, and storminess. Increased CO^2 concentrations lower ocean pH, which in turn lower saturation states of the carbonate minerals calcite, aragonite, and high-magnesium calcite, the materials used to form supporting skeletal structures in many major groups of benthic calcifiers such as corals calcifying macroalgae, benthic foraminifera, molluscs, and echinoderms resulting in smaller size and body weight (Shirayama and Thornton, 2005). It is possible, but untested, that reduced calcification rates would also effect holothurian spicule formation.

Rising temperatures also influence organism's biology, affect dissolved oxygen concentrations in water, and play a direct role in sea level rise and in major patterns of coastal and oceanic circulation. Global climate change, thus has a major potential to affect sea cucumber populations and could in fact, result in the extinction of some species, the alteration of species distributions, and modifications in the flow of energy and cycling of materials within ecosystems. On reefs near low-lying coastal areas, sea level rise would likely increase coastal erosion rates, thus degrading water quality and reducing light penetration.

The coral reefs of New Caledonia, Samoa, Solomon Islands and Vanuatu have been damaged by cyclones in recent years. Cyclone Erica in 2003 destroyed 10–80 percent of live coral cover in some areas of New Caledonia, while Cyclone Heta struck Samoa in 2004, damaging 13 percent of the coral reefs (Lovell *et al.*, 2004).

ACKNOWLEDGEMENTS

The authors would like to thank the following persons for their input, advice and/or provision of information: Alessandro Lovatelli, Aliti Vunisea, Antoine Gilbert, Arsene Stein, Cathy Hair, Chantal Conand, Chris Ramofafia, Franck Magron, Frank Rowe, Joe Konno, Kalo Pakoa, Marcelo Vasconcellos, Marion Henry, Mark Baine, Masahiro Ito, Mecki Kronen, Michael Tekanene, Paul Christian Ryan, Paul Dalzell, Poh Sze Choo, Sam Grant, Samasoni Sauni, Semese Alefaio, Sheila McKenna, Timothy Lam, Timothy Skewes, Veronica Toral-Granda, Warwick Nash, William Sommerville and Yves Samyn. WordFish Center contribution number 1851.

REFERENCES

Adams, T. 1992. Resource Aspects of the Fiji Beche-de-mer Industry. *SPC Beche-de-Mer Information Bulletin,* 4: 13–16.

Adams, T. 1998. The Interface between Traditional and Modern Methods of Fishery Management in the Pacific Islands. *Ocean and Coastal Management,* 40: 127–142.

Adams, T., Aldan, C., Alfred, V., Bertram, I., Bukkurrou, A., Cruz, J., Flores, T., Rosa, F., Seman, R. & Taman, J. 1994. *Assessment of the Northern Marianas Trochus Resource and Recommendations for the Fishery.* Report prepared for South Pacific Commission, Noumea, New Caledonia.

Adams, T., Leqata, J., Ramohia, P., Amos, M. & Lokani, P. 1992. *Pilot Survey on the Status of Trochus and Beche-de-mer Resources in the Western Province of the Solomon Islands with Options for Management.* Report prepared for South Pacific Commission, Noumea, New Caledonia.

Allee, W., Emerson, A., Park, O., Park, T. & Schmidt, K. 1949. *Principles of Animal Ecology.* W.B. Saunders, Philadelphia.

Allen, M., Kinch, J. & Werner, T. 2003. A Basic Stock Assessment of the Coral Reef Resources of Milne Bay Province, Papua New Guinea, Including a Study of Utilization at the Atisanal Level. *In: A Rapid Biodiversity Assessment of the Coral Reefs of Milne Bay Province, Papua New Guinea – Survey II (2000).* G, Allen, J. Kinch, S. McKenna and Seeto, P. (eds.). RAP Bulletin of Biological Assessment No. 29. pp. 56–74. Washington, DC, Conservation International.

Anon. 1997. *L'exploitation des holothuries en Polynésie Française. Papeete, Tahiti, Polynésie Française.* Papeete, Service de la Mer et de l'Aquaculture.

Anon. 1998. *Rapport final relatif à la réalisation d'une étude préalable sur l'exploitation des bêches de mer (Rori): Analyse, par simulation, de la rentabilité économique d'une exploitation des holothuries en Polynésie Française.* Papeete, BIODAX Consulting.

Anon. 1999. *Rapport interne de mission: Exploitation des holothuries (rori) à Fakarava par S. C. A. Hirifa.* Papeete, BIODAX Consulting.

Anon. 2003. *Invertebrate Survey for Airai State.* Report prepared for The Environment Inc., Koror, Republic of Palau.

Australian Fisheries Management Authority. 2004. *Draft Assessment Report – Torres Strait Bèche-de-mer Fishery.* Report prepared for the Australian Fisheries Mangement Authority, Thursday Island, Queensland, Australia.

Australian Fisheries Management Authority. 2005. *Draft Assessment Report – Torres Strait Beche-de-mer Fishery.* Report prepared for the Australian Fisheries Management Authority, Thursday Island, Queensland, Australia.

Baine, M. 2004. From the Sea to the Market Place: An Examination of the Issues, Problems and Opportunities in Unravelling the Complexities of Sea Cucumbers and Trade. *In: Advances in Sea cucumber Aquaculture and Management. A.* Lovatelli, C. Conand, S. Purcell, S. Uthicke, J-F. Hamel and A. Mercier, (eds). pp 119–131. FAO, Rome.

Baker, J. 1929. On the Zonation of Some Coral Reef Holothuria. *Journal of Ecology,* 17: 141–143.

Bakus, G. 1968. Defensive Mechanisms and Ecology of some Tropical Holothurians. *Marine Biology,* 2: 23–32.

Bakus, G. 1973. The Biology and Ecology of Tropical Holothurians. *In. Biology and Geology of Coral Reefs.* O. Jones and R. Endean, (Eds.). 2 Volumes. pp 325–367. New York, Academic Press.

Baquie, B. 1977. *Fishing in Rarotonga.* Unpublished M.A. Thesis, University of Auckland, Auckland, New Zealand.

Battaglene, S. & Bell, J. 2004. The Restocking of Sea Cucumbers in the Pacific Islands. *In: Case Studies in Marine Ranching.* D. Bartley and K. Leber, (eds). FAO Fisheries Technical Paper, No. 429. pp 109–128. FAO, Rome.

Bayliss-Smith, T. 1986. *Ontong Java Atoll: Population, Economy and Society, 1970-1986.* Occasional Paper, No. 9. Fiji, South Pacific Smallholder Project.

Beckett, J. 1977. The Torres Strait Islanders and the Pearling Industry: A Case of Internal Colonialism. *Aboriginal History,* 1(1): 77–104.

Belhadjali, K. 1997. Beche-de-mer Production in Tuvalu. *SPC Beche-de-Mer Information Bulletin,* 9: 2–3.

Bell J., Purcell S., Agudo N., Blazer P., Simutoga M., Pham D. & Della Patrona, L. 2007. Grow-out of Sandfish *Holothuria scabra* in Ponds shows that Co-culture with shrimp *Litopenaeus stylirostris* is not Viable. *Aquaculture,* 273: 509–519.

Bennett, J. 1987. *Wealth of the Solomons: A History of a Pacific Archipelago, 1800–1978.* Pacific Islands Monograph Series, No. 3. Honolulu, University of Hawaii Press.

Beumer, J. 1992. Queensland's Beche-de-mer Fishery. *SPC Beche-de-Mer Information Bulletin,* 4: 12.

Birkeland, C., Green, A., Guilbeaux, M., Donaldson, T., Emilio, D., Kirkendale, L., Mangel, J., Myers, R., Wemg, K. & van Woesik, R. 2000. *Helen Reef Marine Resources in the Year 2000 - Results of Scientific and Community Monitoring Surveys: April 24-May 3, 2000, Hatohobei State, Republic of Palau.* Report prepared for the Helen Reef Marine Resource Management Program, Koror, Republic of Palau.

Bruckner, A., Johnson, K. & Field, J. 2003. Conservation Strategies for Sea cucumbers: Can a CITES Appendix II listing Promote Sustainable International Trade? *SPC Beche-de-Mer Information Bulletin,* 18: 24–33.

Bungitaki, J. & Lindsay, S. 2004. Marine Resource Survey and Assessment of Jaluit Atoll, Republic of the Marshall Islands. *SPC Beche-de-Mer Information Bulletin,* 19: 33–35.

Callaghan, P. 1996. An Economic Model of Sea cucumber Export in Micronesia. *In: Suggestions for the Management of Sea cucumber Resources in Micronesia: Results of the Workshop 'A regional management plan for a sustainable Sea Cucumber Fishery for Micronesia'.* R. Richmond, (Ed). University of Guam Marine Laboratory Technical Report, No. 101. pp 51–65. Mangilao, University of Guam.

Cannon, L. & Silver, H. 1987. *Sea Cucumbers of Northern Australia.* Brisbane, Queensland Museum.

Carroll, T. 2005. Immune from Pessimism: Unicorn Pacific Makes a Mark in Vanuatu. *Pacific Magazine,* August: 13.

Chambers, M. 1989. A Survey of the Beche-de-mer of Vanuatu. *In: The Marine Resources of Vanuatu.* T. Done and K. Navin, (eds.). pp. 107–114. Townsville, Australian Institute of Marine Science.

Chao, S., Chen, C. & Alexander, P. 1994. Reproduction and Growth of *Holothuria atra* (Echinodermata: Holothuroidea) at Two Contrasting Sites in Southern Taiwan. *Marine Biology,* 119: 565–570.

Cheyne, A. 1852. *A Description of Islands in the Western Pacific Ocean.* London.

Clark, H. 1946. *The Echinoderm Fauna of Australia.* Washington, DC, Carnegie Institute of Washington.

Conand, C. 1981. Sexual Cycle of Three Commercially Important Holothurian Species (Echinodermata) from the Lagoon of New Caledonia. *Bulletin of Marine Science,* 31(3): 523–543.

Conand, C. 1988. *Beche-de-mer in New Caledonia: Biology and Fishing.* Paper prepared for the South Pacific Commission's Workshop on Pacific Inshore Fishery Resources, 14–25 March, 1988, Noumea, New Caledonia.

Conand, C. 1989. *Les Holothuries Aspidochirotes du Lagon de Nouvelle-Caledonie: Biologie, Ecologie et Exploitation.* Etudes et Thesses, ORSTROM, Paris, France. (in French – the 'ecology' and 'biology' sections translated into English by the SPC).

Conand, C. 1990. *The Fishery Resources of Pacific Island Countries, Part Two: Holothurians. Fisheries Technical Paper*, No. 272.2. Rome, Food and Agriculture Organization.

Conand, C. 1993. Reproductive Biology of the Holothurians from the Major Communities of the New Caledonian Lagoon. *Marine Biology,* 116: 439–450.

Conand, C. 1996. Asexual Reproduction by Fission in *Holothuria atra*: Variability of some Parameters in Populations from the Tropical Indo-Pacific. *Oceanologia Acta,*19(3–4): 209–216.

Conand, C. 1998. Holothurians (Sea cucumbers, Class Holothuroidea). *In: The Living Marine Resources of the Western South Pacific, Volume 2: Cephalopods, Crustaceans, Holothurians And Sharks. Species Identification Guide For Fishery Purposes.* K. Carpenter and V. Neim, (eds). pp 1157–1191. Rome, FAO.

Conand, C. 2006. Harvest and Trade: Utilization of Sea Cucumber; Sea Cucumber Fisheries; Current International Trade; Illegal, Unreported and Unregulated Trade; By-catch; and Socio-economic Characteristics of the Trade in Sea Cucumbers. *In: Proceedings of the CITES Workshop on the Conservation of Sea Cucumbers in the Families Holothuriidae and Stichopodidae, 1–3 March, 2004, Kuala Lumpur, Malaysia.* A. Bruckner, (ed). NOAA Technical Memorandum, NMFS-OPR-34. pp 51–73. Silver Spring: NOAA.

Conand, C. & Hoffschir, C. 1991. Recent Trends in Sea cucumbers Exploitation in New Caledonia. *SPC Beche-de-Mer Information Bulletin,* 3: 5–7.

Conand, C. & Secretariat of the Pacific Community. 1996. Statistics on Beche-de-mer Exports. *SPC Beche-de-Mer Information Bulletin,* 8: 14–15.

Conand, C., Morel, C. & Mussard, R. 1997. A New Study of Asexual Reproduction in Holothurians: Fission in *Holothuria leucospilota* Populations on Réunion Island in the Indian Ocean. *SPC Beche-de-Mer Information Bulletin,* 9: 5–11.

Convention on the International Trade in Endangered Species of Fauna and Flora. 2006. *Sea Cucumbers.* Paper prepared for the 22nd Meeting of the Animals Committee, 7–13 July 2006, Lima, Peru.

Costa, R. 1995. *Etude de l'abondance relative des holothuries d'intérêt commercial du lagon de Moorea.* Papeete, Service de la Mer et de l'Aquaculture.

Coulter, J. 1847. *Adventures on the Western Coast of South America, and the Interior of California: Including a Narrative of Incidents at the Kingsmill Islands, New Ireland, New Britain, New Guinea, and the other Islands in the Pacific Ocean.* 2 Volumes. London.

Crean, K. 1977. Some Aspects of the Beche-de-mer Industry in Ongtong Java, Solomon Islands. *SPC Fisheries Newsletter,* 15: 36–48.

Dalzell, P. & Adams, T. 1994. *The Present Status of Coastal Fisheries Production in the South Pacific Islands.* 25th Regional Technical Meeting on Fisheries Working Paper, No. 6. Noumea, SPC.

Dalzell, P., Adams, T. & Polunin, N. 1996. Coastal Fisheries in the Pacific Islands. *Oceanography and Marine Biology: An Annual Review,* 34: 395–531.

Dalzell, P., Adams, T. & Polunin, N. 1996. Coastal Fisheries in the Pacific Islands. *Oceanography and Marine Biology: An Annual Review,* 34: 395–531.

Dalzell. P., Lindsay, S. & Patiale, H. 1993. *Fisheries Resources of the Island of Niue.* South Pacific Commission Technical Document, No. 3. Noumea, SPC.

Dance, S., Lane, I. & Bell, J. 2003. Variation in Short-term Survival of Cultured Sandfish (*Holothuria scabra*) Released in Mangrove-Seagrass and Coral Reef Flat Habitats in Solomon Islands. *Aquaculture,* 220: 495–505.

Department of Environment & Heritage. 2004. *Assessment of the Northern Territory Trepang Fishery.* Canberra, Commonwealth of Australia.

Department of Environment & Heritage. 2005. *Assessment of the Western Australian Beche-de-mer Fishery.* Canberra, Commonwealth of Australia.

Department of Environment & Heritage. 2006. National Report: Australia. *In: Proceedings of the CITES Workshop on the Conservation of Sea Cucumbers in the Families Holothuriidae and Stichopodidae, 1-3 March, 2004, Kuala Lumpur, Malaysia.* A. Bruckner, (ed.). NOAA Technical Memorandum, NMFS-OPR-34. pp 207–214. Silver Spring, NOAA.

Department of Fisheries & Marine Resources. No date. *Fisheries Commodity Statement: Beche-de-mer.* Report prepared for the Department of Fisheries and Marine Resources, Port Moresby, Papua New Guinea.

Donnelly, R. 2001. *Socio-Economic Environment and the Effect of the Live Reef Food Fish Trade in Marovo Lagoon, Roviana Lagoon and Ontong Java, Solomon Islands.* Discussion Paper No.5. ACIAR, Canberra

Doty, J. 1977. Fission in *Holothuria atra* and Holothurian Population Growth. Unpublished M.Sc. Thesis, University of Guam, Guam.

Drumm, D. 2004. *Habitats and Macroinvertebrate Fauna of the Reef-top of Raratonga, Cook Islands: Implications for Fisheries and Conservation Management.* Unpublished Ph.D. Thesis, University of Otagao, Dunedin, New Zealand.

Drumm, D. & Lonergan, N. 2005. Reproductive Biology of *Holothuria leucospilota* in the Cook Islands and the Implications of Traditional Fishing of Gonads on the Population. *New Zealand Journal of Marine and Freshwater Research*, 39: 141–156.

Dzeroski, S. & Drumm, D. 2003. Using Regression Trees to Identify the Habitat Preference of the Sea Cucumber (*Holothuria leucospilota*) on Rarotonga, Cook Islands. *Ecological Modelling*, 170: 219–226.

Ebert, T. 1978. Growth and Size of the Tropical Sea Cucumber *Holothuria* (*Halodeima*) *atra* Juger at Enewetak Atoll, Marshall Islands. *Pacific Science,* 32(2): 183–191.

Edward, A. 1997. *Kosrae Sea cucumber Report.* Report prepared for the Kosrae State Government, Federated States of Micronesia.

Eriksson, H. 2006. The Subsistence and Artisinal Sea cucumber Fishery, with Particular Focus on *Stichopus horrens* in Samoa. *In: Sea cucumber Abundance, Diversity and Fishery in Samoa: An Assessment of Lagoon Occurring Species.* Uppsala University Minor Field Study, No.: 122. Pp: 54–73. Report prepared for the Uppsala University, Apia, Samoa.

Etaix-Bonnin, R. 1999. Recent Trends in Beche-de-mer Exports from New Caledonia. SPC Beche-de-mer Information Bulletin, 12: 24.

Fao, B. 1990. Pollution Problems. *SPC Beche-de-Mer Information Bulletin,* 2: 12–13.

Ferdhouse, F. 1999. Beche-de-Mer Markets and Utilization. *SPC Beche-de-Mer Information Bulletin,* 9: 3–9.

Ferdhouse, F. 2004. World Markets and Trade Flows of Sea Cucumber/Beche-de-mer. *In: Advances in Sea Cucumber Aquaculture and Management.* A. Lovatelli, C. Conand, S. Purcell, S. Uthicke, J-F. Hamel and A. Mercier, (eds.). pp 101–118. FAO *Fisheries Technical Paper*, No. 463. Rome, FAO.

Fisk, D. 2002. *Marine Biodiversity Assessment Baseline Survey Report: Aleipata District, May 2002.* Report prepared for the Aleipata Marine Protected Area District Committee and the Samoa Marine Biodiversity Protection and Management Project, Apia, Samoa.

Fisk, D. 2004. *Literature Review and Pilot Baseline Data.* Report prepared for the Niue International Waters Program, Niue.

Fisk, D. 2005. *Community Based Monitoring Trials and Training.* Report prepared for the Niue International Waters Program, Niue.

Fisk, D., Axford, J. & Power, M. 2004a. *Marine Resources of Tokelau Atolls: Fakaofo Marine Resource Management Plan.* Report prepared for the South Pacific Regional Environment Program, Apia, Samoa.

Fisk, D., Axford, J. & Power, M. 2004b. *Marine Resources of Tokelau Atolls: Atafu Marine Resource Management Plan.* Report prepared for the South Pacific Regional Environment Program, Apia, Samoa.

Fisk, D., Axford, J. & Power, M. 2004c. *Marine Resources of Tokelau Atolls: Nukunono Marine Resource Management Plan.* Report prepared for the South Pacific Regional Environment Program, Apia, Samoa.

Fort, G. 1886. *Report on British New Guinea, from Data and Notes by the Late Sir Peter Scratchley, Her Majesty's Special Commissioner.* 2: 953. Melbourne, Government Printers.

Francour, P. 1997. Predation on Holothurians: A Literature Review. *Invertebrate Biology,* 116(1): 52–60.

Franklin, S. 1980. The Reproductive Biology and Some Aspects of the Population Ecology of the Holothurians *Holothuria leucospilota* (Brandt) and *Stichopus chloronotus* (Brandt). Unpublished Ph.D. Thesis, University of Sydney, New South Wales, Australia.

Friedman, K. & Tekanene. M. 2005. White teatfish at Kiribati Sea cucumber Hatchery "Local Technicians Getting Them Out Again". *SPC Beche-de-Mer Information Bulletin,* 21: 31–32.

Friedman, K., Lokani, P., Fale, P., Mailau, S., Ramohia, P. & Ramofafia, C. 2004. *Survey of the Sea Cucumber Resources of Ha'apai, Tonga.* Report prepared for the Tongan Ministry of Fisheries, Tonga'tapu, Tonga.

Friedman, K., Eriksson, H., Pakoa, K., Lasi, F., Solofa A. & Mulipola, A. 2006. *Samoa Sea Cucumber Fishery Status Assessment 2006. PROCFish/C Country Status Report.* Report prepared for the Samoa Fisheries and Uppsala University, Apia, Samoa.

Friedman, K. & Gisawa, L. 2008. Sea Cucumbers: Early retirement or renewal? *Island Business*, March: 48–49

Fufudate, K. 1999. *Field Survey of Sea Cucumber Resources in the Gilbert Islands, the Republic of Kiribati.* Report prepared for the OFCF Hatchery, Kiribati.

Gaudechoux, J. 1993. Statistics on Beche-de-mer Production. *SPC Beche-de-Mer Information Bulletin,* 5: 9–10.

Gentle, M. 1979. Population Ecology of Commercial Beche-de-mer (Echinodermata: Holothuridea) in Fiji. *SPC Fisheries Newsletter,* 18: 13–15.

Green, E. 2003. International Trade in Marine Aquarium Species: Using the Global Marine Aquarium Database. *In: Marine Ornamental Species: Collection, Culture, and Conservation.* J. Cato and C. Brown (eds). pp 31-48. Ames, Iowa State Press.

Hamel J.-F. & Mercier, A. 2000. Cuvierian Tubules in Tropical Holothurians: Usefulness and Efficiency as a Defence Mechanism. *Marine and Freshwater Behaviour and Physiology,* 33: 115–139.

Hamel, J.-F., Conand, C., Pawson, D. & Mercier, A. 2001. The Sea cucumber *Holothuria scabra* (Holothuroidea: Echinodermata): Its Biology and Exploitation as Beche-de-mer. *Advances in Marine Biology,* 41: 129–223.

Harriott, V. 1982. Sexual and Asexual Reproduction of *Holothuria atra* Jaeger at Heron Island Reef, Great Barrier Reef. *Australian Museum Memoir,* 16: 53–66.

Harriott, V. 1984. *Census Techniques, Distribution, Abundance and Processing of Large Sea-cucumber Species (Echinodermata: Holothuroidea) on the Great Barrier Reef.* A Report prepared for the Great Barrier Reef Marine Park Authority, Townsville, Queensland, Australia.

Harriott, V. 1985. Reproductive Biology of Three Congeneric Sea cucumber Species, *Holothuria atra, H. impatiens* and *H. edulis,* at Heron Reef, Great Barrier Reef. *Australian Journal of Marine and Freshwater Research,* 36: 51–57.

Hendler, G. & Meyer, D. 1982. Ophiroids Flagante Delicto and Notes on the Spawning Behaviour of other Echinoderms in their Natural Habitat. *Bulletin of Marine Science,* 32(2): 600–607.

Hill, F. 2006. *Annual Status Report, 2006: Moreton Bay Beche-de-Mer Fishery.* Brisbane, Queensland State Government.

Hoffman, T. 2001. *Reefs of Life to Reefs of Death: The Political Ecology of Coral Reef Health.* Unpublished Ph.D. Thesis, University of California, Berkeley, California, United States of America.

Holland, A. 1994a. The Beche-de-mer Fishery in the Solomon Islands: Recent Trends and Suggestions for Management. *SPC Beche-de-Mer Information Bulletin,* 6: 2–9.

Holland, A. 1994b. *The Status of Global Beche-de-Mer Fisheries with Special Reference to Solomon Islands and the Potentials of Holothurian Culture.* Unpublished M.Sc. Thesis, University of Newcastle, United Kingdom.

Ilek, S. 1991. *A Preliminary Sea cucumber Survey of Reefs in the States of Koror and Airai in Palau.* Report prepared for the Palau Department of Marine Resources, Koror, Republic of Palau.

Ivy, G. & Giraspy, D. 2006. Development of Large-scale Hatchery Production Techniques for the Commercially Important Sea cucumber *Holothuria scabra* var. *versicolor* (Conand, 1986) in Queensland, Australia. *SPC Beche-de-Mer Information Bulletin,* 24: 28–34.

James, R. 1977. *Beche-de-Mer in Solomon Islands.* Regional 9th Technical Meeting on Fisheries Working Paper, No. 17. Noumea, SPC.

Jimmy, R. 1995. Country Statement: Current Management Policies and Problems of the Inshore Fisheries Resources in Vanuatu Sea cucumbers. *In: South Pacific Commission and Forum Fisheries Agency Workshop on the Management of South Pacific Inshore Fisheries: Manuscript Collection of Country Statements and Background Papers, Volume 1.* P. Dalzell and T. Adams, (comps.). Integrated Coastal Fisheries Management Project Technical Document, No. 11. pp: CP. 14. Noumea, SPC.

Jimmy, R. 1996. Current Management Policies and Problems of the Inshore Fisheries Resources in Vanuatu Sea cucumbers. *SPC Beche-de-Mer Information Bulletin,* 8: 11–12.

Kailola, P. & Lokani, P. No date. *Sea cucumbers.* Paper prepared for the National Fisheries Authority, Port Moresby, Papua New Guinea.

Kailola, P., Petelo, W. & Gillett, R. 1995. The Beche-de-mer Fishery in Tonga. *In: South Pacific Commission and Forum Fisheries Agency Workshop on the Management of South Pacific Inshore Fisheries: Manuscript Collection of Country Statements and Background Papers, Volume II.* P. Dalzell and T. Adams, (comps.). Integrated Coastal Fisheries Management Project Technical Document, No. 12. pp: BP. 32. Noumea, SPC.

Kerr, A. 1994. Shallow-water Holothuroids (Echinodermata) of Kosrae, Eastern Caroline Islands. *Pacific Science,* 48(2): 161–174.

Kinch, J. 1999. *Economics and Environment in Island Melanesia: A General Overview of Resource Use and Livelihoods on Brooker Island in the Calvados Chain of the Louisiade Archipelago, Milne Bay Province, Papua New Guinea.* Report prepared for Conservation International, Port Moresby, Papua New Guinea.

Kinch, J. 2002. An Overview of the Beche-de-mer Fishery in the Milne Bay Province, Papua New Guinea. *SPC Beche-de-Mer Information Bulletin,* 17: 2–16.

Kinch, J. 2004a. *The Marine Aquarium Trade in the Solomon Islands, with Specific Notes on the Marau Sound, Guadalcanal.* Report prepared for the Marine Aquarium Council and the Foundation of the Peoples of the South Pacific-International, Suva, Fiji.

Kinch, J. 2004b. *A Review of the Beche-de-Mer Fishery and its Management in Papua New Guinea.* Report prepared for the National Fisheries Authority, Port Moresby, Papua New Guinea.

Kinch, J. 2004c. *The Status of Commercial Invertebrates and other Marine Resources in the Santa Isabel Province, the Solomon Islands.* Report prepared for the United Nations Development Program's Pacific Sustainable Livelihoods Program, Suva, Fiji; and the Isabel Province Development Program, Buala, Santa Isabel Province, Solomon Islands.

Kinch, J. 2005. The Commercial Use of *Thelenota rubralineata* in the Solomon Islands. *SPC Beche-de-Mer Information Bulletin,* 21: 3–4.

Kinch, J., James, M., Thomas, E., Lauhi, P. & Gabiobu, R. 2007. *Socio-economic Assessment of the Beche-de-Mer Fisheries in the Western, Central and Manus Provinces, Papua New Guinea.* Report prepared for the National Fisheries Authority, Port Moresby, Papua New Guinea.

Kinch, J., Mesia, P., Kere, N., Manioli, J. & Bulehite, K. 2006. *Socio-economic Baseline Assessment for the Eastern Marovo Lagoon, Western Province, the Solomon Islands.* IWP-Pacific Technical Report, No.: 35. Apia, SPREP.

King, M. & Faasili, U. 1999. Community-based Management of Subsistence Fisheries in Samoa. *Fisheries Management and Ecology,* 6: 133–144.

Kiribati Fisheries Division. 1995. Kiribati: Country Statement. *In: South Pacific Commission and Forum Fisheries Agency Workshop on the Management of South Pacific Inshore Fisheries: Manuscript Collection of Country Statements and Background Papers, Volume 1.* P. Dalzell and T. Adams, (comps.). Integrated Coastal Fisheries Management Project Technical Document, No. 11. CP. 15. Noumea, SPC.

Koczberski, G., Curry, G., Warku, J. & Kwam, C. 2006. *Village-based Marine Resource Use and Rural Livelihoods: Kimbe Bay, West New Britain, Papua New Guinea.* TNC Pacific Islands Countries Report, No. 5. Report prepared for The Nature Conservancy, Port Moresby, Papua New Guinea.

Kropp, R. 1982. Response of Five Holothurian Species to Attacks by a Predatory Gastropod, *Tonna perdix*. *Pacific Science,* 36(4): 445–452.

Laade, W. 1966. The Islands of Torres Strait. *Bulletin of the International Committee on Urgent Anthropological Research,* 8: 111–114.

Laboute, P. 1987. Mission to the Tokelau Islands to evaluate cyclone damage to Coral Reefs. Orstom Centre, Noumea. 11 p.

Lambeth, L. 1999. *An Assessment of the Role of Women within Fishing Communities in the Republic of Palau.* Report prepared for the Community Fisheries Section, Secretariat of the Pacific Community, Noumea, New Caledonia.

Lambeth, L. 2000. The Subsistence Use of *Stichopus variegates* (now *S. herrmanni*) in the Pacific Islands. *SPC Beche-de-Mer Information Bulletin,* 13: 18–21.

Lane, D. 1999. Distribution and Abundance of *Thelenota rubralineata* in the Western Pacific: Some conservation Issues. *SPC Beche-de-Mer Information Bulletin,* 11: 19–21.

Lawrence, J. 1979. Numbers and Biomass of the Common Holothuroids on the Windward Reff Flat at Enewetak Atoll, Marshall Islands. *Proceedings of the European Colloquium on Echinoderms,* 1: 210–204.

Leqata, J. 2004. *Solomon Islands Country Report.* Paper prepared for the Seminar on Fisheries Development and Environment, 6/01–21/02, Yokohama City, Japan.

Lincoln-Smith, M., Pitt, K., Bell, J. & Ramohia, P. 2000. *Testing the Use of Marine Protected Areas to Restore and Manage Tropical Multi-species Invertebrate Fisheries at the Arnavon Islands, Solomon Islands.* Research Publication, No. 69. Report prepared for the Great Barrier Reef Marine Park Authority, Townsville, Queensland; and the Australian Centre for International Agricultural Research, Sydney, New South Wales, Australia.

Lindholm, R. 1978. *Beche-de-Mer Fishery.* Report prepared for the Department of Primary Industry, Port Moresby, Papua New Guinea.

Lindsay, S. 2000a. *Marine Resource Survey of Lenger Island, Pohnpei State, FSM.* Report prepared for the Conservation Society of Pohnpei, Kolonia, Pohnpei, Federated States of Micronesia.

Lindsay, S. 2001a. *Resource Evaluation of the Current Populations of two Commercially Harvested Sea Cucumber Species (*Actinopyga mauritiana *and* Stichopus chloronotus*), Kosrae State, FSM.* Report prepared for the Kosrae Island Resource Management Program, Kosare, Federated States of Micronesia.

Lindsay, S. 2001b. *Marine Resource Survey of U Municipality, Pohnpei State, FSM.* Report prepared for the Conservation Society of Pohnpei, Kolonia, Pohnpei, Federated States of Micronesia.

Lindsay, S. & Abraham, S. 2004. Resource Evaluation of Current Populations of Two Commercially Harvested Sea cucumber Species (*Actinopyga mauritiana* and *Stichopus chloronotus*), and Recommendations for Management for Kosrae State, Federated States of Micronesia. *SPC Beche-de-Mer Information Bulletin,* 19: 31–33.

Lindsay, S. & McKenna, S. 2006. L'evaluation D'especes Marines Macro-invertebrees Exploitees. In: *Une Evaluation Rapide de la Biodiversite Marien de Recifs Corallines du Mont Panie.* S. McKenna, N. Baillon, H. Blaffart and G. Abrusci, (eds.) RAP Bulletin of Biological Assessment No. 42. pp: 43–52. Washington, DC, Conservation International. (in French, cited from original English version).

Lokani, P. 1990. Beche-de-mer Research and Development in Papua New Guinea. *SPC Beche-de-Mer Information Bulletin,* 1: 8–11.

Lokani, P. 1992. First Results of an Internal Tag Retention Experiment on Sea cucumber. *SPC Beche-de-Mer Information Bulletin,* 4: 9–12.

Lokani, P. & Kubohojam, G. 1993. *Beche-de-Mer Processing and Marketing in Papua New Guinea.* Report to the Department of Fisheries and Marine Resources, Kavieng, New Ireland Province, Papua New Guinea.

Lokani, P., Matoto, S. & Ledau, E. 1996. *Survey of Sea cucumber Resources art Ha'apai Tonga May/June 1996.* Secretariat of the Pacific Community Integrated Coastal Fisheries Management Project, Country Assignment Report. Report prepared for the Secretariat of the Pacific Community, Noumea, New Caledonia.

Long, B., Skewes, T., Dennis, D., Poiner, I., Pitcher, C., Taranto, T., Manson, F., Polon, P., Karre, B., Evans, C. & Milton, D. 1996. *Distribution and Abundance of Beche-de-Mer on Torres Strait Reefs.* Report to Queensland Fisheries Management Authority, Thursday Island, Queensland, Australia.

Lovell, E., Sykes, H., Deiye, M., Wantiez, L., Garrigue, C., Virly, S., Samuelu, J., Solofa, A., Poulasi, T., Pakoa, K., Sabetian, A., Afzal, D., Hughes, H. & Sulu, R. 2004. Status of Coral Reefs in the South West Pacific: Fiji, Nauru, New Caledonia, Samoa, Solomon Islands, Tuvalu and Vanuatu. In: Wilkensen, C. (ed.). *Status of Coral Reefs of the World: 2004.* pp 337–361. Australian Institute of Marine Sciences, Townsville.

Lowden, R. 2005. Management of Queensland Sea cucumber Stocks by Rotational Zoning. *SPC Beche-de-Mer Information Bulletin,* 22: 47.

Massin, C. & Doumen, C. 1986. Distribution and Feeding of Epibenthic Holothuroids on the Reef Flat of Laing Island (Papua New Guinea). *Marine Ecology,* 31(2): 185–195.

Mathews, E. 1995. The Need for Invertebrate Conservation in the Pacific Islands region. *In: Fishing for Answers: Women and Fisheries in the Pacific Islands.* E. Mathews, (ed.). pp. 123–136. Suva, Women and Fisheries Network.

Mathews, E. & Oiterong, E. 1991. *The Role of Women in the Fisheries of Palau.* Division of Marine Resources Technical Report, No. 1. Report prepared for the Palau Division of Marine Resources, Koror, Republic of Palau.

Mathews, E. & Oiterong, E. 1995. Marine Species Collected by Women in Palau, Micronesia. *Micronesica,* 281: 77–90.

McCormack, C. 2005. *Annual Status Report, 2005: Moreton Bay Beche-de-Mer Fishery.* Brisbane: Queensland State Government.

McCormack, G. 1984. *Rori (Sea Cucumbers).* Report prepared for the Cook Islands Conservation Department, Raratonga, Cook Islands.

McElroy, S. 1990. *Beche-de-Mer Survey of the Gilbert Islands.* Report prepared for the Ministry of Natural Resource Development, Tarawa, Kiribati.

McKnight, C. 1976. *The Voyage to Marege: Macassan Trepangers in Northern Australia.* Melbourne, Melbourne University Press.

Mercier A., Battaglene, S. & Hamel, J.-F. 2000a. Periodic Movement, Recruitment and Size-related Distribution of the Sea cucumber *Holothuria scabra* in Solomon Islands. *Hydrobiologia,* 440: 81–100.

Mercier A., Battaglene, S. & Hamel, J.-F. 2000b. Settlement Preferences and Early Migration of the Tropical Sea cucumber *Holothuria scabra. Journal of Experimental Marine Biology and Ecology,* 249: 89–110.

Mladenov, M. & Gerring, P. 1996. *Resource Evaluation of the Sea cucumber (Stichopus mollis) in an Environmentally Sensitive Region of New Zealand.* Paper prepared for the 9th International Echinoderm Conference, August, 1996, San Francisco, California, United States of America.

Moore, C. 2003. *New Guinea: Crossing Boundaries and History.* Honolulu, University of Hawaii Press.

Moore, J. & Marieg, F. 1986. *Yap Fishing Authority Sea cucumber Development Project: Ngulu Atoll, 12 February to 29 April 1986.* Yap Fishing Authority, Kolonia.

Morgan, A. 2000. Aspects of the Spawning in the Sea Cucumber *Holothuria scabra* (Echinodermata: Holothuridea). *Bulletin of Marine Science,* 66: 47–57.

Morgan, A. & Archer, J. 1999. Overview: Aspects of Sea cucumber Industry Research and Development in the South Pacific. *SPC Beche-de-Mer Information Bulletin,* 12: 15–17.

Morrell, B. 1832. *A Narrative of Four Voyages.* New York.

Mulipola, A. 1994. *Current Status of Beche-de-Mer Resource and Exploitation in Western Samoa.* Report prepared for the Research Section, Fisheries Division, Ministry of Agriculture, Apia, Samoa.

Mulipola, A. 2002. *Marine Resources of Samoa.* Samoa's National Report for the World Summit on Sustainable Development, July 2002, Johannesburg, South Africa.

Myint, T. 1996. Beche-de-mer Exports. *PNG National Fisheries Authority Newsletter,* 2(1): 20–21.

Nash, W. & Ramofafia, C. 2006. Recent Developments with the Sea cucumber Fishery in Solomon Islands. *SPC Beche-de-Mer Information Bulletin,* 23: 3–4.

National Fisheries Authority. 2005. *Socio-economic Survey of Small-scale Fisheries in New Ireland Province, Papua New Guinea.* Kavieng, National Fisheries Authority and the Coastal Fisheries Management and Development Project.

National Fisheries Authority. 2006. *Socio-economic Survey of Small-scale Fisheries in Milne Bay Province, Papua New Guinea.* Kavieng, National Fisheries Authority and the Coastal Fisheries Management and Development Project.

National Fisheries Authority. 2007a. *Sea Cucumber Survey: New Ireland Province.* Kavieng, National Fisheries Authority and the Coastal Fisheries Management and Development Project.

National Fisheries Authority. 2007b. *Socio-economic Survey of Small-scale Fisheries in Morobe Province, Papua New Guinea.* Kavieng: National Fisheries Authority and the Coastal Fisheries Management and Development Project.

Northern Province Fisheries Division. 1993. Beche-de-mer Harvesting in the Northern Province of New Caledonia. *SPC Beche-de-Mer Information Bulletin,* 5: 7–8.

Oxley, W., Ayling, A., Cheal, J. & Osborne, K. 2004. *Marine Surveys Undertaken in Elizabeth and Middleton Reefs Marine National Nature Reserve, December 2003.* Report prepared for the Australian Department of Environment and Heritage, Brisbane, Queensland, Australia.

Passfield, K. 1998. *A Report of a Survey of the Marine Resources of Fakaofo Atoll, Tokelau.* Report prepared for the Secretariat of the Pacific Community, Noumea, New Caledonia.

Pauly, G. 2000. Benthic Ecology and Biota of Tarawa Atoll Lagoon: Influence of Equatorial Upwelling, Circulation, and Human Harvest. *Atoll Research Bulletin,* 487: 1–41.

Pawson, D. 1970. The Marine Fauna of New Zealand: Sea cucumbers (*Echinodermata: Holothuroidea*). *New Zealand Department of Science and Industry Research Bulletin,* 201: 1–50.

Polon, P. 2004. The Papua New Guinea National Beche-de-mer Fishery Management Plan. In: *Advances in Sea Cucumber Aquaculture and Management.* A. Lovateli, C. Conand, S. Purcell, S. Uthicke, J.-F. Hamel and A. Mercier, (eds.). pp. 205–217. Rome, FAO.

Preston, G. 1988. *Beche-de-Mer Exploitation in Fiji.* Inshore Fisheries Research Project Country Assignment Report. Noumea, South Pacific Commission.

Preston, G. 1990. Mass beche-de-mer production in Fiji. *SPC Beche-de-Mer Information Bulletin,* 1: 4–5.

Preston, G. 1993. Beche-de-mer. In: *Nearshore Marine Resources of the South Pacific: Information for Fisheries Development and Management.* A. Wright and L. Hill, (eds). pp. 371–407. Honiara, Forum Fisheries Agency.

Preston, G. & Lokani, P. 1990. Report of a Survey of the Sea cucumber Resources of Ha'apai, Tonga. Report prepared for the South Pacific Commission Noumea, New Caledonia.

Preston, G., Vakamoce, V., Lokain, P. & Viala, F. 1989. *Exploitation of the Sea Cucumber Actinopyga miliaris (Blackfish, Driloli) in Northern Fiji.* Reported prepared for the Government of Fiji, Suva, Fiji.

Preston, G., Lewis, A., Sims, N., Bertram, I., Howard N., Maluofenua, S., Marsters, B., Passfield, K., Tearii, T., Viala, F., Wright, D. & Yeeting, B. 1988. *The Marine Resources of Palmerston Island, Cook Islands: Report of a Survey Carried out in September 1988.* Inshore Fisheries Project Technical Document, No. 10. Report prepared for the Secretariat of the Pacific Community, Noumea, New Caledonia.

Purcell, S., Gossiun, H. & Agudo, N. In press. Status and management of the sea cucumber fishery of La Grande Terre, New Caledonia. WorldFish Center Studies and Reviews. The WorldFish Center, Penang, Malaysia.

Purcell, S., Blockmans, B. & Nash, W. 2006a. Efficacy of Chemical Markers and Physical Tags for Large-scale Release of an Exploited Holothurian. *Journal of Experimental Marine Biology and Ecology,* 334: 283–293.

Purcell, S., Patrois, J. & Fraisse, N. 2006b. Experimental Evaluation of Co-culture of Juvenile Sea cucumbers, *Holothuria scabra* (Jaeger) with Juvenile Blue shrimp, *Litopenaeus stylirostris* (Stimpson). *Aquaculture Research,* 37: 515–522.

Purcell, S. & Simutoga, M. 2008. Spatio-temporal and Size-dependent Variation in the Success of Releasing Cultured Sea cucumbers in the Wild. *Reviews in Fisheries Science,* 16: 204–214.

Purcell, S., Gardner, D. & Bell, J. 2002. Developing Optimal Strategies for Restocking Sandfish: A Collaborative Project in New Caledonia. *SPC Beche-de-Mer Information Bulletin,* 16: 2–4.

Purwati, P. 2001. *Reproduction in Holothuria leucospilota in the Tropical Waters of Darwin, NT, Australia.* Unpublished M.Sc. Thesis, Northern Territory University, Darwin, Northern Territory, Australia.

Purwati, P. & Thinh Luong Van, J. 2003. Sexual Reproduction in a Fissiparous Holothurian Species, *Holothuria leucospilota* Clark 1920 (Echinodermata: Holothuroidea). *SPC Beche-de-Mer Information Bulletin,* 18: 33–38.

Qalovaki, S. 2006. National Report: Fiji. *In: Proceedings of the CITES Workshop on the Conservation of Sea Cucumbers in the Families Holothuriidae and Stichopodidae, 1–3 March, 2004, Kuala Lumpur, Malaysia.* A. Bruckner, (ed). NOAA Technical Memorandum, NMFS-OPR-34. pp. 149–152. Silver Spring, NOAA.

Ram, R. (in prep). Impacts of harvest and post harvest processing methods on quality and value of bêche-de-mer in Fiji Islands. Master of Science Thesis. University of the South Pacific.

Ramofafia, C. 2004. *The Sea cucumber Fisheries in Solomon Islands: Benefits and Importance to Coastal Communities.* Report prepared for the Australian Centre for International Agriculture Research, Sydney, New South Wales, Australia.

Ramofafia, C., Battaglene, S. & Bryne, M. 2001. Larval Development in the Tropical Sea Cucumber *Holothuria scabra. In: Echinoderms 2000: Proceedings of the 10th International Conference.* M. Barker, (ed). pp. 369–375. Rotterdam: A.A. Balkema.

Ramofafia, C., Byrne, M. & Battaglene, S. 2001. Reproductive Biology of the Intertidal *Actinopyga mauritiana* in the Solomon Islands. *Marine Biology,* 81: 523–531.

Ramofafia, C., Byrne, M. & Battaglene, S. 2003. Reproduction of the Commercial Sea cucumber *Holothuria scabra* (Echinodermata: Holothuriodea) in the Solomon Islands. *Marine Biology,* 142: 281–288.

Ramofafia, C., Gervis, M. & Bell, J. 1995. Spawning and Early Larval Rearing of *Holothuria atra. SPC Beche-de-Mer Information Bulletin,* 7: 2–6.

Ramofafia, C., Battaglene, S., Bell, J. & Byrne, M. 2000. Reproductive Biology of *Holothuria fuscogilva* (Echinodermata: Holothuroidea) in the Solomon Islands. *Marine Biology,* 136: 1045–1056.

Ramofafia, C., Nash, W., Sibiti, S., Makini, D. & Schwarz, A. 2007. *Household Socio-economics and Beche-de-Mer Resource Use in Kia Community, Isabel Province, Solomon*

Islands (June, 2005). Report prepared for the Australian Centre of International Agriculture Research, Sydney, New South Wales, Australia.

Ramohia, P. 2006. Fisheries Resources: Commercially Important Macroinvertebrates. In: *Solomon Islands Marine Assessment: Technical Report of Survey Conducted May 13 to June 17, 2004.* A. Green, P. Lokani, W. Atu, P. Ramohia, P. Thomas and J. Almany, (eds.). TNC Pacific Island Countries Report, No. 1. pp. 330–400. Brisbane, The Nature Conservancy.

Rees, M., Colquhoun J., Smith, L. & Heyward, A. 2003. *Stocks of Trochus, Beche-de-Mer and Clams at Ashmore Reef, Cartier Reef and Mermaid Reef, northwestern Australia: 2003 Report.* Report prepared for Environment Australia. Australian Institute of Marine Science, Townsville, Australia.

Reichenbach, N., Nishar, Y. & Saeed, A. 1996. Species and Size-related Trends in Asexual Propagation of Commercially Important Species of Tropical Sea Cucumbers (Holothuroidea). *Journal of the World Aquaculture Society,* 27: 475–482.

Richmond, R. 1996a. The Biology and Ecology of Sea Cucumbers. *In: Suggestions for the Management of Sea cucumber Resources in Micronesia: Results of the Workshop 'A regional management plan for a sustainable Sea Cucumber Fishery for Micronesia'.* R. Richmond, (Ed). University of Guam Marine Laboratory Technical Report, No. 101. pp. 7–25. Mangilao, University of Guam.

Richmond, R. 1996b. Introduction and Overview. *In: Suggestions for the Management of Sea cucumber Resources in Micronesia: Results of the Workshop "A regional management plan for a sustainable Sea Cucumber Fishery for Micronesia".* R. Richmond, (Ed). University of Guam Marine Laboratory Technical Report, No. 101. pp. 26. Mangilao, University of Guam.

Roelofs, A. 2004. *Ecological assessment of Queensland's East Coast Beche-de-Mer Fishery.* Report prepared for the Australian Government Department of Environment and Heritage, Canberra, Australia.

Roelofs, A. 2005. *Annual Status Report 2005: East Coast Beche-de-Mer Fishery.* Brisbane, Queensland State Government.

Russell, P. 1970. *The Papuan Beche-de-Mer Trade to 1900.* Unpublished MA Thesis, University of Papua New Guinea, Port Moresby, Papua New Guinea.

Sachithananthan, K. 1971a. *Beche-de-Mer Industry in the South Pacific Islands. I. Market Survey.* First report to the Food and Agriculture Organization of the United Nations, Rome, Italy.

Sachithananthan, K. 1971b. *Beche-de-Mer Industry in the South Pacific Islands. II. Territory of Papua New Guinea.* Second report to the Food and Agriculture Organization of the United Nations, Rome, Italy.

Sebatian, A. & Foale, S. 2006. Evolution of the Artisanal Fisher: Case-studies from Solomon Islands and Papua New Guinea. *SPC Traditional Marine Resource Management and Knowledge Information Bulletin,* 20: 3–10.

Secretariat for the Pacific Community. 1979. *Beche-de-Mer of the Tropical Pacific: A Handbook for Fishermen.* Handbook, No. 18. Noumea, SPC.

Secretariat of the Pacific Community. 2003. *Pacific Island Sea Cucumber and Beche-de-Mer Identification Cards.* Noumea, SPC.

Secretariat of the Pacific Community. 2006a. Fieldwork in the Federated States of Micronesia (FSM). *SPC Fisheries Newsletter,* 117: 30–34.

Secretariat of the Pacific Community. 2006b. Survey Work in Papua New Guinea. *SPC Fisheries Newsletter,* 118: 6–8.

Secretariat of the Pacific Community. 2007a. Fieldwork and surveys in the Cook Islands. *SPC Fisheries Newsletter*, 120: 2–8.

Secretariat of the Pacific Community. 2007b. Fieldwork and surveys in the Cook Islands. *SPC Fisheries Newsletter*, 123: 2–12.

Seeto, J. 1994. *Reproductive Biology of the Sea cucumber Holothuria atra Jaeger 1833 (Echinodermata: Holothuroidea) in Laucala Bay, Fiji, with Notes on its Population Structure and Symbiotic Associations.* Unpublished M.Sc. Thesis, University of Otago, Dunedin, New Zealand.

Sharples, P. 1994. *Pitcairn Island Fisheries Resource Survey – 1994: Observer Trip Report.* Report prepared for the Secretariat of the Pacific Community, Noumea, New Caledonia.

Shelley, C. 1981. *Aspects of the Distribution, Reproduction, Growth and Fishery Potential of Holothurians (Beche-de-Mer) in the Papuan Coastal Lagoon.* Unpublished Master of Science Thesis, University of Papua New Guinea, Port Moresby, Papua New Guinea.

Shelley, C. 1985. Growth of *Actinopyga echinites* and *Holothuria scabra (*Holothuroidea: Echinodermata) and their fisheries potential (as Beche-de-mer) in Papua New Guinea. *Proceedings of the Fifth International Coral Reef Congress, Tahiti, French Polynesia, 27 May–1 June. 5.* pp. 297–302. Papeete, Moorea Antenne Museum-EPHE.

Shelley, C. & Puig, P. 2004. Management of Trepang in the Northern Territory, Australia, and Current Research to Further Improve Understanding of the Fishery. In: *Advances in Sea Cucumber Aquacultureand Management.* A. Lovateli, C. Conand, S. Purcell, S. Uthicke, J.-F. Hamel and A. Mercier, (eds.). pp: 221–230. Rome, FAO.

Shiell, G. 2004. Beche-de-mer – the West Australian Perspective: he Fishing History and Aspects of the Reproductive Biology and Ecology of the Black teatfish, *Holothuria nobilis. In: Advances in Sea Cucumber Aquaculture and Management.* A. Lovateli, C. Conand, S. Purcell, S. Uthicke, J.-F. Hamel and A. Mercier, (eds.). pp. 231–238. Rome, FAO.

Shiell, G. & Uthicke, S. 2005. Reproduction of the Commercial Sea cucumber *Holothuria whitmaei* (Holothuroidea: Aspidochirotida) in the Indian and Pacific Ocean Regions of Australia. *Marine Biology,* 148: 937–986.

Shirayama, Y. & Thornton, H. 2005. Effect of Increased Atmospheric CO^2 on Shallow Water Marine Benthos. *Journal of Geophysical Research,* 110:C09S08.

Skewes, T., Burridge, C. & Hill, B. 1998. *Survey of Holothuria scabra on Warrior Reef, Torres Strait.* Report to Queensland Fisheries Management Authority, Thursday Island, Queensland, Australia.

Skewes, T., Dennis, D. & Burridge, C. 2000. *Survey of* Holothuria scabra *(Sandfish) on Warrior Reef, Torres Strait.* Report to Queensland Fisheries Management Authority, Thursday Island, Queensland, Australia.

Skewes, T., Dennis, D., Jacobs, D., Gordon, S., Taranto, T., Haywood, M., Pitcher, R., Smith, G., Milton, D. & Poiner, I. 1999. *Survey and Stock Size Estimates of the Shallow Reef (0–15 m Deep) and Shoal Area (15–50 m Deep) Marine Resources and Habitat Mapping with the Timor Sea MOU 74 Box, Volume 1: Stock Estimates and Stock Status.* Report prepared for the CSIRO Marine Branch, Brisbane, Queensland, Australia.

Skewes, T., Kinch, J., Polon, P., Dennis, D., Seeto, P., Taranto, T., Lokani, P., Wassenberg, T., Koutsoukos, A. & Sarke, J. 2002a. *Research for the Sustainable Use of Beche-de-Mer Resources in the Milne Bay Province, Papua New Guinea.* Report prepared for the National Fisheries Authority, Port Moresby, Papua New Guinea; and the Australian Centre for International Agricultural Research, Sydney, New South Wales, Australia.

Skewes, T., Dennis, D., Koutsoukos, A., Haywood, M., Wassenberg, T. & Austin, M. 2002b. *Research for the Sustainable Use of Beche-de-Mer Resources in Torres Strait.* Report prepared for the Australian Fisheries Management Authority and Queensland Fishery Service, Thursday Island, Queensland, Australia.

Skewes, T., Dennis, D., Koutsoukos, A., Haywood, M., Wassenberg, T. & Austin, M. 2004. *Stock survey and Sustainable Harvest Strategies for the Torres Strait Beche-de-Mer.* Report prepared for the Australian Fisheries Management Authority and Queensland Fishery Service, Thursday Island, Queensland, Australia.

Skewes, T., Dennis, D., Taylor, S., Haywood, M. & Donovan, A. 2006. *Sustainability Assessment for the Torres Strait Beche-de-Mer Fishery.* Report prepared for the Australian Fisheries Management Authority Torres Strait Research Program, Brisbane, Queensland, Australia.

Smith, A. 1986. Reef Invertebrate Harvesting. *In: Fishing on Guam: Guide to Coastal Resources of Guam.* S. Amesbury, F. Cushing and R. Sakamoto, (Eds.). Volume 2. pp. 68–73. University of Guam: Guam.

Smith, A. 1992. *Federated States of Micronesia Marine Resource Profiles.* FFA Report, No. 17. Report prepared for the Fisheries Forum Agency, Honiara, Solomon Islands.

Smith, L., Rees, M., Heyward, A. & Colquhoun, J. 2001. *Survey 2000: Beche-de-Mer and Trochus Populations at Ashmore Reef.* Report prepared for Environment Australia. Australian Institute of Marine Science, Townsville, Australia.

Smith, L., Rees, M., Heyward, A. & Colquhoun, J. 2002. *Stocks of Trochus and Beche-de-Mer at Cartier Reef: 2001 Surveys.* Report prepared for Environment Australia. Australian Institute of Marine Science, Townsville, Australia.

Smith, R. 1947. *Survey of the Fisheries of the Former Japanese Mandated Islands.* United States Fisheries and Wilidlife Service's Fishery Leaflet, No. 273.

Souter, G. 1963. *New Guinea: The Last Unknown.* Sydney, Angus and Robertson.

Stewart, B. 1993. *Evidence for a Marked Decline of Beche-de-Mer Populations in the Suva and Beqa Areas of Fiji, and a Preliminary Description of a Method of Identifying Beche-de-Mer Individuals based on Characteristic Body Wrinkles.* University of the South Pacific Marine Studies Technical Report, No.1. Suva, USP.

Sumner, R. 1981. A Noisome Buisness: The Trepang Trade and Queensland. *Journal of Australian Studies,* 9: 61–70.

Tiraa-Passfield, A. 1997: The Harvesting of Rori (Sea cucumbers) in Rarotonga, Cook Islands. *SPC Traditional Marine Resource Management and Knowledge Bulletin, 8*: 16–17.

Tonga Ministry of Fisheries. 1995. Kingdom of Tonga Country Report: Status and Management of Inshore Fisheries. *In: South Pacific Commission and Forum Fisheries Agency Workshop on the Management of South Pacific Inshore Fisheries: Manuscript Collection of Country Statements and Background Papers, Volume 1.* P. Dalzell and T. Adams, (Comps.). Integrated Coastal Fisheries Management Project Technical Document, No. 11. pp. CP. 18. Noumea, SPC.

Trianni, M. 2001. *Evaluation of the Resource following the Sea cucumber Fishery of Saipan, Northern Mariana Islands.* Paper prepared for the 9th International Coral Reef Symposium, Bali, Indonesia.

Trianni, M. 2002. Summary of Data Collected from the Sea cucumber Fishery on Rota, Commonwealth of the Northern Mariana Islands. *SPC Beche-de-Mer Information Bulletin,* 16: 5–11.

Tsuda, R. 1997. Survey of Commercially-valued Sea cucumbers in the Saipan Lagoon, CNMI. *SPC Beche-de-Mer Information Bulletin,* 10: 15–17.

Uthicke, S. 1994. Spawning Observations from the Lizard Island Area. *SPC Beche-de-Mer Information Bulletin,* 6: 12–14.

Uthicke, S. 1997. The Seasonality of Asexual Reproduction in *Holothuria (Halodeima) atra, Holothuria (Halodeima) edulis* and *Stichopus chloronotus* (Holothuroidea: Aspidochirotida) on the Great Barrier Reef. *Marine Biology,* 129: 435–441.

Uthicke, S. 2004. Overfishing of Holothurians: Lessons from the Great Barrier Reef. *In: Advances in Sea Cucumber Aquaculture and Management.* A. Lovateli, C. Conand, S. Purcell, S. Uthicke, J.-F. Hamel and A. Mercier, (eds.). pp. 163–171. Rome, FAO.

Uthicke, S. & Benzie, J. 1999. Allozyme Variation as a Tool for Beche-de-mer Fisheries Management: A study on *Holothuria scabra* (Sandfish). *SPC Beche-de-Mer Information Bulletin,* 11: 18–23.

Uthicke, S. & Benzie, J. 2000a. The effect of bêche-de-mer fishing on densities and size structure of Holothuria nobilis (Echinodermata: Holothurioidea) populations on the Great Barrier Reef. *Coral Reefs*, 19: 271–276.

Uthicke, S. & Benzie, J. 2001. Restricted Gene Flow between *Holothuria scabra* (Echinodermata: Holothuroidea) Populations along the North-east Coast of Australia and the Solomon Islands. *Marine Ecology Progress Series,* 216: 109–117.

Uthicke, S. & Benzie, J. 2002. A Genetic Fingerprint Recapture Technique for Measuring Growth in "Unmarkable" Invertebrates: Negative Growth in Commercially Fished Holothurians (*Holothuria nobilis*). *Marine Ecology,* 241: 221–226.

Uthicke, S. & Benzie, J. 2003. Gene Flow and Population History in High Dispersal Marine Invertebrates: mtDNA Analysis of *Holothuria nobilis* (Echinodermata: Holothuroidea) Populations from the Indo-Pacific. *Molecular Ecology,* 12: 2635–2648.

Uthicke, S. & Purcell, S. 2004. Preservation of Genetic Diversity in Restocking of the Sea cucumber *Holothuria scabra* planned through Allozyme Electrophoresis. *Canadian Journal of Fisheries and Aquatic Sciences,* 61: 519–528.

Uthicke, S., Benzie, J. & Ballment, E. 1998. Genetic Structure of Fissiparous Populations of *Holothuria (Halodeima) atra* on the Great Barrier Reef. *Marine Biology,* 132: 141–151.

Uthicke, S., Benzie, J. & Ballment, E. 1999. Population Genetics of the Fissiparous Holothurian *Stichopus chloronotus* (Aspidochirotida) on the Great Barrier Reef, Australia. *Coral Reefs,* 18: 123–132.

Uthicke, S., Conand, C. & Benzie, J. 2001. Population Genetics of the Fissiparous Holothurians *Stichopus chloronotus* and *Holothuria atra* (Aspidochirotida): A Comparison between the Torres Strait and La Réunion. *Marine Biology,* 139: 257–265

Uthicke, S., Welch, D. & Benzie, J. 2004. Slow Growth and Lack of Recovery in Overfished Holothurians on the Great Barrier Reef: Evidence from DNA Fingerprints and Repeated Large-Scale Surveys. *Conservation Biology,* 18(5): 1395–1404.

Vail, L. & Russel, B. 1990. Indonesian Fishermen of Australia's North-West. *Australian Natural History,* 24: 211–220.

Wabnitz, C., Taylor, M., Green, E. & Razak, T. 2003. *From Ocean to Aquaria: The Global Trade in Ornamental Species.* Cambridge, UNEP/WCMC.

Ward, R. 1972. The Pacific Beche-de-mer Trade with Special Reference to Fiji. *In: Man in the Pacific: Essays on Geographical Change in the Pacific Islands.* R. Ward, (ed). pp. 91–123. Oxford, Clarendon Press.

Wilkinson, C. 2004. (ed). *Status of Coral Reefs of the World: 2004.* pp. 337–361. Australian Institute of Marine Sciences, Townsville.

Zoutendyk, D. 1989a. T*rial Processing and Marketing of the Surf redfish (Actinopyga mauritiana) Beche-de-Mer on Rarotonga, and its Export Potential in the Cook Islands.* Report prepared for the Ministry of the Marine Resources, Raratonga, The Cook Islands.

Zoutendyk, D. 1989b. *Beche-de-Mer: Rori of the Cook Islands.* Marine Resources Profile, No. 6. Rarotonga: Ministry of Marine Resources, Raratonga, The Cook Islands.

APPENDIX A

Bêche-de-mer exports (tonnes) from the Western Central Pacific region: 1960–2006

	Micronesia			Polynesia				Melanesia					Australia					
Year	FSM	RMI	Kiribati	Tuvalu	Samoa	Tonga	French Polynesia	Papua New Guinea	Solomon Islands	Vanuatu	New Caledonia	Fiji	Coral Sea	Torres Strait	Great Barrier Reef	Moreton Bay	Northern Territory	Western Australia
1960								2.0										
1961								2.0										
1962								2.0										
1963								2.0										
1964								2.0										
1965								2.0										
1966								2.0										
1967								2.0										
1968								2.0	3.0									
1969								2.0	4.0									
1970																		
1971									5.0									
1972									33.0									
1973									54.0									
1974									18.0									
1975									20.0									
1976									29.0			4.0						
1977								5.3	39.0			17.0						
1978								5.9	34.0			15.0						
1979				1.8				1.3	10.0			11.0						
1980				0.8				2.4	37.0			14.0						
1981				0.1				11.1	8.0			15.0						
1982				0.2				23.0	17.0			16.0						
1983								7.6	9.3	6.0	15.0	33.0						
1984								4.7	44.3	3.0	150.0	53.0						
1985								19.5	13.6	2.0	89.0	66.0						
1986								119.4	134.2	4.0	180.0	229.0						
1987								192.1	146.4	12.0	101.6	640.4						
1988								202.8	147.0	10.0	135.8	717.4						
1989								194.9	87.1	39.0	54.9	365.2						
1990								238.9	118.9		129.4	323.3						
1991			12.2			9.8		626.0	622.4		123.6	319.4						
1992			55.3					655.5	715.4		80.3	402.8						
1993			125.4	0.9		35.4		499.5	316.4		39.5	149.0						
1994			30.0	3.7		61.4		208.8	284.6		79.9	197.0						
1995	2.2		40.0	3.2	29.0	60.2		444.6	219.3		48.0	454.0						
1996	4.7	55.0	74.0		32.0			596.2	113.1	20.0	49.2	666.0		98.5				
1997		29.0	39.0		9.0			505.4	202.9	35.0	56.5	862.0		122				
1998					3.0		2.0	678.8	253.5	25.0	39.1	369.3		81.9				
1999							1.0	394.7	375.7	8.0	49.0	140.7		17.1				
2000			9.0					553.9	160.8		62.0	246.0		2				
2001			14.0			1.0		485.4	374.6	16.0	49.0	245.0	54.4	104.7				
2002			9.0					389.3	177.9	8.0	34.0		29.7	126.6			103.9	71.4
2003		3.0	6.0					488.0	408.7	9.0	52.0				264.0	24.7	278.0	
2004			2.0			1.0		490.8		5.0	81.0			1.2		26.7		
2005								577.0						6.1	319.0	36.3		
2006								611.8										

Source: FSM - Trianni, 2002; RMI – FAO; Kiribati – FAO; Kiribati Fisheries Division, 1995; Tuvalu - Belhadjali, 1997; Conand and SPC, 1996; Belhadjali, 1997; Samoa – FAO, Mulipola, 1994; Tonga – FAO; Tonga Ministry of Fisheries, 1995; French Polynesia – FAO; PNG - Lindohlm, 1978; DFMR, no date; Wright, 1986 cited in Kailola with Lokani, no date; Lokani and Kubohojam, 1993; Lokani, 1990; Myint, 1996; National Fisheries Authority; Solomon Islands – FAO; James, 1977; Gaudechoux, 1993; Leqata, 2004; Division of Fisheries and Marine Resources; Vanuatu – FAO; Preston, 1993; New Caledonia – FAO; Conand and Hoffschir, 1991; Northern Province Fisheries Division, 1993; Conand and SPC, 1996; Etaix-Bonnin, 1999; Fiji – FAO; Preston, 1993; Gaudechoux, 1993; Qalovaki, 2006; Australia – DEH, 2004, 2006; McCormack, 2005; Roelofs, 2005; Hill, 2006; Skewes *et al.*, 2006.

Papua New Guinea: a hotspot of sea cucumber fisheries in the Western Central Pacific

Jeff Kinch
Secretariat for the Pacific Regional Environment Program
Apia, Samoa
E-mail: jeffk@sprep.org

Steve Purcell
The WorldFish Center
Noumea Cedex, New Caledonia
E-mail: SteveP@spc.int

Sven Uthicke
Australian Institute of Marine Science
Queensland, Australia
E-mail: s.uthicke@aims.gov.au

Kim Friedman
Secretariat of the Pacific Community
Noumea Cedex, New Caledonia
E-mail: KimF@spc.int

Kinch, J.; Purcell, S.; Uthicke, S.; Friedman, K. 2008. Papua New Guinea: a hotspot of sea cucumber fisheries in the Western Central Pacific. In V. Toral-Granda, A. Lovatelli and M. Vasconcellos (eds). Sea cucumbers. A global review of fisheries and trade. *FAO Fisheries and Aquaculture Technical Paper*. No. 516. Rome, FAO. pp. 57–77.

SUMMARY

Papua New Guinea (PNG) is one of the largest countries in the Western Central Pacific region and is now the third largest producer of bêche-de-mer in the world, supplying around 10 percent of the global market.

Species of commercial importance recognized by the National Fisheries Authority (NFA) include *Actinopyga echinites, A. lecanora, A. mauritiana, A. miliaris, Bohadschia argus, B. similis, B. vitiensis, Holothuria atra, H. coluber, H. edulis, H. fuscogilva, H. fuscopunctata, H. scabra, H. scabra* var. *versicolor, H. whitmaei, Pearsonothuria graeffei, Stichopus chloronotus, S. herrmanni, S. horrens, Thelenota ananas* and *T. anax*. Other species that are occasionally taken include *A. caerulea, H. leucospilota, S. pseudohorrens, S. vastus* and *T. rubralineata*.

Sea cucumbers are not a common "traditional" food in PNG, hence almost all are exported. They are mainly harvested by hand by free divers, or with spears and lead bombs in deeper water. Night fishing with torches and underwater breathing devices are used now despite their prohibition.

Management of the sea cucumber fishery in PNG was recognized as necessary after catches declined from the fishery's inception in the nineteenth century. Today, the NFA is responsible for the conservation and management of PNG's sea cucumber fishery. It has gazetted the *National Bêche-de-mer Management Plan* in 2001, which aims to maximize the long-term economic benefits from the fishery while ensuring resource and environmental sustainability.

Despite the national management plan, PNG still faces difficulties in enforcement and compliance. The two most continual infringements reported to the NFA involve the illegal buying of bêche-de-mer and seizure in urban centres. There are also occasional discrepancies between export figures and the import figures from Asian markets.

Factors that contribute to management problems include the remoteness of fishers and the limited human and financial resources of provincial fisheries offices. Moreover, export volumes continue to rise as fishers are collecting large quantities of low-value species. There is growing awareness by fishers that sea cucumbers are no longer abundant. Past stock assessments have been largely independent of each other and used differing methodologies and scales, making it difficult to see changes in abundance over time.

Perhaps CITES listing of some sea cucumber species in PNG could help management. However, implementation issues would need to be considered; for example, NFA's policy objectives, administrative capacity, adequate financing and regulatory ability. CITES listings could contribute to enhanced opportunities for technical assistance from regional agencies and promote increased partnerships with importing countries. It could also provide a mechanism for comprehensive and standardized trade and quota reporting. Arguably, the declines in abundances of *H. scabra* and *H. whitmaei* could qualify them for CITES listing.

Recent studies among several Provinces revealed a high economic reliance on sea cucumbers fishing; 12–75 percent of surveyed households gained some or most of their income from this fishery. Fishers currently receive around 65 percent of the export price across all species at the point of sale. Undoubtedly, a collapse of the sea cucumber fishery would have dire social and economic consequences for village-based fishers in PNG. Of all of PNG's existing fisheries, the sea cucumber fishery is one in which good management, appropriate regulation and enforcement could continue to deliver economic benefits, particularly at the rural level.

1. INTRODUCTION

Papua New Guinea (hereinafter referred to as PNG) is the largest country in the Western Central Pacific region and occupies the eastern half of the island of New Guinea, the largest equatorial island in the world (Figure 1). The country includes the islands of New Britain, New Ireland, Bougainville and Manus, and thousands of smaller islands. PNG has a total coastline of approximately 17 110 km, and its coral reefs are among the most diverse in the world. Although all reef types are represented, most are fringing and/or barrier reefs, with an estimated area of 40 000 km^2.

Papua New Guinea is now the third largest producer of sea cucumbers supplying the China Hong Kong Special Administration Region (SAR) and other Asian markets (Conand, 2004; Kinch, 2004), accounting for roughly 10 percent of all bêche-de-mer entering the global market.

Note: To be consistent with the terminology for this fishery, "holothurians" or "sea cucumbers" are used throughout this report when referring to live animals and "bêche-de-mer" is used when referring to the dead animal when processed for commercial purposes. Where older taxonomic classifications have been used in referenced texts or in information provided by colleagues, these have been changed to their new taxonomic determinations.

FIGURE 1
Location of PNG in relation to the Western Central Pacific region

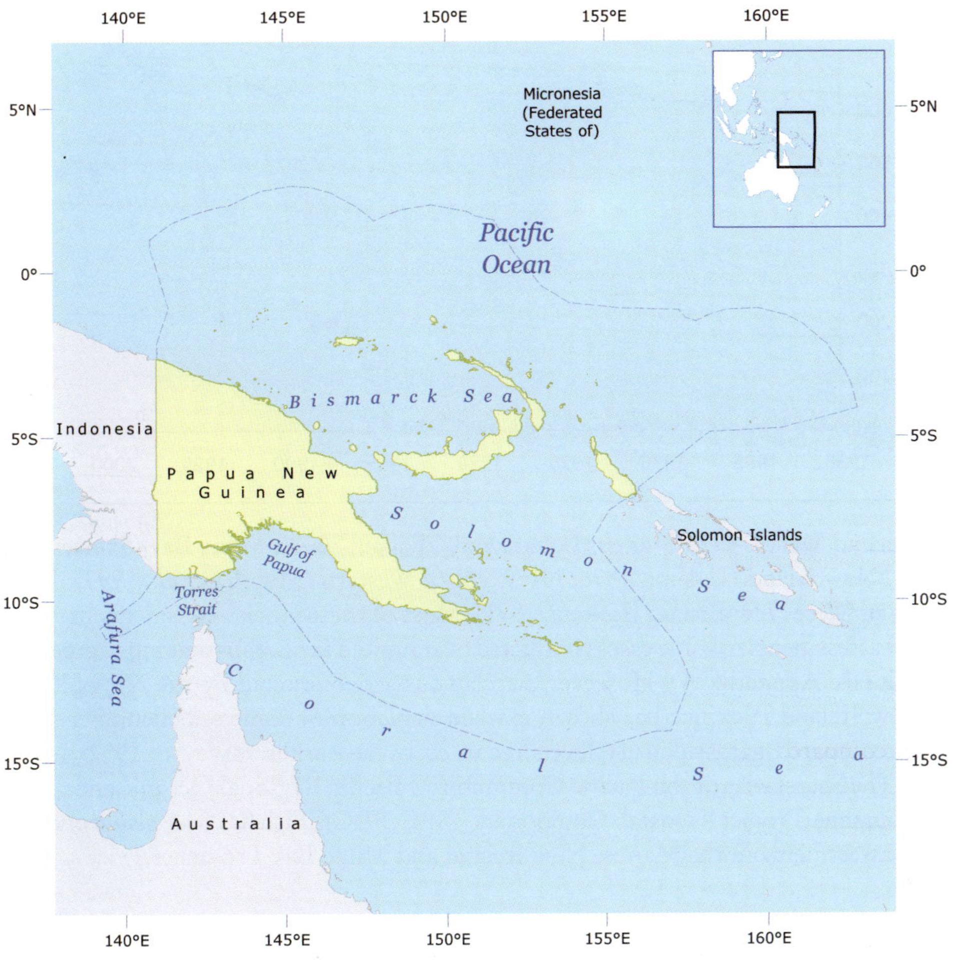

2. BIOLOGICAL AND POPULATION STATUS

2.1 Current species in trade

There are 26 currently known species of sea cucumbers harvested from PNG waters, with most being low-value species. The main species recognized by the PNG National Fisheries Authority (NFA) as having commercial importance include: *Actinopyga echinites, A. lecanora, A. mauritiana, A. miliaris, Bohadschia argus, B. similis, B. vitiensis, Holothuria atra, H. coluber, H. edulis, H. fuscogilva, H. fuscopunctata, H. scabra, H. scabra* var. *versicolor, H. whitmaei, Pearsonothuria graeffei, Stichopus chloronotus, S. herrmanni, S. horrens, Thelonata ananas* and *T. anax.* Other species that are known to be taken include *Actinopyga caerulea, H. leucospilota, S. pseudohorrens, S. vastus* and *T. rubralineata.*

The actual number of species in trade is difficult to determine accurately due to misidentifications by both fishers and companies and because some companies use their own trade names. During a recent assessment of purchasing data provided by the NFA, a total of 53 commercial names were reported in use by companies in PNG, of which 13 had not yet been accredited to known species (Kinch *et al.*, 2007).

2.2 Population status

Population surveys have been conducted in several provinces of PNG, Central (Shelley, 1981), East New Britain (Gisawa, 2002), Madang (Massin and Doumen, 1986; Lokani,

FIGURE 2
Bêche-de-mer exports from Papua New Guinea, 1960–2006

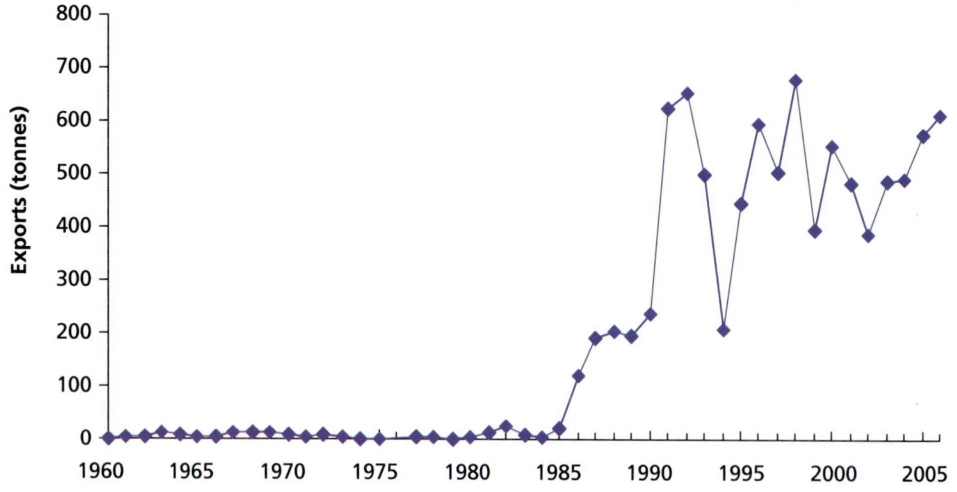

Mobiha and Wafy, 1992), Manus (Lokani and Chapau, 2002), Milne Bay (Skewes *et al.*, 2002), New Ireland (Lokani, 1996; NFA, 2007), Western (Mobiha *et al.*, 1993; Lokani, 2000) and West New Britain (Lokani, 1991). Most of these stock assessments have been one-off assessments in some cases conducted over limited areas, thus making comparisons difficult (see Appendix A). However, a recent survey conducted by the NFA (2007) in the New Ireland Province has shown marked depletion of some sea cucumber species when compared against density levels recorded by an earlier survey in 1992 (Lokani, 1996). The Secretariat of the Pacific Community's Pacific Regional Oceanic and Coastal Development Project Coastal Component (SPC-PROCFish/C) has also conducted recent assessments in the Manus, New Ireland and Milne Bay Provinces.

2.3 Catches

Sea cucumbers are harvested in PNG by hand-collection and free-diving. In deeper water, fisher will dive down and spear them or use the lead bomb (a small harpoon embedded in a lead weight) (Kinch, 2002, 2004). Once arriving at a potential site and sea cucumbers have been located, fishers (including women) will enter the water. If the water is shallow enough, fishers will collect sea cucumbers by wading on the reef. In deeper water, fishers will dive down and spear them or use lead bombs. The use of lights is a common practice and hookah and SCUBA gear has been used in recent years, even though these have been officially banned for numerous years.

Sea cucumber production or catch data is not currently available, and subsequently export data has been used as a determinant of production (Figure 2; Appendix B). Harvesting of sea cucumbers for bêche-de-mer increased dramatically in 1986.

All exports from PNG are as dried bêche-de-mer from capture fisheries. Sea cucumbers do not appear to be a major item on either traditional or modern menus of most PNG communities. However, there are some exceptions. Lokani (1990) notes that some *Actinopyga* species are consumed in West New Britain and some areas in Manus use the toxins of *H. atra* to fish for octopus. In the Milne Bay Province, Trobriand Islanders are also known to consume *H. scabra* (Lindholm, 1978). *Holothuria scabra* is also consumed in the Western Province (Kinch, 2004; Kinch *et al.*, 2007).

2.4 Management of the fishery

The need for the management of the sea cucumber fishery in PNG was recognized from its inception in the nineteenth century due to declining catches. In 1881, a closed season was attempted by the colonial government, but failed to stop illegal harvesting.

In the early twentieth Century attempts were made to manage the sea cucumber fishery under the 1911–1934 *Pearl, Pearl Shell and Bêche-de-mer Ordinance,* which prohibited the harvesting of sea cucumbers between the high water mark and a line drawn parallel to and 800 m distant from the high water mark (Hyndman, 1993; Kinch, 2002, 2004; Kinch *et al.*, 2007; Tom'tavala, 1990, 1992).

In 1992, a *Prohibition of Taking Sedentary Resources* was gazetted under the *Continental Shelf (Living Natural Resources) Act.* Under this gazettal notice, the forerunner of present management strategies was outlined. These included minimum legal dry size limits, prohibitions on the use of hookah and SCUBA, lights at night and ships used in the harvesting of sea cucumbers.

Today, the NFA is responsible for the conservation, management, development and sustainable use of PNG's fisheries under the 1998 *Fisheries Management Act.* The NFA has gazetted the 2001 *National Bêche-de-mer Management Plan* for regulation and management of the sea cucumber fishery (Polon, 2004). Its aim is to ensure that PNG and its people obtain the maximum economic benefit from the fishery, and that fishing is sustainable and has a minimal impact on the marine and coastal environment (Table 1). The *National Bêche-de-mer Management Plan* now over-rides all previous *Provincial Bêche-de-mer Fisheries Management Plans.* Provinces that had management plans previously include Milne Bay (1998), Western Province and Torres Strait (1995), Manus (1997) and New Ireland (2000). The NFA re-emphasized the prohibition on taking sedentary resources at night with the use of lights, and by hookah and SCUBA.

Monitoring and enforcement costs for the sea cucumber fishery in PNG have steadily increased, causing concern for the NFA (Kinch, 2004; Kinch *et al.*, 2007). The two most continual infringements reported to the NFA involve the illegal buying of bêche-de-mer and seizure in urban centres (Kinch *et al.*, 2007) (Table 2). Villagers continue to dive at night and during the closed season, and there has been reported use of hookah and SCUBA (Kinch, 2002; 2004, Kinch *et al.*, 2007).

TABLE 1
Management strategies detailed in the National Bêche-de-mer Management Plan

Permits	Harvest season	Gear type	Total allowable catch (TAC)	Size limits
PNG citizens only Licence for storing or export Reporting requirements attached to licenses	Open season from 16 January to 30 September	Hookah, SCUBA and lights prohibited	TAC for each Province TAC divided into two value groups (high and low)	Minimum legal sizes for most live and dried species

TABLE 2
Notification of infringements to the National Fisheries Authority, 2001–2005

Activity	2001	2002	2003	2004	2005	Total	% of total
Buying of bêche-de-mer		3	3	3	2	11	14.5
Buying undersize		1			1	2	2.6
Foreign involvement				3		3	3.9
Harvesting bêche-de-mer from vessel					1	1	1.3
Harvesting bêche-de-mer in closed season					1	1	1.3
Illegal export by air		1				1	1.3
Illegal shipment of bêche-de-mer to Alotau			1			1	1.3
Illegal shipment of bêche-de-mer to Port Moresby	1	10	11	3		25	32.9
Illegal shipment to West Papua		1				1	1.3
Illegal storage of bêche-de-mer	1	8	6	5	2	22	28.9
Operating outside of home province		1	1			2	2.6
Poaching in Australian waters				1		1	1.3
Use of other export license			1			1	1.3
Use of underwater breathing devices		1	2	1		4	5.2
Total	**2**	**26**	**25**	**16**	**7**	**76**	**100.0**

Although not reflected in the infringement data, poaching by fishers in the Western Province is also of concern for Australian authorities. Villagers from the Western Province regularly move to the Torres Strait Islands for trading and social reasons, and have traditionally claimed fishing rights as far south as the southern end of the Warrior Reef complex which is situated in Australian waters (Kinch *et al.*, 2007).

When the sea cucumber fishery reopened in the Western Province in the early-1990s, Torres Strait Islanders and Australians were not fishing, and harvesting by fishers from the Western Province became prominent on the Warrior Reef complex. During the early 1990s, hundreds of fishers from the Western Province were apprehended in Australian waters, forcing the PNG Government to place a moratorium on fishing. Regular apprehensions are common, and today most fishers time their runs across the border with low tides as this allowed fast collection (Schug, 1995, 1996). The Australian surveillance authority, Coastwatch, sees the Warrior Reef complex as a high risk area, and because incursions/poaching from fishers from the Western Province is ongoing, steps are being undertaken by Australian authorities to improve the effectiveness of future patrol work in the area.

Bêche-de-mer is also moved illegally from the Western Province via Yule Island or Kerema on its way to Port Moresby. Some product is also thought to leave PNG via logging or fishing vessels.

The *National Bêche-de-mer Management Plan* limits the harvest of sea cucumbers and the export of bêche-de-mer. The limits are imposed by setting a Total Allowable Catch (TAC) for each Province (Table 3), intended for a single season. Most TACs in PNG are not based on scientific research. Currently, TACs for some Provinces are set on 70 percent of the total harvestable estimate by dry weight (converted), and from the results of previous stock assessments. For Provinces with no recent stock assessments, the TAC is calculated based on historical catch record.

Sea cucumber species are divided into two groups, high and low value (Table 4). A TAC is set for both groups because the higher value species are more heavily fished than the lower value species. Once the TAC of a value group has been reached, fishing for those species is supposed to stop, though NFA has never enforced a closure by value group. If the allocated TAC for a Province is reached and exceeded by more than 5 tonnes, then the total excess amount should be taken off the next season's TAC. This reduction in TACs has never been enforced by the NFA.

TABLE 3
Papua New Guinea provincial total allowable catches (TACs)

Province	High value (tonnes)	Low value (tonnes)	Total TAC (tonnes)	Previous provincial management plans TACs (tonnes)
Bougainville	20	40	60	
Gulf	(0.5)	(0.5)	(1)	
East New Britain	10	20	30	
East Sepik	7	13	20	
Madang	15	25	40	
Milne Bay	60	80	140	60
Manus	18	32	50	54
Morobe	10	20	30	
NCD and Central	(25)	(55)	(80)	
New Ireland	25	55	80	80
Oro	15	25	40	
West New Britain	20	40	60	
West Sepik	7	13	20	
Western	10	7	17	40
Total	**242.5**	**425.5**	**668**	

Note: Figures in brackets are provisional estimates only. NCD = National Capital District.

TABLE 4
High and low value species and associated live and dry size limits

Low value species	**Live length (cm)**	**Dry length (cm)**
Actinoypga echinites	(25)	(15)
Bohadschia similis	(25)	(7)
Holothuria atra	(30)	(15)
H. coluber		
H. edulis	(25)	(10)
H. fuscopuntata	(45)	(15)
Pearsonothuria graeffei		
Thelenota anax	20	10
High value species		
Actinoypga lecanora	15	10
A. mauritiana	(20)	(8)
A. miliaris	15	10
Bohasdchia argus	20	10
B. vitiensis	20	10
Holothuria fuscogilva	35	15
H. nobilis	22	10
H. scabra	22	10
Stichopus chloronotus	20	10
S. herrmanni	25	10
Thelenota ananas	25	15

Note: Figures in brackets are provisional estimates only.

Movement of bêche-de-mer between Provinces without written permission is prohibited in the *National Bêche-de-mer Management Plan*. Where an exporter's nearest port for export is in another province, or is not functional (or economical), the company may apply for written authorization, from the Managing Director of NFA, to use another port for export. The issue of intra-provincial transfers is an area that needs greater attention by the NFA because they are poorly documented and are not accredited back to the Province of origin (Kinch *et al.*, 2007).

Size limits for harvesting and export of sea cucumbers are set by the *National Bêche-de-mer Management Plan* (Table 4). It also prohibits the trade of undersize bêche-de-mer. However, is doubtful if the minimum legal live sizes limits have achieved the desired goal of protecting individual sea cucumbers until they have spawned at least once. High levels of harvesting persist and undersized animals are still fished, as evident by the large number of undersized product recorded in recent studies by the NFA. A further complicating factor, is that the natural adult sizes of animals differ across regions and habitats (due to location, water quality, depth, differences in temperatures, etc.) making current size limits ineffetual or not respected in some locations.

The appropriateness of minimum legal live sizes limits in PNG (or anywhere in the Western Pacific Region for that matter) as a management tool depends on more studies being conducted on the size at first spawning and location-specific growth rates. The size limits also depend on other variables, such as catchability. Skewes *et al.* (2006) suggests a more sensible approach to reviewing the minimum legal live sizes limits of sea cucumbers is to conduct detailed analyses of egg production (e.g. egg-per-recruit analysis). Minimum legal size limits would therefore be prudent to maximize yield-per-recruit in sea cucumber stocks, since it seems they are often affected by recruitment overfishing. Studies that determine egg-per-recruit would be more beneficial. The trouble with egg-per-recruitment analysis is that it requires data on fecundity at size, some idea about fishing pressure and the population structure, which are currently lacking for most species.

The *National Bêche-de-mer Management Plan* also includes a compulsory closed season each year. This closed season occurs between 1 October and 15 January. If a Province has reached its TAC before the 1 October, then the closed season will commence from the date the TAC was acknowledged to have been reached. Usually Western, Manus and Milne Bay Provinces close early having reached their allotted TACs. Some companies have argued against this imposed closure, citing seasonal weather conditions that make it difficult for fishers to harvest.

The use of hookah and SCUBA, and underwater or surface lights for the fishing of sea cucumbers is prohibited under Section 7d (also refer to G57 National Gazette 4th April 2002). There is no serious enforcement on this regulation, and the use of hookah and SCUBA is a common practice in some provinces. Fishers in all Provinces regularly use lights for the harvesting of sea cucumbers, notably *B. vitiensis* and *A. lecanora* (Kinch, 2004; Kinch *et al.*, 2007).

3. TRADE

Countries that import bêche-de-mer from PNG include Australia, Canada, China Hong Kong SAR, Indonesia, Republic of Korea, Malaysia, New Zealand, Singapore, the People's Republic of China and Taiwan Province of China (Kinch, 2004). Most companies will regularly export to one or two of the same companies, as these are usually their financiers. Papua New Guinea is now the third largest global producer of bêche-de-mer supplying the Chinese Hong Kong SAR and other Asian markets (Conand, 2006; Kinch, 2004, 2006; Kinch *et al.*, 2007).

Records of exports and imports are important elements of management as well as providing documentation along the market chain. There is some cause for concern for understanding the bêche-de-mer trade as PNG exports and imports into China Hong Kong SAR, show anomalies for 2001 and 2004 (Table 5), whereby PNG declared exports are actually smaller in some years then customs records report from China Hong Kong SAR. For example, in 2001 China Hong Kong SAR imported eight tonnes more than PNGs total exports and in 2004 it was over by 45 tonnes. It is also probable that before the moratorium in the Solomon Islands that bêche-de-mer was also being shipped from Bougainville through the Solomon Island's Western Province for onward shipping out of Honiara. It is also more than probable that bêche-de-mer is channeled into the Indonesian Province of West Papua from the West Sepik (Sanduan) and Western Provinces in PNG.

Prices for bêche-de-mer in PNG have increased significantly from 1991 and 2007 (Table 6). Price increases are linked to a multiple of reasons, including new species

TABLE 5
Papua New Guinea bêche-de-mer exports and imports into China Hong Kong SAR

Year	PNG exports (tonnes)	China Hong Kong SAR imports of total PNG exports (tonnes)	Percentage of total PNG exports
1992	655.5	240.0	36.6
1993	499.5	186.2	37.3
1994	208.8	150.3	72.0
1995	444.6	256.0	57.6
1996	596.2	351.6	59.0
1997	505.4	470.1	93.0
1998	678.8	639.8	94.2
1999	394.5	350.3	88.8
2000	553.9	532.0	94.6
2001	485.4	541.1	>100?
2002	389.3	380.6	97.8
2003	488.0	445.5	91.3
2004	490.8	518.3	>100?
2005	577.0	469.0	81.3
2006	611.8	–	–

Note: The last column is the China Hong Kong SAR, imports as a percentage of PNG Exports. All figures supplied by the National Fisheries Authority are indicative only, as the database is continuously updated. When multiple sources have been utilized, the highest value has been incorporated.

TABLE 6
Buying prices (USD/kg) to fishers for bêche-de-mer in 1991 and 2007

Species	Grade	Price (USD/kg)	
		1991	2007
Actinoypga lecanora	A	–	33.30
	B	–	27.25
	C	–	15.15
	D	–	9.10
A. mauritiana	A	–	27.25
	B	–	24.25
	C	–	12.10
	D	–	9.10
A. miliaris	A	2.70	30.30
	B	2.70	27.25
	C	2.70	15.15
Bohadschia argus	A	–	13.65
	B	–	9.10
	C	–	6.05
B. similis	–	–	3.65
B. vitiensis	A	2.70	9.10
	B	2.70	7.60
Holothuria atra	–	3.60	3.65
H. edulis	–	–	3.65
H. fuscogilva	SL	7.20	42.40
	A	7.20	39.40
	B	7.20	30.30
	C	7.20	21.20
H. fuscopunctata	–	–	3.65
H. scabra	SL	–	60.60
	A	16.20	54.55
	B	12.60	48.50
	C	10.80	24.25
	D	–	12.10
H. whitmaei	A	7.20	30.30
	B	7.20	24.25
	C	7.20	15.15
	D	–	9.10
Stichopus chloronotus	A	3.60	33.30
	B	3.60	27.25
	C	3.60	18.20
S. herrmanni	A	1.20	30.30
	B	1.20	27.25
	C	1.20	15.15
	D	1.20	9.10
Thelenota ananas	A	4.50	33.30
	B	4.50	27.25
	C	4.50	18.20
T. anax	–	–	3.65
Pearsonothuria graeffei	–	–	3.65

Note: Grades are determined by length and quality. **"A"** is the highest grade. **SL** = Super Large.
Source: Kinch *et al*., 2007

entering the market, the increased demand by Asian consumers and the rise and fall of the PNG currency. Following recent research, and the high proportion of undersized bêche-de-mer being purchased by exporting companies, the NFA is now considering a three-year moratorium on the sea cucumber fishery, with consultation currently being conducted amongst stakeholders.

From a review of customs declarations for 2006–2007, whereby buying prices were compared to export prices, fishers appear to receive a major proportion of the export value of bêche-de-mer in PNG. On average across all species, fishers received 65 percent

TABLE 7
Average percentage of export value returned to fisher

Species	Grade	Percentage
Actinopyga lecanora	A	74.4
	B	65.7
	C	59.5
A. mauritiana	A	81.1
	B	67.4
A. miliaris		60.4
Bohadschia argus		70.0
B. similis		65.7
B. vitiensis		53.8
Holothuria atra		56.4
H. edulis		63.9
H. fuscogilva	A	85.6
	B	77.6
	C	68.8
H. fuscopunctata		69.4
H. scabra		54.4
	A	69.2
	B	65.1
H. whitmaei	A	73.4
	B	59.4
	C	46.7
Pearsonothuria graeffei		56.4
Stichopus chloronotus		66.8
S. herrmanni	A	65.4
	B	66.6
	C	53.2
Thelenota ananas		63.8
T. anax		59.5
Average		**65.0**

Note: Grades are determined by length and quality. "**A**" is the highest grade. These figures should be treated as indicative only.

Source: Kinch *et al.*, 2007.

of the export price at the point of sale. Certain species had greater returns then others, but the proportionate return to fishers was not very variable (s.d. = 8.7 percent) (Table 7). Generally, higher value species and higher grades within species, provide a proportionately higher share of the export price to fishers than lower value species or lower grade bêche-de-mer. Since large animals yield higher grade bêche-de-mer, it follows that fishing out large and high-value sea cucumbers concomitantly results in fishers harvesting animals of lower value and receiving proportionately less of the export value.

A sobering aspect of the global marketing system for bêche-de-mer is that as the supply of a certain species declines, the demand drives the price up, providing greater incentive for fishers to harvest the remaining vestiges of each species. Alternatively, low pricing by companies forces overexploitation and the harvesting of undersize as fishers attempt to make an adequate profit. Companies in PNG also exert pressure on local fishers to provide bêche-de-mer by giving cash-advances or other materials, and in some cases creating a loan-debt cycle that is difficult to break.

4. SOCIO-ECONOMIC IMPORTANCE TO LOCAL FISHING COMMUNITY

During the lead author's (Jeff Kinch) PhD work at Brooker Island in the Milne Bay Province, bêche-de-mer sales contributed 46.7 percent of all household incomes (Kinch, 1999). A socio-economic survey conducted by the Coastal Fisheries Management and Development Programme (CFMDP) in the Milne Bay Province showed that

FIGURE 3
Juvenile green turtle (*Chelonia mydas*) caught as bycatch in the sea cucumber fishery in Papua New Guinea Western Province

COURTESY OF J. KINCH

21.9 percent of households surveyed, harvested sea cucumbers for income (NFA, 2006). A similar survey in the New Ireland Province showed that 12 percent of households harvested sea cucumbers for income (NFA, 2005). In the West New Britain Province, Koczberski *et al.* (2006) also found the sale of bêche-de-mer to be an important source of income.

A recent socio-economic assessment of the sea cucumber fisheries in the Western, Central and Manus Provinces by Kinch *et al.* (2007) showed that on average, households that harvested sea cucumbers could make between USD 1 000–3 000/year from the sale of bêche-de-mer. In Manus Province, of all households surveyed, 75 percent stated that harvesting sea cucumbers was their most important income stream (Kinch *et al.*, 2007). During this survey, between 24 and 40 percent of all households stated they would move to other forms of fishing and harvesting other marine resources if they could no longer harvest sea cucumbers, and thus possibly starting a spiral of fishing down other valuable species. Of particular concern would be the fishing for sharks for shark fins and sea turtles (Figure 3).

The PNG Government has undertaken a range of financial and economic reforms in recent years, including the Medium Term Development Strategy (2005–2010), the Poverty Reduction Strategy and an Export Oriented Growth Strategy. The majority of this growth is likely to be in export-oriented sectors, such as the sea cucumber fishery because bêche-de-mer is predominantly export-oriented in PNG.

Of all of PNG's existing fisheries, the sea cucumber fishery is the one in which good management, appropriate regulation and enforcement is most likely to continue to deliver economic benefits, particularly at the rural level. Undoubtedly, a collapse of the sea cucumber fishery would have dire social and economic consequences for the rural fishers in PNG.

5. ADDITIONAL THREATS TO SEA CUCUMBER POPULATIONS

As in the Western Pacific Region, the biggest threat to sea cucumber fisheries in PNG is overharvesting (see Dalzell, 1990; Kinch, 2002, 2004; Lokani, 1996; NFA, 2007; Skewes *et al.*, 2002, 2004, 2006). Other impacts on sea cucumber populations and their associated habitats include degradation of habitats around urban centres (Lock, 1986), bleaching events (Davies, Dunne and Brown, 1997) and cyclones (Miller *et al.*, 2004).

6. RECOMMENDATIONS FOR IMPROVING FISHERIES MANAGEMENT AND CONSERVATION

The sea cucumber fishery in PNG has not been well studied, despite its importance in terms of revenue it generates for PNG and the income it provides to a majority of coastal communities. Subsequently, NFA contracted the lead author (Jeff Kinch) to review the sea cucumber fishery and its management (Kinch, 2004) and conduct a socio-economic study (Kinch *et al.*, 2007). Based on those investigations, the following actions for improved management have been recommended (these are in no order of priority or importance):

- To conduct extensive awareness campaigns for coastal and island fishers on the biology and reproductive ecology of holothurians (i.e. spawning, fertilization, larval dispersal and settlement), the dangers of harvesting undersize animals, the importance of protecting them until they reach marketable size or are mature enough to produce the next cohorts, and the susceptibility of sea cucumbers to overharvesting and subsequent decline in yield.
- To conduct research into breeding seasonality, growth rates (due to regional and habitat variations) and size at first spawning to help determine accurate minimum legal live size limits for sea cucumbers in a given area. Size limits can be a useful tool for maximizing the economics of the bêche-de-mer fishery, but use of size limits depends on proper biological information and modeling.
- To carry out research to determine the taxonomic identity of species in trade.
- To develop a standardized and effective assessment sampling protocol for estimating population abundances. This would assist in future replication and permit comparisons between different habitats and regions. It would allow for the collation of a time series of fishery-dependent information, which could be used to determine harvesting impacts within these areas.
- To study the supply and marketing chain, in order to determine the real values of bêche-de-mer along the buying chain, and help accurately determine the equity to the fisher.
- To conduct data collection in order to document the number of bêche-de-mer/grade/kg. This could be used to determine weights for individual grades which could also be used as a measure along with minimum legal dry size limits. The international market sells by weight, rather than size and this could help ameliorate the differences in marketing strategies.
- To determine the conversion ratios from live sea cucumber to dry bêche-de-mer for all species, for converting pooled catch data from one state to another.
- To host workshops and training to encourage value-added processing by coastal and island fishers to the fullest extent possible and to the highest standard possible. This training should be carried out by the industry where possible, and linked to licensing, such that licenses will not be issued unless company employees have been put through this course, and actually carry out extension services.
- To educate fishers and companies on standardized trade names to enable accurate reporting of bêche-de-mer species in trade. The NFA should canvass exporters of differing species in trade and ask for samples that exporters distinguish as

different to the ones that are produced in the SPC *Sea Cucumber and Beche-de-Mer Identification Cards*[1].

- To develop and promote better reporting systems based on suitable logbooks that uses four carbon invoice/receipts with one copy given to the fisher, one copy to the Provincial Fisheries Office, one to the NFA and the book copy retained by the company. This would enable standard collection of catch and effort data when purchasing products from the fishers, and could be converted to a standard unit.
- To investigate equity distribution/reward models to act as an incentive for exporters to add value through better processing and targeting of larger size grades as well as providing a disincentive to the practice of under-declaring export prices.
- To investigate mechanisms for companies to pay a fisheries management levy to NFA to cover the costs of research and management of this fishery. The levy could be a fixed amount/kg exported, rather than a percentage of export value. This would also dissuade exporters from under-declaring their revenues and also to make it less economical to process really low-value product.
- To investigate the impacts of introducing species-specific TACs for heavily-exploited species, or for trials on new commercial species. It is probably better, however, to set zero TACs for depleted species, given that the species value groups are not effectively monitored by the NFA.
- To examine possibilities of co-management of the bêche-de-mer fishery with villages. Assessment criteria could be devised to determine if it would be practically feasible, such as social cohesion, existing management structures, dependency on resources, etc.
- To investigate the possibility of a register of foreign financiers/technical advisors, and further involve them in management discussions. Guidelines could also be developed taking into account the interest of both the non-citizen and the PNG Citizen.
- To limit the number of licences issued for each province. Too many companies make it difficult for the NFA to monitor purchasing and exporting data. Whilst it is generally thought good economic theory that competition raises standards along with prices, this does not appear to have happened as companies compete against each other to secure adequate product, and this leads to increased infringements of some regulations, particularly buying undersize bêche-de-mer. Competition from illegal operators also undermines any incentive licensed operators have to operate legally.
- To monitor intra-provincial transfers and reassign transfers back to province of origin as current export figures are not representative of actual production from a given province.
- To re-introduce buyer's licences that are linked to a specific company. This would assist companies to provide accurate information on catch, size distribution and quality, which could then be used to support management of the fishery.
- To standardize the issuance of exporting licences to a single month, meaning all licenses of intending exporters are submitted during a specific period, followed by inspections of facilities, with all licences given out for the same start month. This would help the NFA in knowing who is legal and who is not, as there is currently considerable confusion over how many legitimate exporters are licensed for each province.
- To establish an industry code of conduct that covers product pricing, size limits, training of fishers in processing methods, and provision of purchasing and exporting data to the NFA. Companies that do not abide by the code of conduct would have their licenses withdrawn.

[1] SPC. 2004. Pacific Island sea cucumber and bêche-de-mer identification cards.

7. EVALUATION OF THE PROS AND CONS OF A CITES LISTING

Bruckner, Johnson and Field (2003) and Sant (2006) give accounts of the process and progress for the possibility of having sea cucumbers listed with the Convention on the International Trade in Endangered Species (CITES). According to these authors, under Resolution Conf. 9.24 (Annex 2a Bi), some sea cucumber species such as some temperate Stichopodidae species meet CITES criteria for Appendix II inclusion because harvesting for international trade is having a detrimental impact on species by exceeding, over an extended period, the level that can be continued in perpetuity.

Papua New Guinea has the most well developed sea cucumber/bêche-de-mer management plan in place, anywhere in the Western Central Pacific region, but as noted above, it is still having difficulties in implementing successful and sustainable management due to a variety of scientific, social, economic and political reasons. Of particular note are the remote locations where coastal and island fishers reside and the limited resources available for monitoring, management and enforcement. Compounding these difficulties is the lack of clear evidence on the surface from aggregate national export data that fishery depletion is actually occurring, as exports of bêche-de-mer continue to increase. This is invariably due to the exploitation of new areas and new species, masking stock depletion of at least some species in some areas. However, stock assessments have not been carried out for most provinces, so there is still substantial uncertainty about this question. There appears to be however, a growing awareness amongst fishers that sea cucumber resources are no longer abundant (Kinch *et al.*, 2007).

In terms of conservation costs and benefits of listing sea cucumber species with CITES, it is clear that a number of implementation issues would need to be addressed in order for benefit to be derived from such a listing in PNG. As noted above, identification of individual species in trade is a problem in PNG. Proper identification is important as is the ability to distinguish taxa in the form they are traded.

Undoubtedly, CITES listing of sea cucumbers in PNG could provide an additional tool to ensure that harvesting to supply international markets is conducted in a sustainable manner. CITES listings could contribute to enhanced opportunities for technical assistance from regional agencies such as the Food and Agriculture Organization (FAO) and SPC, but also draws on financial support from the Australian Centre for International Agricultural Research (ACIAR), and technical support from the Commonwealth Science and Industry Research Organisation (CSIRO) and the Australian Institute of Marine Sciences (AIMS). Finally, CITES listings could help promote increased partnerships with importing countries, and provide a mechanism for comprehensive and standardized trade and quota reporting.

Future prospects for the sea cucumber fishery in PNG and consideration of CITES participation, will however, depend upon several factors including the NFA's policy objectives, administrative capacity, adequate financing and regulatory ability.

It could already be argued that *H. scabra* and *H. whitmaei* could qualify for CITES listing due to their decline over most of their distributional range (Skewes *et al.*, 2002, 2004, 2006).

ACKNOWLEDGEMENTS

The authors would like to thank the following persons for their input, advice and/ or provision of information: Alessandro Lovatelli, Augustine Mobiha, Cathy Hair, Donna Asi, Garry Preston, Leban Gisawa, Marcelo Vasconcellos, Mark Baine, Philip Polon, Timothy Skewes, Ursula Kaly, Veronica Toral-Granda and Warwick Nash.

REFERENCES

Bruckner, A., Johnson, K. & Field, J. 2003. Conservation Strategies for Sea cucumbers: Can a CITES Appendix II listing Promote Sustainable International Trade? *SPC Beche-de-Mer Information Bulletin,* 18: 24–33.

Conand, C. 2004. *Sea Cucumber Biology: Taxonomy, Distribution, Biology and Conservation Status.* A Paper prepared for the CITES Technical Workshop on the Conservation of Sea Cucumbers in the families Holothuriidae and Stichopodidae, March 1–3, Kuala Lumpur, Malaysia.

Conand, C. 2006. Harvest and Trade: Utilization of Sea Cucumber; Sea Cucumber Fisheries; Current International Trade; Illegal, Unreported and Unregulated Trade; By-catch; and Socio-economic Characteristics of the Trade in Sea Cucumbers. *In: Proceedings of the CITES Workshop on the Conservation of Sea Cucumbers in the Families Holothuriidae and Stichopodidae, 1–3 March, 2004, Kuala Lumpur, Malaysia.* A. Bruckner, (ed). NOAA Technical Memorandum, NMFS-OPR-34. pp 51–73. Silver Spring: NOAA.

Dalzell, P. 1990. Beche-de-mer Production from Three Papua New Guinean Atolls between 1982 and 1983. *SPC Beche-de-Mer Information Bulletin,* 1:6–7.

Davies, J., Dunne, R. & Brown, B. 1997. Coral Bleaching and Elevated Sea-water Temperature in Milne Bay Province, Papua New Guinea, 1996. *Marine and Freshwater Research,* 48: 513–6

Gisawa, L. 2002. *A Survey of Some of the Marine Resources of the South Coast Area of the East New Britain of Papua New Guinea.* Report prepared for the National Fisheries Authority, Port Moresby, Papua New Guinea.

Hyndman, D. 1993. Sea Tenure and the Management of Living Marine Resources in Papua New Guinea. *Pacific Studies,* 16(4): 99–114.

Kailola, P. & Lokani, P. No date. *Sea cucumbers.* Paper prepared for the National Fisheries Authority, Port Moresby, Papua New Guinea.

Kinch, J. 1999. *Economics and Environment in Island Melanesia: A General Overview of Resource Use and Livelihoods on Brooker Island in the Calvados Chain of the Louisiade Archipelago, Milne Bay Province, Papua New Guinea.* Report prepared for Conservation International, Port Moresby, Papua New Guinea.

Kinch, J. 2002. An Overview of the Beche-de-mer Fishery in the Milne Bay Province, Papua New Guinea. *SPC Beche-de-Mer Information Bulletin,* 17: 2–16.

Kinch, J. 2004. *A Review of the Beche-de-Mer Fishery and its Management in Papua New Guinea.* Report prepared for the National Fisheries Authority, Port Moresby, Papua New Guinea.

Kinch, J. 2006. National Report – Papua New Guinea. *In: The Proceedings of the Technical Workshop on the Conservation of Sea Cucumbers in the Families Holothuridae and Stichopodidae.* A. Bruckner (ed). NOAA Technical Memorandum NMFS-OPR 44. pp. 218–225. Silver Spring, NOAA.

Kinch, J., James, M., Thomas, E., Lauhi, P. & Gabiobu, R. 2007. *Socio-economic Assessment of the Beche-de-Mer Fisheries in the Western, Central and Manus Provinces, Papua New Guinea.* Report prepared for the National Fisheries Authority, Port Moresby, Papua New Guinea.

Koczberski, G., Curry, G., Warku, J. & Kwam, C. 2006. *Village-based Marine Resource Use and Rural Livelihoods: Kimbe Bay, West New Britain, Papua New Guinea.* TNC Pacific Islands Countries Report, No. 5. Report prepared for The Nature Conservancy, Port Moresby, Papua New Guinea.

Lindholm, R. 1978. *Beche-de-Mer Fishery.* Report prepared for the Department of Primary Industry, Port Moresby, Papua New Guinea.

Lock, J. 1986. *Effects of Fishing Pressure on the Fish Resources of the Port Moresby Barrier and Fringing Reefs.* Fisheries Technical Report, No. 03. Report prepared for the Department of Primary Industries, Port Moresby, Papua New Guinea.

Lokani, P. 1990. Beche-de-mer Research and Development in Papua New Guinea. *SPC Beche-de-Mer Information Bulletin,* 1: 8–11.

Lokani, P. 1991. *Survey of Commercial Sea Cucumbers (Beche-de-Mer) in the West New Britain Province, Papua New Guinea.* Report to the Department of Fisheries and Marine Resources, Kavieng, New Ireland Province, Papua New Guinea.

Lokani, P. 1996. *Fishery Dynamics and Biology of Beche-de-Mer in the Tigak Islands, Papua New Guinea.* Report prepared for the National Fisheries Authority, Port Moresby, Papua New Guinea.

Lokani, P. 2000. *Western Province and Torres Strait Beche-de-Mer Survey.* Report prepared for the National Fisheries Authority, Port Moresby, Papua New Guinea.

Lokani, P. & Chapau, M. 1992. *A Survey of the Commercial Sedentary Marine Resources of Manus Island.* Report to the Department of Fisheries and Marine Resources, Kavieng, New Ireland Province, Papua New Guinea.

Lokani, P., Mobiha, A. & Wafy, A. 1992. (eds.). *Marine Resources Survey of Madang Province.* Report to the Department of Fisheries and Marine Resources, Kavieng, New Ireland Province, Papua New Guinea.

Massin, C. & Doumen, C. 1986. Distribution and Feeding of Epibenthic Holothuroids on the Reef Flat of Laing Island (Papua New Guinea). *Marine Ecology,* 31(2): 185–195.

Miller, I., Sweatman, H., Baine, M., Chatterton, P., Chin, A., Fitzpatrick, B., Gomelyuk, V., Hill, J., Jenkins, A., Jones, G., Kinch, J., Marnane, M., Thorn, R., Malcolm, H. & Smith, L. 2004. Status of Coral Reefs In Australia and Papua New Guinea. In: *Status of Coral Reefs of the World: 2004.* Wilkenson, C. (ed). pp. 303–335. Townsville, Australian Institute of Marine Sciences.

Mobiha, A., Polon, P., Lari, R. & Jogo, S. 1993. *A Survey of the Marine Resources of the Daru Area in Western Province of Papua New Guinea.* Report to the Department of Fisheries and Marine Resources, Kavieng, New Ireland Province, Papua New Guinea.

Myint, T. 1996. Beche-de-mer Exports. *PNG National Fisheries Authority Newsletter,* 2(1): 20–21.

National Fisheries Authority. 2005. *Socio-economic Survey of Small-scale Fisheries in New Ireland Province, Papua New Guinea.* Kavieng, National Fisheries Authority and the Coastal Fisheries Management and Development Project.

National Fisheries Authority. 2006. *Socio-economic Survey of Small-scale Fisheries in Milne Bay Province, Papua New Guinea.* Kavieng, National Fisheries Authority and the Coastal Fisheries Management and Development Project.

National Fisheries Authority. 2007. *Sea Cucumber Survey: New Ireland Province.* Kavieng, National Fisheries Authority and the Coastal Fisheries Management and Development Project.

Polon, P. 2004. The Papua New Guinea National Beche-de-mer Fishery Management Plan. In: *Advances in Sea Cucumber Aquaculture and Management.* A. Lovateli, C. Conand, S. Purcell, S. Uthicke, J.-F. Hamel and A. Mercier. (eds). pp. 205–217. *FAO Fisheries Technical Paper.* No. 463. Rome, FAO. 2004. 425p.

Sant, G. 2006. CITES and Sea Cucumbers. *In: Proceedings of the CITES Workshop on the Conservation of Sea Cucumbers in the Families Holothuriidae and Stichopodidae, 1–3 March, 2004, Kuala Lumpur, Malaysia.* A. Bruckner (ed). NOAA Technical Memorandum, NMFS-OPR-34. pp. 104–121. Silver Spring, NOAA.

Schug, D. 1995. *The Marine Realm and a Sense of Place among the Papuan New Guinean Communities of the Torres Strait.* Unpublished PhD Thesis, University of Hawaii, Honolulu, Hawaii, the United States of America.

Schug, D. 1996. International Maritime Boundaries and Indigenous People: The Case of Torres Strait. *Marine Policy,* 20(3): 209–222.

Shelley, C. 1981. *Aspects of the Distribution, Reproduction, Growth and Fishery Potential of Holothurians (Beche-de-Mer) in the Papuan Coastal Lagoon.* Unpublished MSc Thesis, University of Papua New Guinea, Port Moresby, Papua New Guinea.

Skewes, T., Dennis, D., Taylor, S., Haywood, M. & Donovan, A. 2006. *Sustainability Assessment for the Torres Strait Beche-de-Mer Fishery.* Report prepared for the Australian Fisheries Management Authority Torres Strait Research Program, Brisbane, Queensland, Australia.

Skewes, T., Dennis, D., Koutsoukos, A., Haywood, M., Wassenberg, T. & Austin, M. 2004. *Stock Survey and Sustainable Harvest Strategies for the Torres Strait Beche-de-Mer.* Report prepared for the Australian Fisheries Management Authority and Queensland Fishery Service, Thursday Island, Queensland, Australia.

Skewes, T., Kinch, J., Polon, P., Dennis, D., Seeto, P., Taranto, T., Lokani, P., Wassenberg, T., Koutsoukos, A. & Sarke, J. 2002. *Research for the Sustainable Use of Beche-de-Mer Resources in the Milne Bay Province, Papua New Guinea.* Report prepared for the National Fisheries Authority, Port Moresby, Papua New Guinea; and the Australian Centre for International Agricultural Research, Sydney, New South Wales, Australia.

Tom'tavala, D. 1990. *National Law, International Law and Traditional Marine Claims: A Case Study of the Trobriand Islands, Papua New Guinea.* Unpublished MSc Thesis. Halifax: Department of Law, Dalhousie University.

Tom'tavala, D. 1992. *Yam Houses in the Sea: Marine Claims of the Trobriands Islanders of Papua New Guinea.* Paper presented to the Congress of the Commission on Folk Law and Legal Pluralism, 27–30 August, Wellington, New Zealand.

APPENDIX A

Sea cucumber densities in Papua New Guinea

Species	Density (ind. ha^{-1})	Province	Reference
Actinopyga echinites	1 800.0	Central	Shelly, 1981
	<0.1	East New Britain	Gisawa, 2002
	6.7	Madang	Lokani *et al.*, 1992
	241.3	Madang	Massin and Doumen, 1986
	105.5	Manus	Lokani and Chapau, 1992
	25.0	New Ireland	Lokani, 1996
	119.0	New Ireland	NFA, 2007
	1.8	Western	Lokani, 2000
A. lecanora	<0.1	East New Britain	Gisawa, 2002
	0.9	Madang	Lokani *et al.*, 1992
	<0.1	Milne Bay	Skewes *et al.*, 2002
	3.0	New Ireland	Lokani, 1996
	0.2	New Ireland	NFA, 2007
	2.4	Oro	Anon, 1994
	<0.1	Western	Lokani, 2000
A. mauritiana	13.0	East New Britain	Gisawa, 2002
	38.7	Madang	Lokani *et al.*, 1992
	9.4	Madang	Massin and Doumen, 1986
	9.5	Manus	Lokani and Chapau, 1992
	0.1	Milne Bay	Skewes *et al.*, 2002
	12.0	New Ireland	Lokani, 1996
	1.8	New Ireland	NFA, 2007
	5.0	Oro	Anon, 1994
A. miliaris	2.0	East New Britain	Gisawa, 2002
	4.3	Madang	Lokani *et al.*, 1992
	36.9	Manus	Lokani and Chapau, 1992
	0.1	Milne Bay	Skewes *et al.*, 2002
	15.0	New Ireland	Lokani, 1996
	1.2	New Ireland	NFA, 2007
	57.0	Oro	Anon, 1994
	1.4	Western	Lokani, 2000
A. palauensis	0.2	New Ireland	NFA, 2007
Bohadschia argus	18.0	East New Britain	Gisawa, 2002
	9.5	Madang	Lokani *et al.*, 1992
	24.4	Madang	Massin and Doumen, 1986
	1.3	Milne Bay	Skewes *et al.*, 2002
	5.0	New Ireland	Lokani, 1996
	3.7	New Ireland	NFA, 2007
	8.8	Oro	Anon, 1994
	6.9	Western	Lokani, 2000
B. similis	411.0	New Ireland	Lokani, 1996
	3.2	New Ireland	NFA, 2007
	3.5	Western	Lokani, 2000
B. vitiensis	2.0	East New Britain	Gisawa, 2002
	4.4	Madang	Lokani *et al.*, 1992
	1.0	Milne Bay	Skewes *et al.*, 2002
	136.0	New Ireland	Lokani, 1996
	1.2	New Ireland	NFA, 2007
	5.9	Oro	Anon, 1994
	6.5	Western	Lokani, 2000

APPENDIX A (continued)

Sea cucumber densities in Papua New Guinea

Species	Density (ind. ha^{-1})	Province	Reference
Holothuria atra	38.0	East New Britain	Gisawa, 2002
	80.0	Madang	Lokani *et al.*, 1992
	4 870.6	Madang	Massin and Doumen, 1986
	9.8	Milne Bay	Skewes *et al.*, 2002
	584.0	New Ireland	Lokani, 1996
	40.0	New Ireland	NFA, 2007
	232.9	Oro	Anon, 1994
	143.5	Western	Lokani, 2000
H. coluber	4.6	New Ireland	NFA, 2007
H. edulis	6.0	East New Britain	Gisawa, 2002
	2.2	Milne Bay	Skewes *et al.*, 2002
	6.7	New Ireland	NFA, 2007
	<0.1	Western	Lokani, 2000
H. fuscogilva	3.0	East New Britain	Gisawa, 2002
	3.3	Madang	Lokani *et al.*, 1992
	3.5	Manus	Lokani and Chapau, 1992
	0.4	Milne Bay	Skewes *et al.*, 2002
	23.0	New Ireland	Lokani, 1996
	1.2	New Ireland	NFA, 2007
	2.0	Oro	Anon, 1994
H. fuscopunctata	<0.1	East New Britain	Gisawa, 2002
	18.2	Madang	Lokani *et al.*, 1992
	<0.1	Milne Bay	Skewes *et al.*, 2002
	4.0	New Ireland	Lokani, 1996
	0.1	New Ireland	NFA, 2007
H. hilla	5.5	New Ireland	NFA, 2007
H. leucospilota	<0.1	East New Britain	Gisawa, 2002
	3.3	Madang	Lokani *et al.*, 1992
	<0.1	Western	Lokani, 2000
H. whitmaei	3.0	East New Britain	Gisawa, 2002
	16.8	Madang	Lokani *et al.*, 1992
	9.4	Madang	Massin and Doumen, 1986
	9.8	Manus	Lokani and Chapau, 1992
	0.2	Milne Bay	Skewes *et al.*, 2002
	5.0	New Ireland	Lokani, 1996
	0.9	New Ireland	NFA, 2007
	6.6	Oro	Anon, 1994
H. scabra	2 900.0	Central	Shelly, 1981
	1.0	East New Britain	Gisawa, 2002
	22.2	Madang	Lokani *et al.*, 1992
	19.1	Manus	Lokani and Chapau, 1992
	<0.1	Milne Bay	Skewes *et al.*, 2002
	122.0	New Ireland	Lokani, 1996
	17.0	New Ireland	NFA, 2007
	0.4	Oro	Anon, 1994
	34.9	Western	Lokani, 2000
Pearsonothuria graeffei	3.5	Madang	Lokani *et al.*, 1992
	0.4	Milne Bay	Skewes *et al.*, 2002
	4.0	New Ireland	Lokani, 1996
	6.4	New Ireland	NFA, 2007
	3.3	Western	Lokani, 2000

APPENDIX A (continued)

Sea cucumber densities in Papua New Guinea

Species	Density (ind. ha^{-1})	Province	Reference
Stichopus chloronotus	128.2	Madang	Lokani *et al.*, 1992
	16.0	Manus	Lokani and Chapau, 1992
	3.8	Milne Bay	Skewes *et al.*, 2002
	25.0	New Ireland	Lokani, 1996
	<0.1	New Ireland	NFA, 2007
	51.8	Oro	Anon, 1994
	2.8	Western	Lokani, 2000
S. herrmanni	1.0	East New Britain	Gisawa, 2002
	8.1	Madang	Lokani *et al.*, 1992
	8.6	Manus	Lokani and Chapau, 1992
	0.1	Milne Bay	Skewes *et al.*, 2002
	31.0	New Ireland	Lokani, 1996
	0.3	New Ireland	NFA, 2007
	3.7	Oro	Anon, 1994
	4.1	Western	Lokani, 2000
S. horrens	216.0	New Ireland	Lokani, 1996
	0.4	New Ireland	NFA, 2007
	1.5	Western	Lokani, 2000
Thelenota ananas	12.0	East New Britain	Gisawa, 2002
	8.0	Madang	Lokani *et al.*, 1992
	1.6	Manus	Lokani and Chapau, 1992
	0.5	Milne Bay	Skewes *et al.*, 2002
	8.0	New Ireland	Lokani, 1996
	1.4	New Ireland	NFA, 2007
	1.5	Oro	Anon, 1994
T. anax	6.0	East New Britain	Gisawa, 2002
	6.0	Madang	Lokani *et al.*, 1992
	4.5	Manus	Lokani and Chapau, 1992
	0.6	Milne Bay	Skewes *et al.*, 2002
	1.0	New Ireland	Lokani, 1996
	0.7	New Ireland	NFA, 2007
T. rubralineata	<0.1	New Ireland	NFA, 2007

APPENDIX B

Papua New Guinea bêche-de-mer exports from 1960–2006

Year	Volume (tonnes)	Value (USD)	Reference
1960	1.6	–	Lindholm, 1978
1961	2.4	–	Lindohlm, 1978
1962	4.4	–	Lindohlm, 1978
1963	12.8	–	Lindohlm, 1978
1964	6.3	–	Lindohlm, 1978
1965	4.1	–	Lindohlm, 1978
1966	4.4	–	Lindohlm, 1978
1967	10.5	–	Lindohlm, 1978
1968	11.2	–	Lindohlm, 1978
1969	12.4	–	Lindohlm, 1978
1970–71	6.5	–	Lindohlm, 1978
1971–72	3.9	–	Lindohlm, 1978
1972–73	9.9	–	Lindohlm, 1978
1973–74	4.1	–	DFMR, no date
1974–75	1.2	–	Lindohlm, 1978; DFMR, no date
1975–76	1.7	–	Lindohlm, 1978; DFMR, no date
1977	5.3	–	Lindohlm, 1978
1978	5.9	–	Lindohlm, 1978
1979	1.3	–	DFMR, 1979
1980	2.4	–	Kailola and Lokani, no date
1981	11.1	–	Kailola and Lokani, no date
1982	23.0	–	Kailola and Lokani, no date
1983	7.6	–	Lokani and Kubohojam, 1993
1984	4.7	–	Lokani and Kubohojam, 1993
1985	19.5	–	Lokani and Kubohojam, 1993
1986	119.4	–	Lokani and Kubohojam, 1993
1987	192.1	–	Lokani and Kubohojam, 1993
1988	202.8	–	Lokani and Kubohojam, 1993
1989	194.9	–	Lokani, 1990
1990	238.9	–	Lokani and Kubohojam, 1993
1991	626.0	–	Lokani and Kubohojam, 1993
1992	655.5	–	Myint, 1996
1993	499.5	–	Myint, 1996
1994	208.8	–	NFA database
1995	444.6	3 560 728	NFA database
1996	596.2	5 959 645	NFA database
1997	505.4	5 185 737	NFA database
1998	678.8	8 147 423	NFA database
1999	394.7	4 157 870	NFA database
2000	553.9	5 832 439	NFA database
2001	485.4	5 266 819	NFA database
2002	389.3	5 629 250	NFA database
2003	488.0	6 376 835	NFA database
2004	490.8	7 181 587	NFA database
2005	577.0	9 284 756	NFA database
2006	611.8	11 488 601	NFA database

Note: All figures are indicative only, as the National Fisheries Authority database is continuously updated.

Population status, fisheries and trade of sea cucumbers in Asia

Poh-Sze Choo

The Philippines: a hotspot of sea cucumber fisheries in Asia

Poh-Sze Choo

Population status, fisheries and trade of sea cucumbers in Asia

Poh-Sze Choo
The WorldFish Center
Penang, Malaysia
E-mail: p.choo@cgiar.org

Choo, P.S. 2008. Population status, fisheries and trade of sea cucumbers in Asia. In V. Toral-Granda, A. Lovatelli and M. Vasconcellos (eds). Sea cucumbers. A global review of fisheries and trade. *FAO Fisheries and Aquaculture Technical Paper*. No. 516. Rome, FAO. pp. 81-118.

SUMMARY

The regional review on the population status, fisheries and trade of commercially important sea cucumbers in Asia covers the east and southeast Asian regions including Indonesia, Malaysia, Thailand, Myanmar, Viet Nam, Philippines, Singapore, the Spratly Islands, Japan, Democratic People's Republic Korea, Republic of Korea, Far East Russian Federation, China Hong Kong Special Administrative Region (SAR) and Taiwan Province of China (PC). A total of 52 species are commercially exploited as food with most of them comprising tropical and sub-tropical species from the families Holothuriidae and Stichopodidae, including the genus *Holothuria*, *Actinopyga*, *Bohadschia* and *Stichopus*.

Fisheries in the Asian tropical and sub-tropical waters are multi-species, while the fishery in temperate waters is single species, comprising predominantly only one species, *Apostichopus japonicus*. Fishing and seafaring communities in Asia had been involved in sea cucumber fishing and processing since the sixteenth century. The fresh animals caught were processed into dried forms known as "trepang". Indonesia is the world's top producer of Holothuroidea from the capture fishery. Indonesia, together with the Philippines produced an annual average of 47 percent of the world's Holothuroidea landings, comprising an annual average of 2 572 tonnes (wet weight) between 2000 and 2005. The highest capture fishery producer of the temperate species, *A. japonicus* is Japan, with an average production of 8 101 tonnes per year between 2000 and 2005.

The sea cucumber capture statistics obtained from the Food and Agriculture Organisation (FAO) are recorded in wet weights; for landings in Southeast Asia the figures appear to be grossly underestimated, and the statistics need to be verified as to whether the data reported were actually dried and not wet weight. Except for China, where a substantial amount of sea cucumber production is from aquaculture (an estimated annual production of 10 000 tonnes dry weight), the production in the other Asian countries is derived predominantly, if not exclusively, from capture fisheries.

Apart from gleaning, the most common fishing methods for sea cucumbers include small bottom trawl nets in sandy bottoms, scallop-drag gear in nearshore rocky-bottom habitats, spears, hooks and scoop nets for reefs, and SCUBA and hookah for deeper reef and lagoons.

Overfishing is the main problem contributing to the depletion of sea cucumber resources. Except for Japan, other Asian countries are generally lacking in management measures to conserve and sustain their sea cucumber fisheries. The two most important producing countries, Indonesia and the Philippines do not have management plans specific to sea cucumber conservation. Other threats to sustaining the sea cucumber resources include habitat loss, lack of accurate statistics, global warming and new uncontrolled uses (such as for pharmaceuticals and nutriceuticals) for sea cucumber resources.

1. REGION UNDER STUDY

This paper covers the East and Southeast Asian regions (Figure 1) comprising Indonesia, Malaysia, Thailand, Myanmar, Viet Nam, Philippines, Singapore and the Spratly Islands (in tropical Southeast Asia); China, Japan, Democratic Peoples' Republic Korea, Republic of Korea and Far East Russian Federation (in temperate East Asia) and China Hong Kong Special Administrative Region (SAR) and Taiwan Province of China in the sub-temperate zone. The Spratly Islands are a group of reefs in the South China Sea with sovereign claims from China, Taiwan Province of China, Viet Nam, Brunei, Malaysia and the Philippines.

Generally, information on the biology, ecology, fisheries and trade on sea cucumbers in the Asian region is rather limited. Although China, China Hong Kong SAR and Singapore are important centres for sea cucumber trade, information published in English on trade from these countries were mainly written by researchers based outside the region. It is, however, unclear on how much information on this topic is available in other languages such as Chinese. There are a fair number of publications on sea cucumber from China, Japan, Indonesia, Philippines and Malaysia, but very little information is available from China Hong Kong SAR, Democratic People's Republic of Korea, Myanmar, Thailand, Singapore and Viet Nam, and no information at all (except for some FAO statistics on capture production) from the Republic of Korea.

FIGURE 1

Region under study and countries covered in this review (highlighted in yellow). The region partially encompasses the FAO Statistical Areas 57 (Eastern Indian Ocean), 71 (Western Central Pacific) and 61 (Northwest Pacific)

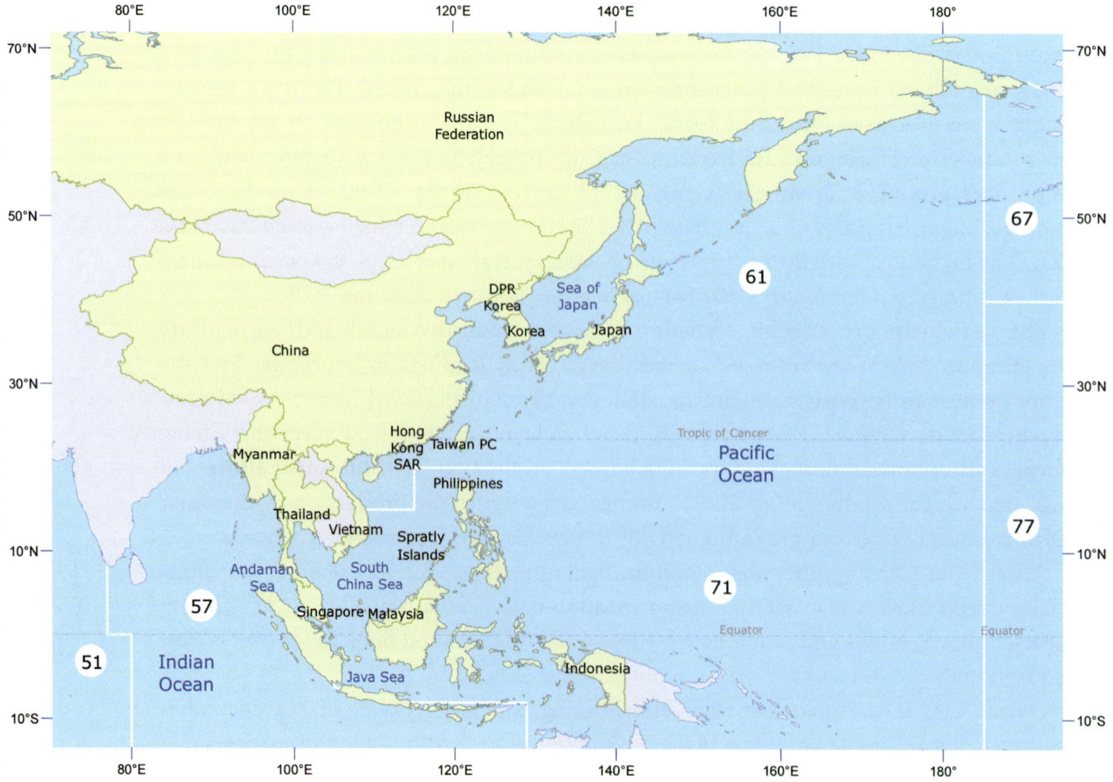

2. BIOLOGICAL AND POPULATION STATUS

2.1 Key taxonomic groups and commercial species

In Asia, sea cucumbers that are commonly fished are derived mainly from the Order Aspidochirotida under two families, Holothuriidae and Stichopodidae. Genuses that are frequently exploited for food include *Holothuria*, *Actinopyga* and *Bohadschia*. Altogether, a total of 125 sea cucumber species were recorded from various references (Chao, 1998; Jeng, 1998; Bussarawit and Thongtham, 1999; Forbes *et al.*, 1999; Flint, 2002; Zaidnuddin, 2002; Tan, 2003; Chen, 2004; Choo, 2004; Mucharin and Putchakarn, 2005; Purwati, 2005; Tan, 2005a, b; Otero-Villanueva and Ut, 2007; Asmedi, undated), with 52 species documented as commercially important (Appendix I). Sven Uthicke (Australian Institute of Marine Science; personal communication) noted some confusion exists in the correct taxonomic names for the teatfish group; this report adheres to the results of a recent morphological and genetic study by Uthicke, O'Hara and Byrne (2004) which suggested that all black teatfish in the areas covered in this review are *Holothuria whitmaei*; it is presumed that *H. nobilis* does not exist in the Asian region covered in this review. Thus, references to *H. nobilis* given in several papers are referred to here as *H. whitmaei*. A country-by-country account on the sea cucumber distribution in the Asian region is given below.

Liao (1997, cited in Chen, 2004) reported that China has about 134 species of sea cucumbers with 28 of them considered edible or with medicinal properties (Zhang, 1954, 1958; Fan, 1979). Twenty seven of these commercial species are listed in Appendix I. Most of the commercial species from China are distributed in the tropical and subtropical regions; only a few species are found in temperate waters (Chen, 2004). *Apostichopus japonicus* is the most commercially important temperate species.

Jeng (1998) recorded a total of 13 species of sea cucumbers in Taiping Island which is one of the largest islands in the Spratly group of islands. Most species found here are also commonly found in the Indo-West Pacific coral reefs (Appendix I). Species like *Actinopyga lecanora*, *Stichopus chloronotus*, *Thelenota rubralineata* and *Synaptula lamperti* have been reported in Taiping Island but had not been recorded from waters off Taiwan Province of China. *A. lecanora* and *T. rubralineata* were rare species and only one specimen of each was found during the survey in Taiping Island.

Chao (1998) listed the presence of 30 species from three Orders and five Families in the shallow waters of Taiwan Province of China. There is no documentation from the FAO statistics that a sea cucumber capture fishery exists in Taiwan Province of China.

In the coastal waters of China, Hong Kong SAR, 14 Holothuroidea species including *H. scabra,* were documented in a paper published in the 1930s (Mortensen, 1934). Xing and Chia (1997) noted that *H. leucospilota* were found in the shallow waters of many areas in China Hong Kong SAR. The author could not find any commercial sea cucumber fishery records for China Hong Kong SAR; sea cucumber trade is dependent on imports from all over the world, a vast percentage of which are re-exported.

Fishing for sea cucumbers in Viet Nam takes place mainly in the waters around the Phu Quoc Archipelago in southern Viet Nam; 25 species of sea cucumbers have been documented here, of which 11 species are exploited for their commercial value (Otero-Villanueva and Ut, 2007).

The sea cucumber fishery in Indonesia is generally artisanal, and is scattered throughout the many Indonesian islands including Lampung, Java, Nusa Tenggara, Sulawesi, Maluku and Irian Jaya (Tuwo and Conand, 1992). There were at least 56 species in the Spermonde Archipelago in Southwest Sulawesi, Indonesia (Massin, 1999). In the Kaledupa Stakeholder Zone in the Wakatobi Marine National Park in Sulawesi, Flint (2002) recorded the presence of 32 species of holothurians of which 26 could be processed into trepang, with *H. hilla* and *H. pervicax* being used as fillers to top up weights during sale transaction.

Aryono (1987) noted the presence of nearly 25 000 sea cucumbers (including *Synapta* sp. and *H. atra*) in a hectare of seagrass meadows in the Seribu Islands. In the coastal waters of Sulawesi, there were probably 30 species of sea cucumber, the most common of which were *H. atra* which measured up to 60 cm in length, *S. chloronotus* (about 30 cm long) and *H. scabra* (Whitten, Mustafa and Henderson, 2002).

Information extracted from Bussarawit and Thongtham (1999) and Mucharin and Putchakarn (2005) indicated the presence of 71 species of sea cucumbers from the waters in Thailand out of which eight were described as commercially important (Appendix I).

A total of 62 species of sea cucumbers belonging to three Orders and five Families from the coral reef areas in Peninsular Malaysia and Sabah have been recorded (Forbes *et al.*, 1999; Zaidnuddin, 2002); out of these 20 were of commercial importance (Choo, 2004). Nineteen commercial species from Malaysia are listed in Appendix I. Commercial species identified only to the genus (*Bohadschia*) level ("lubuyoh tadik" and "mother tadik") are not included in Appendix I. The local names of trepang from Sabah are taken from Filipino migrants (Akamine, J., Nagoya City University, Japan, personal communication).

In the marine park in Pulau Payar, located in the northwest coast of Peninsular Malaysia, Zaidnuddin and Forbes (2000) noted that *H. atra* was the most abundant species, followed by *H. leucospilota*, *S. chloronotus* and *S. horrens.* Zaidnuddin (2002) documented the presence of seven species in Balik Pulau, Penang, which is located in the north–west coast of Peninsular Malaysia. He suggested that the low-value *Paracaudina* sp. found abundantly in the muddy habitats of Balik Pulau and eaten raw by the local population, be more widely promoted as a food source. Boss, Yasin and Tan (1999) completed a preliminary survey of sea cucumbers in Pulau Besar, Johore, located in the southeast coast of Peninsular Malaysia and reported the presence of seven species including five species of *Stichopus* (*S. herrmanni* and four other species which have not been identified), *A. lecanora* and *H. impatiens*. They suggested that the high number of *Stichopus* spp. in Pulau Besar may be due to the variable habitats found there, namely sea grass beds, different types of corals and rocky areas.

Species documented from Singapore waters included those popular in the aquarium trade such as the thorny sea cucumber, *Colochirus quadrangularis* and sea apple, *Pseudocolochirus violaceaus,* found in Pulau Ubin (Tan, 2003, 2005a). Other species recorded from Pulau Ubin included *H. scabra* and *Phyllophorus* sp., while the pink warty sea cucumber, *Cercodemas anceps*, was found in Changi and Pulau Sekudu (Tan, 2005b). However, the species of sea cucumbers stated above were not exploited for commercial purposes in Singapore.

2.2 Biology and ecology of sea cucumbers

Relatively little information is available on the biology and ecology of sea cucumbers in this region. Only a few species of the two orders Aspidochirotida and Apodida found in tropical waters have been studied (Chao, Chen and Alexander, 1995). Among the tropical species with a significant amount of information is the high value species *H. scabra,* and for the temperate species, *A. japonicus*, the latter being one of the best studied species.

Uehara (1991) noted that most benthic sea cucumbers in Asia exhibit two activity patterns: (i) feeding actively the whole day and resting irregularly (examples include *H. atra*, *A. mauritiana* and *H. edulis*); and (ii) demonstrating a cycle of active and inactive periods (examples include *H. scabra*, *B. argus* and *B. bivittata*) where they come out of the sand in the evening, move and feed actively at night and bury themselves under the sand in the morning. Commonly, juvenile sea cucumbers inhabit shallower areas than the adults, and they migrate to deeper areas at a later stage (Uehara, 1991). Most sub-tropical sea cucumber species from Japan reach sexual maturity in about three years and live for more than five years (Uehara, 1991).

TABLE 1
Sea cucumber species from the Balik Pulau District, Penang, Malaysia

Family/Species	Location	Bottom type	Habitat	Abundance
Holothuridae				
H. martensi	Balik Pulau	Muddy with shells	Northwestern area, 3–4 nautical miles from shore with water depth to 30–40 m	<10 Specimens from 1 hour trawl
H. atra	Pulau Kendi	Dead corals	Coral area, shoreward side of reef	<10 Specimens from 1 hour trawl
H. leucospilota	Pulau Kendi	Dead corals	Coral area, shoreward side of reef	<10 Specimens from 1 hour trawl
Stichopodidae				
S. chloronotus	Pulau Kendi	Dead corals	Coral area	1–2 specimen(s) from 1 hour trawl
S. horrens	Pulau Rimau	Mud mixed with shells	Eastern and southern side of Pulau Rimau	1–2 specimen(s) caught with grab
Molpadiidae				
Paracaudina sp.	Border with Perak	Muddy	South of Pulau Pinang, occupying muddy area <10 m deep	>500 specimens; >500 kg from 1hr trawl
Cucumariidae				
Pseudocolochirus spp.	Pulau Rimau	Muddy and rocky	Cage culture area	3 specimens found in discarded netting

Source: Zaidnuddin, 2002.

Zaidnuddin and Forbes (2000) noted that in the Malaysian reef areas, *H. atra* had a preference for soft, sandy bottoms; *S. chloronotus* preferred areas with some boulders mixed with live corals and *S. horrens* occupied areas scattered with boulders of the coral *Porites* spp. Shallow water species inhabiting depths up to 10 m included *S. horrens, H. leucospilota, H. hilla* and *H. pardalis*, while *B. marmorata* and *S. herrmanni* were found in deeper areas. Table 1 shows the substrate preference for seven species of sea cucumber found in the Balik Pulau District, Penang, Malaysia (Zaidnuddin, 2002).

Uehara (1991) noted that most holothurians in Asia are dioecious (having male and female reproductive organs in separate animals but not both together as in hermaphrodites, and sometimes described with the synonym "gonochoric") with external fertilisation. Some species, example *Synaptula hydriformis*, release their larvae through a rupture in the body wall, or more often through a perforation in the wall of the intestine after development in the coelom (Bakus, 1973). For the viviparous species such as *A. africana* and *Scoliodotella lindbergi*, the larvae are housed in the brood pouch (Uehara, 1991).

Chao, Chen and Alexander (1995) conducted a study on the reproductive cycles (including reproductive type, gonad colour, spawning period and gonad morphology) of nine tropical species of sea cucumbers found in the reef flats of the southern tip of Taiwan Province of China. The results are summarised in Table 2.

Sea cucumbers from temperate regions have been found to undergo aestivation – a form of dormancy during the summer months in high temperatures or during the dry seasons. Yuan *et al.* (2007) noted that aestivation has ecological implications and acts as an energy-saving mechanism during the hot period. Aestivation of *A. japonicus* has been widely reported by Chinese researchers (Liu *et al.*, 1996; Sui and Liao, 1988; Yang *et al.*, 2005).

The biology and ecology of the two best studied species in the region (the tropical species *H. scabra* and the temperate species *A. japonicus*) are described in more detail below.

TABLE 2
Reproductive characters of nine species of holothurians from southern Taiwan Province of China. Species of commercial importance in Asia are written in bold letters. Month in parentheses indicates when oocyte diameter was measured

Species	Reproductive type	Oocyte diameter (µm)	Spawning period
A. africana	Gonochoric, fission	350 ± 50 (May)	Mar–Apr
A. echinites	Gonochoric	110 ± 5 (June)	Jun–Jul
H. leucospilota	Gonochoric	120 ± 5 (Jul)	Jun–Sep
H. cinerascens	Gonochoric	100 ± 10 (May)	Apr–Jun
H. difficilis	Gonochoric	75 ± 5 (Aug)	Aug–Sep
S. maculate	Gonochoric	70 ± 5 (Jun)	Jun–Jul
Opheodesma grisea	Gonochoric	100 ± 5 (Jul)	Jun–Jul
Patinapta taiwaniensis	Gonochoric	65 ± 10 (May)	Jun–Jul
Polycheira rufescens	Gonochoric	110 ± 10 (May)	Apr–Jun

Source: Chao, Chen and Alexander, 1995.

2.2.1 Tropical species: Holothuria scabra *(sandfish)*

Holothuria scabra, commonly known as sandfish, is widely distributed throughout the Indo–Pacific, being found roughly between latitudes 30 °N and 30 °S (Hamel *et al.*, 2001). Its range extends from the east coast of Africa and the Red Sea, eastward to the Caroline Islands and Fiji in the Pacific, and from the Amami Islands southward to the Torres Strait (Uehara, 1991). In Japan, the distribution of *H. scabra* is limited to several areas around Okinawa such as Haneji Inland Sea, Nakagusuku Bay, Kin Bay, and in Yakata-katabaru in Onna Village (Uehara, 1991). In Viet Nam, sandfish are found in sandy estuaries or lagoons at depths of 2–25 m or more, often in patches of high concentration (Pitt and Duy, 2003).

Rowe (1969, cited in Hamel *et al.*, 2001) noted the presence of nine species in the subgenus *H. scabra*, with all of them located in the Indo–west Pacific. Sandfish from Khan Hoa Province, Viet Nam, appears to be closer to *H. scabra* in terms of their small size and size at first maturity, with colours ranging from black through dark brown to light beige, often with transverse stripes, and less deeply ridged than those from Oceania (Pitt and Duy, 2003). The subspecies found in Tunku Abdul Rahman Park in Sabah, East Malaysia, was identified as *H. scabra* var. *versicolor* by Forbes *et al.* (1999); the animals which might grow to an estimated 30 cm were found in inner reef flats or near estuaries, and were often buried in sand or covered with leaves. Uehara (1991) noted that *H. scabra* could attain a maximum length of 37 cm and a weight of 600 g.

There is a high degree of polymorphism, which is observed throughout the *H. scabra* geographic range (Conand, 1990). Hamel *et al.* (2001) noted that there are two reasons for the high degree of polymorphism: identification errors and important morphological plasticity over the geographic range. However, Hamel *et al.* (2001) noted that the behaviour and reproductive patterns described for all the species of *H. scabra* in its geographic range differed considerably, suggesting that there might be taxonomic confusion and not all studies were conducted on *H. scabra*.

In countries situated near the Equator, sandfish spawn throughout the year (Agudo, 2006). Tuwo (1999) noted that *H. scabra* from Southwest Sulawesi, Indonesia, had two peak spawning periods, although spawning occured throughout the year; the two post–spawning periods were at the beginning of the dry season (March to July) and at the beginning of the rainy season (November to January). In the Philippines, gonad maturation occured from January to April and from July to October, with main spawning events occurring from May to June and from October to November (Ong Che and Gomez, 1985).

Sandfish, classified as commercially high value species, are easily overexploited since they are commonly found in shallow waters. With the great demand for trepang and the active harvesting from local fishers, there is severe fishing pressure on natural populations of *H. scabra* (Hamel *et al.*, 2001). In Malaysia, Choo (2004) reported on signs of overfishing of the sandfish stock in Sabah, East Malaysia as evident by the decrease in landings and overall size of the animals landed. Otero-Villanueva and Ut (2007) noted that less than 100 kilograms (dry weight) per annum were currently landed from the Phu Quoc Archipelago in Viet Nam, and while conducting over 80 underwater surveys in the area up to a depth of 18 m, they had not encountered sandfish.

2.2.2 Temperate species: **Apostichopus japonicus** *(Japanese spiky sea cucumber)*

A. japonicus (formerly known as *Stichopus japonicus*), a species found along the Asian coast, has long been exploited as an important fishery resource in Far East Russian Federation, China, Japan, Republic of Korea and Democratic People's Republic of Korea (Sloan, 1984, cited in Yang *et al.,* 2005). This species has been successfully cultured on a large commercial scale in northern China. Chen (2004) reported that the total production of *A. japonicus* in China reached over 5 865 tonnes (dry weight) in 2002 against a capture fishery production of 470 tonnes (dry weight) in the same year.

A. japonicus ingests organic matter, bacteria, protozoa, diatoms as well as plant and animal detritus (Zhang, Sun and Wu, 1995, cited in Yang *et al.,* 2005) and reutilises residual food and faeces (Yang *et al.*, 2001, cited in Yang *et al.*, 2005). It becomes inactive when water temperature exceeds 18 °C, and will aestivate at water temperatures about 20–24.5 °C (Sui and Liao, 1988; Liu *et al.,* 1996). Liu *et al.* (1996) noted that in some regions in China, aestivation can last up to four years. From laboratory experiments conducted under controlled conditions, Yang *et al.* (2005) found that large and medium animals (72.3–139.3 g) aestivated at a threshold temperature between 24.5 and 25.5 °C while for small animals, aestivation was observed when the temperature was between 25.5 and 30.5 °C. In general, *A. japonicus* from the more southern areas have a higher threshold temperature for aestivation (Yang *et al.*, 2005).

In Japan, *A. japonicus* are red ("aka namako"), green ("ao namako") or black ("kuro namako") in colour; specimens with different colour morphs fetch different prices and have their own unique tastes (Kan-no and Kijima, 2003). "Aka" and "Ao namako" are the most important morphs for commercial fishing; the three colour morphs are distributed from the surface to a depth of 40 m, over a wide area of the north-eastern Pacific: from Sakhalin Island and Alaska to the Amami Islands (Japan) and the east coast of China (Arakawa, 1990). In Japan, the red type inhabits gravel bed offshore while the green and black types inhabit the sand–muddy bottom inshore (Choe, 1963). Based on their study, Kan-no and Kijima (2003) suggested that the microhabitat differentiation might have resulted in reproductive isolation; however, they noted that the clear genetic independence of the red types could not be explained by differences of microhabitat alone. In a later study using 11 microsatellite markers, Kanno *et al.* (2006) tested the genetic differentiation among the three sympatric colour morphs and showed the strong heterogeneity of the red morph, while no significant difference was observed between the green and black morphs. These results indicated the separate species status of the red morph and supported the population identity of the sympatric green and black morphs (Kanno *et al.*, 2006). A study conducted by Yao *et al.* (2007) on *A. japonicus* from Penglai in Changdao and Lingshandao in Qingdao (China) showed that the genetic diversity of the two stocks was still large but the genetic distance between them was close; a dendogram constructed for the 57 individuals from the two stocks indicated that the genetic structure was unitary for those from Penglai but complex for those from Lingshandao.

2.3 Background of sea cucumber fishery

Sea cucumbers have been fished and traded in the Indo–Pacific regions since the sixteenth century (Akamine, 2004). Akamine (2004) mentioned that Japan traded trepang with China for silk and medicine during the Edo period (1601–1867). About the same time Europeans brought trepang from tropical waters to China in exchange for tea, silk and porcelain (Akamine, 2004). Uehara (1991) reported that dried sea cucumbers were exported in considerable amount to Taiwan Province of China and China from Okinawa, Japan, until about 1940, and those from Haneji in Okinawa (known as "haneji-iriko", comprising *H. scabra*) were especially famous. Currently only small amounts are exported from Okinawa by several fisheries cooperatives. In Asia, sea cucumbers have been reported to be heavily exploited in China, Indonesia, Malaysia, Philippines, Thailand and Viet Nam (Bruckner, 2005).

In the late seventeenth century, the fishing and seafaring communities of Sulawesi became actively involved in sea cucumber gathering which prompted the search for high quality sea cucumber throughout eastern Indonesia and beyond, with most of the trade centred on Makassar in South Sulawesi (Fox and Sen, 2002). In the mid 1880s, fishers from Makassar, Indonesia, were known to have gone as far as northern Australia to fish for sea cucumbers (Purwati, 2005). Exports of seafood products to Singapore from Indonesia have been traced back to the 1800s; in 1830, 180 boats owned by the Bugis were reported to have landed in Singapore bringing with them marine products from eastern Indonesia (Purwati, 2005). Direct trepang trade between China and Indonesia was documented to date back to the sixteenth and seventeenth century (Purwati, 2005).

Butcher (2004) also noted that trepang trade in Southeast Asian countries with China dates back to the 1700s. He reported that Papuan slaves were forced to dive for trepang by people from the Aru Islands. In the latter part of the eighteenth century, British traders began stopping in Jolo in the Philippines where they traded firearms, Indian cloth and opium for trepang which were later sold to the Chinese. At the height of the marine products trade in the 1830s, as many as 68 000 slaves were captured each year and handed to the sultans and chiefs of Jolo who forced the slaves to dive for pearl shell and trepang. During this period, about 600 tonnes of trepang were exported from Jolo each year. The harvesting of huge amounts of trepang in the mid 1800s saw a resurgence of heavy exploitation in the 1980s and 1990s to satisfy the demands from wealthy consumers from Taiwan Province of China, China Hong Kong SAR, southeastern China and the Chinese communities from Southeast Asia (Butcher, 2004).

However, not all countries in the Asian region have a long history of sea cucumber fishery. In Viet Nam, for example, the sea cucumber fishery in the vicinity of the Phu Quoc Archipelago, the mainstay of the country's sea cucumber resources, has only a recent documented history, dating back to the early 1980s (Otero-Villanueva and Ut, 2007). Within a short span of ten years, sea cucumber resources in this region have been over exploited, and sea cucumber fishers are turning to fishing for other organisms for a livelihood. Otero–Villanueva and Ut (2007) reported that in 2004, the average monthly income for a sea cucumber diver in Viet Nam was estimated to be USD 130–194. The sea cucumber fishery in Thailand also has a relatively young history which dates back to the late 1960s (Bussarawit and Thongtham, 1999).

2.4 Species in trade

Based on historical records, the Chinese are probably the first to utilise sea cucumbers as food because of its nutritional and medicinal properties. Chen (2003) noted that eating trepang is a Chinese custom that can be traced back to the Ming Dynasty (1368–1644 BC), and traditionally, they were eaten more for their tonic value than for their taste. Some communities in the Philippines are known to consume sea cucumbers (Trinidad-Roa, 1987; Heinen, 2001) whilst Conand and Byrne (1993) also

TABLE 3

Categorization of sea cucumber species in the international food market according to their commercial value. Common names in parentheses

High value	Medium value	Lower medium value	Low value
H. scabra (sandfish)	*A. lecanora* (stonefish)	*A. echinites* (deep water redfish)	*B. marmorata/ vitiensis* (brown sandfish)
H. fuscogilva (white teatfish)	*A. mauritiana* (good quality surf redfish)	*A. miliaris* (blackfish)	*H. atra* (lollyfish)
H. whitmaei (black teatfish)	*S. chloronotus* (good quality greenfish)	*S. chloronotus* (greenfish)	*H. edulis* (pinkfish)
T. ananas (prickly redfish)	*S. herrmanni* (curryfish)		*H. fuscopunctata* (elephant trunkfish)
			T. anax (amberfish)
			B. argus (tigerfish)

report traditional consumption in Papua New Guinea, Samoa and Fiji. Some species (*S. horrens* and *S. herrmanni*) are utilised commercially for the preparation of medicinal products but less commonly used as food items. In Malaysia, for example, both these species are used in the preparation of traditional medicinal products. Species utilised by humans generally include members of the families, Holothuroidea and Stichopodidae. Fishing down the high value species and the replacement of these species with lower value species is commonly documented in the Southeast Asian region in Malaysia, Thailand, Viet Nam and the Philippines (Bussarawit and Thongtham, 1999; Forbes and Ilias, 1999; Bruckner, Johnson and Field, 2003; Choo, 2004; Gamboa, Gomez and Nievales, 2004; Otero-Villanueva and Ut, 2007). In recent years, commercially less-valuable species like the "worm" species *Synaptula* spp. belonging to the family Synaptidae have been exploited.

Sea cucumbers fished from the tropics are classified into four categories in the international food market according to their commercial values (Ferdouse, 2004). Some species categorized under the four groups are shown in Table 3.

Sea cucumbers are also exploited for the aquarium trade. The most common genera utilised for the aquarium trade are *Euapta, Synapta, Synaptula* or *Opheodesoma* (Toonen, 2002); other species include the thorny sea cucumber (*Colochirus quadrangularis*), yellow sea cucumber (*Colochirus robustus*) and sea apple (*Pseudocolochirus violaceaus*) belonging to the Family Cucumariidae. *T. ananas*, more commonly harvested for food, is sometimes harvested also for the aquarium trade.

The commercial exploitation and trade in sea cucumbers in some countries in Asia are described below.

2.4.1 China

Most of the 27 species of commercially exploited sea cucumbers in China belong to the Order Aspidochirotida, with a few species belonging to the Order Dendrochirotida and the Order Moldavia (Appendix I). Wild fishery catches from China are not recorded in the FAO yearbook on fishery statistics capture production.

2.4.2 Japan

Ohshima (1934, cited in Uehara, 1991) listed 18 species of edible sea cucumbers from the Ryukyu Islands in Japan, among which were: *H. scabra*, *H. atra*, *H. edulis*, *B. argus*, *B. bivittata*, *A. mauritiana* and *T. ananas*. Anon (1989) noted that in Japan, the commercial sub-tropical species, *T. ananas* (locally known as "baika") is the country's largest sea cucumber (maximum length of 70–80 cm) and are found in the waters from Okinawa, Japan, stretching to Micronesia; the temperate species, *Cucumaria frondosa japonica* (locally known as "kinko") is found in cold waters along the coast of the Sea of Japan and north of Ibaraki Prefecture to Hokkaido. *A. japonicus* (called "manamako") is the most common species and is found in shallow waters surrounding Japan.

Parastichopus nigripunctatus (called "oki-namako") grows up to 40 cm, is found along the western coast of Japan at depths up to 160 m. Currently, only small quantities of *P. nigripunctatus* and *C. japonicus* are harvested and processed (Akamine, 2004). There are demands from the Republic of Korea and the United States of America markets for dried *P. nigripunctatus* and its production in Japan has been increasing since 2005 (Akamine, J., Nagoya City University, personal communication).

2.4.3 Democratic People's Republic of Korea

Little information is available on the sea cucumber fishery in Democratic People's Republic of Korea. There is no information on the total number of species exploited; the species commonly exploited for food from the capture fishery is *A. japonicus* which has been reported to be severely exploited (Anon., undated). Attempts to hatchery-reproduce sea cucumber juveniles for restocking programmes and mariculture activities have started in 2001 with the construction of small hatcheries operated by fishing cooperatives (Lovatelli, A., FAO, personal communication)

2.4.4 Indonesia

Thirty-five species of sea cucumber under the order Aspidochirotida and Dendrochirotida are commercially exploited in Indonesia (Appendix I). Commercial species are usually given local names by the fishers, and sometimes, different trepang species are given the same local name, adding to the taxonomic confusion (Purwati, 2005). Statistics on trepang exported are not separated into individual species, and therefore details on production status and landing trends on individual species are not available.

2.4.5 Thailand

Sea cucumbers were initially harvested for local consumption. The export fishery was developed in the beginning of the late 1970s, targeting *H. scabra* and *H. atra* (Bruckner, 2005). Tebchalerm (1984, cited in Bussarawit and Thongtham, 1999) reported that they were harvested commercially in a number of Thai provinces along the Andaman Sea. According to Bussarawit and Thongtham (1999), *H. scabra, H. leucospilota, H. edulis, B. marmorata, T. ananas, S. chloronotus* and *S. herrmanni* were collected commercially by fishermen in Rayong and Chon Buri in eastern Thailand.

2.4.6 Viet Nam

In Viet Nam, sea cucumbers are heavily exploited and processed as trepang which is exported to markets in China, China Hong Kong SAR, Japan, Singapore and Taiwan Province of China (Otero-Villanueva and Ut, 2007). Eleven species of sea cucumbers were harvested commercially in Phu Quoc as shown in Table 4.

TABLE 4
Sea cucumbers harvested commercially in Phu Quoc, Viet Nam

Scientific name	English name	Local name	Average value (VND/kg dry weight) (1USD=15 000 VND)
H. scabra	Sandfish	Đột trắng	500 000–700 000
H. atra	Lollyfish	Đĩa đen, đĩa máu	40 000
H. leucospilota	–	Đĩa mũ	20 000
H. edulis	Pinkfish	Sâu gai	50 000
H. fuscopunctata	Elephant trunkfish	Đột đá, đột da trăn	80 000
S. chloronotus	Greenfish	Sâu biển, đột bê ơ	250 000–300 000
S. herrmanni	Curryfish	Đột ngận đá, đột ngận trường	>300 000
T. ananas	Prickly redfish	Đột điều	800 000
B. marmorata	Brown sandfish	Đột mủ	100 000
B. argus	Tigerfish	Đột da trăn, Sâm vàng	300 000
Pentacta quadragulis		Đột gai đổ	130 000

Source: Otero-Villanueva and Ut, 2007.

Otero-Villanueva and Ut (2007) noted that trepang were traded mainly in Ho Chi Minh City; initially the species traded was *H. scabra* but later the species exploited commercially include also *H. leucospilota, H. edulis, H. atra* and *A. echinites*. With the decline in sea cucumber stocks in the late 1990s, most grocery shops have ceased selling sea cucumbers (Otero-Villanueva and Ut, 2007).

2.5 Population status

Butcher (2004) reported that many Southeast Asian countries had collected trepang for trade with China for many centuries, and during the late 1900s populations of sea cucumbers collapsed in many places because many more fishers were fishing for sea cucumbers and new technology enabled them to harvest far more efficiently.

According to a report by the National Oceanic and Atmospheric Administration (NOAA) (undated), tropical and sub-tropical species that are heavily exploited globally (including Asia) and whose populations are seriously depleted include: *Holothuria fuscogilva*, *H. whitmaei*, *H. scabra* and *T. ananas*, and species that are heavily fished in certain countries including those in Asia are: *A. echinites*, *H. scabra versicolor*, *A. lecanora*, *A. miliaris*, *A. mauritiana*, *S. herrmanni*, *S. horrens* and *S. chloronotus*. Species that may be under potential fishing threats are *B. argus* and *H. fuscopunctata*. The temperate species of sea cucumbers, including *A. japonicus* has also been reported to be depleted in Democratic Peoples's Republic of Korea and Russian Federation (Anon., undated; Konstantinova, 2004).

Sea cucumber population status in some of the countries in this region is reported below.

2.5.1 Indonesia

At Pulau Hoga in Sulawesi, a survey by the Operation Wallacea project indicated that the total holothurian density was about 5 ind./100 m^2 in the reef-crest zone (Anon, 2001), but figures before this study were not available to assess whether the resources have declined. Hoeksema (2004) noted that around the islands of Makassar in South Sulawesi, large sea cucumbers were commonly observed during reef surveys conducted from 1984–1986 in reef slopes deeper than 20 m. Ten years later, after the introduction of hookah diving in this area, the commercially important species have almost disappeared, and the non–commercial species were either absent, rare or only represented by a few small specimens. Fox (2002) noted that in the late 1980s and early 1990s, the Bajaus in Mola, on the island of Wanci in the Tukang Besi Islands shifted to shark fishing in response to the diminishing supply of trepang and the increasing price of sharkfin.

Throughout much of the nineteenth century, the Bajaus, a nomadic fishing community, used to regularly sail to the northern Australian territorial waters to gather trepang without much intervention from the Australian government until the 1960s. In 1974, to prevent unsustainable fishing in Australian waters, the Australian government concluded a Memorandum of Understanding (MOU) with Indonesia (which came into effect on 1 February 1975) that identified five small points (Ashmore Reef, Cartier Islet, Scott Reef, Seringapatam Reef and Browse Islet) on the northwest Australian continental shelf at which traditional Indonesian fishermen were given access to harvest trepang, trochus, abalone, green snail, sponges and all molluscs on the seabed adjacent to these areas (Fox, 2005). A series of sovereignity claims by Australia culminated in a 200 nautical mile Australian Fishing Zone in 1979, which encroached on the traditional fishing grounds of a number of Indonesian fishing communities (Stacey, 2000).

Sea cucumber landings by Indonesian fishers in the area of the MOU (known as the "MOU Box") from 1998 onwards indicated very low catches (maximum recorded catches were 1 000 kilograms dried weight/vessel with median catches around 100 kg dried weight/vessel) even during traditional peak period; a 1999 Commonwealth

Scientific and Research Organization (CSIRO) reeftop survey noted that high value trepang stocks were overexploited in the MOU Box except on Ashmore Reef (Fox and Sen, 2002).

Species heavily exploited in Indonesia include: *A. echinites, A. mauritiana, A. miliaris, B. argus, B. vitiensis, H. atra, H. edulis, H. fuscogilva, H. fuscopunctata, H. whitmaei, H. scabra, H. scabra* var. *versicolor, H. coluber, S. chloronotus, S. herrmanni, T. ananas* and *T. anax* (Tuwo, 2004). Erdmann (1995) showed that the most valuable trepang group, *Actinopyga* spp. are now virtually extinct in the Spermonde Archipelago in southwest Sulawesi, where it was heavily fished by 200–300 households per km of coastline as human population burgeoned and destructive fishing techniques proliferated. *H. scabra* had been overfished in Sulawesi, and in some locations it was regarded as a rare species (Massin, 1999). Massin noted that only few and small specimens were observed during a field trip to Bali and Lombok in 1998. In the Spermonde Archipelago, *S. herrmanni* caught were generally small in size; in a thousand specimens, 36.0 percent were about 16.5 cm in length, 24.5 percent were 27.3 cm, 19.9 percent were 33.8 cm and 10.0 percent were 40 cm (Basri, 1997 cited in Tuwo, 2004). The minimum size for exploitation in Queensland for *S. herrmanni* is 35 cm (Anon, 2004) meaning that only about 10 percent of the Spermonde population is of fishable size based on the Queensland's management measure.

Exploitation of sea cucumbers in Indonesia has increased rapidly during the last decade. Currently, most of the capture sea cucumbers are landed by small- and medium-sized fishing boats of <10 gross tonnage (Tuwo, 2004). These fishing boats increased from 4 to almost 50 boats in 2003 in Barrang Lompo Island, and the catch per unit effort (CPUE) for medium–sized fishing boats in Indonesian waters was low with monthly trips recording about 1 000 specimens versus 2 500 specimens during a five-day trip in Australian waters in 1997 (Tuwo, 2004). Ten years ago, small trawls caught 10–20 specimens per night or 3–7 specimens of *H. scabra* per trawl versus the current catch of 1–2 specimens per night or 0.33–0.66 per trawl (Tuwo, 2004). Trepang fishery in West Nusa Tenggara have suddenly collapsed, dropping from 17.7 tonnes in 1993 to 6.3 tonnes of low quality trepang in 1994 (Fox, 2005).

Tuwo (2004) estimated that the exploitation rate of sea cucumbers in Indonesian waters is more than 0.5, (meaning that 50 percent of the biomass of the stock is harvested annually), which is considered to be the maximum limit for sustainable exploitation. Eyrika (1999) noted that the exploitation rate of *H. scabra* in Saugi island waters was 0.62 in 1999, while in 1997 the exploitation rate of *S. herrmanni* was 0.69 (Tuwo, 2004).

2.5.2 Malaysia

A sea cucumber survey carried out in Sabah, East Malaysia, from July 1996 to December 1998 by Forbes and Ilias (1999) showed that many of the high value species, such as *H. scabra*, were now rare, resulting in increased pressure on the stocks of middle-range and low-value species such as *T. anax, T. ananas* and *Stichopus* spp. Local information also suggested the population and average size of individuals of *T. anax* have decreased.

On the west coast of Peninsular Malaysia, reports from Langkawi Island by Baine and Choo (1999) indicated that there is historical evidence that a fishery once existed for *Stichopus* spp. (local name "gamat") but nowadays, any sea cucumber fishing undertaken is at best part-time and occasional. In 1994, in Pangkor Island located in the west coast of Peninsular Malaysia, in an hour of diving it was possible to collect a large bucket of gamat, however, nowadays, it is difficult to get even three big sea cucumbers (Pankor, 2004).

Studies in marine parks along the east coast of Peninsular Malaysia, where illegal fishing is uncommon, showed healthy populations of sea cucumbers. The density (ind./100 m^2 ± s.d. in brackets) was 4.7 (4.2), 2.8 (2.6) and 11.9 (17.8) in Redang Island, Tioman Island and Tinggi Island, respectively, suggesting low fishing pressure (Harborne *et al.*, 2000). A survey conducted by Coral Cay Conservation (CCC) in the Redang Island Marine Park during March–September 2004 using the CCC baseline transect technique showed that sea cucumbers had the highest overall invertebrate abundance of 0.41 using the semi-quantitative DAFOR[1] ordinal scale (Comley *et al.*, 2004). The survey conducted by CCC indicated that the extraction pressure of sea cucumbers was low in the Redang Island Marine Park (Comley *et al.*, 2004). One of the leading sea cucumber wholesalers in Singapore noted that sea cucumbers were plentiful along the east coast of Peninsular Malaysia because few Malaysians were financially desperate enough to want to collect them (Butcher, 2004).

2.5.3 Thailand

Bussarawit and Thongtham (1999) reported on the overexploitation of sea cucumbers in the Gulf of Thailand and the Andaman Sea. According to the authors, when landings of *H. scabra* had become depleted, less valuable species like *H. atra* and *H. leucospilota* constituted the largest share of trepang produced in Thailand.

2.5.4 Viet Nam

Pitt and Duy (2004) reported that in the Khanh Hoa coastline in Central Viet Nam, small-scale commercial sandfish fisheries still existed in Van Phong and Nha Trang. Fishers in Khan Hoa reported that sandfish population which was once abundant is now depleted (Pitt and Duy, 2003). With the depleting wild resources, some farmers have been trying to rear small sandfish collected from the wild to larger sizes in pond (Pitt and Duy, 2004). In Phu Quoc Island, a sharp decrease in sea cucumber stock during the last five years was reported; *H. scabra* and *S. herrmanni*, once listed as highly abundant are believed to be approaching local extinction (Otero-Villanueva and Ut, 2007).

2.5.5 Russian Federation

Konstantinova (2004) noted that the sea cucumber (species not identified in the article, but likely to be *A. japonicus*) resources in Primorsky Krai, located in the extreme south–eastern region of the Russian Federation, have decreased to 16–20 percent of the stock that existed in the 1960s. The majority of the current stock comprises of specimens of around 40–60 g which were below the marketable size (Konstantinova, 2004).

2.6 Catches

As opposed to the single species sea cucumber fisheries in temperate waters, the fisheries in the Asian tropical and sub-tropical waters are multi-species. In terms of volume, sea cucumber fisheries mostly constitute only a minor portion (<0.2 percent) of the total marine fisheries landings in many of the countries that exploited them (Table 5). The sea cucumber capture landings obtained from FAO are recorded in wet weights; for some landings in Southeast Asia (e.g. Philippines) the figures appear to be grossly underestimated, and the statistics need to be verified as to whether the data reported were actually dried and not wet weight.

Except for China, where a substantial amount of sea cucumber production is from aquaculture, the production in the other Asian countries is derived predominantly, if not exclusively, from capture fisheries. Many tropical areas in the Indo-Pacific region

[1] In the DAFOR abundance rating, 0 number of individuals has a rating of 0; 1–5 individuals rating of 1; 6–20 individuals rating of 2; 21–50 individuals rating of 3; 51–250 individuals rating of 4; and >250 individuals a rating of 5 (Comley *et al.*, 2004).

TABLE 5

Sea cucumber landings in comparison to total landings of fish, crustacean, mollusc in some Southeast Asian and East Asian countries (unless specified, all values are wet weight)

Country/Species	Year	Sea cucumber (tonnes)	Marine fisheries (tonnes)	Sea cucumber landings (%)
Indonesia[1]	1990	1 722	2 285 450	0.075
(Southeast Asia)	2000	3 041	3 811 375	0.080
Holothuroidea	2001	3 517	3 963 422	0.089
(see Appendix I for	2002	3 057	4 038 767	0.076
commercial species)	2003	3 014	4 349 860	0.069
	2004	6 930	4 501 070	0.154
	2005	6 240	3 993 990	0.156
Philippines[1]	1990	4 023	1 623 057	0.248
(Southeast Asia)	2000	730	1 741 079	0.042
Holothuroidea	2001	791	1 813 188	0.044
(see Appendix I for	2002	801	1 899 661	0.042
commercial species)	2003	979	2 036 580	0.048
	2004	1 006	2 071 123	0.049
	2005	762	2 167 889	0.035
Malaysia[2]	1990	1 200	587 875	0.204
(Southeast Asia)	2000	159	1 285 696	0.012
Holothuroidea	2001	165	1 231 287	0.013
(see Appendix I for	2002	139	1 272 105	0.011
commercial species)	2003	100	1 283 256	0.008
	2004	122	1 331 645	0.009
	2005	139	1 209 601	0.011
Republic of Korea[1]	1990	2 491	2 717 000	0.092
(East Asia)	2000	1 419	1 817 854	0.078
Apostichopus japonicus	2001	900	1 984 751	0.045
	2002	833	1 665 730	0.050
	2003	1 281	1 635 366	0.078
	2004	1 154	1 565 035	0.074
	2005	1 135	1 622 462	0.070
Japan[1]	1990	6 426	10 145 435	0.063
(East Asia)	2000	6 975	4 918 742	0.141
Apostichopus japonicus	2001	7 229	4 651 452	0.155
	2002	7 259	4 302 774	0.169
	2003	8 517	4 610 570	0.185
	2004	9 268	4 341 326	0.213
	2005	9 373	4 742 574	0.215
China[3]	2001	358.44 (dry weight)	14 379 457	–
(East Asia)	2002	470.23 (dry weight)	14 305 218	–
Holothuroidea and	2003	–	14 293 783	–
Apostichopus japonicus	2004	–	14 473 187	–
	2005	–	14 792 598	–

Sources: [1] FAO; [2] Choo, 2004 (for 1990, Sabah landed 400 tonnes which included also sea urchins while Peninsular Malaysia landed 800 tonnes. Sea cucumber landings from 2000 onwards comprised landings solely from Sabah and extracted from the Annual Fisheries Statistics, Sabah (Anon. 2000 to 2005); 3. Chen, 2004 (sea cucumber landings in dry weight).

are heavily overfished for holothurians with *H. scabra* the main overfished species (Uthicke and Conand, 2005). Gleaning for sea cucumbers in shallow coastal waters is commonly carried out. Apart from this, the most common fishing methods for sea cucumbers include small bottom trawl nets in sandy bottoms, scallop-drag gear in nearshore rocky-bottom habitats, spears, hooks and scoop nets for reefs, and SCUBA and hookah for deeper reef and lagoon environments (Bruckner, Johnson and Field, 2003).

Because of the very small global landings of sea cucumbers compared to that of other marine species, records on catches are usually not given much attention by the individual governments and therefore may not be accurately documented. Catches are usually not separated by species and sometimes aggregated with other groups of invertebrates or echinoderms. In China, prior to 2001, capture sea cucumber production was classified under the "other" category in the Chinese Fisheries Annual

Statistics (Chen, 2004). The Malaysian sea cucumber catches ceased to be recorded in the FAO fishery statistics yearbook and the Malaysian Annual Fisheries Statistics yearbook after 1993.

Statistics from FAO indicated that Indonesia produced the highest Holothuroidea capture landings in the 2000s, followed by the Philippines. Together, these two countries produced an annual average of 47 percent of the world's Holothuroidea landings between 2000 and 2005 (Table 5). The highest capture fishery producer of the temperate species, *A. japonicus,* is Japan; between 2000 and 2005, an average of 8 101 tonnes per year were produced.

Information on sea cucumber harvesting methods and their landings, and the prices obtained for their harvests from individual countries in the Asian region are described below:

2.6.1 Indonesia

In the nineteenth century, in Sulawesi, fishers used their feet to locate sea cucumbers in shallow waters, while they dived to harvest those found in deeper waters (Butcher, 2004). Fishers from Buton, Salayar and other islands off the south coast of Sulawesi journeyed far out to sea to collect trepang. In the earlier days, fishers free-dived to collect trepang and other invertebrates; since the 1980s, they used compressors to collect their catch (Adhuri and Visser, 2007).

In the Kangean Islands in the Java Sea, women hand-picked sea cucumbers on reef flats during low tide, while men collected them from boats in deeper waters by lowering a weighted, three pronged spear by rope to a point just above a sea cucumber and then releasing the rope to let the weight impale the animal (Butcher, 2004). Whitten, Mustaffa and Henderson (2002) suggested that the inefficient collecting methods have contributed to sustaining the sea cucumber fisheries by preventing overfishing. However, in recent times, with the advent of destructive reef fishing methods (such as blast and cyanide fishing, coupled with the use of hookah and SCUBA gears) invertebrate reef resources in Indonesia have greatly been depleted (Erdmann, 1995). In the Asia-Pacific area, the presence of an estimated 50 000–80 000 indigenous divers, mainly from Indonesia and the Philippines has been reported (Best, 2003).

2.6.2 Malaysia

The sea cucumber fishery in Peninsular Malaysia and in Sabah is artisanal. In Peninsular Malaysia, sea cucumber fishing is very small-scale and is limited to only a few localities, for example in Telok Nipah in Pangkor in the west coast of Peninsular Malaysia where sea cucumbers are harvested for use in traditional medicinal products. The harvesting of *S. horrens* from Pangkor is mainly by hand-picking during low spring tides, and is limited to about 20 days in a month. In Sabah, statistics collected from the mid-1990s onwards showed that 70–80 percent of the total landings are hand–picked (or diving without SCUBA) with the rest landed from trawling. Statistics collected before the mid-1990s indicated that more than 90 percent of the sea cucumber catches were landed by trawl as bycatch (Annual Fisheries Statistics, Sabah, years 1991–2004). Most of the sea cucumbers harvested by trawl gear are collected by trawlers of 10–24.9 and 25–39.9 gross tonnage fishing in waters within 30 nautical miles of the coast (Choo, 2004). Sea cucumbers in Sabah are landed in Semporna, Sandakan, Kudat, Kota Marudu, Kota Belud and Kota Kinabalu; from the mid-1990s there were no landings from Kota Marudu and Kota Belud, indicating that the waters from the surrounding waters from these two areas may have been depleted. The actual volume of sea cucumbers harvested by hand-picking are difficult to capture in official fisheries statistics since gleaners need not report their catches to a central authority; hence sea cucumber landings recorded in the Annual Fisheries Statistics, Sabah, are very likely underestimated and represented predominantly the bycatch from trawling. Statistics available from FAO and also

FIGURE 2
Sea cucumber utilisation in Thailand from 1967–1988

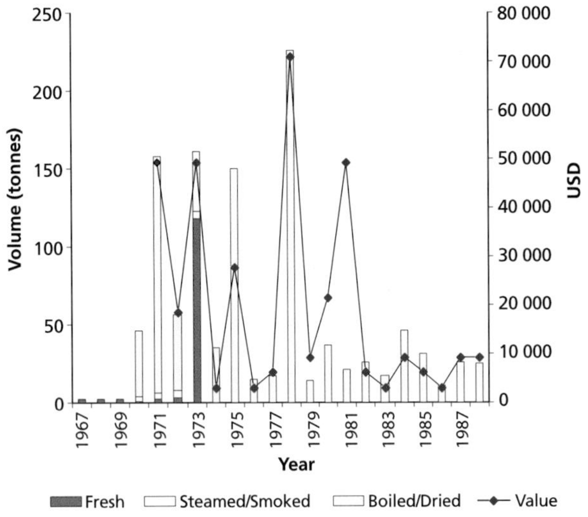

Source: Bussarawit and Thongtham, 1999.

from the Department of Fisheries, Sabah (Table 5) show that sea cucumber landings declined almost ten fold from the 1990s to the 2000s, from an average of around 1 000 tonnes (wet weight) in the 1990s to about 100 tonnes (wet weight) in 2000s .

2.6.3 Thailand

Bussarawit and Thongtham (1999) noted that Thailand supports a very small sea cucumber capture production, and that after 1988 there has been no fishery and trade statistics from the Fisheries Record of Thailand. From 1974 onwards sea cucumbers are processed into the boiled and dried category (Figure 2).

In Thailand, when sea cucumbers were first harvested more than 20 years ago, the most common fishing method was by hand-picking during low tide, but as resources declined, snorkelling to depths of 5–10 m became popular (Bussarawit and Thongtham, 1999). They noted that spears were commonly used by sea gypsies in the Andaman Sea to collect high value species such as *T. ananas, A. echinites* and *H. nobilis*. Sea gypsies often made trips to offshore islands lasting 1–7 days; sea cucumbers caught were gutted and brought back to shore for further processing. SCUBA fishing for sea cucumbers is not practised and fishers prefer to collect them from shallow bays during low tide.

2.6.4 Viet Nam

Otero–Villanueva and Ut (2007) reported overexploitation of the sea cucumber resources in Viet Nam. After the boom in the sea cucumber catches in the late 1980s, the average catch per diver in the 1990s dropped to 20 kg (dry weight) per day for species such as *H. leucospilota, S. herrmanni* and *B. argus* (Otero-Villanueva and Ut, 2007). Due to the sharp decline in stocks since the late 1990s (Table 6), fishers only dive occasionally around Phu Quoc Archipelago to collect sea cucumbers; however, Cambodian stocks in this Archipelago are still abundant and there have been reports of illegal fishing from Vietnamese fishers in Koh Sdach in Koh Kong province and Koh Rong in Sihanouk province (Otero-Villanueva and Ut, 2007). There is a small sea cucumber fishery in Kep and Sihanouk Province, Cambodia, where a wet weight of 470 tonnes and 210 tonnes respectively were landed in 2001 (Gillett, 2004).

Otero-Villanueva and Ut (2007) reported that sea cucumbers were initially caught in shallow waters using large poles operated from boats, but as the shallow water sea cucumber populations decreased, hookah divers replaced pole fishing activity. Sea

TABLE 6
Sea cucumber catches in Phu Quoc Archipelago before and after 1997

Species	Harvests before 1997 (dry weight unless mentioned otherwise)	Harvests after 1997 (dry weight unless mentioned otherwise)
H. scabra	200–500 kg/year	Less than 100 kg/year
H. atra	Many tonnes/year	Less than 1 tonne/year
H. leucospilota	Many tonnes/year	100 kg/year
S. herrmanni	1 tonne/day (fresh weight)	10 kg /day (fresh weight)
Total amount	**5 tonnes/year**	**500 kg/year**

Source: Otero-Villanueva and Ut, 2007.

cucumbers are less cryptic at night and because of stock depletion, hookah divers now collect at night using torches in waters as deep as 30–40 m (Otero-Villanueva and Ut, 2007).

2.6.5 Myanmar

The Burmese sea gypsies (Mokens) spend most of their lives on wooden boats, collecting sea cucumbers and other aquatic organisms from the reefs of the Mergui Archipelago in the Andaman Sea (Project Maje, 2004). Women often dive for shellfish, crabs, shrimp and sea cucumbers (Piprell, 2000). The Mokens do not use modern SCUBA gear and dive with only a mask, fins and a hosepipe acting as a super long snorkel (Northam, 2002). Smoked or dried sea cucumbers are traded with visiting dealers from Mergui town in Myanmar or with those from Ranong in Phuket, Thailand (Piprell, 2000). The amount of sea cucumbers collected by the sea gypsies whose population is estimated to be 2 000–3 000 is not known (Piprell, 2000).

2.6.6 China

Chen (2004) noted that the sea cucumber capture fisheries in Hainan and Guangdong in China comprised tropical and subtropical species such as *T. ananas*, *H. whitmaei*, *H. scabra*, *A. mauritiana*, *B. argus*, *S. herrmanni* and *S. chlononotus*, while in the temperate region of Liaoning and Shandong provinces only *A. japonicus* were exploited while in Zhejiang and Fujian provinces, species like *Acaudina leucoprocta* and *Mensamaria intercedens* were also fished. The wild capture fisheries of *A. japonicus* have declined significantly in the Shandong and Liaoning Provinces from 130–140 tonnes and 130 tonnes (dry weight) in the 1950s to 30–40 tonnes and 26 tonnes, respectively in the 1970s; current landings of wild-caught *A. japonicus* are close to zero (Chen, 2004). Chen (2004) noted that recovery plans for *A. japonicus* are in place together with the establishment of many conservation zones.

Harvesting methods vary from area to area, generally ranging from the use of simple to heavy diving equipment. According to Li (2004) historical records showed that sea cucumber harvesting in the Chinese Islands of Xisha, Zongsha and Nansha in the tropical South China Sea had gone uninterrupted from generation to generation since 1681. In the earlier days, harvesting was by free diving and hand collection, and therefore was restricted to shallower waters up to a depth of 20 m. However, in the 1920s, a tool known as the sea cucumber fork was developed allowing fishers to harvest sea cucumbers from deeper areas. The boats used in sea cucumber fishing are generally 50–80 gross tonnage fitted with a 120–250 hp engine with 12–16 fishers onboard. Each of these vessels are accompanied by three to four smaller boats (15–25 hp; 3–5 gross tonnage) to enable fishers to move easily around the reefs. Each of the smaller boats carries a crew of three, with one crew in charge of steering the boat while the other two are engaged in fishing with the fork equipment. Due to the turbidity of the water, fishing activity is restricted to a depth of 60 m. Each one of the small boats is able to catch up to 2 000 kilograms of sea cucumbers a year. In the northern part of China, temperate species of sea cucumbers are collected using adequate diving equipment (Chen, 2004).

In China, the soaring price of *A. japonicus* has stimulated the development of aquaculture including sea ranching activities. Chen (2004) noted that the output from these two sectors would reach 6 750 tonnes (dry weight) or an equivalent wet weight of 135 000 to 202 500 tonnes. According to FAO statistics the average annual global wild-capture harvest for the same species from Japan and Republic of Korea between 2000 and 2005 was 9 221 tonnes.

2.6.7 Japan

A. japonicus landings from Japan from 1990 to 2005 are shown in Figure 3. Japan's sea cucumber landings were sustained at an average of 8 101 tonnes from 2000

FIGURE 3
A. japonicus landings from Japan in years 1990 to 2005

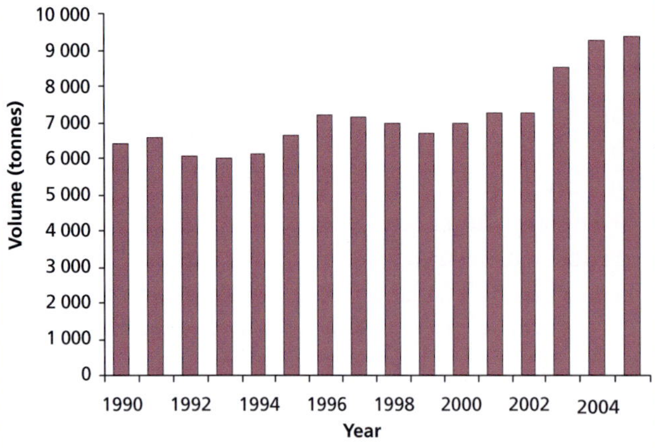

Source: FAO.

to 2005, while the average production between 1990 and 1999 was 6 579 tonnes. Akamine (2004) attributed the steady landings to the several resource management practices by local fisheries cooperative associations including size limits, gear restrictions, closed seasons and stock enhancement.

A. japonicus is harvested by using a dredge net (also known as a beam trawl, according to Kiso, K., Ishigaki Tropical Station, personal communication), hook, clip and twist, spear or dart, or by diving (Akamine, 2004). In the Semposhi Fisheries Cooperative Association (SFCA) in northern Hokkaido, more than 98 percent of the catch is from net dredging which operates mainly in waters that are 40–50 m deep; as of July 2003, there were 11 dredge net operations permitted (Akamine, 2004).

There is a small *S. chloronotus* fishery in Okinawa where a single fisher hand picks and processes the sea cucumbers and sells them in the dried form to Chinese restaurants in Japan (Akamine, J., Nagoya City University, personal communication). The dried *S. chloronotus* are sold between USD 85–90/kg.

2.6.8 Democratic People's Republic of Korea
In the Democratic People's Republic of Korea, the sea urchin and sea cucumber landings had declined from an annual catch of 7 200 tonnes during 1970 to 1975 to about one-third this amount during 1985–1990 (Anon, undated).

2.6.9 Republic of Korea
Capture production of *A. japonicus* from the Republic of Korea (obtained from FAO yearbook on capture fisheries production) showed an average production of about 1 902 tonnes from 1990 to 1999; the average capture production from 2000 to 2005 was 1 120 tonnes.

2.7 Management measures
Except for Japan, countries in the Asian region are generally lacking in management measures to conserve and sustain their sea cucumber fisheries. The two most important producing countries, Indonesia and the Philippines do not have management plans specific to sea cucumber conservation (Bruckner, Johnson and Field, 2003).

The information below describes some of the management measures or proposals for the protection of sea cucumber fisheries.

2.7.1 Japan
In Asia, Japan has the longest history of managing their sea cucumber resources. Mitsukuri (1903) noted that for hundreds of years, the people of Oki Island used to put up loose stone piles in the shallow seas to provide adult *A. japonicus* a place to aestivate and to provide a place for metamorphosing larvae and young juveniles to aggregate. Japan has enforced regulations setting aside certain localities as breeding reserves where stone piles have been constructed, and in these places, sea cucumber fishing was strictly forbidden (Mitsukuri, 1903). The construction of stone piles is still practised in some areas.

The Fisheries Law in Japan, established in 1949, manages fisheries resources with either the "fishery rights" system for stationary fisheries, or the "fishery permit" system for mobile fisheries (Akamine, 2004). The holothurian fishery is regulated by the "fishery rights" system. This "fishery right" is granted by the Sea-area Fishery Adjustment Commission (SFAC). Akamine (2004) noted that no one can freely collect holothurians unless he/she has a "fishery right" and can do so only with hooks, clips and twists or by diving. Fisheries Cooperative Associations in fishing communities are also given the power to regulate fisheries practices: for example, in the Rishiri district located at the northern tip of Hokkaido, the Semposhi Fisheries Cooperative Association (SFCA) does not permit the fishing of *A. japonicus* during the spawning season from May 1 to June 15; with more comprehensive scientific data, the SFCA has proposed to change the closed season from July 20 to the end of September (Akamine, 2004). Further regulations imposed by the SFCA include a minimum catch size of 130 g (wet weight) and an annual catch of not more than 50 tonnes from the Rishiri area.

Japan's aquatic resource recovery plans include the recovery of sea cucumbers in the Ohmura Bay, located in western Kyushu in western Japan where sea cucumbers are caught among other methods by small trawl fishery (Anon., 2005). The Ohmura Bay is a closed bay, and sea cucumber fishery has been carried out for a long time, with *A. japonicus* (red, blue and black varieties) being the target species (Kiso, K., Ishigaki Tropical Station, personal communication). Management measures (effective as of 30 August 2005), established by the Fisheries Policy Planning Department, Fisheries Agency, include closed seasons, minimum harvesting size, mesh size restriction and no–fishing areas (Anon., 2005). Fishing season was reduced to two months (from December to January) from the previous season of five months (November to March), and harvesting sea cucumbers of less than 100 g weight was not permitted (Mitsunaga *et al.*, 2006). Mitsunaga *et al.* (2006) showed that the sea cucumber recovery plan in Ohmura Bay has positive impacts on the sea cucumber fishery: the total resources of *ao* (green) and *aka* (red) *namako* increased to 177 tonnes in 2005 from 150 tonnes in 2004; the density m^{-2} of *ao* (green) and *aka* (red) *namako* in 2005 was 0.24 tonnes versus 0.20 tonnes in 2004, resulting in an increase of 18 percent in 2005.

2.7.2 Democratic People's Republic of Korea

The Oruji Nature Reserve which has an area of 750 hectares and located in Ryongyon and Jangyon counties in the S. Hwanghae Province was established in 1996 mainly for the conservation of sea cucumbers (Anon, undated). Problems facing the management of nature reserves in Democratic People's Republic of Korea include the following (Anon., undated):

- Inadequate and poor enforcement of regulations: currently, most of the newly established nature reserves do not have adequate regulations and do not enforce the Law on Environment Protection which was established in 1986; moreover, there is a lack of coordination among relevant agencies and difficulty in establishing consensus among agencies.
- Poor management capacity: currently, reserves are managed by those who lack the knowledge and expertise. There is no budget allocated for nature reserve management and there are no awareness programmes to educate the local people.
- Weak central agency and the lack of coordination between central agency and local authorities: there is an urgent need to enhance the function and effectiveness of the Ministry of Land Environment Protection and the coordination with local authorities.
- Lack of international connection: some of the reserves are of regional and international importance to a number of countries; the lack of collaboration among these countries is therefore a setback.

2.7.3 Malaysia

In Malaysia, the Establishment of Marine Parks Malaysia Order 1994 acts as a protection against the illegal collection of marine organisms in marine parks. There are no quota system and minimum legal size to protect sea cucumbers from overfishing in non-marine park areas. Waters off Pulau Singa Besar in Langkawi Island off the northwestern coast of Peninsular Malaysia have been earmarked for the establishment of a two-hectare sanctuary in view of the indiscriminate harvesting of sea cucumbers in Langkawi waters (Starmetro Newspaper, 10 January 2007).

2.7.4 Viet Nam

Presently, there are no fishing regulations in placed for the sea cucumber fisheries in Viet Nam (Otero-Villanueva, M., Yancheng Forestry Bureau, Jiangsu Province, personal communication). Viet Nam, including their most important fishing area, the Phu Quoc Archipelago, lacks a management plan for the sea cucumber fishery. Otero-Villanueva and Ut (2007) proposed the following measures for the Phu Quoc Archipelago:

- More detailed studies on the sea cucumber fisheries activities in Phu Quoc Archipelago and the neighbouring Cambodian islands in order to understand the interaction between the CPUE and the yield of commercial sea cucumbers;
- Regulation of all commercial sea cucumber species to prevent overfishing;
- Co–management of the sea cucumber resources; and
- Awareness campaigns to avoid overexploitation of the sea cucumber stocks.

2.7.5 MOU Box: Ashmore Reef

In 1983, the Australian Government declared Ashmore Reef a National Nature Reserve prohibiting the removal of all fauna and flora to a depth of 50 m (Fox, 2002). Fox noted that at the time of this ban, trepang stocks were already depleted near all the reefs and islets in the MOU Box. A survey conducted by the Australian Commonwealth Scientific and Industrial Research Organisation (CSIRO) 15 years after the ban showed that it had resulted in a partial recovery of sea cucumber population (Fox, 2002). Smith *et al.*, (undated) reported that no specimens of *Holothuria timana* were located during a survey in Ashmore Reef in 2000 but this high value species was reported to be in significant numbers in 1988, comprising 11 percent of the catch by Indonesian fishers in the 1980s (Russell and Vail, 1988 cited in Smith *et al.*, undated). *H. whitmaei* was reported to be severely depleted, but *T. ananas* and *H. fuscogilva* were relatively abundant during the 2000 survey by Smith *et al.* (undated).

3. USES

Sea cucumber dishes are almost always prepared by soaking and boiling the dried form (trepang) until they are of the preferred consistency although in some cases, dishes may be prepared from the fresh product. During ancient times when processing methods were limited, sea cucumbers were often processed into trepang and then used later as a food ingredient; recently they have been marketed as jelly, powder, tablets or capsules and consumed as a health supplement (Chen, 2004; Choo, 2004; New Straits Times, 2007).

Chen (2003) considered sea cucumber to be an ideal tonic food; it has higher protein and lower fat level than most foods, and from the Western medical viewpoint, sea cucumber serves as a rich source of the polysaccharide, condroitin sulfate, which is known for its ability to reduce arthritic pain (as little as 3 g/day of dried sea cucumber has been helpful in significantly reducing arthralgia). Chen (2003) also noted that there is a Japanese patent for sea cucumber chondroitin sulfate used for HIV therapy.

In Japan, sea cucumbers are consumed mainly in three forms: fresh, dried and fermented. The dried form is least popular and is categorised into two groups: no more than 30 pieces/600 g and 31–70 pieces/600 g (Anon., 1989). Dried sea cucumbers are used mainly in Chinese restaurants and its quality is judged by the state of the cucumber after reconstitution with water (Anon., 1989). The reconstituted weight should be four to five times the dry weight. Sales of dried sea cucumber have decreased to about three tonnes a year with the decrease being attributed to the unattractive appearance of live sea cucumbers to Japan's younger generation (Anon., 1989). According to Anon. (1989), for the fresh market, the colour, size and origin of sea cucumbers are important, with the red, brown or green variety in the 200–500 g range being the most popular. Fresh or live Japanese sea cucumbers are usually sold for ¥500–1 500/kg (USD 4–13/kg), and are highest during August to September. After depuration in holding tanks, over 12 kilograms of sea cucumbers are packed to meet the net weight of about 10 kilograms upon arrival at market. About 20 percent of the original weight is lost during shipment through water loss. Specialists do not recommend freezing fresh sea cucumbers because they reportedly disintegrate when thawed. The fermented sea cucumber viscera known locally as "konowata", is a delicacy in Japan and is eaten mainly when drinking Japanese whisky or distilled Japanese sake. About eight to ten tonnes of "konowata" are sold at the Tsukiji Central Wholesale Market per year. It is prepared by first thoroughly cleaning the viscera, and then adding salt at an equivalent of 10–15 percent of the total wet weight of the viscera mass. The mixture is stirred frequently for about five hours, drained, and then put in barrels for about a week to age. "Konowata" is found only in specialty stores and the retail price of a glass jar containing about 100–120 g is about ¥3 000 (USD 26) each.

4. TRADE

Global sea cucumber trade meant for the food market is controlled by China, Hong Kong SAR, Singapore and Taiwan Province of China, with China Hong Kong SAR having the largest entrepôt controlling 80 percent of the global import-export sea cucumber trade (Jaquemet and Conand, 1999). The dominance of China Hong Kong SAR is due mainly to its ability to serve as a conduit of goods to the hinterland of mainland China (Clarke, 2002). The lower value products were traditionally shipped to China Hong Kong SAR for re-export to China (Conand and Byrne, 1993).

Producing countries all over the world export trepang to one of these three centres; from where they are re-exported mainly to Chinese consumers worldwide. China Hong Kong SAR, China, Singapore, Malaysia, Taiwan Province of China, Republic of Korea and Japan account for almost 90 percent of the total imports of trepang, with approximately 80 percent of the overall international trade destined initially for China, Hong Kong SAR (Bruckner, Johnson and Field, 2003). From 1996 to 2000, 87 percent of re-exports of trepang from China Hong Kong SAR were designated for China (Clarke, 2002). Due to varied purchasing power, China imports a wide variety and quality of trepang; however, the Chinese market is increasingly demanding good quality products irrespective of the species and consumers are willing to pay better prices for such products (Ferdouse, 2004).

According to Bruckner, Johnson and Field (2003), in 2000 and 2001, Taiwan Province of China imported sea cucumbers from 28 countries and Singapore receives about 50 percent of its imports from China Hong Kong SAR, with Papua New Guinea, Tanzania and Madagascar being the other main suppliers. Two-way trade also exists between the three main exporting centres. China Hong Kong SAR imports peaked in 1991 at 7 885 tonnes and gradually declined to 2 922 tonnes in 1999. From 2000 to 2005, imports averaged at around 4 626 tonnes. Consumption of sea cucumbers in China Hong Kong SAR, peaked in 1991 at 4 456 tonnes. The average consumption decreased from 3 812 tonnes between 1991 and 1994 to 999.9 tonnes between 1995 and 2005. Figure 4 provides the data on sea cucumber imports and re-exports from China Hong Kong SAR, from 1976 to 2005.

FIGURE 4
Sea cucumber (dried, salted and in brine) imports and re-exports from China Hong Kong SAR

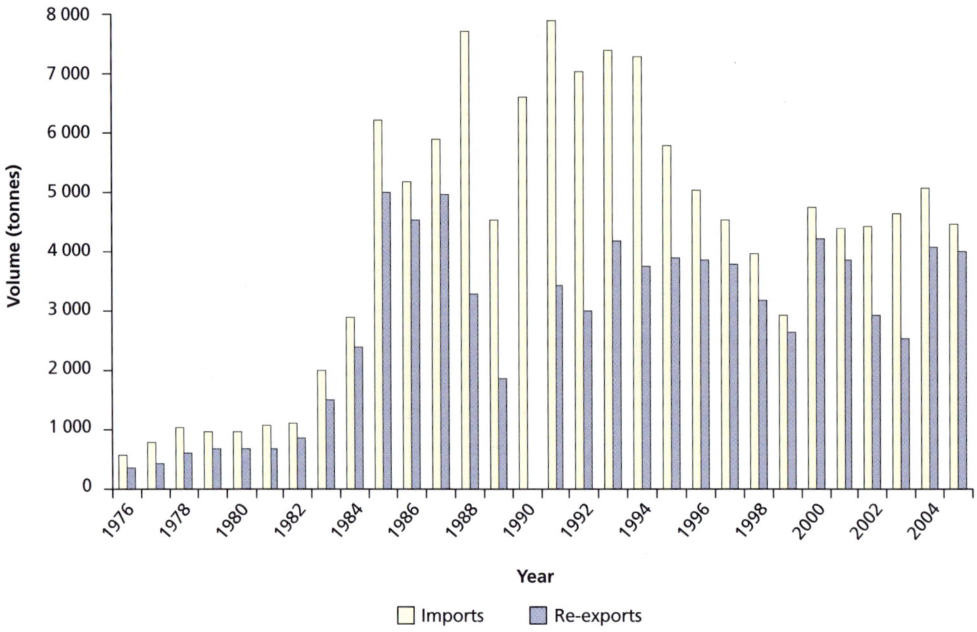

Source: FAO.

Taiwanese consumers prefer the coldwater spiky sea cucumber species such as *A. japonicus* and are extremely quality conscious, preferring the high value species. Singaporeans prefer good quality smoked, cleaned and dried sandfish while Malaysian consumers have a preference for medium quality and medium-size sandfish and teatfish (Ferdouse, 2004). Consumers in China Hong Kong SAR, have preferences for good quality sea cucumber of large and medium sizes (Ferdouse, 2004).

The key factors in determining price are species/type, the size of the dried animal, the degree of dryness of the processed products and the time of year; prices can be 20–30 percent higher just before the Chinese Lunar New Year (Clarke, 2002). Most traders believe that the higher quality goods are those derived from Japan, the Pacific coast of South America, South Africa and Australia while those from Indonesia, the Philippines and China are of lower quality due to species composition and inferior processing techniques (Clarke, 2002).

According to the FAO global statistics on sea cucumber, Indonesia is the largest trepang exporter in the world. About 40–80 percent of the trepangs are exported to China, Hong Kong SAR, with other markets being Japan, Republic of Korea, Singapore, Taiwan Province of China, Malaysia and Australia (Tuwo, 2004). The most important collecting and export centre for marine product trade in Indonesia is Ujung Pandang in Sulawesi, which, in 1981, recorded foreign exports of 2 100 tonnes of reef invertebrates out of which 320 tonnes were trepang (Whitten, Mustafa and Henderson, 2002). The average annual price of Indonesian trepang exported from South Sulawesi from 1996 to 2002 ranged from USD 15.06/kg to USD 1.44/kg (Tuwo, 2004). Apart from the common factors influencing product price (e.g. species composition, size and post-harvest quality), exporters are known also to influence the price of sea cucumbers as they commonly manipulate the price in export documents for the purpose of tax evasion (Tuwo, 2004).

Among the high value species, price trends extracted from INFOFISH Trade News showed that sandfish fetched the highest price (Appendix II). Only the price of one temperate species, *A. japonicus*, has been listed in the INFOFISH Trade News.

The cultured *A. stichopus* fetched almost twice the price as the highest priced grade sandfish.

According to Chen (2004) the retail price of *A. japonicus* has increased dramatically since the 1980s. In 1960, retail price for one kilogram used to be Renminbi (RMB) 18 which increased to RMB500 in 1980 and to the present price of RMB3 000/kg (approximately USD 400).

Apart from the food trade, a large number, possibly hundreds of thousands of sea cucumbers are traded in the aquarium industry, but little data on species, quantities and source countries are available (Bruckner, Johnson and Field, 2003). The volume of sea cucumber used in the medicinal and cosmetic industries is also not known.

5. SOCIO-ECONOMIC IMPORTANCE TO LOCAL FISHING COMMUNITIES

Since historical times, the sea cucumber fishery and trepang processing have been a source of livelihood to many coastal communities. Harvesting, preparing and selling trepang constitute the pillar of the economy of the Bajaus from the Sulawesi Islands since the sixteenth century (Whitten, Mustafa and Henderson, 2002). The most sought-after trepang was painstakingly prepared by the Turijene, a Bajau group based in the Spermonde Archipelago in Sulawesi. The trepang was then sold to the Bugis seafarers for the China trade, which formed the basis of the economic interdependence of these two groups of people.

Up to the present time, in the coastal areas of many of the Southeast Asian countries, the harvesting of sea cucumbers and the processing of trepang constitute an important means of livelihood for the coastal poor. They comprise the sea gypsies from Myanmar, Thailand and Indonesia to the poor fishermen from Sabah, Indonesia and the Philippines who depend on the trepang industry to supplement their meager income. For this reason, increasing attention is being focused on monitoring the trepang trade and on sustainable management of the sea cucumber resources (Clarke, 2002). Tuwo (2004) noted that the exploitation of sea cucumbers in Indonesia is an important contributor to the economic development of the coastal and small island communities; for this reason the sea cucumber population has to be managed properly to sustain its economic and ecological benefits. In Sabah, Biusing (1997) and Choo (2004) noted that sea cucumbers are a valuable source of income for many of the poor rural fishing communities. In Myanmar, the sea gypsies have been reported to be forcibly resettled on land in the late 1990s by Myanmar's military regime, and the increased presence of foreign trawler fleets under joint-ventures with the military regime had also discouraged small-scale local fishing by the sea gypsies (Project Maje, 2004).

H. scabra (size range 3 specimens/kg) were sold at USD 0.12 for each specimen in Saugi Island (Tuwo, 2004). The price received by fishers in Buton in Southwest Sulawesi for their catch, their annual income and the price of sea cucumbers in the international market are shown in Table 7. Only the local names of the sea cucumbers were given by Asmedi (undated) in the table; reference was made to other sources for the scientific names of some of these species. Prices obtained by fishers are one–sixth the international market price, indicating that they are grossly underpaid by the middlemen.

In Phang Nga located in Phuket, Thailand, fishers sold fresh sea cucumbers to processors for prices ranging from Baht 3–35/kg (USD 0.09–1.08/kg); the processors then sold the processed products to middlemen for Baht 130–300/kg (USD 3.6–8.3/kg) (Bussarawit and Thongtham, 1999). Bussarawit and Thongtham (1999) also reported on an entrepreneur from Ranong who supplied fishers with capital and facilities for harvesting sea cucumbers. He would then buy sea cucumbers at a price lower than the market value. Fishers sold their sea cucumbers at prices between Baht 30–50/kg (USD 0.93–1.54/kg) which the middleman resold at Baht 100–200/kg (USD 3.08–6.17/kg).

TABLE 7
Average annual income of fishers and price received by them for their sea cucumber harvest (current USD prices within brackets)

Sea cucumber (local names)	Average price fishers receive per kg dry weight	Average price on the international market	Average annual income of target fishers
STR (low quality mix)	Rp7 000/kg (USD 0.8/kg)	Rp42 000/kg (USD 5/kg)	Rp1 400 000 (USD 164/kg)
Duyung *T. anax*	Rp10 000/kg (USD 1.2/kg)	Rp60 000/kg (USD 7/kg)	Rp2 000 000 (USD 235/kg)
CRDI (medium quality mix)	Rp20 000/kg (USD 2.4/kg)	Rp120 000/kg (USD 14/kg)	Rp4 000 000 (USD 647/kg)
Susu Polos Teatfish	Rp27 500/kg (USD 3.8/kg)	Rp165 000/kg (USD 20/kg)	Rp5 500 000 (USD 647/kg)
Susu Bintik Teatfish	Rp32 500/kg (USD 3.8/kg)	Rp195 000/kg (USD 23/kg)	Rp6 500 000 (USD 3 058/kg)
Gamma *Stichopus* spp.	Rp130 000/kg (USD 15.3/kg)	Rp780 000/kg (USD 92/kg)	Rp26 000 000 (USD 3 058/kg)
SSI (good quality mix)	Rp130 000/kg (USD 15.3/kg)	Rp780 000/kg (USD 92/kg)	Rp26 000 000 (USD 3 058/kg)

Source: JPKP Buton and Yayasan Lintas, Asmedi, undated.

Gleaning for sea cucumbers and other invertebrates constitutes a very important livelihood for many coastal communities in Southeast Asia. Although it is often classified as "informal work", gleaning requires skills and knowledge which are often passed down from one generation to another (LeBlanc, 1997). Gleaning provides a livelihood for the poor and landless – those that cannot afford to be unemployed and must obtain a source of income even if this implies "pitifully low earnings" (LeBlanc, 1997). In the Philippines, it is a highly organised activity which affects a large number of people (LeBlanc, 1997). In periods of fish scarcity, wives supplemented the family income by increasing their gleaning activities collecting species that are not affected by seasonality (Rowe, 1983). However, with the reported overfishing of sea cucumbers in many intertidal areas in the Asian region, gleaning for sea cucumbers will become less important as a source of income compared to gleaning for other invertebrates.

6. CURRENT DEVELOPMENTS

In the Asian region, minimal efforts are being made towards conserving sea cucumber resources through the imposition of fisheries regulations, but some attempts have been made by some countries towards sea cucumber production through aquaculture. Research on the extraction of products from sea cucumber and new product formulation are also being carried out.

Among the tropical species, *H. scabra* is the only species that are currently mass produced in hatcheries. They have been bred mostly on an experimental scale in India, Indonesia, Solomon Islands, New Caledonia, Viet Nam and the Maldives with the reproductive, settling and feeding cycle completed (James, 1996; Battaglene, 1999; Pitt and Duy, 2004; Agudo, 2006). Juveniles measuring 10 to 25 mm in length are transferred to sandy substrata in the field (James, 1996). Polyculture of *H. scabra* with seaweeds and shrimps have also been attempted (Madeali *et al.*, 1993; Daud *et al.*, 1993; Battaglene, Seymour and Ramofafia, 1999).

Among the temperate species, the most successfully cultured is *A. japonicus* with most of the production attributed to China. Work on the culture of *A. japonicus* started in China in the early 1950s, but the breakthrough was achieved only in the 1980s (Chen, 2004).

Three methods (pond and pen culture, and sea ranching) are used in the grow-out culture of *A. japonicus* in China (Chen, 2004). In pond culture, stones or other non-toxic materials are laid in rows or in a pile at the pond bottom, acting as "homes" for the sea cucumbers. Farming sea cucumbers in ponds is popular in China where existing shrimp ponds are modified to meet the ecological needs of sea cucumbers. Pens are located near to the seashore; it has an advantage over pond culture since no pumping of water is required but this method is limited to areas along the coast. Sea ranching experiments were first initiated in the 1950s in Hebei Province and have been found to be an effective method that could help the recovery of wild sea cucumber stocks.

6.1 Indonesia

The species farmed in Indonesia is predominantly *H. scabra*. The potential area for sea farming (cage or pond culture) of sea cucumber in Indonesia is 720 500 ha (Dahuri, 2002, cited in Tuwo, 2004). However, presently only a few of the potential areas have been utilised for farming (Tuwo, 2004). Tuwo (2004) further stated that utilising only 10 percent of the potential farming areas could increase sea cucumber production to about 180 125 tonnes dry weight per year. The four most important areas where grow-out of wild caught animals of *H. scabra* takes place include: Papua, Central Sulawesi, South East Sulawesi and East Kalimantan which recorded a production of 378 tonnes, 200 tonnes, three tonnes and one tonne wet weight per year, respectively (Tuwo, 2004). Since 2002, in Laumbora Bay off Barangka village in Indonesia, cage culture of sea cucumbers using small specimens collected from the wild have been practised by fishers working together with Yayasan Lintas, one of the local NGOs (Asmedi, undated). The introduction of sea cucumber culture in this area has resulted in fishers moving away from destructive fishing practices, and becoming more aware of the importance of conserving the environment. The Buton Coastal Area Development agency and Yaysan Lintas, supported by Oxfam Great Britain since 2002, have approached and negotiated with several exporters of marine commodities to offer better prices to fishers. In turn, the fishers will need to meet the quality standards set by the companies (Asmedi, undated). Sea cucumber cultivation was also recently introduced to 10 villages in the western Seram district (Antara News, 2007).

Mass production of seeds in Indonesia faces many biological and technical problems, the main problem being the lack of broodstock (Handoko *et al.*, 2002, cited in Tuwo, 2004) and the lack of suitable sites, since ponds may be located in acid-sulphate areas or experience low salinity in the monsoon season (Purcell, S., The WorldFish Center, New Caledonia, personal communication,). To optimally utilise 10 percent of the potential farming area, Indonesia needs about 7–14 billion sea cucumber seeds per year, based on a stocking density of 15–20 ind./m^2 (Tuwo, 2004).

6.4 Philippines

The University of the Philippines in Luzon and Mindanao and the National Fisheries Research and Development Institute (NFRDI) of the Bureau of Fisheries and Aquatic Resources (BFAR) are collaborating with the Australian Centre for International Agricultural Research (ACIAR) and the WorldFish Center in a four-year project on "Restocking and sea ranching of sandfish" (Purcell, S., The WorldFish Center, New Caledonia, personal communication). Purcell also mentioned that there is an ongoing sea cucumber research by the Marine Science Institute, University of the Philippines (MSI-UP), genetic studies by BFAR and a project on the distribution and inventory of sea cucumber species by the Department of Science and Technology (DOST). These agencies, together with the Philippine Council for Aquatic and Marine Research and Development (PCAMRD) and the Aquaculture Department of the Southeast Asia Fisheries Development Center (SEAFDEC) are also involved in a National Program on Sea Cucumber Management (Purcell, S., The WorldFish Center, New Caledonia, personal communication).

6.3 Viet Nam

In Cam Ranh, some fishers have been attempting to culture wild-caught small sandfish to larger size in ponds (Pitt and Duy, 2004). Apart from that, the WorldFish Center scientists, working together with their Vietnamese counterparts in Nha Trang, aimed to develop large-scale breeding and rearing methods for commercial culture and/or restocking of sandfish.

Constraints to commercial culture of sandfish in Viet Nam included low prices paid by middlemen, high variability in survival rate at many stages, predation, theft and pond management problems. Positive factors included the wide temperature and salinity tolerance of sandfish, ease of containment, good growth in ponds and pens without the need for adding feed (Pitt and Duy, 2004).

6.4 Malaysia

A local firm in Malaysia, Luxor Network Sdn Bhd, has signed an agreement in 2007 with the International Islamic University in Malaysia for collaboration on research into the properties of gamat (*Stichopus* spp.) as well as research on its breeding and culture (New Straits Times, 2007). The company plans to expand its gamat-based product business overseas in Singapore, Indonesia and member countries of the Organisation of Islamic countries. Luxor Network recorded annual sales turnover of RM60 million (USD 17 391 304) and produces a range of products such as roselle gamat, gamat oral jelly, gamat topical gel, gamat cream, gamat hair shampoo, gamat whitening cream and toothpaste, as well as gamat "tongkat ali" (a plant aphrodisiac).

6.5 China

The largest farm producer of the sea cucumber *A. japonicus* in China is Liaoning Province, where the total area farmed is expected to reach around 51 000 ha with outputs projected to 6 750 tonnes dry weight per year (Chen, 2004). Shandong has become the second largest province to farm sea cucumber in China with a total farming area reaching 17 000 ha with an estimated production of 2 250 tonnes dry weight equivalent to the wet weight production of about 45 000–67 500 tonnes. Hebei province, which produces annually approximately 1 000 tonnes/year dry weight of *A. japonicus*, ranks third (Chen, 2004). Aquaculture production of sea cucumber in the southern provinces is still in its infancy where sandfish, black and white teatfish are considered as choice species for aquaculture development (Chen, 2004). Chen (2004) noted that in the next decade, sea cucumber farming and ranching will become a highly successful industry in China. It will involve an entire industrial supply chain from seed production to processing, the preparation of formulated feeds for culture, the development of special culture techniques and ranching facilities as well as the production of specific chemical products from sea cucumber (Chen, 2004).

6.6 Japan

In the last few years, the Japanese Government, together with the provincial governments has given emphasis to the commercial hatchery production of *A. japonicus* (Akamine, J., Nagoya City University, Japan, personal communication). In 2005, the Iwate Fisheries Technology Center in Kamaishi in the Iwate Prefecture started developing hatchery techniques for *A. japonicus* production as well as sea ranching techniques and breeding them in farms (Yoshimura, 2006). The Center hopes to boost the sea cucumber production in the Iwate Prefecture (which in 2004 had a capture production of 33 tonnes), through aquaculture and sea ranching techniques (Yoshimura, 2006). Some private companies in the Iwate Prefecture have also started commercial aquaculture production of sea cucumbers (Yoshimura, 2006).

6.7 The Russian Federation

In Far East Russian Federation, the first sea cucumber (species not specified but likely to be *A. japonicus*) hatchery valued at USD 1.2 million was established in July 2003 in Kievka Bay in cooperation with the Pacific Fishery and Oceanography Research Institute (TINRO) (Konstantinova, 2004). Aquaculture of sea cucumber in Far East

Russian Federation can apparently produce 10–15 thousands of marketable size sea cucumbers per hectare at an estimated production of 0.5 tonnes/ha.

7. ADDITIONAL THREATS TO SEA CUCUMBER POPULATION

7.1 Illegal fishing

Threats to sea cucumber resources include illegal fishing which dates back many decades in some countries. Indonesian fishers were known to have fished illegally in Australian waters since the eighteenth century. One of the first recorded arrests in Indonesia dates back to 1725 where unauthorized Bajau boats entered Nusa Tenggara illegally to gather trepang (Fox, 2005). Indonesian seafarers were financed and outfitted by the Chinese and other merchants in Makassar in the eighteenth century; they sailed to the north coast of Australia where they collected and processed vast quantities of trepang (Butcher, 2004).

Restriction of illegal trepang fishing was first imposed in 1907 in the form of a license fee which denied Makassan fishers without a license to exploit Australian waters (ACFOA, 2002). The Australian government's first official attempt to regulate Indonesian activity in north Australia occurred in 1968 which later led to the signing of the 1974 MOU between Indonesia and Australia. The MOU permits only traditional Indonesian fishing within the 12 nautical miles territorial sea adjacent to some islands within the MOU Box (Stacey, 2005). Due to illegal overfishing, the reefs in the MOU Box are no longer capable of providing an adequate means of livelihood to those fishers who have previously gathered trepang and trochus (Fox and Sen, 2002).

Stacey (2000) described several measures in the form of apprehension, prosecution, confiscation of boats, catch and equipment, and jail-term for repeat illegal fishing offenders in Australian waters; despite all these measures, illegal Indonesian fishing continues. Indonesian fishers, however, who had been apprehended, detained, sent to jail and deported from Australia often received the status of a "hero" in their homeland, and detention in Australian jails was perceived as enjoyable akin to living in a five-star hotel (Adhuri and Visser, 2007). On the other hand, enforcement measures taken by the Australian Government have been criticised as lax by the local fishing sector in Australia; in the 15 months prior to March 2004, only 168 vessels out of the 1 588 sightings were apprehended (Fish Information and Services, 2004).

The Yomiuri Shimbun (30 December 2006) reported that sea cucumber poaching in Hokkaido and Aomori Prefectures is on the rise as prices of the product has soared because of the demand caused by the rapid expansion of the Chinese economy. The poaching at Aomori Prefecture on 26 July 2006 amounted to 880 kilograms of sea cucumbers worth ¥1.25 million (USD 10 788).

The poaching of sea cucumbers from national parks in Surin Island, Thailand was reported by Bussarawit and Thongtham (1999). Choo (2004) noted that the problems of illegal fishing in Sabah, East Malaysia, will be difficult to solve if the bigger challenge of illegal immigration mainly from the Philippines is not successfully resolved. Filipino fishers were known to fish illegally in Malaysian waters operating at night alone or in pairs, skin-diving for sea cucumbers (Akamine, 2001). Vietnamese fishers have been reported to fish illegally in Cambodian waters after sea cucumber resources in South Vietnamese waters have been depleted (Otero-Villanueva and Ut, 2007).

7.2 Overexploitation

Most of the Asian countries covered in this report have indicated that overfishing is the main problem for the depleting sea cucumber resources. Kim Friedman (Secretariat of the Pacific Community, New Caledonia, personal communication) noted that even the individual size of Asian sea cucumbers is small, with grade-A quality having many more pieces per kg than in the Pacific, and catches are not able to target large mature individuals. Overexploitation of holothurians can have adverse effects since small population sizes may lead to inbreeding and loss of genetic diversity, thus resulting in elevated extinction risks. Most tropical and subtropical holothurians are broadcast spawners and fertilisation success is highly dependent on population density; reduction of population densities by fishing may therefore render remaining individuals incapable of successful reproduction (Bruckner, Johnson and Field, 2003).

7.3 Habitat loss

In Singapore, the main threat to depleting sea cucumber populations is habitat loss due to reclamation or human activities along the coast that pollute the water (Tan, 2003). Sea cucumber collection has also resulted in the trampling of corals resulting in the loss of sea cucumber habitats (Bussarawit and Thongtham, 1999). Pollution has taken a toll on the survival of sea cucumbers in waters off Pangkor and Langkawi Island in Peninsular Malaysia where sedimentation of the coastal waters as a result of land clearing from unsustainable development has polluted the water (Pankor, 2004). Biusing (1997) attributed the severe decline in sea cucumber landings in Sabah to overfishing and the degradation of natural habitats resulting from destructive fishing methods and pollution.

The use of illegal fishing methods has often been documented in the Indo-Pacific region. Fishers who are desperately in need of a livelihood through catching fish and collecting sea cucumbers, have often resorted to the illegal use of explosives in fishing. This was commonly reported in the coastal waters of Palabusa, Uncume and Kapontori Bay in Indonesia (Asmedi, undated; Fox, 2005) and the Philippines (International Marinelife Alliance, 2003).

7.4 Lack of information and inaccurate resources statistics

The under-estimation of the Indonesian sea cucumber statistics due to the limited access of fisheries officers at sea cucumber landing sites have been reported by Tuwo (2004). He estimated that the Indonesian landings may be under-estimated by as much as 25 percent. Sometimes the origin of the fished stock is not clearly elucidated. The sea cucumber landings from Sabah are probably under-estimated due to the complexity of the fishery which has led to constraints in obtaining reliable catch and effort data (Biusing, 1997). In Viet Nam, Otero-Villanueva and Ut (2007) reported that sea cucumbers fished in Cambodian waters in the Phu Quoc Archipelago are usually sold in the Vietnamese side of the Phu Quoc Archipelago and recorded as domestic Vietnamese products, consequently giving the false impression that the Vietnamese side of the Archipelago still has a good sea cucumber supply.

7.5 Inaccurate trade statistics and confusion in commodity codes

In the absence of reliable catch statistics on some aquatic organisms, trade statistics may be used to provide useful information on the exploitation rates and an indicator of resource status. However, there is a tendency for trade figures to be under-reported, linked to reasons such as tax evasion. Moreover, like capture landings statistics, trade statistics are often incomplete or absent. When trading takes place across borders, sometimes no statistics are collected since the trading is clandestine. For instance, trepang trade in Thailand is mainly conducted across borders and does not pass through any customs checkpoint; hence no recent records are available from the Thai Fisheries Statistics Record (Bussarawit and Thongtham, 1999). The sea gypsies from Myanmar are known to sell their sea cucumber harvests to Thai traders in coastal towns such as Ranong and the statistics are therefore not captured in the Burmese fisheries statistics (Piprell, 2000).

The average bias toward under-reporting in the Asian trepang market was greater than 49 percent and there were large discrepancies between the custom's statistics of trepang from China Hong Kong SAR and China (Clarke, 2002). In China, imports of sea cucumber for processing are not included in the customs statistics, thus distorting the actual import quantities (Clarke, 2002). In China Hong Kong SAR, local traders admitted to frequent under–reporting and customs officials do not maintain any inspection programme to discourage this practice (Clarke, 2002).

The use of different commodity codes for the same type of products makes comparison of trade statistics difficult. Clarke (2002) reported that there was a lack of clear relationships between commodity codes between Taiwan Province of China and China Hong Kong SAR. Choo (2004) encountered difficulties with the Standard International Trade Classification (SITC) commodity codes used in Sabah, Malaysia. There were some distinct changes in the category code and the introduction of a new category "fit for human consumption" in 1996 which appeared to have replaced the category "other than fresh, chilled or frozen", hence creating some confusion in the trade statistics.

7.6 New commercial uses for sea cucumber

The consumption of trepang has been reported to be decreasing in the younger population in countries where they have been traditionally consumed (as has been documented in Japan and China Hong Kong SAR and reported by Akamine [2004] and Clarke [2002], respectively). The process of rehydrating trepang is a tedious task and there is a noticeable trend towards the purchase of rehydrated trepang among restaurants, and chefs mentioned that seafood products such as lobster, abalone and fish maw may have overtaken sea cucumber in popularity (Lo, 2004). Although the production of trepang appears to be decreasing (Clarke, 2002), the use of sea cucumbers for new products for medicinal and cosmetic purposes have gained popularity in recent years.

Information from Patent Storm[2] from 2004 to 2006 indicated that there are a number of patent applications by Chinese groups relating to sea cucumbers as nutritional supplements (e.g. CN 1065019). Applications from Japanese groups relate to various carbohydrate moieties from sea cucumber as anticoagulants (e.g. JP 94070085 B2; WO 9008784) and as active components for treating AIDS (e.g. WO 9202231; WO 9009181) while applications from the United States of America (e.g. US 5985330) relate to inhibition of angiogenesis by sea cucumber fractions.

The quantity used for medicinal and cosmetic preparations are poorly documented and the quantity exploited for the aquarium trade is also unclear. Akamine (2005) noted that *H. coluber*, known locally in the Philippines as "patola white" is exported to China to use as fertiliser. Reporting mechanisms should be standardised for sea cucumbers used for medicinal, cosmetic, agricultural and aquarium trade as well as for those used for food.

7.7 Threats of global warming/climate change

Information on the impacts of global warming on sea cucumber fisheries is limited. Global warming may affect sea cucumber fisheries by influencing the stocks and the global trepang supply. Coral reefs, which are important habitats for sea cucumbers may adversely be affected by global warming, and damage to coral reefs will in turn reduce recruitment of sea cucumbers. Higher sea temperature is a major cause of coral bleaching and damage to reef ecosystems. Changing sea temperature and current flows will likely bring shifts in the distribution of sea cucumber stocks with some areas benefiting while others incurring losses.

ACKNOWLEDGEMENTS

I would like to thank V. Toral-Granda, M. Vasconcellos and A. Lovatelli for inviting me to participate in this project and for their comments on the manuscript. I am very

[2] http://www.patentstorm.us/patents/5985330–description.html

grateful to K. Kiso for helping me procure publications and information from Japan. I wish also to thank K. Fukusho for his help with some Japanese translation. Thanks are due to N. Tracey and Dedi Adhuri for forwarding their publications to me, and to Jun Akamine, Kim Friedman, Jeffrey Kinch, Steven Purcell and Sven Uthicke for their comments on the manuscript.

REFERENCES

ACFOA. 2002. *Indonesian "fishers of men"*. Online downloaded 20 December, 2006. www.hrw.org/reports/2002/Australia/australia1202–04html

Adhuri, D.S. & Visser, L.E. 2007. Fishing in, fishing out: transboundary issues and the territorialization of blue space. Special Issue: Transboundary Environmental Issues in Insular Southeast Asia. *Asia–Pacific Forum* No. 36: 112–145. Center for Asia–Pacific Area Studies RCHSS, Academia Sinica, Taipei, Taiwan.

Agudo, N. 2006. *Sandfish hatchery techniques*. Australian Centre for International Agricultural Research (ACIAR), the Secretariat of the Pacific Community (SPC) and the WorldFish Center. 44p.

Akamine, J. 2001. Holothurian exploitation in the Philippines: continuities and discontinuities. *TROPICS* 10(4): 591–607.

Akamine, J. 2004. The status of the sea cucumber fisheries and trends in Japan: past and present. p. 39– 47. *Advances in sea cucumber aquaculture and management*. A. Lovatelli (comp./ed.); Conand, C; Purcell, S.; Uthicke, S.; Hamel, J–F; Mercier, A. (eds.). *FAO Fisheries Technical Paper.* No. 463. Rome, FAO.

Akamine, J. 2005. Role of the trepang traders in the depleting resource management: a Philippine case. In: Indigenous use and management of marine resources. Kishigami, N. and J.M. Savelle. *Senri Ethnological Studies,* 67: 259–278.

Anon. undated. CBD Strategy and Action Plan– DPR Korea (Part II, English version). Online, downloaded 22 November 2006. (http://72.14.209.104/search?q=cache:Umjuw4ZRUpEJ:www.biodiv.org/doc/world/kp/kp–nbsap–01–p2).

Anon. 1989. Japan's sea cucumber harvest and market – Foreign Fishery Developments. Marine Fisheries Review. Online, downloaded 19 November 2006. (http://www.findarticles.com/p/articles/mi_m3089/is_n4_v51/ai_9853197/print).

Anon. 2001. Sustainable management of sea cucumbers in Wakatobi Marine National Park. Online, downloaded January 8, 2007. (http://www.opwall.com/Library/Indonesia/Indonesia%20Marine/Fisheries/2001%20sea20cucumber%20report.htm).

Anon. (annual publications from 1990 to 2005). Annual Fisheries Statistics Sabah. Department of Fisheries Sabah, Kota Kinabalu, Sabah.

Anon. 2004. *Assessment of the Queensland East Coast Beche–de–mer Fishery*. Australian Government Department of the Environment and Heritage, Canberra, Australia. 25pp.

Anon. 2005. Visual Japan's Fisheries September 2005. Fisheries Agency, 1–2–1 Kasumigaseki, Chiyoda–ku, Tokyo 100–8907. (http://www.jfa.maff.go.jp/panf/visual%20japan%20fisheries2005.pdf).

Antara News. 2007. Canadian investor supports sea cucumber farming in west Seram. 8 February. (http://cempaka–marine.blogspot.com/2007/02/canadian–investor–supports–sea–cucumber).

Arakawa, K.Y. 1990. A handbook on the Japanese sea cucumber– its biology, propagation and utilization. (translated by Izumi, South Pacific Commission, Noumea, New Caledonia). Beche–de–mer Information Bulletin, number 3: 8–15, November 1991.

Aryono, B. 1987. *Penelitian tentang kepadatan teripang (Holothuroidea) di perairan karang pantai Bandengan dan perairan karang Pulau Panjang Timur, Jepara*. Fakultas Peternakan UNDP, Semarang.

Asmedi, R. (undated). Sea cucumber: a promising mainstay commodity. Oxfam GB, Indonesia Office. Online, downloaded 2 February 2007. (http://www.oxfamgb.org/eastasia/documents/Indonesia/sea_cucumber_eng.pdf).

Baine, M. & Choo, P.S. 1999. Sea cucumber fisheries and trade in Malaysia. p. 49–63. In: *The conservation of sea cucumbers in Malaysia: their taxonomy, ecology and trade. Proceedings of an International Conference,* 25 February 1999, Department of Agriculture, Kuala Lumpur, Malaysia. Baine, M. (ed.). Printed and published by Heriot–Watt University.

Bakus, G.J. 1973. The biology and ecology of tropical holothurians. p. 325–367. In: *Biology and Geology of Coral Reefs.* Jones, O. A. and Endean, R. (eds.). Academic Press, New York.

Basri. 1997. *Parameter populasi dan laju eksploitasi teripang gama* (Stichopus variegatus) *yang tertangkap oleh nelayan di Pulau Bangko–Bangkoang Kabupaten Pangkep.* Skripsi Fakultas Perikanan UNI, Makassar. 27 pp.

Battaglene, S.C. 1999. Culture of tropical sea cucumbers for stock restoration and enhancement. *NAGA the ICLARM Quarterly,* 22(4): 4–11.

Battaglene, S.C., Seymour, J.E. & Ramofafia, C. 1999. Survival and growth of cultured juvenile sea cucumbers *Holothuria scabra. Aquaculture,* 178: 293–322.

Best, B. 2003. Trade in coral reef animals, algae and products: an overview. p. 73–77. In: *Implications for coral reef management and policy: relevant findings from the 9th International Coral Reef Symposium.* B.A. Best, Pomeroy, R.S. and Balboa, C. M. (eds.). U.S. Agency for International Development, in collaboration with, the World Resources Institute, Conservation International, and the International Society for Reef Studies.

Biusing, R. 1997. Status of the sea cucumber fishery in Sabah. p. 19–35. In: The sea cucumbers of Malaysia: proceedings of a seminar held at the Borneo Marine Research Unit, Universiti Malaysia Sabah, Kota Kinabalu. Baine, M. (ed.). Darwin Initiative funded project. Project partners: International Center for Island Technology, Heriot–Watt University, Orkney, Scotland and Institut Penyelidikan Perikanan, Penang, Malaysia.

Boss, S.Z.R., Yasin, Z. & Tan, A.S.–H. 1999. Species of sea cucumber found in Pulau Besar, Johor, with special emphasis on the genus *Stichopus.* p. 37–41. In: *The conservation of sea cucumbers in Malaysia: their taxonomy, ecology and trade. Proceedings of an International Conference,* 25 February 1999, Department of Agriculture, Kuala Lumpur, Malaysia. Baine, M. (ed.). Printed and published by Heriot–Watt University.

Bruckner, A. 2005. Management and conservation strategies and practices for sea cucumber. In: *The Proceedings of the technical workshop on the conservation of sea cucumbers in the families Holothuridae and Stichopodidae.* Bruckner, A. (ed.). NOAA Technical Memorandum NMFS–OPR 44, Silver Spring, MD. 239 pp.

Bruckner, A.W., Johnson, K.A. & Field, J.D. 2003. Conservation strategies for sea cucumbers: can a CITES Appendix II listing promote sustainable international trade? *SPC Beche–de–Mer Information Bulletin,* 18: 24–33.

Bussarawit, S. & Thongtham, N. 1999. Sea cucumber fisheries and trade in Thailand. p. 26–36. In: *The conservation of sea cucumbers in Malaysia: their taxonomy, ecology and trade.* Proceedings of an International Conference, 25 February 1999, Department of Agriculture, Kuala Lumpur, Malaysia. M. Baine (ed.). Heriot–Watt University, Orkney Campus.

Butcher, J. 2004. *The closing of the frontier: a history of the marine fisheries of Southeast Asia c 1850–2000.* Institute of Southeast Asian Studies, Singapore. 442 p.

Chao, S.M., Chen, C.–P. & Alexander, P.S. 1995. Reproductive cycles of tropical sea cucumbers (Echinodermata: Holothuroidea) in southern Taiwan. *Marine Biology,* 122: 289–295.

Chao, S.M. 1998. *Shallow water sea cucumbers of Taiwan.* National Museum of Natural Science, Taichung, Taiwan. 170 p. (In Chinese).

Chen, J. 2003. Overview of sea cucumber farming and sea ranching practices in China. SPC *Beche–de–Mer Information Bulletin,* 18: 18–23.

Chen, J. 2004. Present status and prospects of sea cucumber industry in China. p. 25–38. In: *Advances in sea cucumber aquaculture and management.* A. Lovatelli (comp./ ed.); Conand, C; Purcell, S.; Uthicke, S.; Hamel, J–F; Mercier, A. (eds.). *FAO Fisheries Technical Paper.* No. 463. Rome, FAO.

Choe, S. 1963. Biology of the Japanese common sea cucumber *Stichopus japonicus* Selenka. Kaibundo, Tokyo.

Choo, P.S. 2004. Fisheries, trade and utilization of sea cucumbers in Malaysia. p. 57–68. In: *Advances in sea cucumber aquaculture and management*. Lovatelli, A.; Conand, C.; Purcell, S.; Uthicke, S.; Hamel, J.–F.; Mercier, A. (eds.). *FAO Fisheries Technical Paper.* No. 463. Rome, FAO.

Clarke, S. 2002. *Trade in Asian dried seafood: characterization, estimation and implications for conservation*. Wildlife Conservation Society Working Paper No. 22 (15 December, 2002).

Comley, J., Walker, R., Wilson, J., Ramsay, A., Smith, I. & Raines, P. 2004. *Malaysia coral reef conservation project: Pulau Redang. Report to the Department of marine Parks, Malaysia, March–September 2004.* Prepared by Coral Cay Conservation, December 2004.

Conand, C. 1990. The fishery resources of Pacific Island Countries, Part 2: Holothurians. *Food and Agriculture Organisation of the United Nations, Rome, Italy*, No. 272.2. Rome, FAO. 143p.

Conand, C. & Byrne, M. 1993. Recent evolution of the world fisheries for sea cucumbers. *Mar. Fish. Rev.*, 55(4): 1–13.

Dahuri, R. 2002. *Membangunan kembali perekonomian Indonesia melalui sector perikanan dan kelautan*. LIPSI, Jakarta. 187 pp.

Daud, R., Tangko, A.M., Mansyur, A. & Sudradjat, A. 1993. Polyculture of sea cucumber, *Holothuria scabra* and seaweed, *Eucheuma* sp. in Sopura Bay, Kolaka Regency, Southwest Sulawesi. In: *Prosiding Seminar Hasil Penelitian*, 11: 95–98.

Erdmann, M. 1995. An ABC guide to coral reef fisheries in Southwest Sulawesi, Indonesia. *NAGA The ICLARM Quarterly*, April 1995, 4–6.

Eyrika. 1999. *Pertumbuhan dan mortalitas teripang pasir* (Holothuria scabra Jaeger) *di Perairan Kecamatan Linkang Tupabbiring Kabupaten Pangkep. Skripsi FIKP Unhas, Makassar.* 35 pp.

Fan, H.Z. 1979. Study on the physical, chemical, toxin and anti-tumor characteristics of mucopolysaccharide. *Communication on Ocean Medicine*, 1: 61–67.

Ferdouse, F. 2004. World markets and trade flows of sea cucumber/beche-de-mer. p. 101–117. In: *Advances in sea cucumber aquaculture and management.* Lovatelli, A.; Conand, C; Purcell, S.; Uthicke, S.; Hamel, J–F; Mercier, A. (eds.). *FAO Fisheries Technical Paper.* No. 463. Rome, FAO.

Fish Information & Services. 2004. Illegal foreign vessels on the increase in local waters, Australia. July 21, 2004.

Flint, D.J. 2002. An assessment of the holothurian fishery in the proposed Kaledupa stakeholder zone (Wakatobi National Park Sulawesi, Indonesia) and the implications for the sustainable management of the fishery. Thesis submitted to the University of Plymouth in partial fulfilment of the requirement for the degree of M. Sc Applied Marine Science, University of Plymouth, Faculty of Science.

Forbes, R. & Ilias, Z. 1999. The taxonomy and ecology of sea cucumbers in Malaysia. p. 42–48. In: *The conservation of sea cucumbers in Malaysia: their taxonomy, ecology and trade.* Proceedings of an International Conference, 25 February 1999, Department of Agriculture, Kuala Lumpur, Malaysia. M. Baine (ed.). Heriot–Watt University, Orkney Campus.

Forbes, R., Ilias, Z., Baine, M., Choo, P.S. & Walbank, A. 1999. A taxonomic key and field guide to the sea cucumbers of Malaysia. Heriot–Watt University. 62 p.

Fox, J. 2002. *In a single generation: a lament for the forests and seas of Indonesia.* Paper prepared for EDEN II Workshop "Sustainability and depletion in island Southeast Asia: forests and fisheries, past and present". Leiden, 20–22 June 2002. Online, downloaded 23 March 2007.

Fox, J. 2005. Fishing resources and marine tenure: the problems of eastern Indonesian fishermen. Institute of Southeast Asian Studies, Singapore. (http://rspas.anu.edu.au/people/personal/foxxj_rspas/In_a_Single_Generation.pdf).

Fox, J.J. & Sen, S. 2002. *A study of socio–economic issue facing traditional Indonesian fishers who access the MOU box.* A Report for Environment Australia. 64 p.

Friedman, K. 2007. Personal communication. Secretariat of the Pacific Community, B.P. D5, 98848 Noumea Cedex, New Caledonia.

Gamboa, R., Gomez, A.L. & Nievales, M.F. 2004. The status of sea cucumber fishery and mariculture in the Philippines. p. 39–47. In: *Advances in sea cucumber aquaculture and management.* Lovatelli, A.; Conand, C.; Purcell, S.; Uthicke, S.; Hamel, J.–F.; Mercier, A. (eds.). *FAO Fisheries Technical Paper.* No. 463. Rome, FAO.

Gillett, R. 2004. The marine fisheries of Cambodia. *FAO/FishCode Review* No. 4. Rome, FAO. 57p.

Hamel, J.–F., Conand, C., Pawson, D.L. & Mercier, A. 2001. The sea cucumber *Holothuria scabra* (Holothuroidea: Echinodermata): its biology and exploitation as beche–de–mer. p. 129–223. In: *Advances in Marine Biology*, Vol. 41. A.J. Southward, Tyler, P.A., Young, C.M. and Fuiman, L.A. (series editors).

Handoko, D., Anindiastuti, S. & Kadek, A.W. 2002. Pemecahan beberapa kendala pada produksi benih teripang pasir (*Holothuria scabra*). *Buletin Budidaya Laut*, 14: 26–32.

Harborne, A., Fenner, D., Barnes, A., Bege, M., Harding, S. & Roxburgh, T. 2000. *Status report on the coral reefs of the east coastof Peninsular Malaysia.* Coral Cay Conservation Ltd.

Heinen, A. 2001. Resource management: between meal and market. *Samudra* April 2001: 13–18.

Hoeksema, B.W. 2004. Biodiversity and the natural resource management of coral reefs in Southeast Asia. p. 49–71. In: *Challenging coasts: transdisciplinary excursions into integrated coastal zone development.* Visser, L. E. (ed.). Mare Publication Series No. 1. Amsterdam University Press.

International Marinelife Alliance. 2003. *Strengthening and institutionalizing destructive fishing reform in the Philippines.* p. 19–26. In: The East Asia–Pacific coral reef conservation initiative 2002–2003, Progress Report February 2003. Honolulu.

James, D.B. 1996. Culture of sea cucumber. In: Artificial reefs and seafarming technologies. K. Rengarajan (ed). *Bulletin of the Central Marine Fisheries Research Institute*, 48: 120–126.

Jaquemet, S. & Conand, C. 1999. The beche-de-mer trade in 1995-96 and an assessment of exchanges between the main world markets. SPC *Beche-de-Mer Information Bulletin*, 12: 11–14.

Jeng, M.–S. 1998. Shallow–water Echinoderms of Taiping Island in the South China Sea. *Zoological Studies* 37(2): 137–153.

Kan–no, M. & Kijima, A. 2003. Genetic differentiation among three colour variants of Japanese sea cucumber *Stichopus japonicus. Fisheries Science,* 69: 806–812.

Kanno, M., Suyama, Y., Li, Q. & Kijima, A. 2006. Microsatellite analysis of Japanese sea cucumber, *Stichopus* (*Apostichopus*) *japonicus*, supports reproductive isolation in colour variants. *Mar. Biotechnol* (NY), 8(6): 672–685.

Kiso, K. 2007. Personal communication. Ishigaki Tropical Station, Seikai National Fisheries Research Institute, Fisheries Research Agency Fukai-Ohta 148–446. Ishigaki, Okinawa 907–0451, Japan.

Konstantinova, I. 2004. Aquaculture in the Russian Far East. International copyright, U.S. & Foreign Commercial Service and U.S. Department of State, 2005. Online, downloaded January 4, 2007. (http://strategis.ic.gc.ca/epic/internet/inimr–ri.nsf/en/gr124181e.html).

LeBlanc, S. 1997. Gleaning in Bais Bay: a case study on an informal sector coastal activity in the Philippines. Thesis submitted in partial fulfilment of the requirements for the degree of Master of Arts in International Development Studies at Saint Mary's University, Halifax, Nova Scotia, Canada. 105pp.

Li, X. 2004. Fishery and resource management of tropical sea cucumbers in the islands of the South China Sea. p. 261– 265. In: *Advances in sea cucumber aquaculture and management*. Lovatelli, A.; Conand, C; Purcell, S.; Uthicke, S.; Hamel, J–F; Mercier, A. (eds.). *FAO Fisheries Technical Paper.* No. 463. Rome, FAO.

Liao, Y.L. 1997. *Fauna Sinica, Phylum Echinodermata, Class Holothuroidea*. Science Press, Beijing. 334 pp.

Liu, Y., Li, F., Song, B., Sun, H. Zhang, X. & Gu, B. 1996. Study on aestivating habit of sea cucumber *Apostichopus japonicus* Selenka: ecological characteristics of aestivation. *Journal of Fishery Sciences of China* 3: 41–48 (in Chinese with English abstract).

Lo, T.H. 2004. Valuation of sea cucumber attributes through laddering. *SPC Beche–de–Mer Information Bulletin*, 20: 34–37.

Madeali, M.I., Tangko, A.M., Pantai, D.E.R. & Marcos, B.P.P.B. 1993. Polyculture of sea cucumber, *Holothuria scabra* and seaweed, *Eucheuma cottonii* in Battoa waters, Polmas Regency, South Sulawesi. *Prosiding Seminar Hasil Penelitian*, 11: 105–109.

Massin, C. 1999. Reef dwelling Holothuroidea (Echinodermata) of the Spermonde Archipelago (South–West Sulawesi, Indonesia). *Zoologisch Verhandelingen*, 329 pp.

Mitsunaga, N., Matsumura, Y., Watanabe, S. & Suzuki, H. 2006. Results of the year 2005 sea cucumber recovery plan. In: Annual Report of Nagasaki Prefectural Fisheries Experimental Station. April 2005–March 2006, p. 50–51.

Mitsukuri, K. 1903. Notes on the habits and life–history of *Stichopus japonicus* Selenka. *Annot. Zool. Jap.*, 5: 1–21.

Mortensen, T.H. 1934. Echinoderms of Hong Kong. *The Hong Kong Naturalist Supplement*, number 3 (February 1934). Online, downloaded 17 January, 2007. (http://209.85.165.104/search?q=cache:7SZVXah3gIQJ:sunzi1.lib.hku.hk/hkjo/view/27/).

Mucharin, A. & Putchakarn, S. 2005. New Thai Record: *Holothuria (Stauropora) discrepans* Semper, 1868. *The Thailand Natural History Museum Journal*, 1(1): 1–8.

New Straits Times. 2007. OIC nations among target marts for Luxor's gamat products. Business Times, New Straits Times, Malaysia. Saturday, May 5, 2007. p50.

NOAA. undated. *Sea cucumber workshop*. NOAA Fisheries: Ecosystem Assessment Division. Online, downloaded November 3, 2006. (http://www.nmfs.noaa.gov/habitat/ead/cucumberworkshop.htm).

Ohshima, H. 1934. Species and scientific names of edible sea cucumbers in Okinawa. Contribution, Dept. of Zoology, Kyusyu University, 67: 139–154. (in Japanese).

Ong Che, R.G. & Gomez, E.D. 1985. Reproductive periodicity of *Holothuria scabra* Jaeger at Calatagan, Batangas, Philippines. *Asian Marine Biology*, 2: 21–30.

Otero-Villanueva, M. & Ut, V.T. 2007. Sea cucumber fisheries around Phu Quoc Archipelago: a cross–border issue between South Viet Nam and Cambodia. SPC Bech–de–mer *Information Bulletin*, 25– February 2007. p. 32–36.

Pankor. 2004. Sea cucumber facts. Langkawi Magazine Nature Section. 16/08/2004. Online, downloaded 27 March 2004. (http://www.best–of–langkawi.com/NATURE–gamut–problem.php).

Patent Storm. 2004–6. Inhibition of angiogenesis by sea cucumber fractions. US patent issued on November 16, 1999. 2004–6 PatentStorm LLC.

Piprell, C. 2000. Endangered Idyll? *The Greater Phuket Magazine*. ArtAsia Press Co., Ltd. Online, downloaded 28 December, 2006. (http://www.phuketmagazine.com/html/PM%20Issues/Vol.14.7/Endangered_Idyll.htm).

Pitt, R. & Duy, N.D.Q. 2003. Breeding and culture of the sea cucumber *Holothuria scabra* in Viet Nam. *Aquaculture Asia*, 8(1): 36–37 & 39.

Pitt, R. & Duy, N.D.Q. 2004. Breeding and rearing of the sea cucumber *Holothuria scabra* in Viet Nam. p. 333–346. In: *Advances in sea cucumber aquaculture and management.* Lovatelli, A.; Conand, C; Purcell, S.; Uthicke, S.; Hamel, J–F; Mercier, A. (eds.). *FAO Fisheries Technical Paper.* No. 463. Rome, FAO.

Project Maje. 2004. *Myanmar "Sea Gypsies" Compendium.* June 2004. Online downloaded on 15 November, 2006. (http://www.projectmaje.org/gypsies.htm).

Purwati, P. 2005. Teripang Indonesia: komposisi jenis dan sejarah perikanan. *Oseana*, 30(2): 11–18.

Rowe, C. 1983. *A survey of fishermen's living condition and assessment of their development potentials in Olympia and dewy Islands in Bais Bay, Negros Oriental, Philippines.* Silliman University, Dumaguete City, Philippines.

Rowe, F.W.E. 1969. A review of the Family Holothuriidae (Holothuroidea: Aspidochirotida). *Bulletin of the British Musuem (Natural History) Zoology*, 18: 119–170.

Russel, B.C. & Vail, L.L. 1988. *Report on traditional Indonesian fishing activities at Ashmore Reef Nature Reserve.* NT Museum of Arts and Scuences Report, Darwin. 179 pp.

Sloan, N.A. 1984. Echinoderm fisheries of the world: A review. *In: Echinodermata (Proceedings of the Fifth International Echinoderm Conference).* P. 109–124. B.E. Keegan & B.D.S. O'Conner (eds.). A.A. Balkema Publishers, Rotterdam, the Netherlands.

Smith, L., Rees, M., Heyward, A. & Colquhoun, J. (undated). *Survey 2000: bêche-de-mer and trochus populations at Ashmore Reef.* Produced by Environment Australia. Australian Institute of marine Science. Online, downloaded on 13 September, 2007. (http://www.environment.gov.au/coasts/mpa/publications/pubs/ashmore–survey.pdf).

Stacey, N. 2000. Pearlers, planes and people of the sea: early Bajo voyages to the north Australian region. *Bulletin of the Australian Institute for Maritime Archaeology*, 24: 41–50.

Stacey, N. 2005. Crossing borders: implications of the Memorandum of Understanding on Bajo fishing activity in northern Australian waters. *The Beagle: Records of Museums and Arts Galleries of the NT Supplement 1*: 229–234.

Starmetro. 2007. Sanctuary for gamat created. M6 Community Wednesday 10 January 2007.

Sui, X. & Liao, Y. 1988. *Culture and enhancement of sea cucumber.* Monograph. China Agriculture Publishing House, Beijing, China (in Chinese).

Tan, R. 2003. *Online guide to Chek Jawa. Sea cucumbers: Class Holothuroidea.* (downloaded 12/20/2006). (http://www.wildsingapore.com/chekjawa/text/p630.htm).

Tan, R. 2005a. *Thorny sea cucumbers (Colochirus quadrangularis) on the shores of Singapore.* (downloaded 12/20/2006). (http://www.wildsingapore.com/wildshores// echinodermata/holothuroidea/colochirus.htm).

Tan, R. 2005b. Pink warty sea cucumbers (*Cercodemas anceps*) on the shores of Singapore. Online, downloaded on 12/20/2006. (http://www.wildsingapore.com/wildshores// echinodermata/holothuroidea/cercodemas).

Tebchalerm, S. 1984. *Edible invertebrates in the Southern part of Thailand.* M.Sc. Thesis, Chulalongkorn University. 418 p.

The Yomiuri Shimbun. 2006. 4 held over poaching sea cucumbers. online, downloaded 03/01/2007. (http://www.yomiuri.co.jp/dy/national/20061230TDY03002.htm).

Toonen, R. 2002. Aquarium invertebrates. *Advanced Aquarist's Online Magazine.* (downloaded 12/20/2006). (http://www.advancedaquarist.com/issues/nov2002/invert.htm), (http://www.cites.org/eng/cop/14/doc/E14–62.pdf).

Trinidad–Roa, M.J. 1987. Bêche–de–mer fishery in the Philippines. *Naga, The ICLARM Quarterly,* October 1987, 15–17.

Tuwo, A. 1999. Reproductive cycle of the holothurian *Holothuria scabra* in Saugi Island, Spermonde Archipelago, Southwest Sulawesi, Indonesia. In: *SPC Beche–de–Mer Information Bulletin*, 11: 9–12.

Tuwo, A. 2004. Status of sea cucumber fisheries and farming in Indonesia. p. 49–55. In: *Advances in sea cucumber aquaculture and management.* Lovatelli, A.; Conand, C; Purcell, S.; Uthicke, S.; Hamel, J.–F.; Mercier, A. (eds). *FAO Fisheries Technical Paper.* No. 463. Rome, FAO.

Tuwo, A. & Conand, C. 1992. Developments in beche–de–mer production in Indonesia during the last decade. SPC *Beche–de–Mer Information Bulletin*, 4: 2–5.

Uehara, M. 1991. Sea Cucumbers. p. 306–313. In: *Aquaculture in Tropical Areas*. Shokita, S. (editor–in–chief). (English version prepared by Yamaguchi, M). Midori Shobo Co. Ltd., Tokyo, Japan.

Uthicke, S. & Conand, C. 2005. Local examples of beche–de–mer overfishing: an initial summary and request for information. SPC *Beche–de–Mer Information Bulletin*, 21: 9– 14.

Uthicke, S., O'Hara, T.D. & Byrne, M. 2004. Species composition and molecular phylogeny of the Indo–Pacific teatfish (Echinodermata: Holothuroidea) bêche–de–mer fishery. *Marine and freshwater Research*, 55: 837–848.

Whitten, T., Mustafa, M. & Henderson, G.S. 2002. *The Ecology of Sulawesi.* (First Periplus Edition). Periplus Editions (HK) Ltd. 754 pp.

Xing, J. & Chia, F.–S. 1997. Heavy metal accumulation in tissue/organs of a sea cucumber, *Holothuria leucospilota*. *Hydrobiologia* 352: 17–23.

Yang, H., Yuan, X., Zhou, Y., Mao, Y, Zhang, T & Liu, Y. 2005. Effects of body size and water temperature on food consumption and growth in the sea cucumber *Apostichopus japonicus* (Selenka) with special reference to aestivation. *Aquaculture Research*, 36: 1085–1092.

Yang, H., Zhou, Y., Wang, J., Zhang, T., Wang, P., He, Y. & Zhang, F. 2001. A modelling estimation of carrying capacities for *Chlamys farreri, Laminaria japonicus* and *Apostichopus japonicus* in Sishiliwan Bay, Yantai. *Journal of Fishery Sciences of China*, 7: 27–31 (in Chinese with English abstract).

Yao, B., Hu, X., Bao, Z., Lu, W. & Hu, J. 2007. Genetic variation in two sea cucumber (*Apostichopus japonicus*) stocks revealed by ISSR markers. *Chinese Journal of Oceanography and Limnology*, 25(1): 91–96.

Yoshimura, H. 2006. Japan's food producers looking abroad for profits. Daily Yomiuri Online: The Daily Yomiuri. Online, downloaded December, 27, 2006. (http://www.yomiuri.co.jp/dy/national/20061227TDY04003.htm).

Yuan, X., Yang, H., Wang, L., Zhou, Y., Zhang, T. & Liu, Y. 2007. Effects of aestivation on the energy budget of sea cucumber *Apostichopus japonicus* (Selenka) (Echinodermata: Holothuroidea). *Acta Ecologica Sinica*, 7(8): 3155–3161.

Zaidnuddin, I. 2002. Sea cucumber species diversity and abundance from the south–western waters of Balik Pulau, Penang. p. 57–62. In: *Workshop on the impact of development on the coastal fisheries off south–west Penang Island.* Choo, P.S., I. Ismail, P.E. Chee and T.T. Chuah (eds). Fisheries Research Institute, Penang, Malaysia.

Zaidnuddin, I. & Forbes, R. 2000. Sea cucumbers of Pulau Payar, Kedah Marine Park Islands. p. 174–177. In: *Proceedings of National Symposium on Pulau Payar Marine Park.* A. Ali, M.A. Arshad, Z. Ilias and S.A. Mohd Sah (eds.). vi, 218p.

Zhang, B., Sun, D. & Wu, Y. 1995. Preliminary analysis on the feeding habit of *Apostichopus japonicus* in the rocky coast waters off Lingshan Island. *Marine Sciences*, 3: 11–13 (in Chinese with English abstract).

Zhang, F.Y. 1954. Edible Sea Cucumber in Xisha Archipelago. *Biology Bulletin*, 6: 21–22.

Zhang, F.Y. 1958. The preliminary report of aquaculture and sea ranching of sea cucumber (*Apostichopus japonicus*). *Journal of Zoology*, 2(2): 65–73.

APPENDIX I

Commercial sea cucumber species recorded from China, Taiwan Province of China, China Hong Kong SAR, Japan, Malaysia, Thailand, Indonesia, Philippines, Viet Nam, Singapore and Spratley Islands (species of commercial importance in each country are marked in red)

	China	Taiwan PC	China HK SAR	Japan	Malaysia	Thailand	Indonesia	Philippines [1]	Viet Nam	Singapore	Spratly
O: Aspidochirotida **F: Holothuriidae**											
Actinopyga lecanora	**x**	x			x	x	**x**	**x**		x	x
Actinopyga mauritiana	**x**	x		**x**	**x**	x	**x**	**x**			x
Actinopyga echinites	**x**	x			**x**	x	**x**	**x**			
Actinopyga miliaris	**x**				x	x	**x**	**x**			
Bohadschia argus	**x**	x		**x**	**x**	x	**x**	**x**	**x**		x
Bohadschia graeffei					**x**	x	x	**x**			x
Bohadschia marmorata	**x**	x			x	**x**	**x**	**x**	**x**		
Bohadschia similis							**x**				
Bohadschia vitiensis						x	x		x		
Bohadschia tenuissima							**x**				
Bohadschia bivittata				**x**					x		
Holothuria atra	**x**	x		**x**	**x**	**x**	**x**	**x**	**x**		
Holothuria coluber					x	x	**x**	**x**			
Holothuria rigida						x	**x**	**x**			
Holothuria pulla								**x**			
Holothuria edulis	**x**			**x**	**x**	**x**	**x**	**x**	**x**		
Holothuria pardalis	**x**	x	x		x	x	**x**				
Holothuria cinerascens	**x**	x				x					
Holothuria moebii	**x**	x	x		**x**	x					
Holothuria whitmaei	**x**	x			x	x	**x**	**x**	x		
Holothuria fuscogilva	**x**				**x**		**x**	**x**	x		
Holothuria difficilis		x				x		**x**			
Holothuria arenicola	**x**	x				x	x				
Holothuria hilla		x			x	x	**x**	**x**			
Holothuria impatiens	**x**	x			x	x	**x**				
Holothuria leucospilota	**x**	x	x		**x**	**x**	**x**	**x**	**x**		x
Holothuria pervicax	**x**	x			x		**x**				
Holothuria conusalba						x	**x**				
Holothuria scabra	**x**		x	**x**	**x**	**x**	**x**	**x**	**x**	x	
Holothuria scabra versicolor					x		**x**	**x**			
Holothuria similis							**x**				
Holothuria fuscopunctata					**x**	x	**x**	**x**	**x**		
Holothuria ocellata					x	x	**x**				
Holothuria fuscocinerea	**x**	x	x		**x**	x		**x**			
Holothuria vagabunda							**x**				
Holothuria vatiensis							**x**				
Pearsonothuria graeffei		x			x		**x**				
O: Aspidochirotida **F: Tichopodidae**											
Stichopus chloronotus	**x**			**x**	**x**	**x**	**x**	**x**	**x**		x
Stichopus sp.					**x**						
Stichopus horrens	**x**	x			**x**		**x**	**x**	x		
Stichopus herrmanni	**x**	x			**x**	**x**	**x**	**x**	**x**		
Stichopus vastus					x		**x**				
Stichopus quadrifaciatus							**x**				
Parastichopus nigripunctatus				**x**							
Thelenota ananas	**x**	x		**x**	**x**	**x**	**x**	**x**	**x**		x
Thelenota anax	**x**				**x**		**x**	**x**			x
Apostichopus japonicus	**x**			**x**							
O: Dendrochirotida **F: Cucumariidae**											
Mensamaria sp.	**x**										
Cucumaria frandosa japonica				**x**		x					
Pentacta quadragulis						x			**x**		
O: Moldavia **F: Caudinidae**											
Acaudina leucoprocta	**x**										
O: Molpadida **F: Molpadiidae**											
Paracaudina sp.					**x**						
No. of commercial species	**27**	**0**	**0**	**11**	**19**	**8**	**35**	**26**	**11**	**0**	0

[1] A more complete list for the Philippines is available in Table 1 in the hotspot paper (this document).

APPENDIX II

Dried sea cucumber price trend

Product	Price (USD/kg) 1 October 2003	Price (USD/kg) 17 October 2005	Price reference & market area	Origin
White teatfish (skin–on), *H. fuscogilva*				
3–5 pcs/kg (Grade A)	23	45	Southeast Asian ports	South Pacific
4–8 pcs/kg Grade B	13	35	Southeast Asian ports	South Pacific
Prickly redfish, *T. ananas*				
6–15 pcs/kg	15	–	Southeast Asian ports	South Pacific
6–20 pcs/kg	–	25	Southeast Asian ports	South Pacific
Black teatfish, *H. whitmaei*				
3–5 pcs/kg (Grade A)	18	–	Southeast Asian ports	Australia
Grade B	10	–	Southeast Asian ports	Australia
4–15 pcs/kg (Grade A)	–	30	Southeast Asian ports	South Pacific
Stonefish, *H. lecanora*			Singapore	Indonesia
10–30 pcs/kg	18	28	Southeast Asian ports	South Pacific
30–55 pcs/kg	–	20	Southeast Asian ports	South Pacific
Sandfish, *H. scabra*				
Grade A	48	–	Singapore	Indonesia
10–30 pcs/kg	56	–	Singapore	South Pacific
15–40 pcs/kg	40	–	Singapore	South Pacific
10–30 pcs/kg	–	66	Southeast Asian ports	South Pacific
30–55 pcs/kg	–	40	Southeast Asian ports	Indonesia
50–90 pcs/kg	–	30	Southeast Asian ports	Indonesia
Greenfish, *S. chloronotus*				
50–120 pcs/kg	25	–	Singapore	South Pacific
50–200 pcs/kg	–	35	Southeast Asian ports	South Pacific
Surf redfish, *A. mauritiana*				
15–35 pcs/kg	11	–	Singapore	South Pacific
15–30 pcs/kg	–	20	Southeast Asian ports	South Pacific
30–50 pcs/kg	–	15	Southeast Asian ports	South Pacific
Tigerfish, *B. argus*				
25–55 pcs/kg	3	–	Singapore	South Pacific
20–50 pcs/kg	–	10	Southeast Asian ports	South Pacific
Brown sandfish, *B. marmorata*	5	–	Singapore	South Pacific
25–100 pcs/kg	–	10	Southeast Asian ports	South Pacific
Curryfish, *S. variegatus*	19	–	Singapore	South Pacific
	–	20	Southeast Asian ports	South Pacific
Elephant trunkfish, *H. fuscopunctata*				
3–8 pcs/kg	5	–	Singapore	South Pacific
3–8 pcs/kg	–	15	Southeast Asian ports	South Pacific
Lollyfish, *H. atra*	1.50	–	Singapore	South Pacific
	–	3	Southeast Asian ports	South Pacific
Chinese Liaofish (farmed)				
50–100 pcs/kg	–	120	Wholesale China	China
100–250 pcs/kg	–	100	Wholesale China	China

Source: INFOFISH Trade News.

The Philippines: a hotspot of sea cucumber fisheries in Asia

Poh-Sze Choo
The WorldFish Center
Penang, Malaysia
E-mail: p.choo@cgiar.org

Choo, P.S. 2008. The Philippines: a hotspot of sea cucumber fisheries in Asia. In V. Toral-Granda, A. Lovatelli and M. Vasconcellos (eds). Sea cucumbers. A global review of fisheries and trade. *FAO Fisheries and Aquaculture Technical Paper*. No. 516. Rome, FAO. pp. 119–140.

SUMMARY

Commercial exploitation of sea cucumbers in the Philippines dates back to the late eighteenth century. Almost all the sea cucumbers harvested in the Philippines are processed into the dried form (trepang or bêche-de-mer) and exported predominantly to China Hong Kong Special Administrative Region (SAR), Singapore, Republic of Korea, Taiwan Province of China and Japan. There are no restrictions on the export of any species of sea cucumbers in the Philippines, even though some species have reached endangered status, nor is there a size restriction on any sea cucumber species for export.

The number of species harvested commercially has increased over the years. There were five commercial species in the 1900s, 24 in the 1980s and currently there are 33 species. Species utilized in the commercial trepang trade has evolved from traditionally high value, low volume species like teatfish and sandfish to the lower value high volume species. Presently the most valuable commercial species are *Holothuria fuscogilva, H. scabra, H. whitmaei, Stichopus chloronotus, S. herrmanni, S. horrens* and *Actinopyga* spp. Many anecdotal reports point to the overexploitation of sea cucumbers which have been confirmed in surveys carried out recently.

Sea cucumber landings and exports statistics from FAO can be confusing and need to be verified. When comparing the FAO data on annual volume produced from capture fishery with the volume of exports, the former is found to be lower (when it should be higher), implying that either the landings or the export data have not been accurately recorded. Acquiring accurate statistics is hindered by two issues: catches from the Philippines are lumped together as Holothuroidea and not separated into individual species; and species are called by various local names. A species can have more than one local name or one local name is shared by two or more species.

Fisheries management in the Philippines in general is regulated by the New Fisheries Code (Republic Act 8550). Currently there are no specific clauses directed at the management of sea cucumbers. The open access nature of the resource encourages overfishing, as leaving them behind means that someone else will collect them. Illegal, unregulated and unreported (IUU) fishing also poses problems. Poaching of fish and invertebrates by foreign fishers in Philippine waters has been documented. As sea cucumber resources become depleted, fishing by Filipino fishers in foreign waters, as well as in Philippine Marine Protected Areas (MPAs) has also become a problem. Although there are more than 150 sites in the Philippines with MPAs, most of these sanctuaries need improved vigilance to enforce the no-fishing boundaries. The lack of enforcement on the use of destructive fishing techniques also caused the destruction of corals and reef inhabitants like sea cucumbers. Various NGOs have established programmes on community-based management and co-management of natural resources in the Philippines, but not all have encountered success.

Currently, sea cucumbers are mainly harvested in one of three ways: commercial fishing targeting solely sea cucumbers; artisanal fishing for sea cucumbers as bycatch; and gleaning in intertidal reef flats during low tide. In many islands and coastal villages in the Philippines, income from sea cucumber fishery used to contribute a significant portion of a family's total income especially where the holothurians were abundant in the intertidal zones. However, current commercial sea cucumber populations in many shallow coastal areas have been reported to be overfished and income derived from sea cucumber gleaning has become less important. Commercial fishers who harvest sea cucumbers from the deeper areas derive a better income compared to those who glean in shallow coastal areas where the resources are already depleted.

Fishers in the commercial and artisanal sectors are recognized to belong to the formal work sector with their livelihood given considerable attention by the government. However, the plight of informal sector workers, like the sea cucumber gleaners, is often ignored and their livelihood given scant consideration.

The Philippine Council for Aquatic and Marine Research and Development (PCAMRD) and several individuals have recommended strategies for the conservation of sea cucumbers in the Philippines. A framework for a national sea cucumber management plan was outlined after the National Forum on Sea Cucumber Fisheries and Management held in June 2007.

1. BIOLOGICAL AND POPULATION STATUS

Commercial exploitation of sea cucumbers in the Philippines dates back to the late eighteenth century; during that period, the Sulu Sultanate was known to have prospered from bêche-de-mer or trepang[1] trade with Spain, Britain and China (Warren, 1985 cited in Akamine, 2001). This trade flourished in the South Pacific during the early nineteenth century and Manila was then an active entrepôt for trepang trade between the Pacific countries and China (Ward, 1972 cited in Akamine, 2001).

Seale (1911) reported the presence of 64 species of Holothurians in the Philippines, with many species bearing different genus names from those reported in the more recent literature, while Domantay (1957) mentioned 99 species and Schoppe (2000), 100 species. Tiu (1981) recorded the presence of 27 species of intertidal holothurians from Mactan and the neighbouring islands in central Philippines.

1.1 Current species in trade and their uses

Almost all the sea cucumbers harvested in the Philippines are processed into trepang and exported predominantly to China Hong Kong Special Administrative Region (SAR), Singapore, Republic of Korea, Taiwan Province of China and Japan (Gamboa, Gomez and Nievales, 2004). A small amount, however, of the harvested sea cucumbers used to be consumed in roasted form by traditional communities in the Busuanga group of islands (Seale, 1911). When sea cucumbers were abundant in Danao Bay, people often brought cooked rice, bananas or root crops to eat together with sea cucumbers found at the beach (Heinen, 2001). They are also consumed by men together with "tuba", a local alcohol (LeBlanc, 1997). The itinerant Boholano divers in Batangas consume some species fresh without the internal organs in salad dishes. The internal organs of

[1] The term trepang is used throughout this document as it is the term normally used in Philippines, Indonesia and other Asian countries.

some *Stichopus* species are fermented and consumed by some traditional communities (Trinidad-Roa, 1987). In the Island Garden City of Samal and nearby communities along Davao Gulf, the body wall of *Holothuria hilla, Stichopus horrens* and *Thelenota rubralineata* are blanched, sliced into bite size, and sold soaked in clean seawater. They are pickled in vinegar and garnished with garlic and onions and are in demand among the Muslims during the Ramadan season (Gamboa, Aurelio and Abreo, 2007). Sea cucumbers are also sold frozen in Davao in Mindanao province (Purcell, S., The WorldFish Center, New Caledonia, personal communication).

Traditionally, sea cucumbers have often been regarded as a functional food and are credited with curative powers to treat high blood pressure and muscular disorders. The cuvierian tubules are used as a crude plaster for minor wounds (Trinidad-Roa, 1987). Akamine (2005a) noted that a species of sea cucumber, *H. coluber*, locally known as "patola white" is reportedly no longer used as food for humans but instead is exported to China and used as a fertilizer.

Collection and processing of sea cucumbers have been reported in almost all the islands in the Philippine archipelago including: Bolinao, Bani and Alaminos in Pangasinan; San Fernando in La Union; San Vicente in Cagayan; Masinloc in Zambales; Polillo Island in Quezon; Calatagan in Batangas; Cebu; Negros Occidental; Surigao del Norte; South Cotabato; and Tawi-Tawi in Sulu (Trinidad-Roa, 1987). Sea cucumbers are also harvested in Mangsee Island in the southern part of the Palawan province and in the Spratly Islands (Akamine, 2001); in the islands of Apid, Digyo and Mahaba off

FIGURE 1
Map of the Philippines showing selected provinces

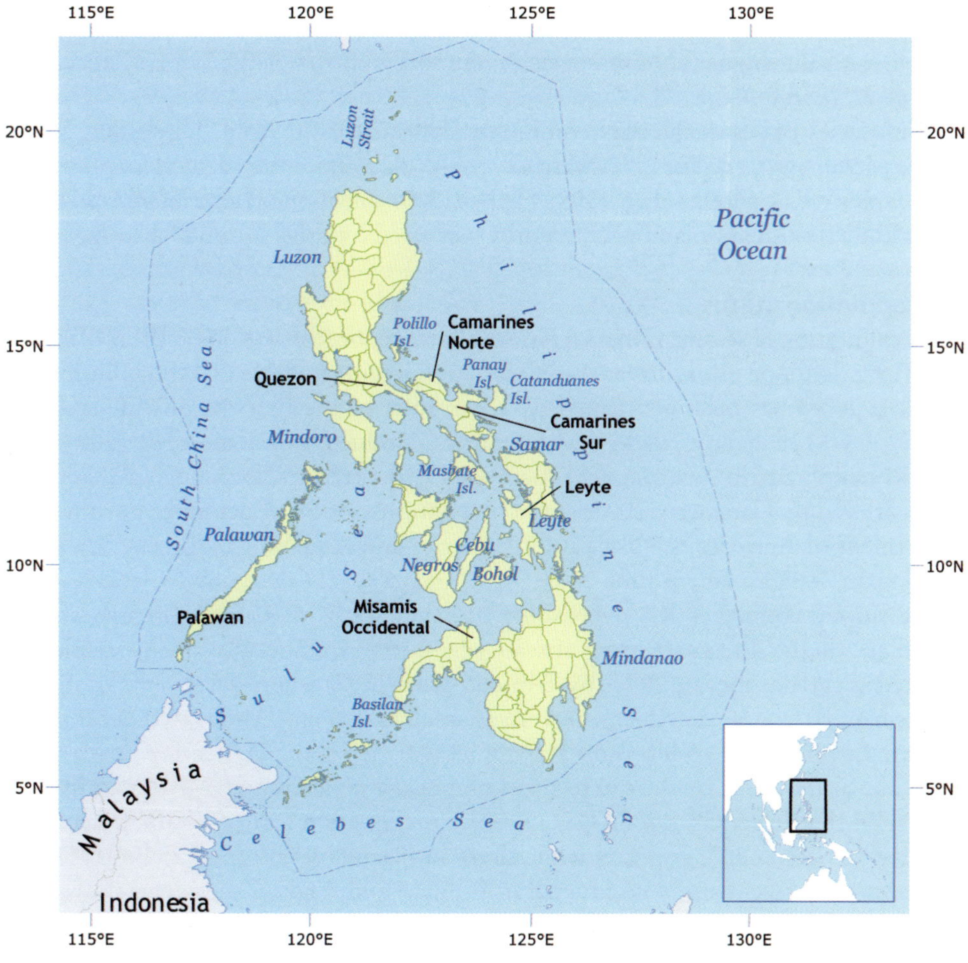

the shore of Inopacan, Leyte (Shoppe *et al.*, 1998) as well as in Cataban Island, central Philippines (Morgan and Panes, 2004). Heinen (2001) reported that sea cucumbers are harvested from Danao Bay located in the province of Misamis Occidental on the island of Mindanao, where a large part of the Bay belongs to the municipal waters of Baliangao. Subsistence and artisanal fisheries are also present in West Central Visayas (Nievales, 2007), Iligan Bay, northern Mindanao (Metillo, Tarranza IV and Arado, 2004) and Davao Gulf, southern Mindanao (Gamboa, Gomez and Nievalaes, 2004). During the 1900s, Seale (1911) noted that trepang supplies for Manila were mainly from Tacloban City in Leyte province, Polillo in Quezon province and Ambos Camarines province. In recent times Zamboanga City in western Mindanao and Puerto Princesa City in Palawan province are the largest entrepôts in the Philippines (Trinidad-Rao, 1987).

The number of commercial species has increased over the years. Seale (1911) listed only five commercial species traded in Manila in the 1900s, with *H. atra* being the most valuable. However, according to Kim Friedman (Secretariat of the Pacific Community, New Caledonia, personal communication), the species might have been wrongly identified as *H. atra* has always been a low value species. Trinidad-Roa (1987) listed a total of 16 commercial species from the Philippines, against 25 species described by Shoppe (2000), 24 species by Akamine (2005a) and 33 species documented in this report (Table 1). In the beginning of 2000s, the most valuable commercial species were *Holothuria fuscogilva*, *H. scabra*, *H. whitmaei*, *Stichopus chloronotus*, *S. herrmanni*, *S. horrens* and *Actinopyga* spp. in descending order of value (Akamine, 2001). Akamine (2001) noted that two unidentified species, known by their Makassar names "tacheritang" and "tundang" were considered as valuable as *H. fuscogilva* and *H. scabra*. Like in other countries in the Southeast Asian region, species utilized in the commercial trepang trade in the Philippines has evolved from traditionally high value, low volume species like teatfish (*H. fuscogilva; H. whitmaei*) and sandfish (*H. scabra*) to the lower value higher volume species, commonly known as the "worm" species (for example *H. leucospilota*).

Five species of sea cucumbers commonly collected from Apid, Digya and Mahaba include *Actinopyga echinites, Bohadschia argus, B. marmorata, Holothuria atria* and *Holothuria rigida* (Schoppe *et al.*, 1998). Heinen (2001) mentioned that *Holothuria difficilis*, locally known as "mani-mani" is commonly harvested by fishers from Danao Bay.

1.2 Population status

Overexploitation of sea cucumbers from anecdotal reports by various authors (Trinidad-Roa, 1987; Schoppe *et al.*, 1998; Akamine, 2001; Heinen, 2001; Gamboa, Gomez and Nievales, 2004) has been confirmed by the results of the first nationwide stock and fishery survey of shallow-water sea cucumbers conducted by Labe *et al.* (2007). Such status is mainly attributed to a very lucrative export trade. The same explanation was earlier offered by Ganchero (2007) at the first nationwide Sea Cucumber Forum held at Department of Agriculture – Bureau of Fisheries and Aquatic Resources (DA-BFAR) in Dagupan, Pangasinan, in June 2007.

The survey confirmed the overfishing of small-sized individuals (Figure 2) which existed in all the 14 study areas spread across the archipelago. Such practice was reported as early as the late 1980s by Trinidad-Roa (1987) where dried samples in Cebu, Pangasinan, Batangas, and Cagayan were as small as 4–5 cm. About two decades after, Metillo, Tarranza and Arado (2004) documented the same practice for northwestern Iligan Bay as indicated by the catch–effort and heavy collection of 5–10 cm individuals. They cited that fishers in Lopez Jaena spent a maximum effort of up to 30 man hours to collect ten commercial species while those in Plaridel spent up to 25 hours for only three. Another recent study conducted in Bolinao and Anda, Pangasinan showed that since 2004, the monthly average landings of *H. scabra* has been around 20 kilograms, way below the 109 kilograms peak in 2002 (Pastor, Catbagan and Ragos, 2007).

TABLE 1
Commercial species currently exploited in the Philippines

Latin name	Common name	No. specimens/kg (average dried weight of specimen in g)	Length of dried specimens (cm)	Size categories
Actinopyga lecanora	Buliq-buliq; Monang; Munang		7.6 6.4 2.5–6.4 <2.5	L M S XS
A. mauritiana	Bakungan; Monang		>7.6	L
A. obesa or *A. miliaris*	Khaki			
A. echinites	Hudhud; Brown beauty; Buli-buli; Khaki; Uwak			
Bohadschia sp.	Lawayan hongkong			
Bohadschia marmorata	Lawayan; Pulutan; Tagukan		10.2 6.4 <6.4	L M S
B. argus	Leopard; Matang Itik; Mat-anan			
B. graeffei	Bulaklak, trompa, pinya			
Holothuria sp.	White beauty			
Holothuria sp. (black colour)	Patola black			
H. scabra	Putian; Cortido; Curtido; Kagisan	15 (67) 20 (50) 40 (25) 60 (17) 80 (13)		XL L M S XS
H. scabra var. *versicolor*	Curtido Bato			
H. fuscogilva	Susuan	3–4 (250–333) 5–6 (167–200) 7–8 (125–143) 8–10 (100–125) 11–15 (67–91)		XL L M S XS
H. whitmaei	Bakungan; Kagisan; sus-an	5–6 (167–200) 7–8 (125–143) 8–10 (100–125 g) 11–15 (67–91 g)		L M S XS
H. atra	Black beauty, Mani		>12.7 10.2–12.7 5.1–10.2	L M S
H. rigida, H. inhabilis	Batunan (fresh for local consumption), Bat-onan			
H. pulla	Patola red			
H. coluber	Patola white; Tambor; Bat-uwak			
H. edulis	Red beauty; Red-black; Hotdog			
H. leucospilota	Patola			
H. fuscopunctata	Sapatos			
H. difficilis	Mani-mani			
H. hilla	Mani-mani			
H. fuscocinera	Labuyuq			
Stichopus horrens, S. herrmanni	Hanginan; "loaf bread"		>7.9 6.4–7.6 5.1–6.4 <5.1	L M S XS
S. chloronotus	Katro kantos			
T. ananas	Tinikan; Pinya-pinya			
T. anax, T. rubralineata	Legs, paag daga			
Neocucumis proteus	Bola-bola			

Source: Schoppe *et al.*, 1998; Schoppe, 2000; Akamine, 2005b; Gamboa, R., University of the Philippines, Mindanao, personal communication.

FIGURE 2
Harvesting of small sea cucumbers in Mindanao

COURTESY OF S. PURCELL

Other local examples of exploitation of the fishery are as follows. According to Heinen (2001), the collapse of the fishery in Danao Bay had become "the symbol of mismanagement and conflict between meal and market" where despite the ban order issued by the mayor, poaching continued. Solandt *et al.* (2001) recorded extremely low temporal and spatial abundance of commercial holothurians in Danjugan Island, Negros Occidental, and did not find a single sea cucumber in North Wall and Hilary's Rock during a 2-year survey. Akamine (2001) believed that the more than 30 years of exploitation in Mangsee Island has resulted in local women no longer involved in fishing and men resorting to hookah diving in search of species in deeper areas.

1.3 Catches

1.3.1 Total catch

The FAO statistics on Philippine's sea cucumber landings and exports can be confusing and need to be verified. A 15-year data on catches and export shows a sharp peak and decline in export quantity in 1985–86 but without corresponding changes in catch volume (Figure 3). The highest catch volume recorded at around 4 000 tonnes in 1990 showed a declining rate thereafter and stayed in a plateau at less than 1 000 tonnes for the last ten years.

Acquiring accurate statistics is hindered by two issues. First, catches from the Philippines are lumped together as Holothuroidea. They are not segregated by species. Furthermore, the country to date has no commercial hatchery for sea cucumbers, thus, all export products are from capture fishery. When comparing the FAO annual volume data derived from capture fishery with the export volumes (see Figure 3), the former is found to be lower (when it should be higher), implying that either the landings or the export data have not been accurately recorded. Assuming that the annual domestic consumption is minimal, and adopting the commonly used wet:dried weight

FIGURE 2
Sea cucumber (Holothuroidea) capture production and trepang exports from the Philippines

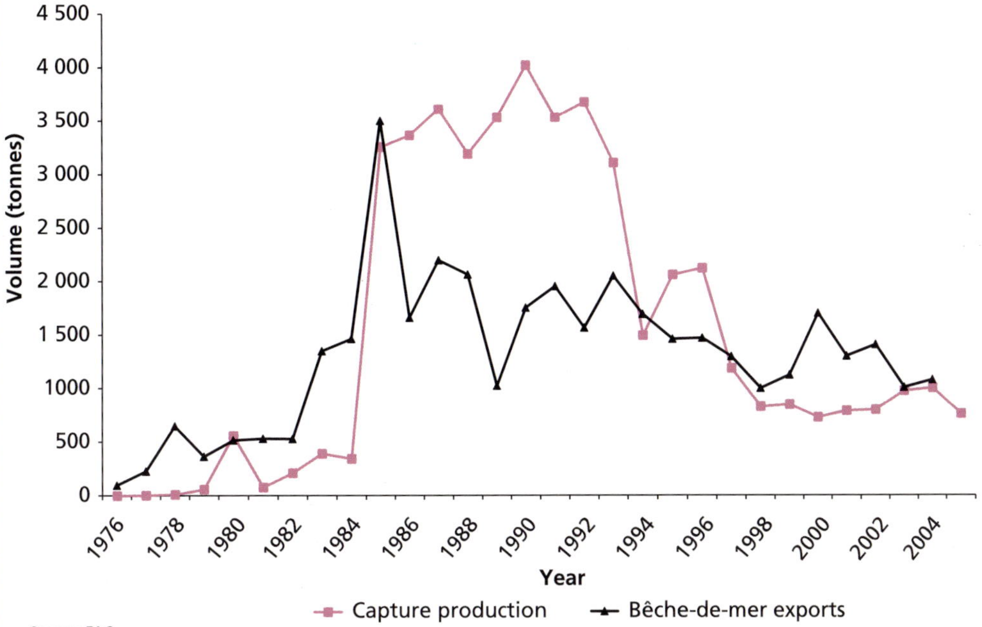

Source: FAO.

conversion factor of 3:1 (FAO, 2000), capture production would correspond roughly to three times the amount of trepang export. However, Kim Friedman (Secretariat of the Pacific Community, New Caledonia, personal communication) commented that very few species give a 30 percent return while most sea cucumbers yield 5 to 12 percent of their wet weight when processed. Interestingly, Ganchero (2007) noted that the Philippine volume-price data obtained from Singapore are much higher than that in national reports indicating a under declaration in the latter.

Second, species are called by various local names. The commercial species shown in Table 1 are difficult to rank. Confusion in taxonomy, as experienced by Gamboa, Gomez and Nievales (2004), is easily apparent in that a species can have more than one local name or one local name is shared by two or more species. Aside from looking at the population status, the nationwide assessment undertaken by the National Fisheries Research and Development Institute (NFRDI) from January 2006 to July 2007 (Labe *et al.*, 2007) was also to address lack of knowledge on nomenclature and taxonomy.

1.3.2 Types of fishing and harvesting techniques

Seale (1911) described the historical harvesting of sea cucumbers by fishers who usually went out at low tide, wading in the shallow waters, dragging a small canoe or "banca" into which they threw all the sea cucumbers which they picked up. Sometimes the fishers harvested from a boat using a long spear with which they gathered the trepang in waters of 3–4 m.

Currently, sea cucumbers are mainly harvested in one of three ways described below: commercial fishing targeting solely on sea cucumbers; artisanal fishing for sea cucumbers as bycatch; and gleaning in intertidal reef flats during low tide.

Sea cucumbers collected through commercial and artisanal fishing are carried out by skin diving or diving with the use of air compressor (Figure 4). Statistics are, however, not available to differentiate the volume of catch resulting from the different forms of sea cucumber harvesting.

FIGURE 4
Woman diver from Mindanao collecting sea cucumbers

COURTESY OF R. GAMBOA

Commercial fishing
The commercial fishers are more organized than the artisanal fishers and gleaners, and they gather sea cucumbers exclusively for direct sales to middlemen (Gamboa, Gomez and Nievales, 2004). Commercial fishing for sea cucumbers in the deeper waters involves the use of motorized vessels (39-foot length keel with a 56 hp engine) which are equipped with echo sounders to locate good habitat for sea cucumbers and have 10–15 crew on board (Akamine, 2001). Fishers spent several weeks out at sea and dived for sea cucumbers with the use of hookah. Ninety-five percent of fishers from Mangsee Island in Palawan were ethnic Bajaus and they usually dived for sea cucumbers in the Spratly Islands.

Akamine (2001) recorded at least 15 commercial vessels fishing in the Spratly Islands in 1998 but in 2000, only a few vessels were still operating. Boat owners advanced the cash for all necessary expenses and deducted them from the catch and divided the net profit among the crew. The former, as boat owner, generally qualified for three shares (for owning boat, engine and echo sounder) and net profit, and handled all transactions without sharing details with the divers (Akamine, 2001).

Harvesting sea cucumbers as bycatch
Harvesting sea cucumbers as bycatch is carried out by traditional fishers whose target fisheries are mainly fish and molluscs. Sea cucumbers encountered along the path are collected. The fishers normally go out alone or in groups of 2–3 and fish by skin diving or with the use of air compressors. Akamine (2001) noted that the fishers may fish early in the morning for sea cucumbers and in the afternoon harvest different kinds of fish and molluscs. In Barangay Bato, Davao del Sur, divers generally fish between 7 pm and 4 am (Gamboa and Junio-Menez, 2007).

Gleaning
Sea cucumbers are also collected by small-scale or artisanal fishers, involving not only men but women as well as children. This activity is carried out during low tide in shallow intertidal reef flats. Gleaning is often classified as "informal" work[2] acting as a safety net for the rural landless (LeBlanc, 1997). Gleaning of sea cucumbers is becoming more and more uneconomical as a source of livelihood because of the decades of uncontrolled overfishing in shallow reef flats that has led to localized depletion of the resource.

1.3.3 Illegal, unreported and unregulated fishing
Poaching of fish and invertebrates by fishers from neighbouring countries in the Philippine coastal waters has often been highlighted as a problem. Illegal fishing in the 1980s by fishers from Malaysia and the Sulu Archipelago in Danao Bay had been

[2] Informal work has been defined by the International Labour Office (cited in LeBlanc, 1997) as "a way of doing things" characterized by: ease of entry; reliance on indigenous resources; family ownership; small-scale operation; labour-intensive and adapted technology; skills acquired outside the formal school systems and unregulated and competitive market.

documented to harvest sea cucumbers by the boatload (Heinen, 2001). Since 2006, 600 Chinese nationals have been arrested for illegal fishing in Philippine waters (The Manila Times, 2007). In 2002, 95 Chinese nationals were detained for poaching 10 sacks of dried sea cucumbers from the Tubbataha Reef Marine Park (a World Heritage Site) off Palawan province (Jimeno, 2007). In December 2006, 32 Chinese, crewmen of the fishing vessel Hoi Wan were arrested by Philippine authorities when their vessel was found loaded with live fish and dried sea cucumbers (The Daily Tribune, 2007).

As sea cucumber resources become depleted in Philippine waters, illegal fishing by Filipino fishers in foreign territorial waters such as in Malaysia, has also been documented (Akamine, 2001), and widespread poaching in Marine Protected Areas (MPAs) in the Philippines had also been noted (Uthicke and Conand, 2005). In 1998, Heinen (2001) documented that large quantities of sea cucumbers were poached from Danao Bay sanctuary by fishers from Tugas village in the Baliangao municipality, landing sacks full of sandfish.

The proliferation of non-Filipino middlemen, usually married to Filipino women, in the last decade was confirmed at the national forum in June 2007 by the trader participants. The traders regard those non-Filipinos as formidable competitors both in terms of financial capital as well as access to domestic and international networks. For example, in Davao City there are at least two foreign traders: a Korean and a Japanese. While both are basically traders of seafood products they are also middlemen and direct exporters of trepang (Gamboa and Junio-Menez, 2007).

1.4 Management of the fishery

Fisheries management in the Philippines in general is regulated by the New Fisheries Code (Republic Act 8550). Currently there are no specific clauses directed at the management of sea cucumbers, although recommendations specific to its management were suggested as early as the late 1980s and early 1990s (see Trinidad-Roa, 1987 and PCAMRD, 1992). For further detail on these recommendations see Section 5.

There are no regulations or monitoring of the sea cucumber fishing activities at the local level and no taxation on the landings; the only payment required is a fee to operate a motorized boat (Gamboa, Gomez and Nievales, 2004). The open access nature of the resource also encourages overfishing, as leaving them behind means that someone else will collect them.

The Philippines is perhaps one of the most advanced countries in the East Asian region with respect to its appreciation and practice of coastal management as a strategy to address problems in the coastal zones (Jacinto *et al.*, 2000). Since the mid-1980s, there have been at least 45 coastal management programmes and projects involving 150 sites all over the Philippines, with MPAs incorporated in many of these plans (Jacinto *et al.*, 2000). Regulations in MPAs include no-fishing zones, regulated fishing zones or a combination of these, as well as open- and closed-season fishing zones (Jacinto *et al.*, 2000). However, enforcement of the regulations is best accomplished in smaller MPAs while the larger ones have only moderate enforcement. A study which surveyed eight MPA sites in Bohol and one in Siquijor in 2003 found a relative lack of sea cucumbers in the study sites reflecting heavy collecting (White *et al.*, 2003). They suggested that the sanctuaries surveyed needed improved vigilance to enforce the no-fishing boundaries.

According to Trinidad-Roa (1987), sea cucumber fishery in the Philippines is uncontrolled and non-selective; if the fishery is to survive, there is an obvious need for appropriate and realistic guidelines to regulate it. Under the Local Government Code, coastal municipalities are empowered to develop their own coastal development plans, but these plans face huge challenges in their implementation due to: (i) the lack of political will in prosecuting illegal fishers; (ii) the lack of support from the community; and, (iii) a shortage of funds for enforcement and where marine police operate on a volunteer basis (Gamboa, Gomez and Nievales, 2004).

Various Non-governmental Organizations (NGOs) have established programmes on community-based management and co-management (working with coastal municipalities) of natural resources in the Philippines with various degrees of success. These initiatives include those in the island of Apo (Negros Oriental province), Balicasag and Pamilacan (Bohol province) and San Salvador (Zambales province) which were established and maintained by organized communities, and are now recognized by coastal resource managers worldwide for their success in using the community-based coastal resource management approach (Oneocean, undated). The Pipuli Foundation formed a community-based management system in 1991 with the establishment of a sanctuary or no-take-zone for the conservation of sea cucumbers and other aquatic organisms in Danao Bay. In that year, local fisher leaders and the NGO began to work on a written resource management plan (Heinen, 2001). The ordinance was well accepted and embraced by the municipal councilors of Plaridel, one out of the six communities around the Bay. In another example, the co-management system for coastal resource management in Banate Bay which was established in 1996 proved very successful, and in 1998, received the Galing Pook national award as one of the most successful development oriented rural organizations in the Philippines (Fernandez Jr., Matsuda and Subade, 2000).

In Sagay, Negros Occidental, the Presidential Proclamation that declared its Marine Reserve was elevated into a Republic Act (law) which gave the local stakeholders more power to implement sustainable fisheries (Gamboa, R., University of the Philippines in Mindanao, Philippines, personal communication). The reserve is administered by a multisectoral Protected Area Management Board (PAMB). It boasts two national awards: the 1997 Gawad Galing Pook Award for innovation and replicability and the 2003 Best Aquatic Resource Management Award. While several species of sea cucumber can be found in the reserve, the fishery is monospecific on *Neocucumis proteus* (*bola-bola*) (Cucumariidae). The regulated harvesting and monitoring of this species is an example of co-management even prior to acquiring scientific baseline information. However, the administrators were quick to realize that handicap and are now making collaboration with academic and research institutions (Dacles, 2007).

Fishers in the commercial and artisanal sectors belong to the formal work sector with their livelihood given considerable attention by the government. However, the plight of informal sector workers, like the sea cucumber gleaners, is often ignored and their livelihood given scant consideration. LeBlanc (1997) reported that the implementation of the Philippines Local Government Code in 1991 provided for changes in Bais Bay use and access patterns, with many gleaners deprived of their gathering activities in intertidal areas which were converted to mussel and oyster farms.

2. TRADE

2.1 Main export destination, trade route, illegal trade

The Philippines is one of the important sea cucumber fishing countries that have no pertinent national management or regulatory measures for sea cucumbers (Labe, 2007). Consequently, there are no restrictions on the export of any species of sea cucumbers even though some species have reached endangered status (although not officially declared as such), nor is there a size restriction on any sea cucumber species for export. All export consignment conforming to trade regulations are therefore allowed to leave the country legally. Philippines is a signatory to the World Trade Organization (WTO) but according to Salayo (2000) fish and its products are not classified as sensitive agricultural products and therefore do not fall under the WTO Agreement on agriculture. However, fishery trade is covered by the general rules of the General Agreement on Tariffs and Trade (GATT).

FIGURE 5
Trade relationship between the different suppliers and buyers in the sea cucumber trade in the Philippines

China Hong Kong SAR is the major direct importer of Philippine trepang. There, the products are either further processed or repackaged after which almost half are "exported" to China at higher prices. The Republic of Korea ranks second and Singapore third while North American and Western European countries with sizeable Oriental populations are minor markets (Ganchero, 2007). For the trade route, Akamine (2001, 2005a) noted that in the early 2000s there were four major trepang exporters, all of whom operated from Manila, and the exporters were either ethnic Chinese or who had a spouse with Philippine citizenship. Apart from sea cucumber, these exporters also traded other dried marine products like sharkfin and dried sea horse whose main market is China.

Akamine (2001, 2005a) indicated that wholesale dealers often classify trepang into five categories: (i) those that can be stored without further treatment; (ii) those that require a half day of sun drying; (iii) those that require more than one day of sun drying; (iv) those that require smoking; and (v) those that require further treatment. Further treatment includes procedures which require rinsing away salt, removing odour, spoiled meat, scorched parts, sand in the body, and layer of lime used in processing.

Figure 5 shows the relationship among the major importing countries and the various suppliers, buyers and fishers in the sea cucumber supply chain in the Philippines. The recent visible proliferation of non-Filipino middlemen in key cities outside Metro Manila poses strong competition among local middlemen (Gamboa, R., University of the Philippines in Mindanao, Philippines, personal communication).

Given the above flow of sea cucumber trading, some big-time middlemen finance local contacts creating more competition among local traders (Gomez, 2007). This can potentially drop the buying price making the local fishers the real losers. Gomez (2007) recommended examining the value chain and identifying activities critical to buyer satisfaction or market success. This can serve as guide in improving the value of the sea cucumbers especially for local stakeholders.

Table 2 compares the volume and value of the Philippines exports to the world exports of sea cucumbers. The value index, calculated as the ratio between the percent value and percent volume of exports from the Philippines compared to the world exports, provides an indication of the relative value of the exported products. For instance, the indices calculated from 1983 to 2004 are well below one, indicating that the trepang exported are low value species or the processed products are of low quality. In contrast, Sri Lanka[3], which exports a small volume of higher value products, have value indices for 2003 and 2004 of 2.45 and 0.99 respectively. These statistics form another warning sign for the need to put in place national regulations to increase the economic benefits from sustainable sea cucumber fisheries.

[3] Sri Lanka exported 2 tonnes (USD 31 000) of sea cucumbers in 2003 and 36 tonnes (USD 103 000) in 2004.

TABLE 2
Export volume and value of sea cucumber (dried, salted or in brine) from the Philippines

Year	Philippines		World		Percentage world production		Value index
	Volume (tonnes)	Value (USD x1 000)	Volume (tonnes)	Value (USD x1 000)	Volume	Value	% Value / % Volume
1983	1 349	2 008	5 116	10 651	26.37	18.85	0.72
1984	1 463	1 145	6 214	11 648	23.54	9.83	0.42
1985	3 499	2 993	13 040	19 854	26.83	15.08	0.56
1986	1 659	1 772	10 787	18 947	15.38	9.35	0.61
1987	2 195	2 532	12 179	24 770	18.02	10.22	0.57
1988	2 062	2 364	12 481	36 397	16.52	6.50	0.39
1989	1 022	1 465	5 372	20 310	19.03	7.21	0.38
1990	1 752	3 253	7 661	30 145	22.87	10.79	0.47
1991	1 952	3 560	9 292	40 991	21.01	8.69	0.41
1992	1 609	3 216	10 350	41 045	15.55	7.84	0.51
1993	2 049	3 986	23 216	115 386	8.83	3.45	0.39
1994	1 692	4 120	22 688	123 895	7.46	3.33	0.45
1995	1 459	4 803	17 939	87 006	8.13	5.52	0.68
1996	1 469	4 827	21 383	120 925	6.87	3.99	0.58
1997	1 297	4 505	18 776	103 636	6.91	4.35	0.63
1998	NA	NA	–	–	–	–	–
1999	1 125	3 653	17 850	85 774	6.30	4.26	0.68
2000	NA	NA	–	–	–	–	–
2001	NA	NA	–	–	–	–	–
2002	1 407	2 386	4 420	14 642	31.8	16.30	0.51
2003	1 009	2 095	3 264	14 715	30.91	14.23	0.46
2004	1 079	2 176	4 928	21 080	21.90	10.32	0.47

Source: Landings in volume and values obtained from FAO.

2.2 Prices to fishers and exporters

Trinidad-Roa (1987) reported that fresh animals harvested in the 1980s were sold by fishers to middlemen with prices varying from 0.50 pesos (USD 0.01) to 1.50 pesos (USD 0.03) per piece for less preferred species like *H. atra* and *S. chloronotus* (however, currently *S. chloronotus* is sold for USD 90/kg in Japan according to Jun Akamine (Nagoya City University, Japan, personal communication), while more preferred species like *H. whitmaei, T. ananas* and *A. mauritiana* were sold between 7 pesos (USD 0.15) to 15 pesos (USD 0.32) per piece; for *A. echinites* and *A. miliaris*, the middle price range was 3 pesos (USD 0.06) to 5 pesos (USD 0.11) per piece. LeBlanc (1997) reported that gleaners sold Balat-bagisan or *H. scabra* at 5 pesos (USD 0.11) per kg, while Batuli or *H. leucospilota* were sold by gleaners for 0.10 pesos (USD 0.002) per piece.

Sea cucumbers are processed into trepang usually by middlemen who then sell them to wholesalers who would grade the products. High value trepang is categorized according to individual weight estimated by hand or according to length as measured against the middle finger (Akamine, 2001). Low value species are not categorized into weight or length classes. Middlemen and traders have often been considered as the main beneficiaries in the sea cucumber trade at the expense of the fishers. Most recently, good quality *H. scabra* in China Hong Kong SAR can fetch up to USD 110/kg (Ganchero, 2007). This is a specific point where value chain analysis, as pointed out by Gomez (2007), can help fishers optimize their share of profit in this international trade.

Fishers are not well compensated for their catches. In the 1900s, fishers received a price which was 30 percent lower than the retail price of sea cucumbers in Manila (Seale, 1911), whilst Gamboa, Gomez and Nievales (2004) reported that fishers received only 12–20 percent of the export price when they sold *H. scabra* bought at 187 pesos/kg (USD 4/kg) and then exported at 888 pesos/kg (USD 19/kg) to 1 075 pesos/kg (USD 23/kg).

Wholesale prices of almost all species of sea cucumbers have increased, especially the high value species. Akamine (2005a) noted that *H. scabra* has shown a steady increase in price in Puerto Princesa City over the years between 1998 to 2002, but prices of low

TABLE 3
Wholesale prices of trepang in Puerto Princesa City from 1998 to 2002

Species	Size: Number of specimens/kg (mean dried weight of specimen)	Price (USD/kg)				
		1998	1999	2000	2001	2002
H. scabra	XL: 15 (67 g)	29.7	35.0	36.7	37.3	40.4
	L: 20 (50 g)	22.8	27.5	31.1	29.4	36.5
	M: 40 (25 g)	16.0	18.8	24.4	21.6	23.1
	S: 60 (17 g)	9.1	11.3	16.7	13.7	15.4
	XS: 80 (13 g)	6.9	8.8	12.2	12.7	13.5
H. fuscogilva	XL: 3–4 (250–333 g)	21.7	30.0	35.6	35.3	35.6
	L: 5–6 (167–200 g)	20.5	27.5	34.4	33.3	34.6
	M: 7–8 (125–143 g)	17.1	22.5	26.7	21.6	24.0
	S: 8–10 (100–125 g)	12.6	15.0	17.8	15.7	16.3
	XS: 11–15 (67–91 g)	9.1	12.5	12.4	9.8	10.6
S. horrens	L: >7.9 cm	12.6	20.0	21.1	21.6	28.8
S. herrmanni	M: 6.4–7.6 cm	9.1	12.5	14.4	15.7	20.2
	S: 5.1–6.4 cm	6.9	10.0	11.6	11.8	14.4
	XS: <5.1 cm	4.1	6.3	6.7	5.9	6.7
Actinopyga spp.	L: >7.6 cm	14.8	20.0	24.4	21.6	25.4
	M: 6.4 cm	10.3	13.8	15.6	15.7	19.6
	S: 2.5–6.4 cm	8.0	11.3	11.6	11.8	12.5
	XS: <2.5 cm	5.7	10.0	6.7	7.8	9.6
H. whitmaei	L: 5–6 (167–200 g)	14.8	17.5	26.7	19.6	23.1
	M: 7–8 (125–143 g)	12.6	15.0	22.2	17.6	19.2
	S: 8–10 (100–125 g)	10.3	11.3	17.8	13.7	15.4
	XS: 11–15 (67–91 g)	9.1	8.8	11.1	9.8	9.6
A. mauritiana	L: >7.6 cm	8.2	11.3	14.4	12.7	13.5
	M: 6.4–7.6 cm	5.0	7.5	11.1	8.8	8.7
	S: 3.8–6.4 cm	3.7	6.3	8.0	5.5	5.4
	XS: 2.5–3.8 cm	2.3	3.0	4.0	2.4	2.3
A. echinites		9.6	11.3	15.6	12.7	13.5
B. argus		5.3	7.0	8.4	8.2	8.3
Bohadschia spp.	L: >10.2 cm	3.7	5.5	6.9	5.9	6.0
	M: 6.4 cm	2.7	5.0	6.2	5.3	5.4
	S: <6.4 cm	1.8	3.0	4.0	3.3	3.3
H. edulis	L: >12.7 cm	2.3	3.3	5.3	4.7	4.6
	S: 5.1–12.7 cm	–	–	4.9	3.9	4.1
H. atra	L: >12.7 cm	2.5	4.0	5.3	3.9	4.2
	M: 10.2–12.7 cm	1.6	2.1	3.1	2.4	2.7
	S: 5.1–10.2 cm	0.7	1.0	2.2	1.6	1.5
H. leucospilota	–	1.8	3.3	4.9	3.9	4.0
T. anax	–	3.4	4.3	4.9	3.7	4.0
H. fuscopunctata	–	1.8	2.8	2.9	2.7	2.9
B. graeffei	–	1.4	2.1	2.4	1.8	2.1

Source: Akamine, 2005a.

value species (*H. leucospilota, T. anax, H. fuscopunctata* and *B. graeffi*) appeared to be more stable (Table 3). Prices listed in Table 3 were for class A top quality products whilst class B products consist of semi-dried, out of shape, or poorly processed products of high value species (*H. fuscogilva, H. whitmaei, H. scabra, S. herrmanni, S. horrens* and *Actinopyga* sp.). These were classified as "rejects" and sold at 40 percent of the prices of class A products (Akamine, 2001). On the other hand, rejects of "bola-bola" in Sagay are bought at 50–70 percent lower than the prices for healthy looking pieces (Dacles, 2007).

3. SOCIO-ECONOMIC IMPORTANCE TO LOCAL FISHING COMMUNITY

In the Philippines, like in many other Southeast Asian countries, sea cucumber fishing, along with fishing for molluscs and other aquatic organisms, provides an income to many poor coastal fishers. In Mangsee Island, commercial sea cucumber fishing in deep waters is carried out not at subsistence level, but as an activity to supplement the fisher's income (Akamine, 2001). In many islands and coastal villages in the Philippines, income from sea cucumber fishery used to constitute a significant portion of a family's total income especially where the holothurians are abundant in the intertidal zone (Trinidad-Roa, 1987). In Barangay Bato, Sta. Cruz, Davao del Sur sea cucumbers, even as bycatch can contribute as much as 41 percent to the daily take home income of fishers (Gamboa, Aurelio and Abreo, 2007). However, current commercial sea cucumber populations in many shallow coastal areas have been reported to be overfished and income derived from sea cucumber gleaning has therefore become less important. In Bais Bay, in a survey carried out from July to November 1993, and in January 1994, sea cucumbers comprised only 1.2 percent of the total value of organisms gleaned during that period (LeBlanc, 1997).

Fishers who harvest sea cucumbers from the deeper areas derive a better income compared to those who glean in shallow coastal areas where the resources are now scarce. Although gleaning is an important harvesting method, there is no official statistics to indicate the volume of sea cucumbers collected by this method. Gleaning is usually adopted by the self-employed, poor coastal fishers who cannot afford to be without work and livelihood. LeBlanc (1997) noted that those who must find employment are likely to do so by entering the flexible income sector or the informal sector, and gleaning for sea cucumbers and other shellfish in the Philippines contributes to an important means of livelihood for the poor and landless. LeBlanc (1997) noted that in Bais Bay located on the island of Negros, a total of 25 223 gleaners (39 percent women, 35 percent men and 26 percent children) were recorded from July 1993 through to January 1994. They shared the coastal waters with an estimated 3 077 coastal fishers many of whom were involved in shellfish and seaweed cultivation. A study conducted by Schoppe *et al.* (1998) in the islands of Apid, Digyo and Mahaba off Leyte revealed that gleaning is a year round activity and 70 percent of the gleaners were females, 30 percent males with children in the age group between 7 to 15 years representing almost 40 percent of all gleaners. In Danao Bay, Heinen (2001) reported that 30 percent reef gleaners were males and gleaning was mainly carried out by women. In Apid and Mahaba, 85 percent and 62 percent of gleaners respectively, worked for subsistence purposes with the surplus catch being sold in the local market. Reef gleaning on the islands ranked second in the list of economic activities conducted to augment income derived from marine resources; 3.7 percent of the gleaners concentrated on collecting sea cucumber resources, with echinoderms comprising 16.6 percent of the catch in Apid, 10 percent in Digyo and 8.4 percent in Mahaba (Schoppe *et al.*, 1998).

LeBlanc (1997) classified the social roles of gleaning in Bais Bay into three categories:

1. Gleaners from households for which collecting shellfish and other invertebrates was the main income generating activity all year round;
2. Gleaners from households for which collecting shellfish and other invertebrates was not the most important income generating activity;
3. Gleaners from households involved in shellfish and other invertebrate collection for their own consumption.

Gleaning proved not to be a small and unstructured activity, but rather highly organized fishing affecting large numbers of people (LeBlanc, 1997). LeBlanc (1997) further noted that gleaning requires skill and knowledge to know where and when to find large quantities of the targeted species, and these skills are passed down from the adults to the children of the neighborhood.

Access to gleaning areas has traditionally been open to everyone, but this is slowly changing (LeBlanc, 1997). Gleaners' access to collection sites is being increasingly compromised by changes in the Philippines legislation; the Philippines Local Government Code indirectly produced changes in both access and use patterns in the coastal zone. Gleaners face steep competition when sites are no longer accessible when coastal areas were converted to mussel or oyster farms (LeBlanc, 1997).

4. ADDITIONAL THREATS TO SEA CUCUMBER POPULATIONS

Other reasons driving the decline of the sea cucumber resources in the Philippines include the fact that it is a common pool resource with a high demand and fetching a lucrative price. The absence of fishing regulations and poor understanding of its ecology also contributes to its overfishing (Morgan and Panes, 2004). There are various management regimes to sustain the sea cucumber resources, ranging from open access to co-management with local municipalities and to community-based management. Very few of these systems, however, seem to work because of the lack of enforcement on regulations and political will to prosecute wrong-doing. Other factors include the vulnerability of sea cucumbers which are easily harvested because of their sessile and defenseless nature, the low larval recruitment, the intermediate time of 2–6 years to reach maturity, as well as the dependence on high animal density for successful reproduction (Morgan and Panes, 2004; Uthicke, Welch and Benzie, 2004; Bruckner, Johnson and Field, 2003).

The use of destructive fishing techniques for finfish (such as cyanide, blast-fishing, hookah), destroys corals and reef communities. Hookah fishing of sea cucumbers is unsustainable because it enables fishers to stay longer in deeper waters thus contributing to increased fishing pressure of an already depleted resource. In "muro-ami" fishing, an encircling net is used with blocks of stone or cement slabs which are lowered into reefs to break up the corals (thus destroying sea cucumber habitats) and scare the fish into the nets.

In Polillo Islands, the greatest direct threat to the reefs and associated organisms was attributed to the use of explosives and poisonous substances (LeBlanc, 1997). Gleaning also contributes to the destruction of corals, and human impacts in gleaned areas are often evident from broken corals caused by stepping, and breaking of corals in order to gain access to the sea cucumbers (Schoppe *et al.*, 1998).

5. RECOMMENDATIONS FOR IMPROVING FISHERIES MANAGEMENT AND CONSERVATION OF SEA CUCUMBER POPULATIONS

Trinidad-Roa (1987) noted that the most practical method to regulate sea cucumber fishery is to impose a size regulation allowed for exports, and to restock depleted reefs with seeds produced from a hatchery. Experiences by fishers and traders reveal that size regulation is not necessarily enforceable (Gamboa, R., University of the Philippines in Mindanao, Philippines, personal communication). For example, a local trader buys undersized pieces out of benevolence to fishers who wasted time and effort in fishing and processing those catches. Hatchery production, on the other hand, is relatively expensive, takes years before real benefits can be noticed and need sustained community involvement.

The following strategies suggested by PCAMRD in 1992 are hereto revisited and commented by Gamboa (University of Philippines in Mindanao, Philippines, personal communication) using available information:

1. Establishment of catch quotas and minimum size limits for exported species.
   Although the Philippines has no national regulatory and monitoring system in place, at the local level the harvesting of *N. proteus* (*bola-bola*) in Sagay Marine Reserve observes catch quota and size limits, among others. The regulated fishery started only in 2004 and monitoring is in progress. Two visible successes at this early stage are: i) keeping off non-local residents from the fishing ground; and ii) income to the city and incentives to "barangays" (smallest political unit) as a result of profit-sharing (Dacles, 2007).
2. Establishment of closed season for harvesting during the breeding season.
   In the Sagay Marine Reserve this measure was relatively easy to enforce because the fishery is confined to one species, *N. proteus*, despite the presence of other commercial (but less valuable) species.

3. Rotational harvesting in reef areas. With collection areas divided into sections and each section opened to harvesting during certain periods on a rotational basis.
   A majority of the sea cucumber fishing grounds around the archipelago consists of several commercial species. One knowledge gap to this strategy is local information on the reproductive biology of those species. This review found information available only for three species:
   a) *H. scabra* is a continuous breeder but with reproductive pulses in June and December (Ong Che and Gomez, 1985);
   b) *H. pulla* spawns only in October (Batoy, Alino and Pocsidio, 1998);
   c) *H. coluber* in December (Batoy, Alino and Pocsidio, 1998).
   According to Gamboa, dissertations and technical reports available in universities and offices of government fisheries and marine bureaus may help fill some of the knowledge gaps about the biology, ecology and taxonomy of other commercial species.
4. Designation of permanent survey sites for monitoring the harvesting pressure and seasonal variation of sea cucumber populations.
   The national survey made by the National Fisheries Research and Development Institute can help identify those sites. There is likewise a need to standardize monitoring tools for easy comparison of data.
5. Establishment of reserve areas to help in the recruitment of stocks.
   This is perhaps the most popular strategy because of the presence of a national legal framework under the National Integrated Protected Areas System (NIPAS) Act. However, two common handicaps at the barangay level is the lack of properly trained enforcers and a sound monitoring and reporting system. Poaching and lack of political will are two other problems. Successful poaching can be attributed to insufficient logistic and financial support while the lack of political will often occurs in municipalities whose political leaders and their extended families are themselves engaged in fishing trade.
6. Ban on the use of SCUBA for harvesting except for species that are not found in intertidal regions and are found only in deep waters where the use of SCUBA is necessary.
   The reality is that SCUBA is too expensive for marginal fishers but hookah is affordable and does not need any formal licensing. The ban on the latter is at the discretion of the local government. A big common handicap in its implementation is defining political boundaries in the coastal waters and policing the area especially at night.
7. Monitoring the volume of harvests and sales/exports to assess catches from natural stocks. Existing stocks in exploited and unexploited areas should be monitored for effective management of the resource.
   This recommendation suffers again from the lack of a national regulatory system in place. Furthermore, the monitoring officers at the municipality including those at the Bureau of Fisheries, the Coast Guard, and the Immigration Department need to have some training on the taxonomy of the many species as well as a common Identification Guide to fresh and dried species (Gamboa, R., University of the Philippines in Mindanao, Philippines, personal communication).
8. Artificial propagation of commercial species to provide seeds for mariculture and restocking of depleted reefs.
   Sea cucumber culture was pilot tested at the Bolinao Marine Laboratory of the University of the Philippines Marine Science Institute in 2001 (Gamboa and Juinio-Menez, 2007). Follow through attempts attained 15–33 percent survival to early juvenile stage, and grow-out experiments demonstrated that field cages provided better survival when larger-sized juveniles were used (Gamboa, Gomez

and Nievales, 2004). In 2005, Nievales (2007) successfully produced juveniles of both *H. scabra* and *B. marmorata* var. *marmorata* in experimental scale at the University of the Philippines Visayas. These efforts, so far, have yet to reach the grow-out phase.

In 2006, the Department of Science and Technology (DOST) – PCAMRD in collaboration with the Australian Centre for International Agricultural Research (ACIAR) – The WorldFish Center and NFRDI took cognizant of the need for concerted research efforts on sea cucumber fisheries. They provided financial support to a national programme on hatchery production and resource management was initiated by the University of the Philippines (Marine Science Institute, Visayas and Mindanao). The 3-node project took off the next year simultaneous with that at the NFRDI. In the next 3–6 years the programme will try to: a) meet the needs for R&D on juvenile production and management of *H. scabra*; b) train fisher groups on sea ranching, restocking, processing and marketing; and c) develop incentives for responsible management of the said fishery (Gamboa and Juinio-Menez, 2007).

A commercial-scale multi-species hatchery for high value finfish and invertebrates such as abalone, sea cucumber, sea urchin (USAID, 2007) is about to be completed in Bongao, Tawi-Tawi, southwestern Mindanao. It is a joint project by the provincial government, BFAR, the Western Mindanao Community Initiatives Project (WMCIP) and Growth with Equity in Mindanao (GEM–USAID). The hatchery is meant to supply quality seed stock for mariculture parks which are being promoted by BFAR nationwide (BFAR, 2007). Kim Friedman (Secretariat of the Pacific Community, New Caledonia, personal communication), however, commented that the various sea cucumber culture systems are not market ready even for *H. scabra* and will take many years before other species come on line. Even when these species are farmed successfully, management of the natural stocks is still needed as total dependence on hatchery produced seed will be very expensive.

Sea cucumber farming using wild broodstock has been attempted by several growers who were participants at the National Forum (Gamboa, R., University of the Philippines in Mindanao, Philippines, personal communication) The problems encountered included:

- Lack of knowledge in the right stocking density and grow-out practices in general;
- Uncooperative and less enthusiastic cooperative members;
- Low production;
- Market access and strategies.

Some of the key points raised at the National Forum were:

- The need to regularly maintain the pens;
- The significance of an organized group and their contribution;
- The need to coordinate with "barangay" officials;
- The need for proper monitoring and documentation.

Among local traders, the problems raised were as follows:

- Dwindling stock due to destruction of reefs;
- Knowledge of processing techniques;
- Prices depend on up-line traders;
- Proliferation of non-Filipino traders/middlemen.

Sea cucumber species are now currently under review by the Convention on International Trade in Endangered Species of Wild Fauna and Flora (CITES) to examine whether its enlistment as an endangered species under CITES would help in the conservation and management of the sea cucumber resource and fishery. The international technical workshop on the conservation of sea cucumbers in the families

Holothuriidae and Stichopodidae convened by CITES in March 2004 in Malaysia recommended that commercially important species under these two families could be enlisted under Appendix II. Several benefits to the CITES enlistment include: curtail of illegal trade and harvest; awareness rising amongst stakeholders; comprehensive measures to comply with CITES provisions and perpetuation of sustainable fisheries (Toral-Granda, 2006). Under Appendix II listing, sea cucumber trade can be managed to yield the greatest sustainable benefits to fishers, exporters and importers, while they can be sustainably maintained to serve their important ecological roles (Bruckner, Johnson and Field, 2003).

In the Philippines, when an animal or plant species is listed under CITES, the national legislation does not permit the trading of the species, irrespective of the Appendix the species fall under. Although listing sea cucumbers under CITES may give them some respite from fishing, the ban may affect the livelihood of the poor people who are dependent on sea cucumber fishery. A plan to provide alternative livelihood to those affected by the ban must be in place to ensure that fishers dependent on sea cucumber fishery will not be left without livelihood. Toral-Granda (2006) suggested that including sea cucumber species in the Appendices of CITES may curtail trade and harvest. On the contrary, the listing of sea cucumbers under CITES may actually enhance illegal fishing and trade especially for the high value species whose black market price may sky rocket and make it very lucrative for fishers and traders to take risk and break the law. Enforcement of fisheries regulations has always been a problem in the Philippines; placing sea cucumbers under CITES control may not be an effective means to prevent fishing. Without strict enforcement on harvesting regulations, CITES listing may encourage smuggling and illegal trade for sea cucumbers that were caught illegally.

Some difficulties which need to be resolved before addressing whether it is appropriate to enlist sea cucumbers under CITES include: taxonomic uncertainties within the families, ability to distinguish taxa in the form they are traded, adequacy of biological information for making non-detriment findings, and ability to make legal acquisition findings (Bruckner, Johnson and Field, 2003). Bruckner, Johnson and Field (2003) also mentioned that it is possible to identify most of the common species that are traded as live animals for home aquaria and other uses, based on the gross morphology but it is very difficult to determine the dried products. With the listing in CITES, enforcement and custom officials in both exporting and importing countries will need to familiarize themselves with the identification of trepang which may not be so easy, especially to the less-scientific and sometimes, uninitiated personnel.

Akamine (2005b) suggested that international intervention is not the only way to save depleting resources and is against enlisting sea cucumber under CITES. He noted that the management of trepang resources requires commitment and support from trepang traders. Akamine (2005b) further noted CITES intervention will undermine cultural diversity and deny certain communities their historical preferences for trepang consumption and rob them of the use of sea cucumbers for medicinal purposes.

Other ways of conservation, such as regulating the catch by introducing community-based management where fishers are invited to play an active role in the decision-making process can be an alternative to CITES enlistment. Such community-based fisheries management and co-management efforts have proved successful in some small-scale pilot projects for example in the project implemented in Danao Bay and Banate Bay (Heinen, 2001; Fernandez, Matsuda and Subade, 2000). Akamine (2005b) suggested that all levels of stakeholders, especially trepang traders (whose trade may run from one generation to another) because of their vast knowledge (scientists are able to benefit from their knowledge and experiences) should be involved in issues on sea cucumber management and conservation. Akamine (2005b) also claimed that without actual knowledge of the trepang market, it is impossible to plan a management programme for more than 20 commercial species. La Viña (2002) noted that in the

Philippines, local governments must support community initiatives and the national government must ensure that community efforts are supported. He further noted that community-based resource management approach should not be grounded merely on the improvement of management of resources by reinforcing control and enforcement mechanisms through greater participation, but emphasis should be placed on the rationale of equity and justice which are necessary conditions for the attainment of environmental sustainability in the Philippines.

6. TOWARDS A NATIONAL SEA CUCUMBER MANAGEMENT PLAN

Ruth Gamboa (University of the Philippines, Mindanao, personal communication) provided the following information highlighted in a National Forum on Sea Cucumber Fisheries and Management which was held in Dagupan City, Pangasinan in June 2007. The forum was a first step towards a nationwide initiative to promote responsible and sustainable management of the sea cucumber fisheries.

The salient issues highlighted are:

<u>Production</u>

- Degradation of habitat and declining catch;
- Lack of scientific information and technical know-how on the biology, taxonomy, production and marketing;
- Unavailability of juveniles for grow-out due to absence of commercial hatcheries.

<u>Post-harvest</u>

- Poor handling and processing methods.

<u>Marketing and trade</u>

- Absence of a formal marketing system in local trade.

At the end of the forum a framework for a national sea cucumber management plan was agreed upon. Its objectives and programmes are:

General objectives:

- To assess the status of sea cucumber fishery and industry and develop management recommendations for national policies that can help ensure sustainability of the resource and equitable benefits to stakeholders;
- To formulate a multi-sectoral action plan towards a comprehensive management strategy for Philippine sea cucumber resources.

Specific objectives:

- To provide an overview of the current state of knowledge and practices related to the fisheries;
- To facilitate information sharing and leveling of understanding among various stakeholders;
- To identify key issues and strategic actions to improve the management of the fisheries and culture practices.

Programme areas include:

- Management of capture fisheries
- Grow-out culture
- Post-harvest
- Marketing

The participants decided on a one-year Action Plan to finalize the framework and help address the issues raised.

ACKNOWLEDGEMENTS

I would like to extend my appreciation and thanks to V. Toral-Granda, M. Vasconcellos and A. Lovatelli for inviting me to contribute to this project. My special thanks go to

Ruth Gamboa, whose comments I appreciate very much and for her contribution to Section 6. I would also like to thank J. Akamine, K. Friedman, J. Kinch and S. Purcell for their comments on the manuscript.

REFERENCES

Akamine, J. 2001. Holothurian exploitation in the Philippines: continuities and discontinuities. *TROPICS*, 10(4): 591–607.

Akamine, J. 2005a. Role of the trepang traders in the depleting resource management: a Philippine case. *In*: Indigenous use and management of marine resources. Kishigami, N. and J.M. Savelle. *Senri Ethnological Studies*, 67: 259–278.

Akamine, J. 2005b. International intervention is not the only way to save depleting resources. *Journal of Chinese Dietary Culture*, 1(2): 1–30.

Batoy, C., Alino, P.M. & Pocsidio, G.N. 1998. Reproductive development of *Holothuria pulla* and *Holothuria coluber* (Holothuroidea: Echinodermata) in Pamilacan Island, Central Philippines. *Asian Fisheries Science*, 11: 169–176.

BFAR. 2007. Online, downloaded 01 December, 2007. http://www.bfar.da.gov.ph/news/RPtotapChina_Japanmarkets(50707).htm.

Bruckner, A.W., Johnson, K.A. & Field, J.D. 2003. Conservation strategies for sea cucumbers: Can a CITES Appendix II listing promote sustainable international trade? *SPC Beche-de-Mer Information Bulletin*, 18: 24–33.

Dacles, T.P. 2007. *Sagay Marine Reserve, Sagay City, Negros Occidental.* Paper presented at the National Forum on Sea Cucumber Fisheries Manangement. Dagupan, Pangasinan, Philippines. June 6–9, 2007.

Domantay, J.S. 1957. *The zoogeographical distribution of the Indo-West Pacific litoral Holothuroidea.* Proceedings Eighth Pacific Science Congress, 1953, Vol. 3: 417–455.

FAO. 2000. Conversion factors landed weight to live weight. *FAO Fisheries Circular* No. 847, Revision 1. Rome, FAO. 176 pp.

Fernandez Jr., P.R., Matsuda, Y. & Subade, R.F. 2000. Coastal area governance system in the Philippines. *Journal of Environment & Development*, 9(4): 341–369.

Gamboa, R., Aurelio, R. & Abreo, N.A. 2007. *Local fishing and trading of sea cucumber in Davao Gulf, Philippines.* 9th National Symposium in Marine Science. Program and Abstracts (Paper Presentation). Iloilo, Philippines, 24–26October 2007.

Gamboa, R., Gomez, A.L. & Nievales, M.F. 2004. The status of sea cucumber fishery and mariculture in the Philippines. p. 69–78. In: *Advances in sea cucumber aquaculture and management.* Lovatelli, A., Conand, C., Purcell, S., Uthicke, S., Hamel, J.-F. and Mercier A. (eds). *FAO Fisheries Technical Paper*. No. 463. Rome, FAO. 2004. 425p.

Gamboa, R. & Juinio-Menez, M.A.J. 2007. Pilot-testing the laboratory production of the sea cucumber, *Holothuria scabra* (Jaeger), in the Philippines. *Philipp. Scient*, 40: 111–121.

Ganchero, A. 2007. *Overview of trade in sea cucumber resources and products.* Paper presented at the National Forum on Sea Cucumber Fisheries Management, Dagupan, Pangasinan, Philippines, 6–9 June 2007.

Gomez, A.L.V. 2007. *Value chain analysis for sea cucumber marketing.* Paper presented at the National Forum on Sea Cucumber Fisheries Management. Dagupan, Pangasinan, Philippines. 6–9 June 2007.

Heinen, A. 2001. Resource management: between meal and market. *Samudra*, April 2001: 13–18.

Jacinto, G.S., Aliño, P.M., Villanoy, C.L., McManus, L.T. & Gomez, E.D. 2000. Chapter 79: The Philippines. p. 405–423. In: *Seas at the Millennium: an environmental evaluation, Vol. II Regional Chapters: the Indian Ocean to the Pacific.* Sheppard, C.R.C. (ed.). Pergamon, Elsevier Science Ltd.

Jimeno, A.R.L. 2007. The rape of Tubbataha Reef. *Manila Standard Today MST Online.* Downloaded January 30, 2007. http://ecologyasia.com/news-archives/2007/jan-07/mst_070115_1.htm.

Labe, L. 2007. *Review of national and international policies, management options and instruments pertinent to sea cucumber resources.* Paper presented at the National Forum on Sea Cucumber Fisheries Management. Dagupan, Pangasinan, Philippines. 6–9 June 2007.

Labe, L., Acera, L.K., Romena, N. & Manululu, V. 2007. *Stock and fishery investigation of commercial Aspidochirotid Holothurians (Holothuriidae and Stichopodidae) in the Philippines.* 9th National Symposium in Marine Science, Program and Abstracts (Paper Presentation). Iloilo, Philippines. 24–26 October 2007.

La Viña, A.G.M. 2002. Community–based approaches to marine and coastal resources management in the Philippines: a policy perspective. p. 91–142. In: Institutional issues and perspectives in the management of fisheries and coastal resources in Southeast Asia. Torell, M. and A.M. Salamanca (eds.). *ICLARM Technical Report* 60.

LeBlanc, S. 1997. Gleaning in Bais Bay: a case study on an informal sector coastal activity in the Philippines. Submitted in partial fulfillment of the requirements for the degree of Master of Arts in International Development Studies at Saint Mary's University, Halifax, Nova Scotia, Canada. National Library of Canada, Ottawa. vi + 105pp + Appendices.

Metillo, E., Tarranza, D. IV & Arado, J. 2004. Notes on the species composition, relative abundance and fishery of sea cucumbers (Echinodermata, Holothuroidea) in Northwestern Iligan Bay, Philippines. *Philipp. Scient*, 41: 155–175.

Morgan, S. & Panes, H.M. 2004. *Gold rush: the rise and fall of a sea cucumber fishery in the central Philippines.* Presentation in: The Graduate student conference on "Fisheries and marine Ecosystems: integrating science and management", April 30– May 2, 2004. British Columbia, Canada.

Nievales, M.F. 2007. *Sea cucumber fishery and culture in the West Central Visayas.* Paper presented at the National Forum on Sea Cucumber Fisheries Management. Dagupan, Pangasinan, Philippines. June 6–9, 2007.

Oneocean. Undated. The Philippine experience in coastal resource management. Online, downloaded 3 September, 2007. http://www.oneocean.org/about_crmp/where_to.html

Ong Che, R.G. & Gomez, E.D. 1985. Reproductive periodicity *of H. scabra* Jaeger at Calatagan, Batangas, Philippines. *Asian Mar. Biol*, 2: 21–30.

Pastor, D., Catbagan, T. & Ragos, C. 2007. *Characterization of the sea cucumber fishery in Boolinao and Anda, Pangasinan through trade records and catch volumes.* 9th national Symposium in Marine Science, Program and Abstracts (Paper Presentation). Iloilo, Philippines. October 24–26, 2007.

PCAMRD. 1992. Primer on sea cucumbers. *Philippine Council for Aquatic and Marine Research and Development– Department of science and Technology (PCAMRD–DOST) Technology* No. 16. 4pp.

Salayo, N.D. 2000. *International trade patterns and trade policies in the Philippine fisheries.* Discussion Paper Series No. 2000–15. Philippine Institute for Development Studies. Downloaded February 20, 2007. 41 pp.

Schoppe, S. 2000. Sea cucumber fishery in the Philippines. *SPC Beche–de–Mer Information Bulletin*, 13 : 10–12.

Schoppe, S., Gatus, J., Milan, P.P. & Seronay, R.A. 1998. Gleaning activities on the islands of Apid, Digyo and Mahaba, Inopacan, Leyte, Philippines. *Philipp. Scient*, 35: 130–140.

Seale, A. 1911. The fishery resources of the Philippine Islands, part IV: miscellaneous marine products. *The Philippine Journal of Science XII*, D. General Biology, Ethnology and Anthropology, Vol. VI, December 1911, No. 6: 284–320.

Solandt, J.L., Slater, J., Beger, M. & Harborne, A.R.H. 2001. *Danjugan Island, Negros Occidental, Philippines: Reef Check Report 1997–1999. Danjugan Island Summary Report 4.* The Philippines Reef and Rainforest Conservation Foundation Inc., Negros Occidental, Philippines. 25 pp.

The Daily Tribune (Philippines). 2007. *Chinese "poachers" seek new investigation into case.* 26 January 2007. Online, downloaded on 29 January 2007. http://www.drugpolicycentral.com/bot/xarticle/tribune.net.ph4400.htm.

The Manila Times. 2007. *Poaching now "organized crime"*. Online, downloaded on 13 January 2007.
http://www.illegal–fishing.info/item_single.php?item=news&item_id=1211&approach.

Tiu, A.S.T. 1981. The intertidal holothurian fauna (Echinodermata: Holothuroidea) of Mactan and the neighbouring islands, central Philippines. *The Philippine Scientist*, 18: 45–119.

Toral–Granda, V. 2006. *Summary of FAO and CITES Workshops on sea cucumbers: major findings and conclusions.* Convention on international trade in endangered species of wild fauna and flora. Twenty–second meeting of the animals committee, Lima (Peru), 7–13 July 2006. AC22 Inf.1, 5 pp.

Trinidad–Roa, M.J. 1987. Bêche-de-Mer fishery in the Philippines. *Naga, The ICLARM Quarterly,* 15–17 October 1987.

USAID. 2007. *Sustainable aquaculture of high–value species in Mindanao seen by year end.* Growth with Equity in Mindanao Program–USAID. Online, downloaded 24 April 2007. http://www.mindanao.org/news/0603/17.htm

Uthicke, S., Welch, D. & Benzie, J.A.H. 2004. Slow growth and lack of recovery in overfished holothurians on the Great Barrier Reef: evidence from DNA fingerprints and repeated large-scale surveys. *Conservation Biology*, 18: 1395–1404.

Uthicke, S. & Conand, C. 2005. Local examples of beche-de-mer overfishing: an initial summary and request for information. *SPC Beche-de-Mer Information Bulletin*, 21: 9–14.

Ward, R.G. 1972. The Pacific Beche-de-Mer trade with special reference to Fiji. In: *Man in the Pacific Islands: Essays on Geographical Change in the Pacific Islands.* Ward, R.G. (ed.). 91–123. Oxford: Oxford University Press.

Warren, J.F. 1985. *The Sulu zone: The dynamics of external trade, slavery, and ethnicity in the transformation of a Southeast Asian maritime strait.* Quezon City: New Day Publications. (Originally published from Singapore University Press, 1981).

White, A.T., Stockwell, B., Meneses, A., Ovenden, M., Tesch, S., Apurado, J. & White, E. 2003. *Summary Field Report: Coral Reef Monitoring in Bohol and Siquijor, Philippines. 11–30 April 2003.* The Coastal Conservation and education Foundation, Inc. and Coastal Resource Management Project, Cebu City. 126 pp.

Population status, fisheries and trade of sea cucumbers in Africa and the Indian Ocean

Chantal Conand

Seychelles: a hotspot of sea cucumber fisheries in Africa and the Indian Ocean region

Riaz Aumeeruddy and Chantal Conand

Population status, fisheries and trade of sea cucumbers in Africa and the Indian Ocean

Chantal Conand
Université de la Réunion
Saint Denis, France
E-mail: conand@univ-reunion.fr

Conand, C. 2008. Population status, fisheries and trade of sea cucumbers in Africa and the Indian Ocean. In V. Toral-Granda, A. Lovatelli and M. Vasconcellos (eds). Sea cucumbers. A global review of fisheries and trade. *FAO Fisheries and Aquaculture Technical Paper*. No. 516. Rome, FAO. pp. 143–193.

SUMMARY

The region covered in this review is very diverse, including four FAO Fishing Areas and 30 countries. Sea cucumber fisheries presently exist in the Western Indian Ocean (WIO), where 16 countries have been documented, and in the Eastern Indian Ocean (EIO), with two countries covered in this review. The fisheries are for the dried product ("trepang" or "bêche-de-mer") which is consumed by Chinese populations and have a long history dating back to the ninteenth century.

Nearly thirty species are presently exploited (23 Holothuriidae; 6 Stichopodidae), with commercial value varying among species. Several differences in species composition between the Indian and tropical Pacific region have been shown recently, such as the teatfish *H. whithmei* found only in the Pacific, and *H. notabilis* and *H. spinifera* in the Indian Ocean.

The main information on the population status, reproductive biology and ecology of the commercially important species is synthesized in the present document. In 12 out of the 30 countries in the region the resource appears to be overexploited or fully exploited. Sea cucumbers are harvested and processed in different ways throughout the region, varying from small-scale, artisanal to semi-industrial activities. Globally, and according to FAO statistics, the region produces at least 1/3 of the world dried sea cucumber products.

There are several national management measures, including total bans; however, these seem to be insufficient for a sustainable use of the resources. The trade is characterised by exports from the producer countries, imports in "intermediate" (e.g. Yemen, Dubai) and final markets, where the key role of China Hong Kong SAR is most apparent. Illegal trade remains a problem in many countries.

The socio-economic aspects essential in small-scale fisheries are presented for several countries. Finally, several current projects in fisheries, or aquaculture as an alternative measure, are detailed.

In conclusion, the need for co-management, the improvement of the export statistics and the implementation of sustainable use are discussed.

1. REGION UNDER STUDY

The region covered for the present regional review (Figure 1) is very diverse and had to be separated into several sub-areas, following FAO Fishing Areas (FAO, 2005a). Table 1 lists the countries in each sub-area and points out the countries for which information on sea cucumber fisheries is available and countries that, although information is lacking, there may be commercial exploitation of sea cucumber species.

As in the tropical Pacific (Conand, 1986, 1990) the fishery for sea cucumbers in the Indian Ocean, for exports to the Asian markets, has a long history. For example, in India it is reported by Hornell (1917), in Madagascar by Petit (1930), in eastern Africa by Sella and Sella (1940) and in the Seychelles, Marguerite (2005) dates exports back to 1894. Following the FAO review in Lovatelli *et al.* (2004), an important recent source of information is the status report on sea cucumber fisheries from the Western Indian Ocean (WIO) prepared for the Western Indian Ocean Marine Science Association (WIOMSA) by Conand and Muthiga (2007a).

In the Africa and Indian Ocean region, the species commercially fished are benthic. Some species live on hard substrate (rocks and coral reefs), but most inhabit soft bottoms, either on the surface, or buried (temporarily or permanently) in the sediment. The common characteristics of the commercial species are: (i) abundance in rather shallow waters; (ii) large mean size of the specimens; (iii) diurnal habits; (iv) thickness and stiffness of the tegument that ensures a good processed trepang (bêche-de-mer). The first three characteristics make the commercial species very vulnerable to overexploitation.

An introduction is given on the four areas and the sub-areas of each main area, with general characteristics. Given the characteristics of the commercial species, they are found in the different shallow benthic environments, generally distant from estuaries as

FIGURE 1

Region under study and countries covered in this review (highlighted in yellow). The region encompasses the FAO Statistical Areas 51 (Western Indian Ocean), 57 (Eastern Indian Ocean), 34 (Eastern Central Atlantic) and 47 (Southeast Atlantic)

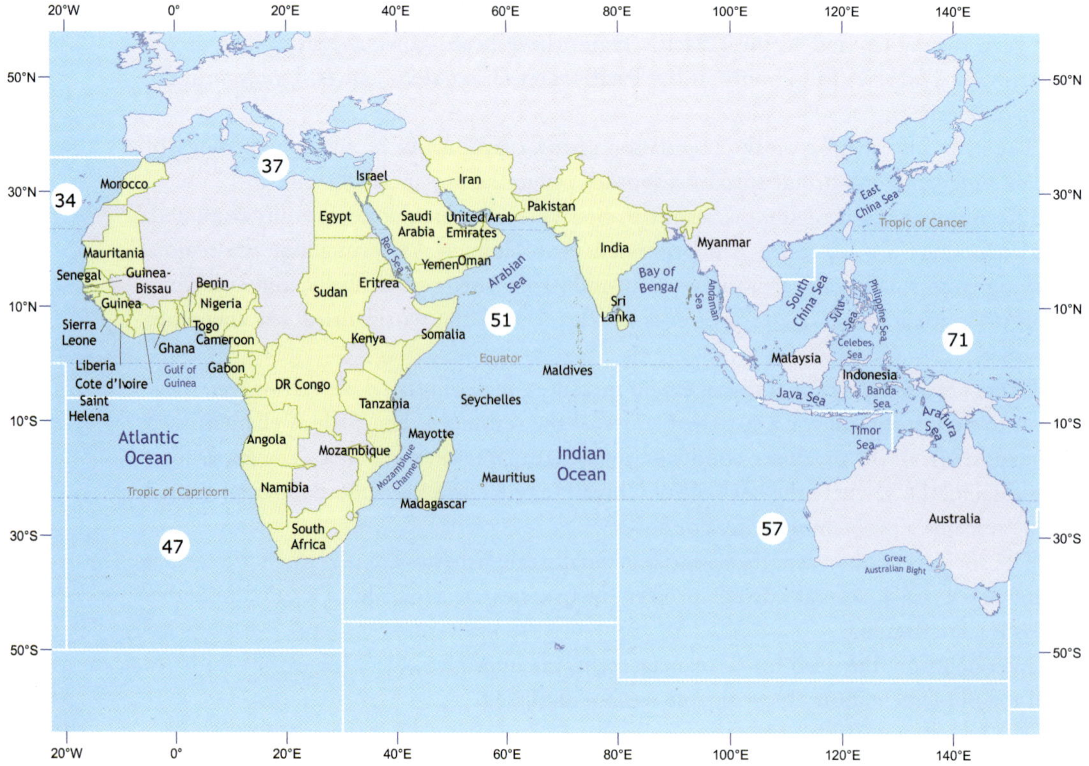

TABLE 1
Countries in the Africa and Indian Ocean region organized according to FAO fishing areas and sub-areas. Countries with documented sea cucumber fisheries are in italics and bold

FAO Area	FAO Sub-Area		Country	Main references
Western Indian Ocean (51)	51.1	Red sea	***Egypt***	Lawrence *et al.*, 2004; Ahmed and Lawrence, 2007
			Sudan	–
			Eritrea	Tewelde and Jeudy de Grissac, 2007; Kalaeb *et al.* 2008
			Saudi Arabia	–
			Israel	–
			United Arab Emirates	–
	51.2	Gulf	Djibouti	–
	51.3	Western Arabian Sea	***Oman***	Al-Rashdi, Al-Busaidi and Al-Ramadi, 2007
			Yemen	–
			Iran IR	Tehranifard *et al.*, 2006
	51.4	Eastern Arabian sea	Pakistan	–
			India	James, 1982, 1994b, 2001
			Chagos	Spalding, 2006
			Maldives	Reichenbach, 1999
	51.5	Somalia Kenya Tanzania	***Somalia***	Lovatelli, 1995
			Kenya	Marshall, Milledge and Afonso, 2001; Muthiga, Ochiewo and Kawaka, 2007; Samyn, 2003
			Tanzania	Marshall, Milledge and Afonso, 2001; Mmbaga and Mgaya, 2004; Mgaya and Mmbaga, 2007
	51.6	Madagascar, Mozambique Channel	***Madagascar***	Conand *et al.*, 1997, 1998; Mara *et al.*, 1997; Conand, 1999; Rasolofonirina, 2007
			Comoros	Samyn, Vanden-Spiegel and Massin, 2005, 2006
			Mayotte	Pouget, 2004, 2005; Conand *et al.*, 2005
	51.7	Islands	***Mauritius***	Arakaki and Fagonee, 1996; Conand and Muthiga, 2007b
			Réunion	Conand, 2003; Conand and Mangion, 2002; Conand and Frouin, 2007
			Seychelles	Aumeeruddy *et al.*, 2005; Aumeeruddy, 2007, 2008
	51.8	Mozambique	***Mozambique***	Marshall *et al.*, 2001; Hill, 2008
Eastern Indian Ocean (57)	57.1	Bay of Bengal	***India***	see 51.4
			Sri Lanka	Adithya, 1969; Terney Pradeep Kumara, Cumaramathunga and Linden, 2005
			Myanmar	see Choo, this volume
			Malaysia	see Choo, this volume
	57.2	Northern	Indonesia	see Choo, this volume
	57.3	Central		–
	57.4	Oceanic		None
	57.5	Western Australia	Australia	see Kinch *et al.*, this volume
	57.6	Southern Australia	Australia	see Kinch *et al.*, this volume
Eastern Central Atlantic (34)	34.1	Northern coastal	Morocco	No information available
			Mauritania	No information available
			Senegal	No information available
	34.2	Northern oceanic		No information available
	34.3	Southern coastal	Guinea Bissau	No information available
			Guinea	No information available
			Sierra Leone	No information available
			Liberia	No information available
			Ivory coast	No information available
			Ghana	No information available
			Togo	No information available
			Benin	No information available
			Nigeria	No information available
			Cameroon	No information available
			Equatorial Guinea	No information available
	34.4	Southern oceanic		No information available
				No information available
Southeast Atlantic (47)	47.1	Western coastal	Gabon	No information available
			Congo	No information available
			Rep Congo	No information available
			Zaire	No information available
			Angola	No information available
			Namibia	No information available
	47.2	Agulhas coastal	South Africa	No information available
	47.3	Southern oceanic		No information available
	47.4	Tristan da Cunha		No information available
	47.5	St Helena and Ascencion		No information available

most echinoderms are stenohaline organisms (intolerance to changes in salinity), with the possible exception of the sandfish *Holothuria scabra* (Conand, 1989; Hamel *et al.*, 2001); coral reefs, grass beds, or lagoons are favourable environments (Conand, 1989, 1990; Conand and Muthiga, 2007a).

Western Indian Ocean (FAO 51)

The Western Indian Ocean is divided in eight sub-areas and includes 24 countries. Only 16 countries had available documented information on sea cucumbers. Sea cucumber fisheries occur traditionally in most of them. However, it has changed during the last decade and show depletion of the stocks in many countries (Conand, 2004a, 2006b; Conand and Muthiga, 2007a). This has created awareness for management which have been raised at different stages and has had more or less success.

Eastern Indian Ocean (FAO 57)

The Eastern Indian Ocean is divided into six sub-areas and includes six countries. The sea cucumber fisheries are less important and less documented. Some will be presented in the Asia regional review (as Bay of Bengal) (Choo, this volume) and some are parts of Australia and the Pacific Island Countries and Territories (PICTs) (Kinch *et al.*, this volume).

Eastern Central Atlantic (FAO 34)

The Eastern Central Atlantic encompasses the areas of the tropical Atlantic on Africa's northwest coast. It is divided into four sub-areas and includes 14 countries. A few studies have been conducted on the inventories of Echinoderms, including sea cucumbers, but there is presently no sea cucumber fishery or information available in any of the countries.

Southeast Atlantic (FAO 47)

The Southeast Atlantic is divided into five sub-areas surrounding the southern part of the African continents, and includes seven countries. Many studies have been conducted on the inventories of sea cucumber species, mostly in the Republic of South Africa (Thandar, 2007), but currently no information is available on cucumber fishery and aquaculture.

This regional review will therefore concentrate on the Indian Ocean where information has been accessible for several countries. The information reported in this review is mostly based on articles from the 26 published Bêche-de-mer Information Bulletins (SPC 1970 to 2007) (www.spc.int/coastfish/), CITES meetings (Bruckner, 2006; Toral-Granda, 2007b), the FAO Workshop on Advances in Sea Cucumber Aquaculture and Management (ASCAM report; Lovatelli *et al.,* 2004), the ongoing Marine Science for Management (Masma) programme of the Western Indian Ocean Marine Science Association (WIOMSA) (Conand *et al.,* 2006; Conand and Muthiga, 2007a), and personal unpublished information and data.

2. BIOLOGICAL AND POPULATION STATUS

2.1 Key taxonomic groups

Within the region the key taxonomic group targeted for commercial purposes are the families Holothuriidae and Stichopodidae from the order Aspidochirota. These families comprise 100 percent of the catch. Table 2 presents the main 30 species fished, arranged taxonomically and with their commercial importance ranked from high (1) to low (4). These categories, sometimes different from previous documents (Conand, 1999, 2006a, b) and may well change in the future with the level of exploitation and the country concerned.

TABLE 2

Main commercial sea cucumber species in Africa and Indian Ocean. Value: 1 = high commercial value; 2 = medium commercial value; and 3 = low commercial value

Family	Genus	Species	Common name	Value
Holothuriidae	*Actinopyga*	*echinites*	Deep water redfish	2
		lecanora	Stonefish	3
		mauritiana	Surf redfish	2
		miliaris	Blackfish	2
	Bohadschia	*atra*	Tigerfish	3
		marmorata	Brownfish	3
		vitiensis	Brownfish	3
		subrubra	Leopardfish	3
	Pearsonothuria	*graeffei*	Flowerfish	3
	Holothuria	*atra*	Black lollyfish	3
		coluber	Snakefish	3
		cinerascens	–	3
		edulis	Pink lollyfish	3
		impatiens	–	4
		leucospilota	–	3
		notabilis	–	3
		nobilis	Black teatfish	1
		fuscogilva	White teatfish	1
		"pentard"	Flower teatfish	1
		fuscopunctata	Elephant trunkfish	2
		scabra	Sandfish	1
		scabra versicolor	Golden sandfish	1
		spinifera	–	3
Stichopodidae	*Stichopus*	*chloronotus*	Greenfish	2
		herrmanni	Curryfish	2
		horrens	–	3
		quadrimaculatus	–	4
	Thelenota	*ananas*	Prickly redfish	1
		anax	Amberfish	2

Source: Conand and Muthiga, 2007a.

This list was based on Bruckner (2006) and Toral-Granda (2006, 2007a, 2007b); however, some species have been added given the recent data from the Masma project in the WIO region (Conand and Muthiga, 2007a), which shows new species entering trade (i.e. *Holothuria* sp. "pentard" which is dominant, yet undescribed, in the Seychelles fishery [Aumeeruddy, 2007]; *H. notabilis* in Madagascar [Rasolofonirina, 2007] and *H. spinifera* fished in Sri Lanka and other islands [James, 1989]).

2.2 Biology and ecology of sea cucumbers

Due to the large number of species and countries covered in this review, a synthesis of the ecological and reproductive information available for each of the sea cucumber species in the region is presented in Table 3. More detailed information on a species basis is provided in the following sections.

Clark and Rowe (1971) and Richmond (1997) presented the general geographic distribution of the holothurians in the region. Recent papers on more specific localities have been published by Samyn (2003), Muthiga, Ochiewo and Kawaka (2007) for Kenya, Conand (2003), Conand and Mangion (2002), Conand and Frouin (2007) for Réunion, Rasolofonirina (2007) for Madagascar, Aumeeruddy (2007) for Seychelles, Marshall, Milledege and Afonso (2001) for Mozambique and Tanzania. Conand and Muthiga (2007b) have also summarised other recent data for the WIO (Conand and Muthiga, 2007a).

TABLE 3

Summary of ecological and reproductive information available for sea cucumber species in Africa and Indian Ocean. Geographical distribution – given the large number of countries and species, two categories are defined: W = wide distribution and R = restricted to a few countries within the region; NA: no information available

Genus	Species	Geographical distribution	Habitats	Substrate	Depth range (meters)	Ecological role	Other notes of interest on behaviour	Sexual reproduction	Size at Sexual Maturity	Larval development type	Fission (season and rate)
Actinopyga	*echinites*	W	Reef flats and seagrass	Coral and rubble	0–10	Deposit-feeder		Warm season	45 g		
	lecanora	W	NA	NA	NA						
	mauritiana	W	Outer reef flats	Coral	0–5	Deposit-feeder		Warm season	23 cm TL		
	miliaris	W	Reef flats and seagrass	Sand	NA	Deposit-feeder					
Bohadschia	*atra*	W	Seagrass	Sand	NA	Deposit-feeder					
	marmorata	W	Back reef, seagrass	Sand	0–20	Deposit-feeder	Burying				
	vitiensis	W	Back reef, seagrass	Sand	0–20	Deposit-feeder	Burying	Warm season			
	subrubra	W	Rubble	Sand	0–30	Deposit-feeder	Buried during day				
Pearsonothuria	*graeffei*	W	Living coral	Hard substratum		Deposit-feeder					
Holothuria	*atra*	W	Back reef, reef flats	Sand and Rubble	0–10	Deposit-feeder, bioturbation, productivity		Warm season			X cool season
	coluber	R	Back reef	Rubble	NA						
	cinerascens	W	Outer reef	Hard	0–3	Filter-feeding					
	edulis	W	Seagrass, lagoons	Sand	0–30						
	leucospilota	W	Back reef, reef flats	Sand and rubble	0–10	Deposit-feeder, bioturbation, productivity		2 seasons: February – May	180 g (total weight)		X, low rate
	notabilis	R (Madag.)	Lagoon, seagrass	Sand	0–10	Deposit-feeder					
	nobilis	W	Reef flats, slopes	Rubble	0–40	Deposit-feeder					
	fuscogilva	W	Slopes, lagoons	Sand	10–50	Deposit-feeder					
	"pentard"	R (Seych.)	Lagoons	Sand	10–50				1 500 g		
	fuscopunctata	W	Lagoons, seagrass	Sand	0–20	Deposit-feeder					
	scabra	W	Back reef,	Muddy sand	0–5	Deposit-feeder, bioturbation, productivity	Burying	November to April; mature individuals all year		Planctotrophic 2 weeks; epibiont juveniles 6 weeks	
	scabra versicolor	W	Lagoons	Sand	0–30	Deposit-feeder					
	spinifera	R India, Sri Lanka	Lagoons		NA			Warm season			
Stichopus	*chloronotus*	W	Reef flats	Rubble	0–5	Deposit-feeder					
	herrmanni	W	Lagoons, seagrass	sand	0–30	Deposit-feeder					
	horrens	W	Lagoons, seagrass	Sand, rubble	0–5	Deposit-feeder	Nocturnal	Warm season			X, cool season
	quadrimaculatus	R	Reef flats	rubble	0–5	Deposit-feeder	Nocturnal				
Thelenota	*ananas*	W	Coral slopes	hard	5–35	Deposit-feeder					
	anax	R	Lagoons	sand	10–40	Deposit-feeder					

Family Holothuriidae

Genus *Actinopyga*

Actinopyga echinites

Actinopyga echinites (the common name is deep water redfish, which does not correspond to its ecology) is a medium-sized species whose tegument varies in colour from light brown to orange. It has numerous papillae dorsally and the anus is ringed by five calcareous teeth as for the other species of this genus. The Cuvierian tubules are rarely expelled. It is a common species on the shallow reef flats of the region. On Réunion reefs, it is mostly found on outer reef flats with oxygenated waters, with a maximum density of 3 ind. m^{-2} (Conand and Mangion, 2002).

Some traits of its population biology (densities, reproduction, biometry) have been studied from Réunion as part of the Masma (WIOMSA) regional project. This first study in the region shows that the sexual reproduction is annual during the warm season and the weight at first maturity is at 45 g (Kohler, Conand and Gaudron, 2007). This species lives exposed on the hard substrate, without burying.

Actinopyga lecanora

Actinopyga lecanora (common name stonefish) is also a medium-sized species whose tegument makes it look as a stone. It is found in many countries in the region, but specimens are relatively rare and its biology has never been studied.

Actinopyga mauritiana

Actinopyga mauritiana (common name surf redfish) is relatively large, with a mean total length of 20 cm and up to 35 cm. Its medium weight is 380 g in Réunion (Conand and Mangion, 2002). It is common in most countries of the region. Its colour is variable according to locations, from uniform brown to dark brown; some specimens have a brown median band and are whitish laterally. The trivium is generally white. There are different morphs in the Indian Ocean and Pacific.

The habitat is restricted to the outer reef flats under the influence of the surf, where the individuals are exposed and hold strongly to the hard bottom (Conand, 1989). There is a narrow transition zone where *A. echinites* and *A. mauritiana* can be found together.

Several studies on the biology have been conducted in the Pacific (Conand, 1989, 1993; Ramofafia, Battaglene and Byrne, 2001) where the sexual reproduction occurs during the warm season. In Egypt (Red Sea) the reproduction has been studied and takes place between spring and summer; the spawning is partial and the gonads are reabsorbed at the end of the season; individuals are mature at 23 cm total length; after artificially induced fission the regeneration is generally slow (Gabr *et al.*, 2004).

Actinopyga miliaris

Actinopyga miliaris (common name blackfish) is a species with black tegument and characteristic five anal teeth. This species averages 20–25 cm total length. The individuals live in shallow waters up to 20 m on seagrass beds or rubble (Conand, 1989). Despite the high densities of some populations, there are no references on the biology of this species. They are found in Comoros, Mayotte, Madagascar, Seychelles and Kenya, but not in the Mascareigne Archipelago (Réunion, Mauritius, Rodrigues). They are actively fished in Kenya with *A. miliaris* contributing to 17 percent of the total catches in the two main landing beaches (Muthiga, Ochiewo and Kawaka, 2007). This species lives exposed on hard bottoms or in seagrass beds, without burying (Conand, 1989).

Genus *Bohadschia*

The species are of medium-low commercial value. Their Cuvierian tubules are very abundant and sticky which makes the processing difficult. They are usually collected when overexploitation has diminished the abundance of other commercial species.

Bohadschia atra

Bohadschia atra was recently taxonomically described (Massin *et al.*, 1999) and although no official records existed it was already commercially exploited. This shows that even large species, exploited in some places are probably still undescribed. It had probably been misidentified previously as *B. argus* and the fishermen call it tigerfish, the same name as *B. argus*. It is a black species with brown spots, numerous Cuvierian tubules. Since its description, it has been observed in several countries and it is a rather common species inhabiting shallow waters on sandy bottoms. Individuals do not bury themselves under the substratum.

Bohadschia subrubra

This species recently received interest as a commercial species (Conand 1999), and was re-described from Madagascar by Massin *et al.* (1999) (first described by Quoy and Gaimard in 1833 from Mauritius). It has an external appearance similar to *B. atra*, but its bivium is brown to orange with black areas that makes it easy to distinguish. Its Cuvierian tubules are expelled rapidly in large quantities. It is also a large species and may grow up to 35 cm. It is common in the region and populations are abundant; no studies have yet been conducted on its biology.

Several other *Bohadschia* species are not so easy to distinguish (Samyn, 2003) and there are probably yet some misidentifications between *B. marmorata, B. vitiensis, B. cousteaui* and *B. bivittata*.

Bohadschia vitiensis and *Bohadschia marmorata*

Bohadschia vitiensis has recently been identified by C. Massin (Département des invertébrés, Institut royal des sciences naturelles, Belgium, personal communication) from Réunion specimens and from Comoros (Samyn, Vanden-Spiegel and Massin, 2006) where *B. marmorata* has not been found.

The morphology of these species is rather similar; the body is cylindrical, up to 30 cm in length. The bivium is light brown dorsally spotted by brown dots in the tube feet, the trivium yellow-white; transverse banding could characterize some *B. marmorata*. The abundant large Cuvierian tubules are rapidly ejected when disturbed. *B. vitiensis* lives in rather dense populations in shallow back reef areas, where the individuals burrow into the anaerobic sand during the night and emerge at the surface of the oxygenated sand in the morning around 10 A.M. at Réunion (Conand, unpublished data). The characteristic behaviour during sexual reproduction (individuals in an upright position expelling sperm or oocytes out of swollen gonopores), has been described from Réunion during the warm season (Gaudron, 2006). Sexual reproduction also occurs during the warm season in Madagascar (Rasolofonirina, 1997).

Bohadschia vitiensis is cited for Kenya by Muthiga, Ochiewo and Kawaka (2007), while *B. marmorata* by Samyn (2003) and for the Seychelles by Aumeeruddy (2007). In Mauritius, *B. marmorata* is cited by Arakaki and Fagonee (1996). Its distribution and abundance in Mauritius in Trou d'Eau Douce has been described by Muller (1998) and in La Preneuse and Baie du cap sites by Luchmun *et al.* (2001). A taxonomic revision of the genus *Bohadschia* is still needed.

Genus *Pearsonothuria*

Pearsonothuria graeffei

Its previous name was *Bohadschia graeffei.* It is a species easy to distinguish given the elongated body, mottled grey and cream dorsally and with three bands of tube feet ventrally; the black oral tentacles bordered with a pale zone are conspicuous. The Cuvierian tubules are present but not expelled. The average length is 35 cm. It lives in shallow reefs, often feeding day and night on detritus found on corals. The species does not burrow. It is found in most countries of the region. Little is known on its biology. The juvenile of this species is different from the adult and very characteristic, looking like Phyllidiidae nudibranchs. The commercial value of this species is low as the tegument is thin; a fishery should therefore not be encouraged.

Genus *Holothuria*

Holothuria atra

It is the most abundant and frequent holothurian in all countries within the region (common name: black lollyfish), and therefore one of the most studied species despite its low commercial value (Conand, 1996; Jaquemet, Rousset and Conand, 1999; Uthicke, Conand and Benzie, 2001; Conand and Mangion, 2002; Conand, 2004a; Mangion *et al.*, 2004; Taddei, 2006). It is a black cylindrical species with the body often covered with sand, except in circular areas, with most specimens living in back-reef areas. Populations from the outer reef flat are composed of larger individuals uniformly covered in sand and presenting wart-like protrusions on the dorsal area. The populations have mostly been studied from the tropical Pacific (see references in Conand 1990; Conand and De Ridder, 1990). A few studies from the Indian Ocean show the characteristics (density, size) of the populations of this species; fission is an important mode of reproduction. In Réunion *H. atra*, showed higher densities in La Saline reef than *H. leucospilota* another common species in the site. The mean value in the reef was 0.68 ± 1.67 ind. m^{-2}. *H. atra* was dominant in back reefs (1.44 ± 2.46 ind. m^{-2}). However the distribution of this species over the reef complex is highly heterogeneous. Maximum values of 5.1 ± 0.1 to 6.6 ind. m^{-2} (Fabianeck and Turpin, 2005; Conand and Mangion, 2002) have been recorded in one back reef station, whereas no individuals were observed from other back reefs. In inner reef flats, densities were inferior to 0.1 ind. m^{-2}.

Conand (1996) had pointed the existence of populations with different biometric parameters, with larger individuals and low densities on the outer reef flats and smaller individuals in back reef and inner reef flat stations; these are also characterised by high fission rates. Conand (2004c) analysed data collected from 1993 to 2000 and found stable densities over the long term (average 4.7 ± 0.3 ind. m^{-2}), which could be partly explained by this particular mode of reproduction.

The nutrition of *H. atra* has recently been studied, as the species is a major component of the soft-bottom reef communities (Conand and Mangion, 2002; Taddei, 2006). The nutrition of the two dominant species, *Holothuria atra* and *H. leucospilota,* has first been studied at two sites, one eutrophicated and one oligotrophic (Mangion *et al.*, 2004). A relation has been established between the enrichment level of the sites and holothurian densities: holothurians are abundant (densities up to 3 ind. m^{-2}) in eutrophic areas whereas low densities characterise oligotrophic areas. Gut content analysis showed that the organic matter ingested from the sediment was used with 10 percent efficiency for both species (Mangion *et al.*, 2004). The Carbon:Nitrogen ratio decreased along the gut showing organic matter degradation. This shows the ability of these holothurians to break down the organic matter from the sediment and to make it easily available for other organisms (Mangion *et al.*, 2004). Holothurians from the oligotrophic reference station appeared to discriminate soft bottom patches with high nutritional value. In contrast, no selectivity was observed in the eutrophicated stations. *H. atra* also shows a significantly higher assimilation than *H. leucospilota*. Data suggest that efficiency of assimilation decrease with the eutrophication gradient. More recently, Taddei (2006) evaluated globally the role of the soft-bottom compartment during the transfer of matter and energy in the reefs. The holothurian biomass constituted a major element that could reach 7.92 g dry weight. m^{-2} in the most productive areas. The soft-bottoms played a key role in catching the organic matter provided by the back-reef. This was deeply influenced by the high hydro-dynamism which modulated the loss of mater and energy of the reef. These losses were however limited by the action of the holothurians, which stored organic matter such as biomass and probably enhanced local production.

The sexual reproduction of *H. atra* has attracted considerable research (Conand, 1981, 1982, 1989, 1993, 2006a). The reproductive biology (sexual and by fissiparity), population variations and genetics for this species have been described (Uthicke, Conand and Benzie, 2001; Conand, Uthicke and Hoareau, 2002; Uthicke and Conand, 2005b). The characteristics of the different stages in fission and regeneration have been described. Asexual reproduction (Figure 2) is seasonal on the back reef, with a higher fission rate during the cool season. Regeneration lasts about five to six months. Fission is much higher in the back-reef zones than on reef flats (Conand, 1996), as is also the case in other mixed populations of this species, for example on the Great Barrier Reef (Uthicke, 2001). The population showed stable density and weight distributions, and seems to have attained the optimum density in relation to the abiotic and biotic back-reef conditions, at a euthropic station of a Réunion Island reef (Conand, 2004b). These results are relevant to the assessment of species productivity which is affected by fission in addition to growth and recruitment.

The population genetics has been compared between the Torres Strait (Pacific) and Réunion to investigate connectivity between populations separated by large geographic distances (Uthicke, Conand and Benzie, 2001). This is the first study on genetics of a sea cucumber from the Indian Ocean. This comparison has shown that a high gene flow exists between the regions. Nevertheless, pooled populations within each region were significantly different from those of the other region. Despite the importance of asexual reproduction, the potential for widespread dispersal mediated by sexually produced larvae is large.

H. atra moves while ingesting the sediment at a rate of 30 cm h^{-1}; it is able to ingest the sediment under the surface, but it does not burrow (Taddei, 2006).

FIGURE 2

Fission in *H. atra* (a) at 1/3 of the length from the mouth and *H. leucospilota* (b) at 1/5 of the length

COURTESY OF C. CONAND

Holothuria coluber

Holothuria coluber (common name: snakefish) is a low commercial value species, sometimes fished in Madagascar. The body is an elongated dark cylinder up to 50 cm in length. The dorsal podia are large and stout yellow or white, arranged in irregular rows while the ventral ones are dispersed on the trivium. The oral tentacles are pale yellow. They live in calm shallow waters, on sandy bottoms with rubble. The species is often covered with sand but does not burrow. The posterior part of the body is generally hidden under blocks, making it look as *H. leucospilota.* However, its rigid tegument and the colour of the papillae make it different from the former species.

Holothuria cinerascens

Holothuria cinerascens is a low commercial value species, sometimes fished in Madagascar. The species live in crevices, in shallow water reefs or beach rocks, with a strong hydro-dynamism. The dorsal podia are large and stout but few while the ventral ones are very numerous. The oral tentacles are pelto-digitated and the species filters water which is an exception among the aspidochirotids. The body is a short cylinder, brown, measuring from 12–18 cm in length; it is burrowed or hidden, only the mouth and tentacles surfacing with an appearance of a cauliflower (local name given in Madagascar). Very little is known on this species.

Holothuria edulis

Despite its species name (*edulis* means "edible") the pink lollyfish is of low commercial value and little is known apart its distribution. It looks like *H. atra*, but the colour is different, from pink to grey on the bivium, mostly pink on the trivium. The species attains up to 40 cm of length. It is reported to occur in several countries. The species is often covered with sand but does not burrow.

Holothuria leucospilota

Holothuria leucospilota is a conspicuous long black species very frequent and abundant in many countries of the region. In Réunion, the individuals were present in all back reefs and inner flat stations, with total mean density of 0.59 ± 0.38 ind. m^{-2}. Highest densities were in back reefs (mean 0.84 ± 0.38 ind. m^{-2} versus 0.39 ± 0.26 ind. m^{-2} in inner reef flats), with a maximum density of 1.4 ± 0.7 ind. m^{-2}. It is fished in several countries despite its low value.

Fission in *H. leucospilota* (see Figure 2) has been described for the first time in Réunion (Conand, Morel and Mussard, 1997). The morphology and anatomy of the different stages of the regeneration have been detailed and allow their observation in the field. The fission rates are much lower than for other species such as *S. chloronotus* or *H. atra*. Fission has now been observed on other reefs (Purwati and Luong Van, 2003; Drumm and Loneragan, 2005).

The modalities of the sexual reproduction have been followed in Réunion by monthly sampling. (Kohler, 2006; Gaudron, Kohler and Conand, 2007) and will allow comparisons with other WIO countries. Two peaks of spawning were found in Réunion, the first in February marked by a sharp decrease in the gonadal index and numerous individuals in post-spawning stage, the second peak in May. It reaches SOM at a weight of above 180 g. It is also studied in Kenya to allow a regional synthesis on this species (Kawaka and Muthiga, 2007).

As it is so common and abundant in the region, the effects of its nutrition have been evaluated on the reefs of Réunion as an average of 88.8 g dry weight.$ind^{-1}.d^{-1}$ of sediment in the two sites studied (Mangion *et al.*, 2004). Holothurians from the oligotrophic reference station appeared to discriminate soft bottom patches with high nutritional value. In contrast, no selectivity was observed in the eutrophicated stations. Assimilation by *H. leucospilota* shows a lower rate than *H. atra* (Taddei, 2006). It moves while ingesting the sediment at a rate of 30 cm h^{-1}, without daily pattern and with periods without movement; *H. leucospilota* ingests the surface sediment with its large tentacles, but it does not burrow (Taddei, 2006).

Holothuria notabilis

This species, first described by Ludwig (1875), has been added to the list of commercial species, as it is presently one of the main species exploited in the south of Madagascar (Rasolofonirina, 2007), where it has been determined recently (Massin, C., Département des invertébrés, Institut royal des sciences naturelles, Belgium, personal communication). It lives in seagrass beds where the population biomass had already been estimated at 6 000 tonnes/ha (Mara *et al.*, 1997). This species has not been observed in other countries in the region. The holothurian fauna of Madagascar, based on Cherbonnier's (1988) studies, is the richest in the region. Its reproductive biology is presently under study (Rasolofonirina, R., Université de Toliara, Madagascar, personal communication).

Teatfish *Holothuria (Microthele)*. There are two common teatfish species *H. nobilis* and *H. fuscogilva*; a third one, the "pentard", as it is called by the Seychelles fishers may need to be added to this group. These large species typically present lateral processes on the body wall known as "teats".

Holothuria nobilis

The common name of *Holothuria nobilis* is black teatfish despite more or less developed white to golden patches on the body wall. It has recently been separated from the Pacific black teatfish (*H. whitmaei*) by Uthicke, O'Hara and Byrne (2004) and in a near future other species may probably differ for the two regions, as (i) more attention is being paid to holothurians and, (ii) an increased use of genetic tools will allow better identifications.

H. nobilis is one of the species with the highest commercial value and therefore overexploited. It is a large species, up to 50 cm in length, loaf-shaped. The body is usually covered by sand but it does not bury. The Cuvierian tubules are present, but not expelled. The species occurs in a variety of coral reef habitats, in dispersed populations, at low densities. For example, only a few dispersed specimens have been seen on Réunion reefs, raising the question of successful egg fertilization if spawning and recruitment occurs (so far never detected).

Holothuria fuscogilva

Holothuria fuscogilva (common name: white teatfish) has been described by Cherbonnier (1980) from New Caledonia, but the fishers from the tropical Pacific had traditionally used different names for this white teatfish as well as the black teatfish (*H. nobilis* now *H. whitmaei*). It is a large species (mean length 40 cm, maximum up to 60 cm) with a variable coloration from yellow to grey and often presents brown patches. It inhabits deep habitats, found up to 40 m depth, but its densities are generally low (Conand, 1989). Despite the taxonomic uncertainties (Samyn, 2003), it appears that it is now accepted as different from the black teatfish. However, it remains to be demonstrated if it is the same species in the Pacific and the Indian Ocean. Uthicke, O'Hara and Byrne (2004) using mtDNA sequences have shown that a large intraspecific sequence divergence exists indicating potential for presence of several cryptic species in a white teatfish complex.

It is fished in Comoros, Kenya, Madagascar and Seychelles, but now found much deeper than before, as it is highly overfished in coastal waters.

In the Maldives, Reichenbach (1999) assessed its ecology from three habitats. It is the dominant species in the lagoon floor and is also abundant in passes. The species recruits in shallow seagrass beds, as observed in Fiji (Conand, 1989) and then moves to deeper waters. The species is often covered with sand but does not burrow. Size at maturity was found to be the same as in New Caledonia (Conand, 1981, 1993) around 1.5 kilograms, which is a very late maturity. The growth is also probably very slow; they are long-lived and therefore sensitive to overexploitation.

Holothuria "pentard"

This third teatfish (common local name: "pentard") is the main species collected and exported by Seychelles (Aumeeruddy, 2007; Aumeeruddy *et al.*, 2005; Aumeeruddy and Conand, 2007) (Figure 3). It is also found in the Comoros (Samyn, Vanden-Spiegel and Massin, 2006). Future studies will help decide if it is another species or simply a variety of the black teatfish. Very little is known apart from its presence in the catches.

Holothuria fuscopunctata

Unlike the other *Holothuria (Microthele)*, *Holothuria fuscopunctata* (common name: elephant trunkfish) has a low commercial value. It is yet a large species up to 60 cm length. The colour is characteristic, with a dorsal body wall yellow to light brown scattered with the darker spots of the tube feet. The body wall is thick. No Cuvierian tubules. It has been observed in Madagascar, Kenya, Seychelles, Comoros and Mayotte, but not in the Mascareigne Archipelago (Réunion, Mauritius, Rodrigues). It lives on coral sand in shallow waters, up to 30 m. The populations are not as abundant in the Indian Ocean as in the Pacific and individuals are rare and scattered. The species is often covered with sand but does not burrow.

Sandfish *Holothuria scabra* and *H. scabra* var. *versicolor*

The two sandfish *Holothuria scabra* and *Holothuria scabra versicolor* (Conand, 1986, 1989, 1998b, 1999; Hamel *et al.*, 2001*)* have also raised considerable interest in recent years. These species have a high commercial value and rank probably number one on the world bêche-de-mer market. Genetic studies have recently allowed proper identification of the species. The results of the DNA and allozyme analyses show that they are distinct species (Uthicke, Purcell and Blockmans, 2005). In this review we still use the names *H. scabra* and *H. scabra* var. *versicolor.*

Holothuria scabra

Holothuria scabra (common name: sandfish) is largely distributed and probably supports most of the tropical captures of sea cucumbers for bêche-de-mer. It shows

FIGURE 3
Dried (a) and salted (b) specimens of "Pentard" teatfish from the Seychelles

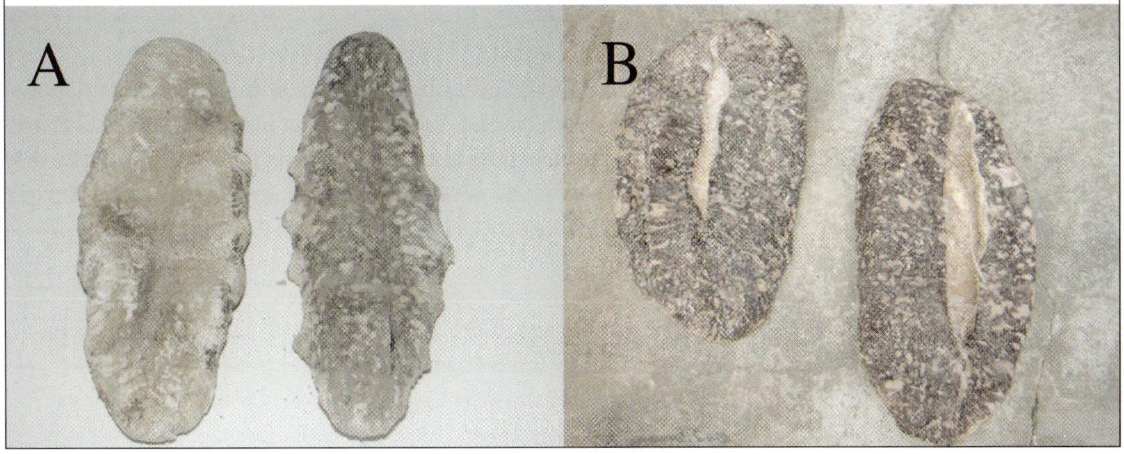

COURTESY OF R. AUMEERUDDY AND C. CONAND

some variability in colour between the Pacific and Indian Oceans. In the Indian Ocean it is most often grey to dark with small yellow patches or horizontal bands. Its lateral wrinkles are characteristic. It shows a preference for muddy sands and is often found in the vicinity of mangroves. Many studies (Kithakeni and Ndaro, 2002; Rasolofonirina *et al.*, 2005) and a recent synthesis (Hamel *et al.*, 2001) have been conducted on this species, which is also used for stock enhancement or aquaculture in different places (India, the Solomon, Madagascar, Indonesia). In India aquaculture experiences started well before other countries (James, 1994a, 1994b, 1996, 2004a, 2004b; James *et al.*, 1988). Such development then started in the Pacific with the support from The Worldfish Center in the Solomon Islands (Ramofafia, 2002; Mercier, Battaglene and Hamel, 1999) and New Caledonia (Purcell, 2004). More recently, the interest emerged in Madagascar where the aquaculture of this species has progressed with assistance from the Belgian cooperation (Jangoux *et al.*, 2001; Rasolofonirina, Mara and Jangoux, 2004; Rasolofonirina and Jangoux, 2005).

The reproduction of this species has been studied in Tanzania by Kithakeni and Ndaro (2002) and in Kenya by Muthiga, Kawaka and Ndirangu (2007). Rasolofonirina *et al.* (2005) detailed the gonad anatomy and the tubule's development from monthly samplings carried out at Toliara (Madagascar) from 1998 to 2001; these are synchronous within an individual but gametogenesis was not synchronous in the population; most individuals were mature or spawning between November and April. The burying behaviour of the adults has been described by Yamanouchi (1956) and of the juveniles by Mercier, Battaglene and Hamel (1999, 2000).

Holothuria scabra var. *versicolor*

Holothuria scabra var. *versicolor* (common name: golden sandfish) differs from *H. scabra* by a number of characters, including a larger mean size and a deeper habitat in New Caledonia (Conand, 1986, 1989, 1998a, b). The dorsal tegument is highly variable from beige to black, with many specimens having black patches. The papillae and tube feet are also more developed. Its seasonal reproductive cycle differs from *H. scabra.* The genetics has been studied by Uthicke, Purcell and Blockmans (2005). In the WIO, the reproduction has been studied from Madagascar; spawning is annual during the warm season and gametogenesis is synchronous in the population (Rasolofonirina, 1997). The burying behaviour, observed in New Caledonia (Conand, 1989), is not described from the Indian Ocean.

Holothuria spinifera

In India, the species *Holothuria spinifera* (common name: brown sandfish) is abundant (James, 2001). The species reaches 30 cm in length. In the Gulf of Mannar and Palk Bay it is found on clean sands where it burrows. In Sri Lanka it has the local name "Gal-atta" or "Weli-atta".

Family Stichopodidae

The main commercial Stichopodidae are less numerous than the Holothuriidae and belong to two genera *Stichopus* and *Thelenota*. Their characteristics are: a body square-shaped or trapezoidal in cross section, Cuvierian tubules always absent, gonad in two tufts appended on each side of the dorsal mesentery and dominant spicules in branched rods and C- and S-shaped rods.

Genus *Stichopus*

The genus *Stichopus* has a bivium covered by tubercules and papillae at least on its sides; the spicules develop as tables, branched rods and C and S rods.

Stichopus chloronotus

Stichopus chloronotus (common name: greenfish) is a rather small species with a firm body of quadrangular shape. Its green tegument gives origin to its common name. The ventral mouth is surrounded by a row of 20 stout tentacles. It lives in very shallow areas of the coral environment, with some populations attaining high densities. Some populations show a high rate of asexual reproduction by fission. In Réunion, *S. chloronotus*, exhibits two spawning seasons, similar to other populations previously studied on the Great Barrier Reef (GBR, Australia) (Franklin, 1980; Uthicke, 1997). However, this population is quite unusual in terms of the high densities recorded (3.7 ind. m^{-2}), low average specimen weights (between 35 and 80 g in the two populations sampled between 1998 and 2001) (Conand, Uthicke and Hoareau, 2002) and high rate of asexual reproduction (Conand *et al.*, 1998). Spawning has recently been observed at Étang Salé site by Barrère and Bottin (2007); fission occurs during the cool season (Conand, Uthicke and Hoareau, 2002). The population genetic has been compared between Pacific and Indian Ocean sites, showing that asexual reproduction is an important feature in most populations, over a wide geographic range, but also that the potential for widespread dispersal mediated by sexually produced larvae is large (Uthicke, Conand and Benzie, 2001). More recently, amplified fragment length polymorphism (AFLP) analysis have indicated the importance of both asexual and sexual reproduction in this holothurian in the Indian and Pacific Ocean. Although asexual reproduction is important for the maintenance of populations, large distance dispersal of sexually produced larvae provides the genetic link between populations (Uthicke and Conand, 2005b). It is a very common species throughout the region. It does not burrow.

Stichopus herrmanni

*Stichopus herrmanni (*formerly called *variegatus*) (common name: curryfish) is a medium-sized species (mean length about 35 cm), but larger specimens have been observed. The firm body wall disintegrates easily when collected, as in other *Stichopus.* Its colour, yellow to greenish gives it its name of curryfish. The bivium is covered by irregular conical warts arranged in eight longitudinal rows, with smaller papillae in between. It occurs in reefs and lagoons, in seagrass beds, rubbles and muddy-sand bottoms. It is not rare in Kenya and Madagascar. The sexual reproduction takes place during the warm season in the Pacific. The only study in the Indian Ocean by Tehranifard *et al.* (2006) in Kish Island (Islamic Republic of Iran) also shows a summer spawning. The authors have calculated that the average body length at first maturity is 310 mm and the diameter of mature oocytes is 200 µm. This species does not burrow and is able to move rapidly when disturbed.

Other Stichopus species are quite difficult to distinguish and need a taxonomic revision: *S. horrens, S. monotuberculatus, S. quadrifasciatus, S. quadrimaculatus* and *S. naso*.

Stichopus horrens

It is presently heavily exploited in Madagascar (under the name "smurf"), but yielding a rather low quality product (Rasolofonirina, 2007). The body is square in cross section and the length is up to 40 cm. Dorsal papillae are conspicuous. The reproductive biology is presently studied as part of the Masma (WIOMSA) project (Rasolofonirina, R., Université de Toliara, Madagascar, personal communication); it will be the first study for this species in the Indian Ocean. Fission has recently been described in Japan (Kohtsuka, Arai and Uchimura, 2005). The species is mentioned from several countries of the region and seems to have a wide distribution, as it currently illegally fished in the Galapagos Islands (Hearn and Pinillos, 2006). It does not burrow, but hides under rocks during the day.

Stichopus monotuberculatus

In Réunion, the species previously called *S. horrens* has been identified as *S. monotuberculatus* (Massin, C., Département des invertébrés, Institut royal des sciences naturelles, Belgium, personal communication). It is probably also found in Comoros, Madagascar and other places. In length it averages 35 cm. The species is nocturnal and hides contracted under rocks or rubbles during the day, which explains why it has not been observed all that frequently.

Genus *Thelenota*

Thelenota ananas

Thelenota ananas (common name: prickly redfish) is a large species (maximum length 85 cm; mean 45 cm) found in the reef environment. It is fished in several countries as it has a high commercial value. Although it is still found in storage rooms of collectors and middlemen from Madagascar and the Seychelles (Aumeeruddy and Conand, 2007), this species is now becoming increasingly rare. Its bright orange-red to brown in colour and the large papillae give it a pineapple appearance and make it very vulnerable to fishing. Its densities are generally low. It lives in coral reef environment, mostly on outer slopes. It is observed in Comoros, Mayotte (including Geyser Banks, see Mulochau, Conand and Quod, 2007), Madagascar, Kenya and Seychelles. It is very rare in Réunion. The potential fecundity is low and sexual maturity late in New Caledonia (Conand, 1989, 1993), compared with other species from the Holothuriidae (from studies in New Caledonia by Conand, 1981, 1993) which makes it vulnerable to overexploitation. Few studies have been made on its biology. It does not burrow.

Thelenota anax

Thelenota anax (common name: amberfish) is even larger (mean length 60 cm; up to 90 cm) than *T. ananas*. The body is squarish. Despite its medium commercial value, it is exploited in several countries. It lives exposed on white coral sand at the base of the relatively deep reef slopes. Nothing is known about its biology. It is cited from Comoros, Madagascar, Seychelles and Kenya, and more recently from South Africa, but not from the Mascareigne Archipelago or the Red Sea. Nothing is known about its biology in the region. It does not burrow.

A conclusion from the above summary is that the biology of most species has received less attention in the Indian Ocean than in the Pacific where many studies started in the 1980s. The regional Masma project from WIOMSA (Conand *et al.,* 2006; Conand and Muthiga, 2007a) will shortly contribute new data on the reproductive and fishery biology of several commercial species.

Globally the main ecological role of these detritus-feeders is bioturbation of the sediments and building organic matter from bacteria and micro-phytobenthos. More attention is currently given to their specific role in tropical ecosystems (Taddei, 2006; Wolkenhauer, S., Australian Commonwealth Scientific and Research Organization [CSIRO], Australia, personal communication).

Genetic analysis is only starting in the region (Uthicke and Benzie 2003; Uthicke, Conand and Benzie, 2001; Uthicke and Conand, 2005b; Kerr *et al.*, 2005); some studies have been concentrating on comparisons with Australian populations of *H. atra, S. chloronotus* and *H. nobilis* (Uthicke, Conand and Benzie, 2001; Uthicke and Conand, 2005b) as little is known on the connectivity among populations.

Several species can reproduce by fission, but they also reproduce sexually with a seasonal periodicity (Conand, Uthicke and Hoareau, 2002). Recruitment from sexual reproduction is still poorly understood, as the juveniles are not found for most species, apart for a few anecdotal observations (Shiell, 2004a, b, 2005). Burying is known from several species, but needs scientific investigations to understand its rhythms and its effects. Aestivation is not observed for any of the tropical Indo-Pacific species.

2.3 Background of sea cucumber fishery

Following the data presented during two recent and important workshops (Lovatelli *et al.,* 2004; Bruckner, 2006), the review will first concentrate on the WIO where a recent synthesis for the regional Masma (WIOMSA) project has been prepared (Conand and Muthiga, 2007a).

Kenya

In Kenya the sea cucumber fishery is (Muthiga, Ochiewo and Kawaka *in* Conand and Muthiaga, 2007a) primarily artisanal as many other coastal fisheries along the Kenyan coast, however it does contribute to the livelihoods of fisher households (Muthiga and Ndirangu, 2000; Marshall, Milledge and Afonso, 2001; Beadle, 2005; Conand and Muthiga, 2007a). The fishery has been in existence since the early 1900s as shown by Kenya National Archives (KNA) and Government reports (1985, 1991). Many species have local names, indicating the cultural identification of this fishery by the local communities. Sea cucumbers are currently collected either as bycatch by spear fishermen and other gleaners or by fishers that target only sea cucumbers using snorkel and mask or SCUBA equipment. The gear and boats are usually provided by dealers who purchase the processed product, while the collectors carry out the processing and drying at the landing beaches. The local level dealers in turn sell their products to a few exporters based in the coastal capital of Mombasa that export to China Hong Kong SAR. Information on the trade and fishing of sea cucumbers is poorly documented and primarily consists of records from the National Archives, the Fisheries Department, and recent assessments and studies by Muthiga and Ndirangu (2000), Marshall, Milledge and Afonso (2001) and Beadle (2005). The start of the sea cucumber trade is thought to coincide with the influx of Asian nationals in the 1900s. For example, by 1918 there were fisheries concessions in Malindi, Mombasa and Kipini in northern Kenya (KNA Coast 1917–1925a). Concessions were granted for three to five years at a time and Chinese fishers and dealers dominated the trade. The death of a concessionaire meant his licence ended but his family members and other interested parties could take up the concession. The fishery was then considered very lucrative. Today harvesting is primarily done by a small number of fishers concentrated in a few villages in Lamu (northern Kenya) and Kwale (southern Kenya) districts. Forty species of sea cucumbers have been reported along the Kenyan coast and of these 17 species are currently harvested.

Réunion

Holothurians have never been exploited in Réunion, because the reefs are so small that the resource would be depleted within a few days or weeks and the valuable species are not abundant (Conand and Frouin *in* Conand and Muthiga, 2007a). Some projects presented to local administrations have always been discouraged, as the scientists rejected them as a precautionary approach.

Madagascar

In Madagascar sea cucumber harvesting is a traditional and permanent activity in coastal regions, especially near coral reefs (Conand *et al.*, 1997; Rasolofonirina, Mara and Jangoux, 2004; Rasolofonirina *in* Conand and Muthiga, 2007a). This activity steadily expanded from the early 1990s and it is currently rather intense on the west coast of Madagascar (Petit, 1930; Rasolofonirina and Conand, 1998). Fishers collect various resources on the reef flat, during the low tide, such as shellfish, urchins, octopus and sea cucumbers (Rosa, 1997; Rakotonirina, 2000). Presently, more than thirty species are exploited. The harvested species, however, vary according to the market price, the international demand and their availability. *Holothuria scabra, H. fuscogilva, H. nobilis, Stichopus horrens* and *Thelenota ananas* are the main collected species. *Holothuria notabilis* which has recently been described for Madagascar (Massin, C., Département des invertébrés, Institut royal des sciences naturelles, Belgium, personal communication) is also intensively fished.

Seychelles

In Seychelles, the fishery has developed recently and has been closely followed by the Seychelles Fishing Authority (SFA) through an FAO project (Aumeerudy *in* Conand and Muthiga, 2007a). The Seychelles has been chosen as the hotspot presentation (see Aumeeruddy and Conand, this volume, for more details). Sea cucumbers in Seychelles have been fished for more than a hundred years, with reports of bêche-de-mer exports dating back to the late 1800s (Marguerite, 2005). However, the quantities fished were fairly low and it is only in the late nineties that the fishery has seen a rapid development. This is due to several factors, the main one being the high demand for bêche-de-mer on the international market and higher prices offered for the product (Aumeeruddy and Payet, 2004). The fishery has evolved from a collector-type, whereby fishers collect sea cucumbers on foot, to a more sophisticated one where most of the harvesting is done by divers using SCUBA gear. A lot of fishers have entered the fishery during the last eight years, and the fishery which was uncontrolled and unregulated, had to be put under management. The national fisheries authority first placed a number of management measures in 1999 in view of the local depletion of some species. The measures adopted included a licensing system for fishing and processing sea cucumbers, a quota on the number of fishing licenses allocated annually, and limits of four divers for each fishing license. Before this recent interest for the fishery, very little was known about the holothurians from Seychelles. A synthesis by Clark (1984) based on previous publications by Clark and Rowe (1971) listed 115 Echinoderm species, detailed 151 species, including 35 sea cucumbers and gave some brief information on the ecology of the most common ones.

Tanzania

Sea cucumbers constitute an important marine resource for Tanzania (Semesi *et al.*, 1998; Marshall, Milledge and Afonso, 2001; Mgaya and Mmbaga *in* Conand and Muthiga, 2007a). There has been a rapid expansion of sea cucumber exploitation at some sites of Tanzania (Mgaya, Muruke and Semesi, 1999). The sea cucumber fishery developed without baseline biological data and without any monitoring. Therefore, to-date, Tanzania has unknown and unquantified sea cucumber resources, though the fishery provides income to local collectors and generates export earnings (Mmbaga and Mgaya, 2004). The fishery is largely artisanal with a small commercial operation monopolised by few exporters. Local exploitation occurs year-round on reefs close to the shore sheltered from prevailing winds. The peak in the collection of sea cucumbers falls in the period October to December and April to May when winds are usually light, and trips can be made to the outer reefs. The level of exploitation differs from shore to shore depending on fisher experience, number and category of fishers, fishing techniques used and season. Sea cucumbers are purchased by a number of traders based in Dar-es-Salaam, Tanga and Zanzibar from where they are exported to eastern Asia, mainly to China mainland, China Hong Kong SAR and Singapore (Marshall, Milledge and Afonso, 2001). The increase in export of bêche-de-mer was observed from 1980s (< 200 tonnes per annum) to 1992 (617 tonnes) while they have continued to decline thereafter (Marshall, Milledge and Afonso, 2001). These authors report further that the number of official bêche-de-mer exporters on the Tanzania mainland has decreased from 23 tonnes in 1993 to 8 tonnes in 1997 and that the exporters have largely halted exports as a result of the declining profits. A list of 15 species currently harvested dates back to the synthesis by Marshall, Milledge and Afonso (2001) and includes *Actinopyga miliaris*, *Bohadschia* sp., *Stichopus variegatus* (now *herrmanni*) and *Bohaschia argus* (probably *B. atra*).

Comoros

The fishery of Comoros is briefly addressed in Samyn, Van den Spiegel and Massin (2005). They express their deepest "concern with regards to the present, rather blind, overexploitation of sea cucumbers in the Comoros" giving a few qualitative observations to support this. They have recently produced a detailed description (in French) of the different species and their habitat (Samyn, VandenSpiegel and Massin, 2006). The harvesting and processing is controlled by Chinese immigrants.

Mayotte

In Mayotte, several brief studies have shown that the same problems are encountered (Pouget, 2004, 2005; Conand *et al.,* 2005). Mulochau, Conand and Quod (2007) in a quantitative evaluation have shown that in the different habitats from Geyser Bank (off Mayotte), the sea cucumbers are presently very few and only from a few species; it is highly probable that they have now been fished.

Mauritius

The Mauritius (including the islands of Rodrigues and Saint Brandon) holothurian fauna has attracted attention recently, but a few species as *Actinopyga mauritiana* had been described long ago by Quoy and Gaimard (1833). Michel (1974) listed 23 species of Aspidochirotida and Arakaki and Fagonee (1996) gave a list of species from several countries including Mauritius. The distribution of a few species has been studied (Luchmun *et al.,* 2001) and unpublished reports present further data on the distribution of the main species *A. echinites, H. atra* and *H. leucospilota* (Muller, 1998). The echinoderms from Rodrigues have been sampled by the Shoals of Capricorn and studied by Rowe and Richmond (2004) who identified 29 holothurians. Other studies on the distribution have been recently conducted (Mrowicki, 2007). In Saint Brandon the main holothurian abundances have recently been reported (Laxaminaraya, A., Indian Council of Agricultural Research [ICAR], India, personal communication). The exploitation in Mauritius started on a trial basis in late 2005 and was continued by six licensed operators with exports of around 80 tonnes (Kadun, S., Mauritius Government, Mauritius; Laxaminaraya, A., ICAR, India, personal communication).

Mozambique

It has not been possible to get recent information from Mozambique, except on the present fishery from Querimbas Archipelago (Hill, 2008). The detailed report by Marshall, Milledge and Afonso (2001) remains the main review for this country where only "minimal data on the fishery" and "illegal exports to Tanzania" exist. The dozen species exploited are the same as in the other countries of the region. The mean annual quantities exported for the decade 1980–90 were around 65 tonnes, but they have dropped in the 1990s. Exports are to Taiwan Province of China, China Hong Kong SAR and Singapore. The report by Marshall, Milledge and Afonso (2001) presents several management initiatives and recommendations for different localities, as for example in Inhaca, in 1990, a closed season from December through to March, a minimum size of 20 cm, rotational exploitation arrangements, while in 1995 for Inhambane the closure of the fishery (which fishers continued to exploit due to lack of control). From recent information, most of the sea cucumber resources in the central and the southern regions of the country are depleted, with the exception of those in the Bazaruto Archipelago National Park (Motta, H., Mozambique, personal communication). The species most captured in these two regions is *H. scabra*. In the north the mammy holothurians (teatfish) may still be available but only at some depth, for which fishers need to dive in deeper waters. A recent study by Hill (2008) shows that the fishery is still active in the Querimbas Archipelago. Mozambique is certainly one of the countries in the region where the sea cucumber fishery should be carefully followed, aquaculture developed and management of wild resources encouraged.

Egypt

The bêche-de-mer fishery in Egypt has been described by Lawrence *et al.* (2004). It began in 1998, in the southern part of the country. It was initially performed by trawling boats. It expanded dramatically and the Red Sea Governorate initiated a ban in 2001. Illegal fisheries therefore developed along the coast. Under the pressure from the fisheries authorities the ban was lifted in 2002. As the preliminary observations on the stocks showed a rapid decline of the resources, a new ban was decreed in 2003 and new decisions were expected based on the results of the stock assessment. Ahmed and Lawrence (2007) have recently updated the situation. Following the initial study, they have re-visited some of the sites to determine whether there was any evidence of stock recovery following the 2003 fishery ban. Four sites were assessed using the belt transect method applied in the original study. The sites were selected based on their accessibility, initial levels of stocks and levels of exploitation. A further six sites were assessed visually to determine the presence or absence of commercial species. It appeared that four years after the ban on the fishery, there is some evidence of the return of selected commercial species to some of the sites, but no evidence of stock recovery.

Eritrea

Recent information on sea cucumber fisheries has been reported by Tewelde and Jeudy de Grissac (2007) and Kalaeb *et al.* (2008). Sea cucumbers have been harvested for more than forty years by local artisanal fishers, but only in small quantities. The artisanal fishers collected sandfish (*Holothuria scabra*) by skin diving using a "nadur" or home-made mask. The catches are gutted, dried and marketed to Aden (Yemen). Exploitation of sea cucumber on a larger scale started in 2000 with the products exported to Asia. Ten species are presently collected and the largest quantity is from *Holothuria atra* (lolly fish) locally known as "Lega" which is found in abundance in shallow waters. Deeper species are now harvested by divers using an air compressor or SCUBA or by artisanal trawling.

India

In India this old artisanal activity is reported to have been introduced by Chinese long time ago (Hornhell, 1917). During the last two decades considerable scientific

TABLE 4
Commercial sea cucumber species exploited in Eritrea

Species	Common name	Local name	Commercial value
Holothuria scabra	Sandfish	Hedra beyda	1
H. fuscogilva	White teatfish	Abu habhab abyed	1
H. nobilis	Black teatfish	Abu habhab aswed	1
H. atra	Lolly fish	Lega	3
H. edulis	Pink fish	Abu sanduk tina	2
Thelonota ananas	Prickly redfish	Abu mud	2
Actinopyga miliaris	Black fish	Abu shelalik	2
A. mauritiana	Surf redfish	Abu sanduk hager	3
Stichopus herrmanni	Curry fish	Hamra	3

Source: Tewelde and Jeudy de Grissac, 2007.

studies have been conducted on the taxonomy, the ecology and the exploitation of sea cucumbers; a comprehensive list of references is given in James (2004b). Subsequent studies focused on hatchery techniques (James, 1994a). Starting in 1994, the need for management of the resources was recognized as overexploitation was noted as the fishery typically focused on successive species, first *H. scabra*, then *A. echinites*, followed by *A. miliaris* (James and James, 1994). The Government of India has banned collection, processing and export of all sea cucumbers from several regions since 2002 (James, 2004a). Illegal fisheries, however, still continue with "huge quantities of *Stichopus herrmanni* processed every year near Tuticorin during the months of May to July. Two lakhs (Indian numbering system equal to one hundred thousand) of sea cucumbers are collected every day during this period. No seed is produced since 2000 since the scientists are unable to collect broodstock due to the fishing ban. There are some indications that the ban may be lifted in the near future if certain conditions are met" (James, D.B., India, personal communication). Recently "the National Fishers Forum (NFF) has sought the Prime Minister's intervention in lifting the ban as this had adversely affected the livelihood of thousands of fishers in Tamil Nadu. The Forum said there was no fear of depletion of this fish as it was caught by diving and its longevity was estimated to be 10 years. The ban on shark and some items was lifted following protests by the fishing community, but no exemption was made in the case of sea cucumber, the Ministry had constituted a committee to review the ban, but there has been no response since then. The Forum has also sought lifting the ban on *Holothuria atra* and *H. spinifera* species. The NFF activists arrived in the capital and would sit on an indefinite dharna near Parliament until the ban would be lifted"[1].

Maldives

Maldives has experienced a short term fishery with a boom-and-bust cycle after only a few years (Reichenbach, 1999). Some projects for mariculture had been set up. No recent information has been found.

Sri Lanka

In Sri Lanka sea cucumber collection "is rooted in antiquity, when Arab and Chinese merchants employed local inhabitants of western, northern and eastern maritime regions, to gather and cure the animals for them" (Adithiya, 1969). Terney Pradeep Kumara, Cumaranathunga and Linden (2005) have recently described the present fishery. In the western, northern and eastern waters the activity flourished with increased demands from East Asian countries. As shallow areas were fished out, the availability of SCUBA gear in the 1990s enabled deeper habitats to be exploited. Published Sri Lanka Customs Department statistics (Terney Pradeep Kumara, Cumaranathunga and Linden, 2005) for the whole Island shows steep rises and declines in quantities exported (see Figure 7). These fluctuations appear to correspond with discoveries of new sea cucumber grounds and their depletion as a result of unrestricted and intensive collection. In the southern parts, along the coastline from Negombo to Dondra the fishery commenced only about ten years ago. Regrettably, overexploitation with no effective management measures has led to its complete collapse. Sea cucumbers are locally known as "muhudu kekiri" or "atta" and are exported to other Asian countries. There are nearly 200 known species in the ocean around Sri Lanka. About 75 species have been shown to be present in shallow waters while nearly 50 species can be collected from the intertidal region (Clark and Rowe, 1971). Although sea cucumbers were abundant, they were not harvested in the south until buyers from Singapore created a demand. They bought mainly two species, *Holothuria edulis* and *Holothuria atra* for very low prices. Although they paid only one rupee (USD 0.01) per animal, fishers earned a considerable amount of money because

[1] Source: The Hindu (http://www.hinduonnet.com/2007/08/29/stories/htm).

of the abundance of the organism in the shallow coastal waters. The price increased up to five rupees each as supplies dwindled. Harvesting at first was done by hand while wading or using snorkel gear in shallow water. As the shallow areas were fished out, SCUBA gear was used to exploit increasingly deeper sea cucumber beds. Over the past three to four years the sea cucumber fishery in the shallow areas off southern Sri Lanka declined rapidly and collapsed. The fishers and divers of these areas then turned to distant sea cucumber beds. Although fishers from Negombo on the western coast started the fishery in the southern part of the Island, fishers/divers from the southern fishing towns of Mirissa and Dondra now dominate. The sea cucumber fishery in the south of the Island is by opportunistic tuna fishers. Whenever exploitable sea cucumber beds are located they switch to that; and when the beds are fished out, revert to tuna fishing. The duration of each spell of activity depends on the sea cucumber population and the number of boats and divers that participate in the fishery. The catch is washed and stored on ice. SCUBA divers complain that they now need to dive to increasing depths to harvest holothurians, leading to accidents, and also that they are forced to look for new fishing grounds.

Oman

Oman has re-started its sea cucumber fishery very recently (Al-Rashdi, Al-Busaidi and Al-Rassadi, 2007). Mahout Bay in the Gulf of Masira (Arabian Sea) is the main area of sea cucumber fishery in Oman and the fishing dates back to 1960s, when Mahout Island was considered a travelling port to East Africa and India. The islanders used to collect sea cucumber and exchange it for food provisions from abroad. The fishery was then discontinued in 1970s as the port was no longer in operation. However, the fishery re-commenced in 2003 due to the demand of bêche-de-mer from traders from the United Arab Emirates. Intensive fishing was practiced throughout 2004 and 2005 when the fishery was totally unregulated. The sea cucumber in Oman forms a minor fishery of the traditional sector with almost 50 percent women involvement. The fishery is located in Mahout Bay and supported by the sandfish, *Holothuria scabra* which is still abundant in seagrass beds. The local name of the species is "Feik Albahar" meaning "sea jaw". The sandfish are hand picked by walking in the shallow areas during low tide between late November and May. Sandfish is traditionally processed to a dried form after gutting, boiling and cleaning.

Chagos Archipelago

The Chagos Archipelago, in the central Indian Ocean, were first inhabited in the late eighteenth century, and by the mid-twentieth century there was a small economy based around copra (dry coconut flesh), but the resident population was removed by the British in the early 1970s. With the exception of Diego Garcia, where there is now a large US military base, the islands have remained uninhabited for over 30 years. Surveys of the coral reefs during the 1970s revealed a remarkably pristine reef fauna. Fisheries patrols, largely to monitor the licensed tuna fishery, began in the 1990s. At the same time the first observations of illegal sea cucumber fishing were made. An expedition in 1996 revealed that reef shark populations had declined by some 85 percent linked to the illegal fisheries (Spalding, 2006).

South Africa

According to the author no information exists or is readily available on sea cucumbers exploitation in South Africa where the rich holothurian fauna was recently reviewed by Thandar (2007).

Most reports on sea cucumber fisheries are still qualitative and the situation in the region is not very clear particularly for a number of countries. Probably uncontrolled overexploitation is widespread and occurs in many areas.

2.4 Species in trade

Currently, the only use of sea cucumbers in the region is for the export of bêche-de-mer (Conand, 2001, 2004a; Bruckner, 2006; Conand and Muthiga, 2007). Table 2 lists 30 species with their commercial importance ranked as follows: 1 = high commercial value; 2 = medium commercial value; and 3 = low commercial value.

The processing is generally done artisanally in this region, following the methods already described for sandfish and the other species (Anonymous, 1994; Conand, 1999); in the Seychelles some processors operate large plants with considerably large drying rooms (Pinault and Conand, 2007). Currently the fishers have to fish further away from the traditional grounds than they used to. Sea cucumbers tend to eviscerate after collection they need to preserve their catches in salt as described for Indonesia by Tuwo and Conand (1992). This practice occupies less room in the boat and enables fishers to go out on longer fishing trips. In India some processing techniques have been refined as for example the removal of the skin of the sandfish which suppresses the need of the burying them in sand (James, 1994b).

2.5 Population status

The population status is a important parameter which can be estimated through several methods: (i) scientific evaluations of densities, or biomasses; (ii) fishery data on catches, efforts and CPUE; (iii) trends in these indicators; and (iv) retro-calculations from export statistics (Conand, 1989). Concerning the scientific evaluations, the method used and the spatial scale of the study are important factors to take into consideration. Some density data have been transformed to numbers of individuals per 100 m^2 to allow comparisons. Some examples will be presented, despite few data available and no synthesis has been carried out. The lack of historical data in many countries reflects the lack of interest on this resource until recently. Table 10 summarizes the population exploitation status in each country in the region. Recent information on abundance has been collected from WIO countries through the Masma project (Conand and Muthiga, 2007) and is presented below.

In **Kenya** the relative abundance of sea cucumbers, calculated from transects, was highly variable ranging between 0.3 to 7 ind. 100 m^{-2} and averaging ~3.5 ind. m^{-2} (Muthiga and Ndirangu, 2000; Muthiga, Ochiewo and Kawaka, 2007). The density of individual species was generally low (0.01 to 1.5 ind. m^{-2}). In general, the highest densities of sea cucumbers were found in shallow water habitats, primarily reef lagoons and channels (around 6 ind. 100 m^{-2}), as well as in Marine Parks; however, the factors that control variability have not been yet sufficiently explored.

In **Réunion** quantitative data on holothurian densities can be found in Conand and Mangion (2002), Conand (2003) and Fabianek and Turpin (2005) (Conand and Frouin, 2007; Frouin *et al.*, 2007), however, more data is needed for a good understanding of the structure and function of holothurian guilds in Réunion reefs. *H. leucospilota* and *H. atra* are the dominant species. Most studies were done in the Saint-Gilles/La Saline reef which is the most important in Réunion (almost 40 percent of reef surface in Réunion). Back reef areas (characterized by dominant sandy bottoms) and inner reef flats (dominated by hard substrata) were extensively sampled. *H. atra* showed higher densities in this reef than *H. leucospilota*. The mean value in the reef was 0.68 ± 1.67 ind. m^{-2}. *H. atra* was dominant in back reefs (1.44 ± 2.46 ind. m^{-2}). The distribution of this species over the reef complex was highly heterogeneous. Maximum values of 5.1 ± 0.1 to 6.6 ind. m^{-2} (from Fabianeck and Turpin, 2005; and Conand and Mangion, 2002, respectively) or a mean around 580 ind. 100 m^{-2}, has been recorded in one reef station (Planch'Alizes). Jaquemet, Rousset and Conand (1999) found 4.8 ± 0.53 ind. m^{-2}. Conand (2004c) analysed data collected from 1993 to 2000 and determined a density over this long term (4.7 ± 0.3 ind. m^{-2} or 470 ind. 100 m^{-2}). In inner reef flats the densities were <10 ind. m^{-2}. In contrast, *H. leucospilota* was present in all back reef and inner flat stations, with total mean density of 0.59 ± 0.38 ind. m^{-2}. Highest densities were in back reefs (mean 0.84 ± 0.38 ind. m^{-2} versus 0.39 ± 0.26 ind. m^{-2} in inner reef flats), with a maximum density of 1.4 ± 0.7 ind. m^{-2} in one station (Planch'Alizés). Other holothurian species have been observed in the field, but few quantitative data are available. *Stichopus chloronotus* showed a total mean density of 0.34 ± 0.69 ind. m^{-2} with no individuals found in inner flats. Mean density in back reefs was 0.78 ± 0.91 ind. m^{-2} (Fabianeck and Turpin, 2005). Conand *et al.* (1998) found 3.7 ind. m^{-2} in Trou d'Eau, a station where the sexual reproduction and fission were studied. Large size *Actinopyga echinites* and *A. mauritiana* on the outer reef flats and reef front respectively, are also frequent species (Conand and Mangion, 2002).

In **Madagascar** an investigation of the abundance and distribution of the different exploited species was carried in 1996 on the Toliara Great Barrier Reef (southwest of Madagascar) (Mara *et al.*, 1997; Rasolofonirina *in* Conand and Muthiga, 2007a). The results indicate the biomass of edible species on different habitats of the reef (Table 5). The maximum biomass and density observed at the time was for *Holothuria notabilis* with 0.06 ind. 100 m^{-2} (or 60 kilograms) on the spatially-limited zone of the dune in the seagrass habitat. This species was not exploited at that time, but currently it is the most exploited one, which has resulted in a considerable decrease of its biomass.

For **Seychelles,** Aumeeruddy and Conand (this volume) and other reports (Aumeeruddy *et al.*, 2005; Aumeeruddy, 2007) present density and biomass estimates for large areas in the Seychelles (using a stratified sampling according to depth and substrates). Due to the extended size of Seychelles' EEZ (1.4 millions km^2), a recent stock survey focused on two main fishing grounds, i.e. the Mahé Plateau and the Amirantes which together cover an area of approximately 48 305 km^2. During the surveys, around 20 species of large sea cucumbers were recorded. The overall average density in the study area was 19.78 ind. ha^{-1} which equates to a standing stock of approximately 95 millions individuals corresponding to an estimated live weight of about 98 600 tonnes. The highest density of commercial sea cucumbers was found in the shallow strata (<20 m) while the deep strata (>50 m) had the lowest density. In the intermediate depth strata (20–50 m), the sea cucumber density was of an intermediate level, but due to the large size of the specimens, it had the bulk of the overall standing stock. Figure 3 presents the mean densities for the main species.

TABLE 5
Biomass of the main holothurians per biotope in the Toliara Great Barrier Reef, southwest Madagascar. The unit in kg/ha is chosen given the low biomass of most species

Species	Common name	Biomass (kg/ha)				
		Inner slope	Seagrass	Microatoll	Detrital fringe	Outer flat
Actinopyga echinites	Trokena		0.018		1.125	
A. mauritiana	Fotsitreka				1.159	0.117
A. lecanora	Zangabato	0.09		0.299		
Bohadchia vitiensis	Falalijaka	33.38	2.151			
Holothuria atra	Stylo noir	2.617	0.618	3.7	0.884	1.316
H. cinerascens	Zanga fleur				1 520	
H. edulis	Stylo rouge	0.602	0.095	0.515		
H. excellens	Delave	1.380	0.274	0.124		
H. leucospilota	Zanga kida	13.302	1.004	4.243	1.515	
H. notabilis	Dôrilisy		6 000			
H. nobilis	Benono	0.1	0.046			
H. scabra	Zanga fotsy	0.298	3.774			
H. scabra var. *versicolor*	Zanga mena	0.802	1.207			
Stichopus horrens	Smurf	0.228	0.137	0.218	0.113	0.033
S. herrmanni	Trakitera	0.636	1.022	0.048		
Total		**53.44**	**6 010.3**	**9.15**	**1 524.80**	**1.47**
Percentage		**0.70**	**79.09**	**0.12**	**20.07**	**0.02**

Source: Mara *et al.*, 1997 in Rasolofonirina 2007; Common names are from Conand (1999).

FIGURE 4
Abundance of the main commercial species in the Seychelles

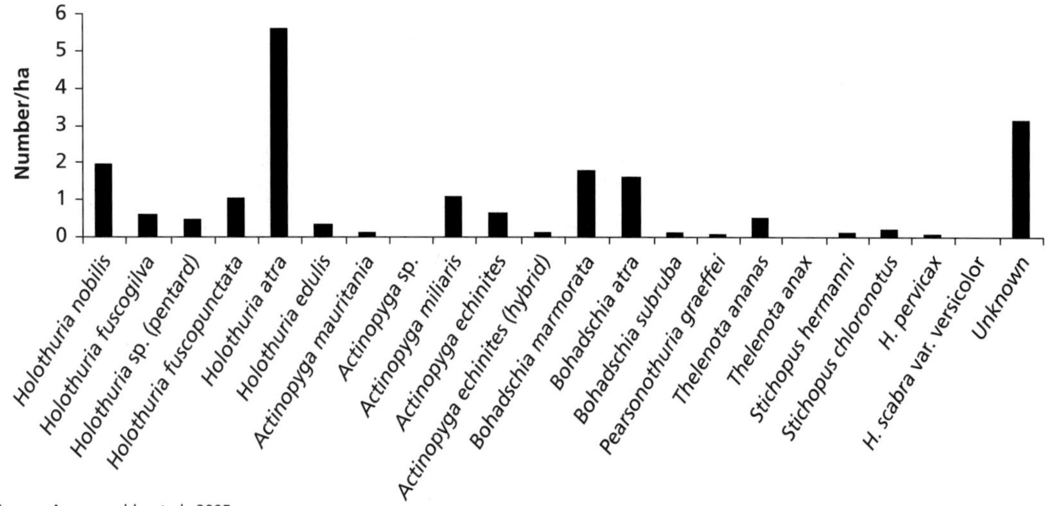

Source: Aumeeruddy *et al.*, 2005.

As in other tropical countries, *H. atra* is the most frequent and abundant species with a mean density of nearly 0.06 ind. 100 m^{-2}; the three teatfish species are quite abundant with a mean density of 0.02 ind. 100 m^{-2}. It is worth noting that there is a need to harmonize stock monitoring methodologies in order to produce results and data which are comparable, as different methods will produce dissimilar findings in terms of densities and biomass.

Concerning other countries, data are mostly lacking or are too old, but they give at least an idea on the state of the populations, as in **Maldives** where *H. fuscogilva* has been heavily exploited (Reichenbach, 1999). Lawrence *et al.* (2004) provided detailed density estimates for different areas in **Egypt**. *Holothuria atra* was by far the most abundant followed by *A. mauritiana*. A second survey conducted four years following the official fishing ban, has shown that there was some evidence of the return of some commercial species, but no evidence of stock recovery (Ahmed and Lawrence, 2007).

Catch and effort data are necessary to calculate catch-per-unit-effort (CPUE) values in a given fishery. Nevertheless, a value of CPUE is sometimes used by scientists as an indicator of abundance, when the densities are not directly accessible, as in the case of very large zones. The counts of the different holothurian species are therefore done during a temporal interval rather than over a unit surface area, as it was done for the crown of thorns (*Acanthaster* sp.) in New Caledonia (Conand, 1985). Overall, CPUE values are not yet available on a regular basis for the countries in the region, except for the **Seychelles** where the fishery is regularly monitored (Aumeeruddy and Conand, this volume). The data presented on the main species by Aumeeruddy (2005) shows that the CPUE, given in number of specimens per diver per day, has decreased from 2001 to 2006 (Table 6); the exception of the "pentard" is explained by an increase of effort on this species.

TABLE 6
Number of the main commercial sea cucumber species in the Seychelles collected daily by one diver from 2001 to 2006

Year	Black teatfish	Sandfish	White teatfish	Prickly red	Pentard	Other spp.
2001	7,15	0,20	29,09	4,86	4,83	5,95
2002	4,75	0,52	30,06	4,67	7,32	29,78
2003	3,22	0,01	9,97	6,09	18,68	27,15
2004	1,82	0,12	7,98	2,38	11,51	10,12
2005	1,52	0,01	6,04	2,26	11,01	12,88
2006	1,05	0,19	3,94	1,43	16,22	10,08

Source: Aumeeruddy *et al.*, 2005.

Based on a socio-economic survey in **Eritrea** the mean CPUE by day by diver is 1.68 kilograms dried sea cucumber (Tewelde and Jeudy de Grissac, 2007); given the mean crew per boat of 4.68, the mean number of days of the trip of 23 the average boat catch in dry weight is 175 kilograms.

In conclusion, the unit used for catches, effort and CPUE should be standardized in the future to get a more accurate regional view. The relations between fresh and dried weights, as shown in Aumeeruddy and Conand (2007) are useful to compare different sets of data.

2.6 Catches

In the region the only use of the wild sea cucumbers resources is for bêche-de-mer for the export markets (Conand, 2004a; Conand, 2006b *in* Bruckner, 2006b). Aquaculture has been at an exploratory level in Madagascar from 2000 (Jangoux *et al.*, 2001), but the encouraging results obtained from the Belaza hatchery (Rasolofonirina, Mara and Jangoux, 2004; Lavitra *et al.*, 2007) may well encourage the commercial production of sea cucumber (Rasolofonirina, R., Université de Toliara, Madagascar, personal communication).

The fishery is artisanal in most countries, but at a larger scale in the Seychelles. As a result there are generally no reliable data on the catches as the sea cucumbers are collected by untrained fishers from scattered villages. It is however possible to obtain a general picture of the catches by analyzing the statistics of the processed products, which being for export market require official export permits. In most countries the sea cucumbers (and other marine species) are generally collected at low tide wadding on the reefs and snorkelling. Since the reef flats have been mostly depleted, fishers now dive with SCUBA.

Table 7 presents FAO production data from different countries for the last 12 years. The data shows that (i) seven countries in the FAO Area 51, compared to only 3 countries in Area 57 report sea cucumber catches; and (ii) the WIO accounts for over a third of the world's total sea cucumber landings. The regional data presented are mere indications as probably several countries in the region are exploiting sea cucumbers without declaring exports or simply lumping sea cucumber with other dried export items such as shark fins.

More detailed information on present and historical catches in selected WIO countries is presented below (Conand and Muthiga, 2007a).

TABLE 7

Sea cucumber production by countries and fishing areas and percentage of total world production (in tonnes)

Country	1994	1995	1996	1997	1998	1999	2000	2001	2002	2003	2004	2005	Average 1994–2005	Percentage world total
Egypt	-	-	-	-	-	-	20	139	2 310	527	15	5	**502.7**	*3.4*
Madagascar	5 400	5 400	5 400	5 400	1 446	1 500	1 500	1 500	500	500	500	500	**2 462.2**	*1.8*
Maldives	66	94	145	318	85	54	205	226	191	239	182	117	**160.2**	*1.1*
Tanzania	1 591	1 460	1 644	1 527	1 800	189	372	340	65	75	10	10	**756.9**	*0.2*
Kenya	41	55	15	41	38	15	30	13	68	27	28	18	**32.4**	*0.2*
Yemen	102	-	-	-	-	1	-	-	14	10	380	400	**151.2**	*1.0*
Mozambique	0	6	54	7	2	8	12	11	10	4	1	1	**9.7**	*0.1*
FAO Area 51	7 200	7 015	7 258	7 293	3 371	1 767	2 139	2 229	3 158	1 382	1 116	1 051	**4 075**	*27.8*
Indonesia	548	227	269	338	630	689	903	697	649	870	360	624	**567.0**	*3.9*
Sri Lanka	92	100	150	272	203	170	145	90	150	170	280	260	**173.5**	*1.2*
Malaysia	-	-	-	-	-	-	-	-	-	-	-		**-**	*-*
FAO Area 57	640	327	419	610	833	859	1 048	787	799	1 040	640	884	**741**	*5.1*
World total [1]	16 246	15 556	17590	15 295	13 613	12 596	16 020	11 964	15 243	18 578	13 017	9 878	**14 633**	*100*

[1] Excluding data from Japan and the Republic of Korea as data is reported in fresh weight.

Source: FAO, 2005b.

FIGURE 5
Kenyan annual bêche-de-mer production

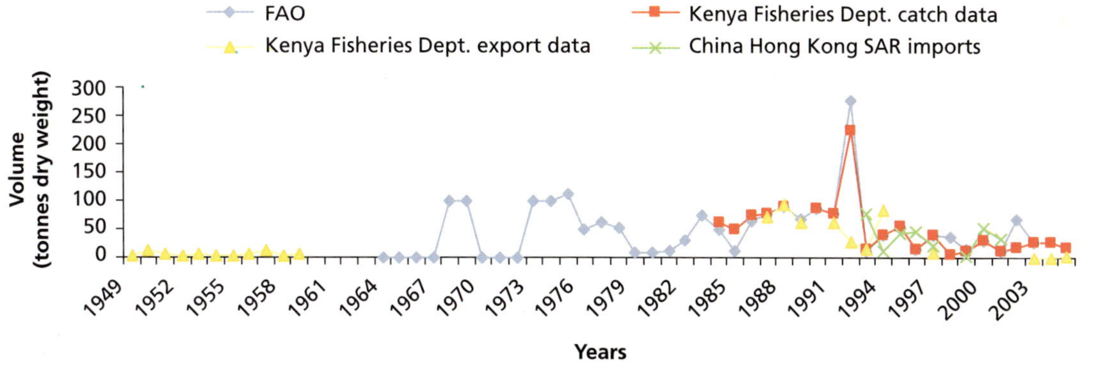

Source: Kenya Fisheries Department; FAO; Kenyan imports into China Hong Kong SAR (Infofish Trade News) (from Muthiga, Ochiewo and Kawaka, 2007).

Kenya (from Muthiga, Ochiewo and Kawaka, *in* Conand and Muthiga, 2007a). Historical and current production statistics on sea cucumbers are difficult to obtain as they are inadequately collected and recorded often resulting in landing and export statistics that often fail to tally. Comparison between data from FAO (FAO, 2005b), the Kenyan Fisheries Department and Kenyan imports into China Hong Kong SAR show a variable trend of production (Figure 5). In general, the catches were fairly low in the early 1950s averaging between 3 and 12 tonnes; they increased to about 80 tonnes between the late-1960s and mid-1990s. The production has averaged about 20 tonnes in the last decade. In Kenya, although more than seventeen species of sea cucumbers are harvested, *H. fuscogilva* dominates the catch (Muthiga, Ochiewo and Kawaka, 2007).

TABLE 8
Evolution of trepang exports from Madagascar

Year	Exports from Madagascar (tonnes)
1987	60.6
1988	119.4
1989	113.8
1990	202.6
1991	545.3
1992	423.2
1993	356.6
1994	539
1995	317.2
1996	307.4 ; (340.5)[3]
1997	(150.9)[1,2]; (331)[2]; (161.5)[3]
1998	322.5, (6.5)[2]
1999	326.6
2000	389.8
2001	355.2
2002	986.9
2003	204.5
2004	299.9

Sources: [1] Report SIR-PRH/MPRH (Inter-Regional Servlce of Fishery); [2] Veterinary Service (COS); [3] Statistics Services in Fishing and Marine Resource Office.

Madagascar (from Rasolofonirina *in* Conand and Muthiga, 2007a). Concerning the historical records, the first exports recorded from Madagascar were in 1920 with about 40 tonnes of trepang from three species (Petit, 1930). Then exports varied annually from 50 to 140 tonnes. Since 1990, the harvesting of sea cucumbers greatly increased. Exports of trepang reached nearly 600 tonnes in 1991 and 1994 (or about 6 000 tonnes in fresh weight) and about 980 tonnes in 2002. After 2002 official data showed a significant decline in exports. Since 1990 signs of overexploitation have been observed and included: (i) the declining quality of the product; (ii) the decrease in product size; (iii) the use of illegal fishing equipment; (iv) strong competition between collectors (Conand *et al.,* 1998); (v) fishing outside territorial waters; and (vi) the collection of juveniles (Conand, 1999).

In **Seychelles** (from Aumeeruddy *in* Conand and Muthiga, 2007a) bêche-de-mer has been exported for more than a century. For a long period and until 1999, the fishery was unregulated and thus no catch and effort data were collected. Figure 6 shows the limited yearly exports from 1894 to 1999 (<10 tonnes) and the sudden increase in 1999.

FIGURE 6
Bêche-de-mer exports from the Seychelles

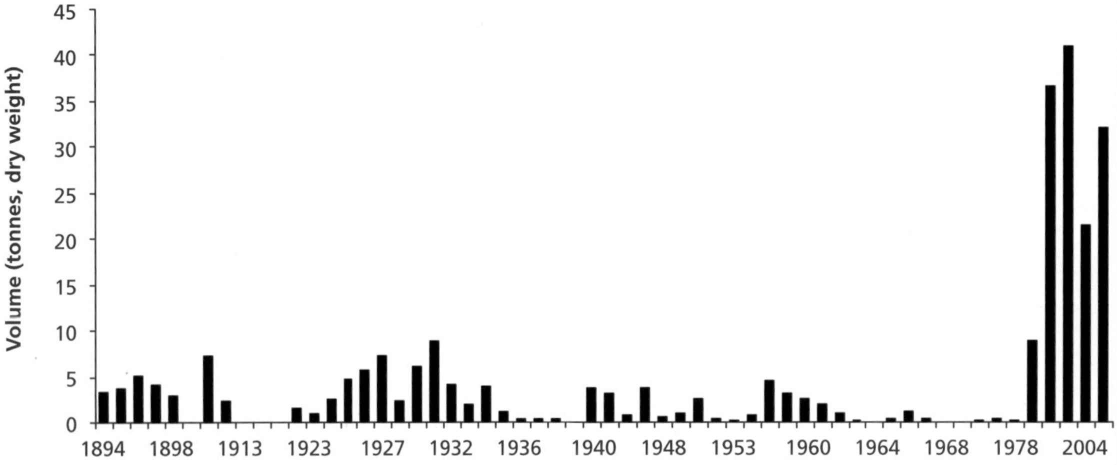

Source: Aumeeruddy, 2007.

Traditional markets are countries from Southeast Asia (China, China Hong Kong SAR, Malaysia and Singapore). Aumeeruddy and Conand (this volume) report an upward trend in the catch of the main targeted species since 1999. Fishers have been targeting mainly six high-value species (*Holothuria nobilis*, *H. fuscogilva*, *H. scabra*, "pentard", *Thelenota ananas* and *Actinopyga mauritiana*). The very high increase in catch between 2004 and 2005 is worrying, considering that the number of fishing licenses (25) has remained the same. It is possible that the reporting of catch which was an issue in the previous years improved in 2005. However one of the main reasons for the high increase in catches in 2005 is the change in fishing practice adopted by fishers. Several fishers who have small fishing units are now transferring their catch at sea to a mothership which then brings back the sea cucumbers to the processors. They can thus stay longer at sea, and the fishing effort has increased considerably with the same number of fishing licenses.

From these examples, it appears that catches in the region have increased strongly during the 1990s and then have dropped in some countries due to overexploitation. Countries have been aware of the situation and some of them are trying to take measures for a better management, as is the case in Seychelles.

FIGURE 7
Export of bêche-de-mer from Sri Lanka

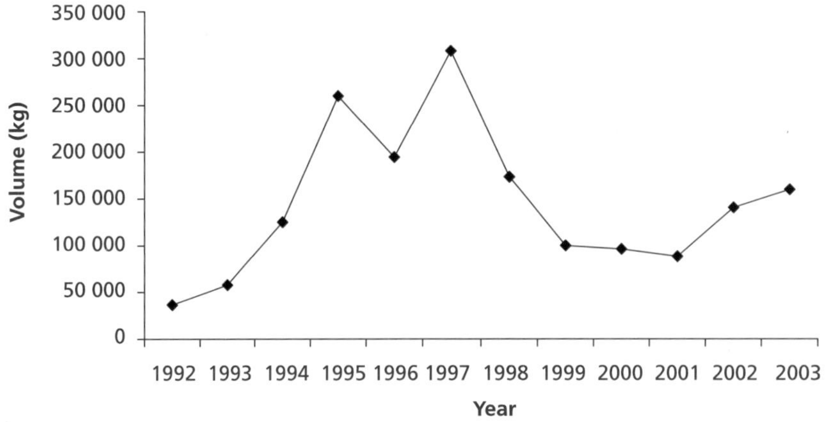

Sources: External trade statistics of the customs; Terney Pradeep Kumara, Cumaranathunga and Linden, 2005.

TABLE 9
Official sea cucumber harvest in Eritrea from 1998 to 2006, gutted and dried weight (tonnes)

Year	1998	2000	2001	2002	2003	2004	2005	2006
Harvest	0.258	11.66	80.584	451.89	283.58	379.90	278.00	241.77

Source: Ministry of Fisheries and National Fisheries Corporation (Tewelde and Jeudy de Grissac, 2007).

Data on exports from some countries as **Sri Lanka** (Terney Pradeep Kumara, Cumaranathunga and Linden, 2005) (Figure 7) can be used to infer catches. They also appear to have peaked in the 1990s with around 300 tonnes and decreased to 150 in 2003. These data slightly differ from the FAO statistics (cf Table 7).

The recent report by Tewelde and Jeudy de Grissac (2007) gives the annual data on the products prepared from the catches in **Eritrea**, from 1998 to 2006. In 2007 the fishery was closed to prepare a new management plan. Two government bodies follow the planning regulation, management, exploitation and marketing of fisheries resources: Ministry of Fisheries (MOF) and National Fisheries Corporation (NFC); they will be presented in section 2.7. The highest production occurred in 2002, after three years of increase. In the last three years the harvest dwindled. However, the decrease could be interpreted as a result of increase of illegal trade to Yemen. The fishermen are attracted by the high price of sea cucumber in Yemen and the secure supplies of fuel, food items and spare parts. The survey showed the mean catch per unit effort (CPUE) per day for collector is 1.68 kilograms.

Taking into account the available information from the different countries on densities, CPUE, catches and other more or less anecdotal information, the status of sea cucumber resources in the region can be summarized in the Table 10.

It appears that in 12 out of the 30 countries covered in the region, the resources are presently over-exploited (at least for the main commercial species) or fully-exploited. Even in the countries where the fishery has recently started (or re-started), resources are been depleted rapidly, given the use of scuba the favourable fishing grounds. This "bust and boom" cycles are sometimes difficult to identify in the catch statistics. In conclusion, with an increased market demand in the 1990s, the sea cucumber fishery increased in the different countries of the region, then dropped slightly in the 2000s. Because bêche-de-mer is mainly an export product, the volume of catches can only be inferred from the export statistics in countries that have no capacity to set up the collection of catch statistics. Careful consideration to the units of export used (pieces numbers or weights, fresh, in salt or dried) is rather important when estimating catches from exports.

2.7 Management measures in place

Unlike for marine finfish, there are no large-scale assessment of the stocks of sea cucumbers, except for the Seychelles (Aumeeruddy *et al.*, 2005). Management measures have therefore not been based on scientific data and have sometimes caused conflicts of interest. Several examples of national management have been presented in Lovatelli *et al.* (2004) and Bruckner (2006). Recent information concerning the WIO is presented in Conand and Muthiga (2007).

Despite the long history of the sea cucumber exploitation in **Kenya** (Muthiga, Ochiewo and Kawaka *in* Conand and Muthiga, 2007a), this fishery has evolved without much intervention from the management authority. Early records of management interventions include the removal of royalties in 1959, originally imposed to control collection and to revitalize sea cucumber trade. Along the Kenyan coast only a few studies have been conducted on some aspects of the fishery (Muthiga and Ndirangu, 2000; Samyn and Vanden Berghe, 2000; Marshall *et al.*, 2001; Samyn, 2003; Beadle, 2005). Most have raised concerns about the viability of this fishery especially as trade records indicate a rapid growth of the fishery from the 1960s to 1990s. However, the lack of historical information makes it difficult to discern whether production peaks

TABLE 10
Exploitation status of sea cucumber resources, by area and country. Countries with documented sea cucumber fisheries are in italics and bold

Western Indian Ocean		Overexploited	Fully exploited	Moderate	Underutilized	NA or other regional review
Red Sea	***Egypt***	X				
	Sudan					NA
	Eritrea		X	X		
	Saudi Arabia					NA
	Israel					NA
	UAE					NA
Aden Gulf	Djibouti					NA
Western Arabian Sea	***Oman***					
	Yemen					
	Iran IR					
Eastern Arabian Sea	Pakistan					NA
	India	X				
	Chagos	X				
	Maldives	X				
Somalia, Kenya, Tanzania	Somalia					NA
	Kenya	X				
	Tanzania	X				
Madagascar, Mozambique channel	***Madagascar***	X				
	Comoros	X	X			
	Mayotte	X				
Islands	***Mauritius***		X	X		
	Réunion				X	
	Seychelles		X	X		
Mozambique	***Mozambique***	X				
Eastern Indian Ocean						
Bay of Bengal	***India***	X				
	Sri Lanka	X				
	Myanmar					Poh Sze, this volume
	Malaysia					Poh Sze, this volume
Northern	Indonesia					Poh Sze, this volume
Western Australia	Australia					Kinch *et al.*, this volvume
Southern Australia	Australia					Kinch *et al.*, this volume

NA = no information available

and troughs are due to stock depletions, gear improvements, global markets or other socio-economic factors impacting the coastal communities.

The resource management problems in **Madagascar** (Rasolofonirina *in* Conand and Muthiga, 2007a) are related to overfishing, legislation and administration. Various actions have been taken since 1990 by the scientists, professional organisations and the government to monitor the fishery. While legislation on sea cucumber exploitation exists in Madagascar, they are not effectively applied or even not applicable to all species exploited. Although the use of diving tanks is forbidden, this gear type is commonly used. A major problem is the lack of reliable statistics. The different administrations ignore the exact quantity of product that the country exports, and exporters wishing to avoid paying corresponding taxes and consequently do not declare the exact quantity of trepang. Implementation of sustainable management methods will only take place through actions involving the various participants in the fishing sector, from the fisherman to exporters (Conand, 2001, 2004a).

The national PRE/COI/UE programme, funded by the European Union (EU) and coordinated in Madagascar by the National Committee (CN-MAD), has helped the efforts of the Fishery Department and the professionals to manage the resource. The National Association of Sea Cucumber Producers (ONETH, in French) was created in 1996 through a pilot operation to implement the sustainable management of sea cucumber resources and exploitation (Conand, 1998a; Conand *et al.*, 1998). In order to reduce over-exploitation of sea cucumbers, they planned to study the current status of the resource, the production and export statistics, to create a quality management manual for this product (Conand, 1999) and to experiment with sea cucumber farming. ONETH encountered several problems in the years following its creation. It is now active again but still faces difficulties in the implementation of sustainable fisheries for the resource (Rasolofonirina, R., IH-SM, Université de Toliara, Madagascar, personal communication).

The information presented for **Tanzania** is based on Magaya and Mbaga *in* Conand and Muthiga (2007a). In this country, the new Fisheries Act of 2003 and the Fisheries (General Amendment) regulations of 1997 (under the old Fisheries Act of 1970) recognise three types of licence fees paid annually by dealers and exporters. They include, (i) a local individual or company with approved shore-based fish processing facilities, (ii) a local individual or company without approved shore-based fish processing facilities, and (iii) foreigner individual or company with approved shore-based fish processing facilities. Foreigner–individual or company without approved shore-based fish processing facilities cannot have licenses to operate. There is no specific legislation in Tanzania that refers to sea cucumbers. The biggest problem facing Tanzania, however, is not the development of appropriate legislation, but rather hoe to develop capacity to enforce regulations, including those that may be developed to protect bêche-de-mer fisheries. Legislation is useless without its three counterparts: political will, education and enforcement. However, education and enforcement are expensive, hence casting a shadow of doubt on compliance by the fisher folk to existing regulations.

Seychelles (Aumeeruddy and Conand, this volume; Aumeeruddy, 2007) offers an important example of management since the beginning of the recent fishery development. In 1999, the need for fisheries management was felt and measures to control the fishery were done through the Fisheries Regulations (1999). A management plan was also prepared, and is based on the results of the resource assessment done on a large spatial scale (Aumeeruddy *et al.*, 2005). Stakeholders were consulted several times, and the management plan was completed in August 2005 during a final stakeholders' workshop. This plan is based on a mix of input controls (limited number of fishing licenses) and output controls (Total Allowable Catch [TAC]) for each commercial species) (Payet, 2005). The TAC was calculated based on the maximum sustainable yield (MSY) for each species (Aumeeruddy *et al.*, 2005). The total TAC for all species has been calculated at 1 707 tonnes landed weight (gutted), from which the high value species represent 425 tonnes landed weight. An Advisory Management Committee, composed of representatives of government departments, professionals involved in the sea cucumber industry (boat-owners, divers, processors and exporters) and environmental NGOs was set up to oversee the implementation of the management plan. Discussions have not been finalised yet, and several other issues are being taken into consideration (number of fishing licenses, closed season). The timely provision of catch and effort data by fishers and processing data by processors remains a problem which makes it difficult to assess the level of catch and will be even more difficult when TACs are introduced. Currently, the accuracy of the data provided is not checked. Recent interviews carried out with the different actors of the "fishery system" (Pinault and Conand, 2007), showed that the holothurian fishery is very difficult to assess and to effectively monitor catches at fishermen level. It seems more efficient and easier to intervene at the processor level, by monitoring the export stocks and the undersized sea cucumbers.

In **Eritrea** (Tewelde and Jeudy de Grissac, 2007), the Ministry of Fisheries (MOF) and the National Fisheries Corporation (NFC) are the two government bodies that follow the planning, regulation, management and exploitation of fisheries resources. The main responsibilities of the MOF are (i) planning and implementing national fisheries policies and regulations; (ii) fisheries research and statistics collection; (iii) fisheries management design and implementation; (iv) monitoring, control and surveillance; (v) fish quality control; and, (vi) infrastructure and community development. The NFC was established in 2004 as an autonomous agency of the Government with the objective to develop and promote a profitable and long term export fishing industry. There is a Memorandum of Understanding between the MOF and NFC concerning the sea cucumber fishing activities where the NFC is given the authority to establish sea cucumber fishing enterprises. In addition to its own industrial development programs, the NFC undertakes trade of fishing products including dry products such as sea cucumber produced by local artisanal and commercial fishermen. Finally, the Corporation allocates quotas, given by MOF, to locals as well as chartered industrial fishing foreign vessels or companies. Presently no law directly regulates the harvesting and management of sea cucumber. However the MOF, in an effot to ensure that marine resources are not endangered by exploitation, has instroduced a closed season (from November to March), fishing quotas, and a minimal size limit of 5 cm. The fishing quotas of sea cucumber are 500 tonnes per year of dried sea cucumber (this quota has not been reached until now). In addition MOF established areas protected from fishing of sea cucumber and prohibited the operation of fisheries within four miles away from fishing villages. The fishery was closed in 2007 in order provide opportunity to spawning and will make some reorganization in the sea cucumber fisheries that will benefit the country's economy.

Commercial fishermen do not have business licenses but enter in contract agreements with the NFC. Licenses are issued for the boats owned by the commercial fishermen and working permits for sea cucumber non-Eritrean divers and assistants. MOF does not ask diving certificate when issuing working permits for the divers, and no regular checking is made for the diving equipment.

The MOF acts as regulatory as well as an economic stakeholder through collecting licensing and registration fee, royalty and resources utilization fee. The NFC exploits the marine resources through government-owned firms, private commercial fishermen and artisanal fishermen. The commercial fishermen enter into contract agreements with the NFC to exploit fisheries resources (37 presently, but less than 25 working). The contract contains the following terms and conditions: (i) licenses for all vessels working in the collection of the products; (ii) specific fishing quotas for the fishermen; (iii) calculation and payment of registration, royalty, and utilization fees; fishing, marketing and collection, and selling of the products to NFC; (iv) sea cucumber prices; (v) closed season and areas; (vi) transshipment and illegal trade; (vii) grade of the products; (viii) employment; (ix) reporting, inspection; (x) size limit; (xi) quality of the products; and (xii) revocation, suspension and termination. The NFC does not buy products from the fishermen who do not have a contract. NFC collects sea cucumber from the commercial fishermen and exports it. It set the prices that are paid to the commercial fishermen based on the export prices. NFC also procures the fishing gears to fishermen who regularly supply sea cucumbers and provides supporting letters to commercial fishermen to allow them get food supplies and to assist them in getting permits to employ foreign divers given the shortage of local fishermen.

Bans on the whole fishery, or on the use of some gears, have been chosen in different countries where signs of over-exploitation had been observed, but they have often been abandoned under the fishers and exporters pressure. There was a ban on the use of SCUBA in the **Republic of Maldives** to protect the stocks of *H. fuscogilva* and *T. ananas* (Reichenbach, 1999). No recent information has been provided. **Egypt** is an example of a country introducing and lifting bans (Lawrence *et al.*, 2004; Ahmed and Lawrence, 2007).

In **Oman,** the situation of the *H. scabra* fishery's management is now changing (Al-Rashdi, K., Marine Science and Fisheries Center, Sultanate of Oman, personal communication). For the first time in the country, a set of regulations have been set up and are in place now, including the identification of the amount produced, the determination of the quantity exported, the total amount fished and the price from the fisheries statistics year book. Before this, no systematic data collection existed as this fishery was not of official concern, and data were included among unidentified fish. Currently only *H. scabra* is commercially exploited; most of the product is sent to the United Arab Emirated (UAE) where they are dried for (bêche-de-mer) or frozen (a smaller quantity than dried one).

Sea cucumber fishing in Oman was not subjected to any management and control regulation. This is related to lack of research on biological and ecological information of Oman sandfish resources. A few general fishery management restrictions are practiced, such as the restriction of SCUBA diving for any fishery in Oman; sea cucumber collection is mainly restricted to hand-picking while walking at low tide or via free diving at depth less than 2 m. The presence of unexploited populations below the present depth limits has been considered a possible buffer against overfishing, however the presence or extent of such stocks have not been demonstrated. Although there is no specific closed season for sea cucumber, fishers usually collect them for six months, with peak fishing from January to March. Moreover, all fishers involved in sandfish fishing are originally from local community of Mahout Area, and restricted to traditional fisheries only. From the above description of the fishery coupled with the walk survey using transect (Al Rashdi, Al-Busaidi and Al-Rassadi, 2007), several issues have been raised in the management point of view: (i) because of lack of information, collection of data on landing, effort, CPUE, density and processed production of *H. scabra* should be initiated; (ii) the Directorate of Fisheries at Al-Wusta (Central) Region should collect the above data by providing the local traders with export permission certificate, showing the name of the trader, age, area of the trader, form of product (dried, frozen, live or salted), quantity of the product, origin of the product, country/area to which the product to be exported, name of border, date of permission; (iii) the fish inspectors at borders should be trained on sea cucumber species and bêche-de-mer to facilitate the Directorate of Fisheries Statistics to group sea cucumber data separately in the annual statistical book; (iv) research in various aspects of biology, ecology and stock assessment and enhancement and marketing should be initiated; (v) exploration of new fishing grounds and application of marine protected areas (MPA) in some remote sea cucumber sites are also needed to minimize the fishing pressures in limited grounds and to protect brooders respectively; (vi) marketing structure should be studied to develop the socioeconomic part of this fishery; (vii) initiation of feasibility studies on sandfish aquaculture in Oman is important; (viii) capacity building programmes (e.g. training, study tours, participation in national and international meetings and workshops on sea cucumbers) should be conducted and supported by governmental organizations.

3. TRADE

Data for several countries have already been presented in section 2.6, as catches are often evaluated from exports. The principal sea cucumber export destinations are China, China Hong Kong SAR, Singapore, Malaysia and Taiwan Province of China with some reciprocal exchanges between them, which complicate the synthesis (Conand and Byrne, 1993; Conand, 2004a; 2006b). The destination of the exports is known for some countries. In the **Seychelles** the recent exports appears to be mostly to China Hong Kong SAR (Table 11) (from Aumeeruddy and Conand, this volume).

TABLE 11
Exports of dried sea cucumbers (kg) from the Seychelles and percentage of total export volume to the main market destinations

	China Hong Kong SAR	Malaysia	Singapore	Others
2001	4 662	2 387	1 729	0
2002	22 805	8 995	2 170	2 625
2003	30 467	13 085	1 075	0
2004	12 555	0	2 605	0
2005	22 858	2 415	6 030	1 725
2006	46 794	8 742	5 762	240
Total	**140 141**	**35 624**	**19 371**	**4 590**
%	70	18	10	2

TABLE 12
Annual dry sea cucumber exports (kg) from Sri Lanka to the main market destinations

	Singapore	Taiwan Province of China	China Hong Kong SAR	Total
1996	73 266	27 457	69 803	170 526
1997	–	–	–	307 578
1998	88 959	68 330	46 424	203 713
1999	30 905	45 112	22 001	98 018
2000	16 479	39 626	29 530	85 635
2001	19 739	48 649	14 205	82 593
2002	25 519	50 593	40 057	116 169
2003	47 223	44 866	40 746	132 835

Source: External trade statistics of the customs (From Terney Pradeep Kumara, Cumaranathunga and Linden, 2005).

Another example is taken from **Sri Lanka** where the major export destinations of sea cucumbers are Singapore, Taiwan Province of China and China Hong Kong SAR, with Singapore being the dominant buyer from Sri Lanka since 1999. Taiwan is the second biggest market and China Hong Kong SAR is the third (Table 12) (Terney Pradeep Kumara, Cumaranathunga and Linden, 2005). The mean annual export to these three major countries since 1997 is around 50 tonnes dry weight.

As suitable shallow habitats for sea cucumbers are restricted in the southern part of Sri Lanka and following the collapse of this fishery in the south, fishermen from these areas have been compelled to seek other sites for their fishery, in distant locations (e.g. the Chagos Archipelago, the Laccadive Islands and the Andaman Islands). Customs statistics for the whole country suggests that import of sea cucumbers in Sri Lanka commenced in 1996, with small quantities from the Maldives (collected by Sri Lankans) for processing and re-export. The imported quantity gradually increased to 23 609 kilograms in 2000. The source of imports has not been established, but probably reflects clandestine operations. This makes the interpretation of the available data difficult.

The trade routes in **Oman** are particular as generally almost all its bêche-de-mer production is exported to the United Arab Emirates (UAE) for re-export to the international markets. However, lately (2000/2001) import of Oman bêche-de-mer products was made by China Hong Kong SAR (Bruckner, Field and Johnson, 2003) with lower quantities comparing with UAE import at the same market and same time. The production from Oman rarely appears in the international bêche-de-mer statistics because small quantities are exported directly into international markets or to the UAE. Local price paid to collectors varies from 10 Omani Rials (OR) (approximately USD 26) per kilogram of dried sandfish to 50 OR, depending on the size of sea cucumber and season of occurrence.

Table 13 presents the 2004 imports by China Hong Kong SAR from the countries of the region and the total for the period 1999–2005. It appears that (i) the main supplier is Madagascar; (ii) countries from the Red Sea and western Arabian Sea, as Yemen and the UAE which are not well documented, are major suppliers; (iii) the percentage of the region is important and increasing with 23 percent in 2004, more than the mean

TABLE 13
China Hong Kong SAR imports of sea cucumber from Africa and the Indian Ocean region between 1996–2005

Country of origin	Quantity (tonnes)		Total
	2004	2005	1996–2005
Madagascar	176	246	1 982
Yemen	479	301	1 031
Tanzania	95	59	1 019
Sri Lanka	107	78	619
Maldives	21	8	348
Mozambique	42	33	330
South Africa	15	5	309
Kenya	22	23	205
United Arab Emirates	140	15	194
Djibouti	9	0	148
Seychelles	18	40	100
Mauritius	4	0	76
Egypt	18	22	58
India	21	2	54
Ethiopia	12	12	24
Saudi Arabia	9	–	13
Oman	4	4	10
Saudi Arabia	9	3	13
Comoros	0.7	0.3	1.6
Sudan	–	0.6	0.6
Total region	**1 201**	**851**	**6 533**
Total (all imports)	**5 070**	**4 479**	**44 000**
Region/total	*0.24*	*0.20*	*0.15*
Value HKD (million)	778	869	–
Value USD (million)	100	112	–

Source: Census Statistic Department, China Hong Kong SAR, 2005.

TABLE 14
Imports of sea cucumber into China Hong Kong SAR from Oman and United Arab Emirates

Year	Oman		United Arab Emirates	
	Quantity (tonnes)	Value (USD 1 000)*	Quantity (tonnes)	Value (USD 1 000)
1996			3	19
1997			22	70
2000	0.96	14.25	10.85	161
2001	0.49	7.26	40.62	602

* Estimated values based on UAE values during the same period

Source: After Bruckner, Johnson and Field, 2003; Ferdouse, 2004 *in* Al Rahsdi, Al-Busaidi and Al-Rassadi, 2007).

value of 15 percent for the period 1999–2005. Table 14 presents the reported imports of bêche-de-mer into China Hong Kong SAR, from UAE and Oman.

The illegal, unregulated and unreported (IUU) fisheries for sea cucumbers are a problem mostly encountered where countries have no regulations or little control. A first example will be taken from the **Chagos** presented by Spalding (2006). The United Kingdom government efforts to police these waters have increased since 1996 and the waters are now patrolled year-round by a Fisheries Protection Vessel while regular sorties to the islands are made by British military personnel. This has led to increases in prosecution of illegal fishing vessels and crew; however, rates of arrival of new illegal vessels remain high. It seems likely that, as fish stocks become increasingly impacted elsewhere around the Indian Ocean margin then the risks associated with illegal fishing in Chagos waters become increasingly worth taking. Reports point to an illegal sea cucumber fishery in Chagos that has been in operation from Sri Lanka for several years. This is supported by direct observations and discussions with fishers.

Large quantities of sea cucumber from Chagos waters were found in Sri Lanka ports (Spalding, 2006). The start of the fishery in the years leading up to 2000 would tie in well with the recorded decline in national stocks in Sri Lanka that began around this time. In April 2005 a camp of illegal Sri Lankan fishers was observed in Chagos, on the northern end of Eagle Island, a Strict Nature Reserve. About ten fishermen were present. They had arranged a series of drying platforms on wide sandy beach area at this end of the island, with strong black plastic sheeting to cover the sea cucumbers during rainy days. Around 6 000 sea cucumbers were observed drying. These were not identified, although it would appear that they included several species. These fishers were reported to the British officials on Diego Garcia. The fishers were arrested, duly tried and fined, and on payment of the fine were allowed to return (without their catch) to Sri Lanka. Since this date, over a period of five months, a further three vessels and crew have been caught and each fined amounts ranging from £6 000–10 000 (USD 10 000–17 000). Some, but not all were found with diving equipment, which would greatly extend their harvesting capability and would allow them to access some of the higher value species. To be added to the cost of these fines must be considered the costs of mounting an expedition to fish in Chagos waters. The journey is made in relatively small vessels, using considerable quantities of fuel. An expedition may take 1–2 months. Such an expedition is clearly worth undertaking when the chances of success are high. This is heightened by the growing market value of these animals. The total value of the haul observed drying in April 2005 could have been as high as USD 60 000–80 000, and this by no means represents the complete intended catch. Perhaps the capture of four such ventures in the last five months will begin to have a deterrent effect (Spalding, 2006).

Another example is given by Tewelde and Jeudy de Grissac (2007) for **Eritrea**. The decline in catches is partially explained by the illegal trade practiced by the local fishermen with Yemen. Apart from official trade records, traders and fishermen claimed that there is a significant amount of illegal (not recorded) trade to Yemen. Fishermen are attracted by the higher prices of sea cucumber and low price of food and fuel in Yemen. For instance, according to fishermen, the price of *H. scabra* (dry) per kilogram is about three times higher in Yemen than in Eritrea for the same products (ERN 296 to 536). In order to avoid illegal trade across the Red Sea the Eritrean Ministry of Fisheries has a plan to reduce royalty for shore-based fishermen, but this has still to be implemented.

A prices analysis is also necessary, but never easy, as there are several buying levels from the fishers to the exporters (Conand, 2004a, 2006b).

In **Seychelles** some sea cucumber prices are given in the hotspot report (Aumeeruddy and Conand, this volume) extracted from the export documents or collected from interviews made in May 2007 (Pinault and Conand, 2007). The price depends on the species, the size and the grade of the product; it is not easy to obtain the pricing structure from the fishers or the processors. The high valued species such as white teatfish, pentard and black teatfish fetch respectively a maximum of USD 17.3, USD 13.8 and USD 8.6 per piece for the best grade and large size dried individuals.

Although **Oman** produces sea cucumber since 1995, the Directorate of Statistics groups sea cucumber production along with unidentified fishes (Al Rahsdi, Al-Busaidi and Al-Rassadi, 2007). The prices paid by local traders to the collectors vary by sizes and season. However, the average price paid from 2000 to 2004 was roughly OR 10 per 100 pieces of live sea cucumber (1 Omani Rial = USD 2.61). During January–March 2005, the price jumped to OR 45–50 for 100 live sea cucumber with an average price of OR 30. All fishing and processing expenditures, are made by the traders, these include transportations, fuel, mask, gas, salt and generator. This is also the case in other countries as for example Eritrea (Tewelde and Jeudy de Grissac, 2007). In Oman, when the bêche-de-mer is ready, it is stored in warehouse in bulk unsorted. Just two or three days before loading, they are sorted into different sizes and packed in polypropylene bags. Presently, four major local traders are actively involved in marketing of bêche-de-mer. They have agents in UAE purchasing the products. Prior to the fishing season, the agents visit the area and give instructions to the traders on the required processing techniques and negotiate the price.

In conclusion, the trade in bêche-de-mer differs considerabely among the countries in the region. A more careful analysis, based on the main markets (China, China Hong Kong SAR, Malaysia, Singapore and Taiwan Province of China,) is necessary. Data on imports, exports and re-exports will have to be collected and processed. Trade between countries, as shown on the example of Oman and the UAE, complicates the analysis, as is the reciprocal trade between the markets of Singapore, China Hong Kong SAR and Taiwan Province of China (Conand, 2004a, 2006b).

4. SOCIO-ECONOMIC IMPORTANCE TO LOCAL FISHING COMMUNITIES

Understanding the socio-economic aspects of sea cucumber fisheries are essential in the case of small-scale fisheries (McClanahan and Pet-Soede, 2000; McClanahan and Sheppard, 2000; Cesar *et al.*, 2002; De La Torre-Castro, 2006). It also constitutes an important objective of the regional WIO Masma project (Conand *et al.*, 2006; Muthiga and Conand, 2006). The project foresees the application of the same methods, using specific questionnaires and interviews, in selected sites of the different countries (De La Torre-Castro *et al.*, 2007). A conceptual model of sea cucumber resource dynamics in the WIO countries has been developed (De La Torre-Castro *et al.*, 2007) based on the previous knowledge on these fisheries (Conand, 2001, 2006b). This model encompasses multiple levels and considers the main structures of the sea cucumber fishery such as fishing and collection grounds, resource users and other stakeholders involved in the fishery (e.g. fishers, middlemen and importers), the links between stakeholders, villages and countries, and the associated management initiatives at different levels. This framework is promising when analyzing the sea cucumber national fisheries from a holistic perspective, considering both social and ecological interactions. The results are still preliminary for a few countries.

In **Kenya** (Ochiewo and De La Torre-Castro, 2007), the study has been conducted in villages located along its southern coast. The results indicate that sea cucumber collectors are mainly men who fish in the sub-tidal areas, between three and ten metres deep. These fishers do not use SCUBA and fishing is heavily done during the north-east monsoon season when the sea is calm and water is clear. About 50 percent of the sea cucumber fishers also collect other marine products such as octopus. Sea cucumbers are sold fresh to local dealers (middlemen) who process them and sell to more prominent middlemen in the neighbourhood of Mombasa. The fishers occasionally borrow money from dealers especially when they fail to catch sea cucumbers. This in turn makes them in debt to the dealers who lend them money. Sea cucumber fishing is regulated by the Fisheries Department that issues fishing licences. According to the law, fishers pay Kenya Shillings (KES) 100 (1 KES = USD 0.015) for the fishing licence annually. However, apart from the main fisheries legislation, there is no special law or policy that is devoted specifically to govern the sea cucumber fishery.

In **Madagascar**, the fishery is very active; when a site is overexploited, then fishers move temporarily to another place along the coast, set up a new village, and dive either with artisanal mask and fins or with SCUBA. These settlements sometimes raise conflicts with traditional villages.

The photos in Figure 8, taken in June 2007, illustrate this activity. It is worth observing that catches are composed of large specimens, many of the high commercial value species (*H. fuscogilva, T. ananas* and *Bohadschia* spp.); these fishers were diving for the first time in this site and will move again when the catches decline.

FIGURE 8

Sea cucumber fishery in northwestern Madagascar, 2007. A: the skin divers and their canoe; B: fishing sea cucumbers; C: processing in a temporary village; D: multispecies catch

COURTESY OF G. ANCEL, UNIVERSITÉ DE LA RÉUNION

The situation in the **Seychelles** is presented in Marguerite (2005) and Aumeeruddy (2007). Interviews of the different stakeholders from boat owners to international traders of sea cucumbers were conducted during a recent survey of the Masma project (Pinault and Conand, 2007). Through this socio-economic study, it was estimated that the holothurian fishery is employing about 100 people in the Seychelles, and from the Marguerite's survey (2005) which focussed on the economical dimension, it was estimated that 120 to 125 households were, to various degrees, dependent on the sea cucumber harvesting and processing business. This fishery amounts to between 9–10 percent of the total employment in the artisanal sector. Boat owner data tend to indicate that the harvesting activity is not profitable with the prices declared during the interviews, especially for the bigger vessels. Increasing prices for "pentard" gives more realistic results but it appears that the boat owner fishing strategy is more commanded by the high prices proposed by the processors for one or two species than by a real profitability evaluation. The Amirantes sector seems to present the more profitable fishing grounds with an interesting relation between the distance to cover and the money to earn. Processors interviews tend to indicate that the trade in dried sea cucumber is unprofitable. A comparison between the export incomes based on the InfoFish (http://www.infofish.org) prices and the expenditures of processors to buy the products to fishermen, indicates that the processors and exporters would be making important losses. On the other hand, the continuous interest in both harvesting and processing would suggest that the sea cucumber business is still profitable

At present, **Sri Lankan** fishermen from the south are exploiting sea cucumber beds in distant parts of the Indian Ocean (Terney Pradeep Kumara, Cumaranathunga and Linden, 2005). Although fishermen from Negombo on the western coast started the fishery in the southern part of the Island, fishermen/divers from the southern fishing towns of Mirissa and Dondra now dominate. There are around thirty-five boats engaged in the sea cucumber fishery along the southern coast. Fishermen use multi-day operating craft (MDOC) and Global Positioning Systems (GPS) to navigate far from shore. The lengths of the boats range from 35 to 50 feet and they use inboard 45 Hp engines. These boats are usually four to six years old and modified to accommodate 10 to 12 people. The crew consists of the skipper of the boat, divers, a cook, a compressor operator and an electricity generator operator. At times these boats operate in three or four units: by this strategy they are able to maximise their profits by carrying fewer support personnel (such as cooks and compressor operators) and a larger number of divers. The sea cucumber fishery in the south is by opportunistic tuna fishermen. Whenever exploitable sea cucumber beds are located they switch to that; and when the beds are fished out, revert to tuna fishing. The duration of each spell of activity depends on the sea cucumber population and the number of boats and divers that participate in the fishery.

A few data have also been presented by Al Rashdi, Al-Busaidi and Al-Rassadi (2007) for **Oman**. Sea cucumbers are harvested by hand picking during low tides, mainly spring low tides. There are five main *H. scabra* fishing grounds well-known to the collectors. To reach barrier flats, collectors use motorized boats, which usually belong to the traders. The latter often contracts a group of collectors, mostly women, to buy their sea cucumber catches, providing them with free services such as transportation (boats or vehicles), drivers, mask and food. About three boats each by four main traders are usually used in fishing operation with 7–10 collectors in each boat, while the vehicle may take two round trips with 10 collectors in each trip. The time required to get to the fishing grounds ranges from 15 to 60 minutes. Fishers work individually; they collect and store sea cucumbers separately in a large flour bags or a plastic bucket. Free diving (mask only) is infrequently used and is restricted to men only. The number of monthly fishing days was estimated to vary from 10 to 20. The estimated number of fishers increased from 100 in 2004 to 200 in 2005, which could indicate an increase in demand for bêche-de-mer and high income for the fishers and traders. The average fishing hours per working day (unit of effort) is 3–4 hours with a collection of 100 live sandfish approximately. Among the collectors, women represented the largest share of about 50 percent, while men and children accounted for 30 and 20 percent, respectively. The fishery for sea cucumber is usually linked with the shrimp fishing season in the area. Once the shrimp fishing activity decreases, the fishing season of sandfish commences. The sea cucumber fishing activity begins in late November and reaches its peak between January and March and then gradually decreases to May. The fishing season is also influenced by traditions of Mahout Community and sea conditions. As most of the Mahout communities are Bedouins, they seasonally move out of Mahout to adjacent areas particularly in the summer (June-August) when the sea condition is rough.

In **Eritrea** (Tewelde and Jeudy de Grissac, 2007) the situation is complex, with four big traders companies who employ fishermen (Eritrean and foreigners) and sell the products to NFC through auctions. Table 15 shows the main traders organization and the dried sea cucumber supplied to NFC in 2005 and 2006. The number of diving tanks and air compressors shows the importance of SCUBA diving. This fishing practice results in frequent decompression accidents (i.e. paralysis or death) because fishers are not well trained on Scuba diving.

TABLE 15
Main traders, fishermen and dried sea cucumber harvesting in Eritrea

Trader	Divers (fishermen) origin					No. boats	No. dive tanks	No. air compressor	Tonnage 2005	Tonnage 2006
	Eritrea	Somalia	Yemen	Egypt	Tanzania					
1	230	–	–	–	–	30	80	40	166.02	85.61
2	350	30	161	–	–	–	–	24	54.42	43.05
3	40	–	–	70	85	6	100	–	7.27	4.1
4	8	30	3	–	–	12	120	3	1.49	7.07
Other	–	–	–	–	–	–	–	–	32.34	94.62
Total									**278.37**	**241.84**

Source: Tewelde and Jeudy de Grissac, 2007.

This presentation clearly shows that although most sea cucumber fisheries are generally small in size, they are important for many communities in the Africa and Indian Ocean region.

5. CURRENT PROJECTS/DEVELOPMENTS ON FISHERIES, BIOLOGICAL STUDIES, AQUACULTURE VENTURES UNDERWAY

In the region there are several projects underway investigating commercial holothurians. The Masma (WIOMSA) Regional Project for WIO, which started in 2005, (Conand *et al.*, 2006; Conand and Muthiga, 2007a) has as main goals to: (i) increase the understanding of the status of sea cucumbers and their management including their potential for aquaculture; (ii) provide key skills and information for management including identification skills and information on reproduction and recruitment of key commercial species that is crucial for fisheries management; (iii) improve the knowledge of the management systems and the gaps in knowledge that will form the basis for any management plans; (iv) increase the knowledge of the impact of the fishery on the socio-economic status of coastal communities. It is intended to be multidisciplinary, with a close collaboration between biologists, ecologists and social scientists. Training in sea cucumber taxonomy, biology and fisheries will provide the capacity for monitoring and evaluating the effectiveness of the management systems currently in place. The project will focus on Kenya, Madagascar, Réunion, Seychelles and Tanzania. The main intended outputs are: species inventories and distribution, assessment of the impacts of MPAs, understanding of the biology of the main species, socio-economy and management of sea cucumber fisheries, and the training of practitioners and students, as well as guidelines for the collection of catch statistics. Several national contributions are underway (see abstracts for the WIOMSA's symposium, Durban, October 2007), but the regional synthesis will be prepared at a later stage.

Several countries, including Egypt (Ahmed and Lawrence, 2007), Eritrea (Tewelde and Jeudy de Grissac, 2007; Kalaeb *et al.*, 2008), Oman (Al-Rasdi, Al-Bussaidi and Al-Rassadi, 2007), are giving a special attention to sea cucumbers resources and fishery through different projects. It is important that their results and reports are disseminated. The socio-economy of sea cucumber fisheries also deserves a special attention in the Middle East region.

Aquaculture is considered as a solution for sustainable management of this resource, to reduce the anthropic pressure on the natural population and to restock wild stocks. Since 1998, some successful aquaculture experiences have been conducted in Tuticorin, south-eastern coast of India, on the sandfish *H. scabra* (James, 1994a, b, 1996, 2004a; James *et al.*, 1988; James *et al.*, 1995).

In Madagascar (Rasolofonirina, 2007), a programme was set up with the help of the Belgian Cooperation resulted in the construction of a sea cucumber hatchery and farm in Toliara region (Conand, 1997; Conand *et al.*, 1998; Jangoux *et al.*, 2001; Rasolofonirina, Mara and Jangoux, 2004; Rasolofonirina and Jangoux, 2005; Lavitra *et al.*, 2007; Rasolofonirina *in* Conand and Muthiga, 2007). Recent information has been communicated by I. Eeckhaut, R. Rasolofonirina and M. Jangoux (Université libre de Bruxelles, Bruxelles, personal communication) on "Madagascar Holothurie" (MH S.A.) the first trade company based on sea cucumber aquaculture in Madagascar. They also provide a brief history of the project. In 1999, a sea cucumber mariculture project undertaken in Madagascar arose from alarming reports of widespread overexploitation of the natural sea cucumber populations. The work was funded by the Belgian University Corporation for Development (CUD) and involved the Universities of Brussels and Mons (both from Belgium) and the Malgachian University of Toliara. It first consisted in building a sea cucumber hatchery on the setting of the Toliara Marine Sciences Institution (IH.SM). The hatchery was functional in 2003 and currently produces tens of thousand of juveniles of the valuable *H. scabra*. Its main section consists of a 120 m^2 air-conditioned building with six rooms for growing seaweed, rearing larvae, caring for the broodstock and undertaking microscopic and computer analyses. The aquaria of the hatchery are connected to a saltwater pumping station, whose reservoir fills up at high tide and whose water pours into a 30 m^3 settling pond. Decanted water is then sterilised by repeated applications of UV before being used in the larvae rearing tanks.

The second phase of the project, launched in 2004, consisted in setting up a sea cucumber farm to grow-out juveniles until they reach a marketable length. The farm was erected at Belaza, 20 km south from Toliara. This setting, which borders a mangrove, fulfils the ecological requirements of *H. scabra*. The farm allows bringing sea cucumber juveniles produced by the hatchery to a marketable size and weight (more than 20 cm and 300 g) in 10 to 12 months. The method for growing sea cucumber includes three successive phases, each requiring specialized infrastructures (internal aquaria, external tanks and sea fences) related to the animal sizes. Aquaria are set up in a 120 m^2 air-conditioned building. They each contains 200 l of filtered sea water where juveniles grow up to a length of 2 cm. Sea cucumbers are then transferred in 25 000 l external tanks where they are maintained until they reach a length of 5 cm. This size is adequate for allowing individuals to survive in sea fences. Thousands of sea cucumbers are currently produced by the farm.

In March 2008 when the funds from the Belgian CUD ended, private firms have formed a partnership involving a spin-off from Belgian universities, the IH.SM and private firms to form the first trade company based on sea cucumber aquaculture in Madagascar. The expected production of the trade company in the next five years is of 250 000 sea cucumbers per year. Development of cooperatives in villages will be essential for the development of the sea cucumber mariculture. The farm productivity is indeed not limited by the number of juveniles bred in the hatchery (a single pair of genitors, one male and one female, produce tens of thousands of fecundated eggs), but rather by the surface of the available farming structures (surface of enclosed spaces). The trade company would have to supply various local organizations (groups of fishermen, coastal villages) so that they could be trained to operate the farming. If the experience is positive, it could rapidly extend to the entire west coast of Madagascar when a reliable mean of transportation of juveniles is set up (enhancement of various coastal sites for the growing of specimens).

In the other countries, there are probably ongoing projects, but the author has not been able to get further information. The Manual on Sandfish Hatchery Techniques by Agudo (2007) and other publications from the Pacific (Purcell, Blokmans and Agudo, 2006) will certainly be useful for the countries who want to find alternative ways to protect their natural sea cucumber resources.

6. ADDITIONAL THREATS TO SEA CUCUMBER POPULATIONS

A general indirect threat to sea cucumber populations worldwide is the degradation of the habitats. This is more important in the tropics and the overall decrease of the coral reefs worldwide is widely recognized. Bryant *et al.* (1998) in a publication entitles "Reefs at Risk" gave a general picture of the threat of the reefs worldwide. The indicator (low, medium and high threat) is a composite of four coastal risk factors. Table 16 indicates that in the Middle East and Indian Ocean more than half of the reefs are with medium and high threats. From these regions, Tanzania and the Comoros have both high biodiversity and high threat from human activities.

The coral reef holothurians are particularly vulnerable. For India (Gulf of Mannar) James (1982) mentioned that several species such as *S. chloronotus*, which were abundant in 1927, had disappeared as a result of habitat destruction. Aumeeruddy and Conand (this volume) have analysed the situation for the Seychelles which is also applicable to other countries in the region with decreasing CPUE.

TABLE 16
Total area of reefs at risk

Region	Total reef area (km²)	Threat category (%)			Coastal population
		Low	Medium	High	density/km²
Middle East	20 000	39	46	15	24
Indian Ocean	36 100	46	29	25	135
Global	255 300	42	31	27	101

Source: Bryant *et al.*, 1998.

The main direct threat for sea cucumbers is overexploitation for the production of of bêche-de-mer. Species with high economic value (see Table 2) are more threatened, but as their numbers decrease other species with lower commercial value are now declining. The major consequence is the depletion of sustainable breeding populations to allow natural replenishment of populations. In Malaysia other uses are made from sea cucumber, for medicines, ("gamat oil" or "gamat water") and a variety of balms. Pharmaceutical research for new products is also ongoing (Baine and Choo, 1999), but this is not reported in other countries of the region.

Illegal fisheries are probably quite important in several countries of the region. The example of the Chagos is characteristic (Spalding, 2006). These islands represent one of the few remaining wilderness areas of the Indian Ocean, and provide an invaluable reserve of natural reef communities, and may well play a wider regional role in the movements of species and genetic material to other reefs across the ocean. The government of the United Kingdom has expressed its concern and will be further seeking to halt this illegal fishery through improved detection and arrest as well as through diplomatic approaches with other Indian Ocean states.

Improving sea cucumber fishery management and conservation is presently very urgent. The efforts of international and national agencies should lead to international, regional and national recommendations and application of new management measures (see Bruckner, 2006; Aumeruddy and Conand, this volume).

ACKNOWLEDGEMENTS

I would like to thank V. Toral-Granda, A. Lovatelli and M. Vasconcellos for inviting me to contribute to the sea cucumber project and improving the manuscript. This review has greatly benefited from input by numerous colleagues and revision by S. Uthicke and D. Herrero-Pérezrul. The work of the Masma team from the WIOMSA sea cucumber project in the WIO has provided much information; several contributors of the Bêche-de-mer Bulletin published by SPC, have responded to my request for up-dated information.

REFERENCES

Adithya, L.A. 1969. Bêche-de-mer in Ceylon. *Spolia Zeylan. Bull. Natl. Mus.*, Ceylon, 31(2): 405–412.

Agudo, N. 2007. Manuel d'écloserie d'holothuries de sable. SPC, ACIAR: 44 pp. (www.spc.int/coastfish/rapport/worldfish_F/BDM_Manuel.pdf).

Ahmed, M.I. & Lawrence, A.J. 2007. The Status of Commercial Sea Cucumbers from the northern Red Sea Coast of Egypt. *Beche-de-Mer Information Bulletin,* 26: 14–18.

Al-Rashdi, K., Al-Busaidi, S. & Al-Rassadi, I. 2007. Status of the sea cucumber fishery in the Sultanate of Oman. *Beche-de-Mer Information Bulletin,* 25: 17–21.

Anon. 1994. Sea cucumbers and beche-de-mer of the Tropical Pacific. SPC, Noumea, Handbook 18. 51 pp. (http://www.spc.int/coastfish/Fishing/BDM_HdBook18/HdBook18E.htm).

Anon. 2001. Illegal beche-de-mer fishing in the Seychelles, Indian ocean. Le Quotidien (La Réunion, 11 May 2001). *Beche-de-Mer Information Bulletin*, 15: 39.

Arakaki, Y. & Fagonee, I. 1996. Corals and Echinoderms of the western Indian Ocean islands, Mauritius, Madagascar and Mahe (Seychelles). *Bull. Mei Univ.*, 2: 113–125.

Aumeeruddy, R. 2007. Chapter 5. Sea cucumbers in Seychelles. *In: Commercial Sea Cucumbers: A Review for. the Western Indian Ocean.* Conand and Muthiga (eds). WIOMSA Book Series No. 5.

Aumeeruddy, R. & Conand, C. 2007. Seychelles' sea cucumber fishery: data on processed products and other parameters. *Beche-de-Mer Information Bulletin,* 26: 19–25.

Aumeeruddy, R. & Payet, R. 2004. Management of the Seychelles sea cucumber fishery: status and prospects. *In: Advances in sea cucumber aquaculture and management.* A. Lovatelli, C. Conand, S. Purcell, S. Uthicke, J.-F. Hamel and A. Mercier, (eds). *FAO Fisheries Technical Paper.* No. 463. Rome, FAO. 2004. 425p.

Aumeeruddy, R., Skewes, T., Dorizo, J., Carocci, F., Coeur de Lion, F., Harris, A., Henriette, C. & Cedras, M. 2005. Resource assessment and management of the Seychelles sea cucumber fishery. FAO Project Number: TCP/Seychelles/2902(A), November 2005. 49 pp.

Baine, M. & Choo, P. 1999. Sea cucumber fishery and trade in Malaysia. *In: The conservation of sea cucumbers in Malaysia: Their ecology, taxonomy and trade.* M. Baine (ed.). 49–63.

Barrère, A. & Bottin, C. 2007. Observations in situ de la reproduction sexuée de *Stichopus chloronotus* sur un récif frangeant de La Réunion (Océan indien). *SPC Beche-de-Mer Information Bulletin,* 25: 37–38.

Beadle, R. 2005. Management, conflict and compliance: The case of the Kenyan sea cucumber (Echinodermata: Holothuroidea) fishery. M.Sc. Thesis in Tropical Coastal Management University of Newcastle upon Tyne.

Bruckner, A.W., Johnson, K.A. & Field, J.D. 2003. Conservation strategies for sea cucumbers: can a CITES Appendix II listing promote international trade? *Beche-de-Mer Information Bulletin,* 18: 24–33.

Bruckner, A.W. (ed) 2006. The Proceedings of the Technical workshop on the conservation of sea cucumbers in the families Holothuridae and Stichopodidae. *NOAA Technical Memorandum* 44, 239 pp.

Bryant, D., Burke, L., McManus, J. & Spalding, M. 1998. *Reefs at risk. A map-based indicator of threats to the world's coral reefs.* WRI, ICLARM, WCMC, UNEP, 56 pp.

Cesar, H., Pet-Soede, L., Westmacott, S. & Mangi, S. 2002. Economic analysis of coral bleaching in the Indian ocean – Phase II. in *Coral Reef Degradation in the Indian Ocean.* Linden O., Souter D., Wilhelmsson D. and Obura D. (eds.). Report 2002. CORDIO, Kalmar. 251–262.

Cherbonnier, G. 1980. Holothuries de Nouvelle-Calédonie. *Bull. Mus. Natl. Hist. Nat.*, Paris, 4e Ser., 2(3): 615–667.

Cherbonnier, G. 1988. *Echinodermes: Holothurides. Faune de Madagascar.* ORSTOM 70, 292 pp.

Clark, A.M. 1984. Echinodermata of the Seychelles. *In*: *Biogeography and Ecology of the Seychelles Islands.* Stodart D.R. (ed.). *Monographiae biologicae*, 55: 83–102.

Clark, A.M. & Rowe, F.W.E. 1971. *Monograph of the Shallow-Water Indo-West Pacific Echinoderms.* Bath, England, Pitman Press.

Conand, C. 1981. Sexual cycle of three commercially important holothurian species (Echinodermata) from the lagoon of the New Caledonian. *Bull Mar Sci*, 31: 523–543.

Conand, C. 1982. Reproductive cycle and biometric relations in a population of *Actinopyga echinites* (Echinodermata: Holothuroidea) from the lagoon of New Caledonia, Western Tropical Pacific. *In: Echinoderms. Proceeding of the International Conference* Tampa Bay. J.M. Lawrence (ed.), Balkema, Rotterdam: 437–442.

Conand, C. 1985. Distribution, reproductive cycle and morphometric relationships of *Acanthaster planci* in New Caledonia, Western Tropical Pacific. *In*: *Echinoderms: Proceeding of the 5th International Conference,* Galway, Balkema, Rotterdam: 499–506.

Conand, C. 1986. Les ressources halieutiques des pays insulaires du Pacifique. Deuxième parti : Les Holothuries. *FAO Doc. Tech. Pêches*, 272.2: 108 p.

Conand, C. 1989. Les Holothuries Aspidochirotes du lagon de Nouvelle-Calédonie: biologie, écologie et exploitation. Etudes et Thèses, ORSTOM, Paris : 393 pp.

Conand, C. 1990. The fishery resources of Pacific island countries. Part 2: Holothurians. *FAO Fisheries Technical Paper*. N. 272.2: 143 pp.

Conand, C. 1993. Reproductive biology of the characteristic holothurians from the major communities of the New Caledonia lagoon. *Mar. Biol.*, 116: 439–450.

Conand, C. 1996. Asexual reproduction by fission in *Holothuria atra*: Variability of some parameters in populations from the tropical Indo-Pacific. *Oceanologica Acta*, 19(3): 209–216.

Conand, C. 1997. Gestion durable de la filière holothurie à Madagascar. Séminaire International sur les récifs Nosy Bé, 7 pp.

Conand, C. 1998a. Overexploitation in the present world sea cucumber fisheries and perspectives in mariculture. In: *9th International Echinoderm Conference, San Francisco*. R. Mooi and M. Telford, (eds), pp. 449–454. A.A. Balkema, Rotterdam.

Conand, C. 1998b. Holothurians. *In FAO species identification guide. The marine living resources of the Western Central Pacific. Vol. 2 cephalopods, crustaceans, holothurians and sharks,* K Carpenter and V. Niem (eds.). 1157–1190.

Conand, C. 1999. *Manuel de qualité des holothuries commerciales du Sud-Ouest de l'Océan Indien.* PRE/COI: 39 pp.

Conand, C. 2001. Overview of sea cucumbers fisheries over the last decade – what possibilities for a durable management? *In: Echinoderms 2000.* Barker (ed.). Swets and Zeitlinger: 339–344.

Conand, C. 2003. Les Echinodermes de La Réunion, Annexe 1; 129 pp. *In* : *Biodiversité du milieu récifal réunionnais: échinodermes, poissons et hydraires.* Conand C. Chabanet P. and Gravier-Bonnet N., Rapport au Conseil Régional.

Conand, C. 2004a Present status of world sea cucumber resources and utilization: an international overview: 13–23 *In: Advances in sea cucumber aquaculture* and *management*. Lovatelli A., Conand C, Purcell S., Uthicke S., Hamel J.-F. and Mercier A. (eds). *FAO Fisheries Technical Paper*. No. 463. Rome, FAO. 2004. 425 pp.

Conand, C. 2004b. Convention on International Trade in Endangered Species of Wild Fauna and Flora (CITES): conservation and trade in sea cucumbers. *SPC Beche-de-Mer Information Bulletin*, 20: 3–5.

Conand, C. 2004c. Monitoring a fissiparous population of Holothuria atra on a fringing reef on Réunion Island (Indian Ocean). *SPC Bêche-de-Mer Information Bulletin,* 20: 22–25.

Conand, C. 2006a. Sea cucumber biology: taxonomy; distribution; biology; conservation status. pp: 33–50. In The Proceedings of the CITES workshop on the conservation of sea cucumbers in the families Holothuriidae and Stichopodidae. Bruckner, A.W. (ed.). *NOAA Technical Memorandum*, 244 pp.

Conand, C. 2006b. Harvest and trade: utilization of sea cucumbers; sea cucumber fisheries; current international trade; illegal, unreported and unregulated trade; bycatch; socio-economic characteristics of the trade in sea cucumbers: 51–73. *In:* The Proceedings of the CITES workshop on the conservation of sea cucumbers in the families Holothuriidae and Stichopodidae. Bruckner, A.W. (ed). *NOAA Technical Memorandum*, 244 pp.

Conand, C. & Byrne, M. 1993. A review of recent developments in the world sea cucumber fisheries. *Mar. Fish. Rev.*, 55(4): 1–13.

Conand, C. & De Ridder, C. 1990. Reproduction asexuée par scission chez *Holothuria atra* (Holothuroidea) dans des populations de platiers récifaux. In: *Echinoderm Research*, De Ridder, Dubois, Lahaye et Jangoux (eds.), Balkema, Rotterdam: 71–76.

Conand, C. & Frouin, P. 2007. Chapter 3. Sea cucumbers in La Réunion. *In: Commercial Sea Cucumbers: A Review for the Western Indian Ocean.* Conand and Muthiga (eds). WIOMSA Book Series No. 5.

Conand, C. & Mangion, P. 2002. Sea cucumbers on La Réunion Island fringing reefs: diversity, distribution, abundance and structure of the populations. *SPC Bêche-de-Mer Information Bulletin*, 17: 27–32.

Conand, C. & Muthiga, N. (eds.). 2007a. *Commercial Sea Cucumbers: A Review for the Western Indian Ocean.* WIOMSA Book Series No. 5.

Conand, C. & Muthiga, N. 2007b. Chapter 7 – Conclusion and recommendations. *In: Commercial Sea Cucumbers: A Review for the Western Indian Ocean..* Conand C. and Muthiga N. (eds.). WIOMSA Book Series.

Conand, C., Galet, L.N., Randriamiarana, H., Razafitseheno, G. & De San, M. 1997. Sea cucumbers in Madagascar: Difficulties in the fishery and sustainable management. *SPC Bêche-de-Mer Information Bulletin*, 9: 4–5.

Conand, C., Morel, C. & Mussard, R. 1997. A new case of asexual reproduction in holothurians: Fission in *Holothuria leucospilota* populations on Réunion Island in the Indian Ocean. *SPC Bêche-de-Mer Information Bulletin*, 9: 5–11.

Conand, C., Jerome, A. Dijoux, N. & Garryer, J. 1998. Fission in a population of *Stichopus chloronotus*, on Réunion Island, Indian Ocean. *Bêche-de-Mer Information Bulletin*, 10: 15–23.

Conand, C., De San, M., Refeno, G., Razafitseheno, G., Mara, E. & Andrianjatovo, S. 1998. Difficulties and Sustainable management of the sea cucumber fishery sector in Madagascar. *SPC Bêche-de-Mer Information Bulletin*, 10: 7–9.

Conand, C., Uthicke, S. & Hoareau, T. 2002. Sexual and asexual reproduction of the holothurian *Stichopus chloronotus* (Echinodermata): a comparison between La Réunion (Indian Ocean) and east Australia (Pacific Ocean). *Invert. Reprod. Develop.*, 41(1–3): 235–242.

Conand, C., Dinhut, V., Quod, J.-P. & Rolland, R. 2005. Sea cucumber inventory in Mayotte, South West Indian Ocean. *SPC Bêche-de-Mer Information Bulletin*, 22: 19–22.

Conand, C., Muthiga, N., Aumeeruddy, R., De La Torre Castro, M., Frouin, P., Mgaya, Y., Ochiewo, J. & Rasolofonirina, R. 2006. A three years regional project on sea cucumbers in the South-Western Indian Ocean: National and regional analyses to improve management. *SPC Bêche-de-Mer Information Bulletin*, 23: 11–15.

De la Torre-Castro, M. 2006. *Human and seagrasses in East Africa: a social-ecological systems approach.* Ph.D. Thesis, Stockholm University, 62 pp.

De la Torre-Castro, M., Ochiewo, J., Mmbaga, T. & Pinault, M. 2007. A framework addressing socioeconomics and management of the sea cucumber resources in the Western Indian Ocean considering multiple scales. *SPC Bêche-de-Mer Information Bulletin*, 25: 22–25.

Drumm, D.J. & Loneragan, N.R. 2005. Reproductive biology of *Holothuria leucospilota* in the Cook Islands and the implications of traditional fishing of gonads on the population. *New Zealand J Mar Freshwater Research*, 39: 141–156.

Fabianek, F. & Turpin, A. 2005. Distribution des populations d'holothuries peuplant le récif de St Gilles/La Saline, à l'île de La Réunion. Report. MBPE, Univ. de La Réunion, 33 pp.

Franklin, S.E. 1980. The reproductive biology and some aspects of the population ecology of the holothurians *Holothuria leucospilota* (Brandt) and *Stichopus chloronotus* (Brandt). Ph.D. Thesis, Univ. of Sydney, 253 pp.

FAO. 2005a. *Yearbook of fishery statistics.* Rome

FAO. 2005b. FAO Fishery Information Data and Statistics Unit. Databases: Capture Production 1950–2005. Rome: FAO, 2005 (http://www.fao.org/fi/statist/fisoft/fishplus.asp).

Ferdouse, F. 2004. World markets and trade flows of Sea cucumber / beche de mer. *In: Advances in sea cucumber aquaculture* and *management.* Lovatelli A., Conand C., Purcell S., Uthicke S., Hamel J.-F. and Mercier A. (eds.). *FAO Fisheries Technical Paper.* No. 463. Rome, FAO. 2004. 425p.

Frouin, P., Conand, C., Burgos, A. & Hollinger, C. 2007. Spatial patterns of holothurian populations on the shallow reefs in Réunion Island. 5th Wiomsa scientific Symposium. Durban.

Gabr, H., Ahmed, A., Hanafy, M., Lawrence, A., Ahmed, M. & El Etreby, S. 2004. Mariculture of sea cucumber in the Red Sea - the Egyptian experience. 373–384. *In: Advances in sea cucumber aquaculture* and *management.* Lovatelli A., Conand C., Purcell S., Uthicke S., Hamel J.-F. and Mercier A. (eds.). *FAO Fisheries Technical Paper.* No. 463. Rome, FAO. 2004. 425p.

Gaudron, S. 2006. Observation of natural spawning of *Bohadschia vitiensis. Bêche-de-Mer Information Bulletin*, 24: 54.

Gaudron, S., Kohler, S. & Conand, C. 2008. Reproduction of the sea cucumber *Holothuria leucospilota* in the Western Indian Ocean: biological and ecological aspects. *Invert Reproduction Dev.*, 51(1): 19–31.

Government of Kenya. 1985. Wildlife (Conservation and Management) Act, Cap 376 of the Laws of Kenya. Government Printer, Nairobi.

Government of Kenya. 1991. The Fisheries Act Cap 378 of the Laws of Kenya: Revised Edition (1991). Government Printer, Nairobi.

Hamel, J.F., Conand, C., Pawson, D.L. & Mercier, A. 2001. Biology of the sea cucumber *Holothuria scabra* (Holothuroidea : Echinodermata) and its exploitation as beche-de-mer. *Adv. Mar. Biol.*, 41: 129–223.

Hearn, A. & Pinillos, F. 2006. Baseline information on the warty sea cucumber *Stichopus horrens* in Santa Cruz, Galápagos, prior to the commencement of an illegal fishery. *Bêche-de-Mer Information Bulletin*, 24: 3–10.

Hill, N. 2008. Information on sea cucumber fishery from Mozambique, Querimbas Archipelago. *Bêche-de-Mer Information Bulletin*, 27: 16.

Hornhell, J. 1917. The indian bêche-de-mer industry, its history and recent revival. *Madras Fish. Bulletin*, 11(1): 119–150.

Jaquemet, S., Rousset, V. & Conand, C. 1999. Asexual reproduction parameters and the influence of fission on a *Holothuria atra* sea cucumber population from a fringing reef on Réunion Island (Indian Ocean). *Bêche-de-Mer Information Bulletin*, 11: 12–18.

James, D.B. 1982. Ecology of intertidal echinoderms of the Indian seas. *J. Mar. Biol. Ass. India*, 24: 124–129.

James, D.B. 1989. Bêche-de-mer. Its resources, fishery and industry. *Mar. Fish. Inform. Serv.*, 92: 35 pp.

James, D.B. 1994a. Experiments on rearing of the juneviles of *Holothuria scabra* Jaeger produced in the Hatchery. *Cent. Mar. Fish. Res. Inst.*, 1–9.

James, D.B. 1994b. Improved methods of processing holothurians. *In*: Proc. Natl Workshop on beche-de-mer. K. Rehgarajan and D.B. James (eds.). *Bull. Cent. Mar. Fish. Res. Inst.*, 46: 71–75.

James, D.B. 1996. Culture of sea cucumbers. *Bull. Cent. Mar. Fish. Res. Inst.*, 48: 120–126.

James, D.B. 1999. Hatchery and culture technology for the sea cucumber *Holothuria scabra*, in India. *Naga, ICLARM* 22: 12–16.

James, D.B. 2001. Twenty sea cucumbers from Seas around India. *Naga, ICLARM* 24: 4–8

James, D.B. 2004a. Captive breeding of the sea cucumber *Holothuria scabra*, from India: 385–395. *In: Advances in sea cucumber aquaculture* and *management*. Lovatelli A., Conand C., Purcell S., Uthicke S., Hamel J.-F. and Mercier A. (eds.). *FAO Fisheries Technical Paper*. No. 463. Rome, FAO. 2004. 425p.

James, D.B. 2004b. A bibliography on Indian sea cucumbers. *Tech. Bull. 2*, Chennai-600.

James, D.B. & James, P.S.B.R. 1994. *Handbook on Indian sea cucumbers.* Central Marine Fisheries Research Institute special publication 59: 1–46.

James, D.B., Rajanpandian, B., Baskar, B.K. & Gopinathan, C.P. 1988. Successful induced spawning and rearing of the holothurian Holothuria (metriatyla) scabra Jaeger at Tuticorin. *Mar. Fish. Infor. Serv.*, 30–33.

James, D.B., Gandhi, A.D., Palaniswamy, N. & Rodrigo, J.X. 1995. Hatchery techniques and culture of the sea cucumber *Holothuria scabra. SPC Bêche-de-Mer Bulletin,* 7: 28–29.

Jangoux, M., Rasolofonirina, R., Vaitilingon, D., Ouin, J.M., Seghers, G., Mara, E. & Conand, C. 2001. A sea cucumber hatchery and mariculture project in Tulear, Madagascar. *SPC, Beche de Mer Information Bulletin,* 14: 2–5.

Kalaeb, K., Ghirmay D., Semere, Y. & Yohannes, F. 2008. Status of sea cucumber fishery in Eritrea and recommendations for a sustainable future. Preliminary assessment of Eritrea's sea cucumber fishery. *SPC, Beche-de-Mer Information Bulletin,* 27: 8–12.

Kawaka, J.A. & Muthiga, N.A. 2007. Annual Reproduction in the Indo-Pacific Sea Cucumber *Holothuria leucospilota* as a Response to Variability in the Environment. 5th Wiomsa scientific Symposium. Durban October 2007.

Kerr, A., Janies, D., Clouse, R., Samyn, Y., Kuszak, J. & Kim, J. 2005. Molecular phylogeny of coral-reef sea cucumbers based on 16S mitochondrial ribosomal DNA sequence. *Mar. Biotech.,* 7: 53–60.

Kithakeni, T. & Ndaro, S.G.M. 2002. Some aspects of Sea cucumber, *Holothuria scabra* (Jaeger, 1935) along the coast of Dar-es-Salaam. *Western Indian Ocean Journal of Marine Science,* 1(2): 163–168.

Kohler, S. 2006. Reproduction sexuée de *Holothuria leucospilota* à La Réunion. Report. MBPE. Universite de La Réunion. 25 pp + annexes.

Kohler, S., Conand, C. & Gaudron, S. 2007. Reproductive biology of sea cucumbers from La Réunion: a contribution for a regional management of the fishery. 5th Wiomsa scientific Symposium. Durban (abstract).

Kohtsuka, H., Arai, S. & Uchimura, M. 2005. Observation of asexual reproduction by natural fission of *Stichopus horrens* Selenka in Okinawa Island, Japan. *SPC Beche-de-mer Information Bulletin,* 22: 23.

Lavitra, T., Rasolofonirina, R. Jangoux, M. & Eeckhaut, I. 2007. Effect of food quality and settling density on growth and survival of epibiotic juveniles of the sea cucumber *Holothuria scabra.* 5th Wiomsa scientific Symposium. Durban.

Lawrence, A.J., Ahmed, M., Hanafy, M., Gabr, H., Ibrahim, A. & Gab-Alla, A.A. 2004. Status of the sea cucumber fishery in the Red Sea – the Egyptian experience. pp. 79–90. *In*: *Advances in sea cucumber aquaculture* and *management.* Lovatelli A., Conand C, Purcell S., Uthicke S., Hamel J.-F. and Mercier A. (eds.). *FAO Fisheries Technical Paper.* No. 463. Rome, FAO. 2004. 425p.

Lovatelli, A. 1995. EC rehabilitation programme for Somalia: artisanal fisheries. Final report. European Commission, Nairobi (Kenya). EC Somalia Unit. 209p.

Lovatelli, A., Conand, C., Purcell, S., Uthicke, S., Hamel, J.-F. & Mercier, A. 2004. Advances in sea cucumber aquaculture and management. *FAO Fisheries Technical Paper.* No. 463. Rome, FAO. 2004. 425p.

Luchmun, J., Mangar, V., Mosaheb, J., Terashima, H. & Yamamoto, M. 2001. Holothurian distribution in the lagoon at La Preneuse and Baie du Cap, Mauritius, Indian Ocean. *Rep. Mar. Ecol; Res. Inst.,* 3: 13–25.

Ludwig, H. 1875. Beitrage zur Kentniss der Holothurien. *Zool. Inst. Würzburg,* 2(2): 77–118.

Mangion, P., Taddei, D., Frouin, P. & Conand, C. 2004. Feeding rate and impact of sediment reworking by two deposit feeders *Holothuria leucospilota* and *Holothuria atra* on a fringing reef (Réunion Island, Indian). *In: Echinoderms: München.* Heinzeller and Nebelsick (eds.), Taylor and Francis, London: 311–317.

Mara, E., Rasolofonirina, R., Rabesandratana, H., Rakotoarinivo, W., Ravelo, I. & Conand, C. 1997. Etude de la pêcherie aux holothuries et propositions de mesures d'aménagement dans le Sud-ouest de Madagascar. Report IH.SM/ONE/BM, Toliara. 40 pp.

Marguerite, M. 2005. Capacity building in the Resource Assessment and Management of the Seychelles holothurian fishery–Socio-economic analysis. FAO Project Number: TCP/Seychelles/2902(A). 55 pp.

Marshall, N.T., Milledge, A.H. & Afonso, P.S. 2001. *Trade Review: Stormy Seas for Marine Invertebrate: trade in Sea Cucumbers, Seashells and Lobsters in Kenya, Tanzania and Mozambique.* TRAFFIC East/Southern Africa, Nairobi, Kenya. 70 pp.

Massin, C., Rasolofonirina, R., Conand, C. & Samyn, Y. 1999. A new species of *Bohadschia* (Echinodermata, Holothuroidea) from the Western Indian Ocean with a redescription of *Bohadschia subrubra. Bull inst. Royal Sciences Belgique,* 69: 151–160.

McClanahan, T.R. & Pet-Soede, L. 2000. Kenyan coral reef fish, fisheries and economics - trends and status after the 1998 coral mortality. *In* S. Westmacott, H. Cesar and L. Pet-Soede. Amsterdam, Institute of Environmental Studies, Free University: 79–100.

McClanahan, T.R. & Sheppard, C.R. (Eds). 2000. *Coral Reefs of the Indian Ocean: Their Ecology and Conservation.* New York, Oxford University Press.

Mercier, A., Battaglene, S.C. & Hamel, J.-F. 1999. Daily burrowing cycle and feeding activity of juvenile sea cucumbers *Holothuria scabra* in response to environmental factors. *Journal of Experimental Marine Biology and Ecology* 239:125–156.

Mercier, A, Battaglene, SC. & Hamel, J.-F. 2000. Settlement preferences and early migration of the tropical sea cucumber *Holothuria scabra*. Journal of Experimental Marine Biology and Ecology 249:89–110.

Mgaya, Y.D., Muruke, M.H.S. & Semesi, A.K., 1999. The sea cucumber and molluscs fishery in Bagamoyo. *In: Coastal Resources of Bagamoyo District, Tanzania.* University of Dar-es-Salaam, Dar es Salaam, Tanzania, pp. 65–72. Howell, K.M. and Semesi, A.K. (eds).

Michel, C. 1974. Notes on marine biology studies made in Mauritius. *The Mauritius Institute Bull.,* 7(2): 287 pp.

Mmbaga, T. & Mgaya, Y. D. 2004. Studies on sea cucumber in Tanzania and the gaps towards resource inventory and management 193–203. *In: Advances in sea cucumber aquaculture and management* A. Lovatelli, C. Conand, S. Purcell, S. Uthicke, J.-F. Hamel and A. Mercier (eds.). *FAO Fisheries Technical Paper.* No. 463. Rome, FAO. 2004. 425p.

Mgaya, Y.D. & Mmbaga, T.K. 2007. Chapter 6. Sea cucumbers in Tanzania. *In: Commercial Sea Cucumbers: A Review for the Western Indian Ocean.* Conand and Muthiga (eds). WIOMSA Book Series No. 5.

Mrowicki, R. 2007. The Abundance and Distribution of Holothurians (Echinodermata: Holothuroidea) in the Shallow Coastal Lagoon of Rodrigues, Mauritius. *SPC Beche-de-mer Information Bulletin,* 25: 43.

Muller, C. 1998. The role and distribution of holothurians (Echinodermata: Holothuroidea) in shallow coastal lagoon, Mauritius. M.Sc. Thesis, University of Bangor, North Wales, UK, 100 pp.

Mulochau, T., Conand, C. & Quod, J.-P. 2007. Les holothuries et autres échinodermes du Banc de Geyser (Mayotte – Océan Indien). *SPC Bêche-de-Mer Information Bulletin,* 26: 7–13.

Muthiga, N.A. & Ndirangu, S. 2000. Village based larviculture and stock enhancement of sea cucumbers (Echinodermata: Holothuroidea) on the Kenyan coast. pp. 66. Final Technical Report, Biodiversity Support Fund.

Muthiga, N.A. & Conand, C. 2006. Regional sea cucumber project: Proceedings of the start-up workshop, Mombasa Kenya, 26–29 January 2006. 20 pp. (on line www.wiomsa.org)

Muthiga, N.A., Ochiewo, J. & Kawaka, J. 2007. Chapter 2. Sea cucumbers in Kenya. *In: Commercial Sea Cucumbers: A Review for the Western Indian Ocean.* Conand and Muthiga (eds.). WIOMSA Book Series No. 5, pp. 8–20.

Muthiga, N.A., Kawaka, J.A. & Ndirangu, S. 2007. The Reproductive Biology of the Commercial Sea Cucumbers *Holothuria fuscogilva* and *H. scabra* along the Kenyan coast. 5th Wiomsa scientific Symposium. Durban, October 2007.

Ochiewo, J. & De La Torre-Castro, M. 2007. The social and economic features of the sea cucumber fishery in Kenya. 5th Wiomsa scientific Symposium. Durban.

Payet, R. 2005. Management Plan of the Seychelles Sea cucumber Fishery. Final Report of Project No. FAO/TCP/SEY/2902(A), September 2005. 46 pp.

Petit, G. 1930. L'industrie des pêches à Madagascar. *Faune des Colonies Françaises.* Mart G. et P. Colon (eds.). 392 p.

Pinault, M. & Conand, C. 2007. Contribution to the Seychelles' sea cucumber Masma project. Report from a mission made in May 2007. Unpub. report 19 pp. (on line www.wiomsa.org).

Pouget, A. 2004. Sea cucumber fisheries in the Mayotte reef system, Indian ocean *SPC Bêche-de-Mer Information Bulletin,* 19: 35–38.

Pouget, A. 2005. Abundance and distribution of holothurians on the fringing reef flats of Mayotte. *SPC Bêche-de-mer Information Bulletin,* 21: 22–25.

Purcell, S. 2004. Criteria for release strategies and evaluating the restocking of sea cucumbers. pp. 181–191. *In: Advances in sea cucumber aquaculture* and *management.* Lovatelli A., Conand C, Purcell S., Uthicke S., Hamel J.-F. and Mercier A. (eds.). *FAO Fisheries Technical Paper.* No. 463. Rome, FAO. 2004. 425p.

Purcell, S., Blockmans, B. & Agudo, N. 2006. Transportation methods for restocking of juvenile sea cucumber, *Holothuria scabra. Aquaculture,* 251: 238–244.

Purwati, P. & Luong-Van, J.T. 2003. Sexual reproduction in fissiparous holothurian species, *Holothuria leucospilota. SPC Beche-de-Mer Information Bulletin,* 18: 33–38.

Quoy, J.R. & Gaimard, J.P. 1833. Zoologie: Zoophytes. *In: Voyage de la corvette l'Astrolabe sous le commandement de J. Dumont d'Urville,* Paris: 390pp.

Rafalimanana, T. 1997. Rapport de l'enquête historique sur l'exploitation des holothuries à Madagascar. Ministère de la pêche, ONET, PRE/COI/UE. Report. 20 pp.

Rakotonirina, J.S. 2000. Evaluation de l'activité de pêche à pied sur les platiers du récif frangeant entre Fiherenamasay et Tsifota (Région de Toliara, Sud-Ouest de Madagascar). DEA, IH.SM, Université de Toliara, Madagascar.

Ramofafia, C. 2002. Reproduction and development of *Holothuria* and *Actinopyga* species (Echinodermata: Holothuroidea) from Solomon Islands: Implications for aquaculture. Ph.D. Thesis, Sydney University.

Ramofafia, C., Battaglene, S.C. & Byrne, M. 2001. Reproduction of the intertidal sea cucumber *Actinopyga mauritiana* in the Solomons islands. *J. Mar. Biol. Ass. UK,* 81: 523–531.

Rasolofonirina, R. 1997. Ecologie, Biologie et Exploitation de deux espèces d'Holothuries Aspidochirotes, *Bohadschia vitiensis* et *Holothuria scabra* var. *versicolor* sur le Grand Récif de Toliara (Sud-ouest de Madagascar). Mémoire de D.E.A. en Océanologie appliquée. IH. SM, Université de Toliara. 84 p.

Rasolofonirina, R. 2007. Chapter 4. Sea cucumbers in Madagascar *In:* Commercial Sea Cucumbers: A Review for the Western Indian Ocean. Conand and Muthiga (eds). WIOMSA Book Series No. 5.

Rasolofonirina, R., Mara, E. & Jangoux, M. 2004. Sea cucumber fishery and mariculture in Madagascar: A case study of Toliara, south-west of Madagascar. pp. 133–149. *In: Advances in sea cucumber aquaculture and management.* A. Lovatelli, C. Conand, S. Purcell, S. Uthicke, J.-F. Hamel & A. Mercier (eds.). *FAO Fisheries Technical Paper.* No. 463. Rome, FAO. 2004. 425p.

Rasolofonirina, R. & Jangoux, M. 2005. Apparition et Développement des éléments squelettiques chez les larves et les juvéniles épibiontes de *Holothuria scabra. SPC Beche-de-Mer Information Bulletin,* 22: 6–10.

Rasolofonirina, R., Vaïtilingon, D., Eeckhaut, I. & Jangoux, M. 2005. Reproductive Cycle of Edible Echinoderms from the South-Western Indian Ocean. II: The sandfish *Holothuria scabra* (Jaëger, 1833). *Western Indian Ocean Journal of Marine Science,* 4: 61–75.

Rasolofonirina, R. & Conand, C. 1998. Sea cucumber exploitation in the Toliara region of south-west Madagascar. *SPC Beche-de-Mer Information Bulletin,* 10: 10–13.

Reichenbach, N. 1999. Ecology and fishery biology of *Holothuria fuscogilva* (Echinodermata: Holothuroidea) in the Maldives. *Bull. Mar. Science,* 64: 103–113.

Richmond, M. 1997. *A guide to the seashores of eastern Africa and the Western Indian Ocean Islands.* SIDA, 448 pp.

Rosa, J.C. 1997. Etude de la pêche à pied sur les platiers des récifs frangeants entre les villages de pêcheurs de Maromena, Befasy et Beheloka (Sud-Ouest de Madagascar). DEA, IH.SM, Université de Toliara, Madagascar.

Rowe, F.W.E. & Richmond, M. 2004. A preliminary account of the shallow water echinoderms of Rodrigues, Mauritius, western Indian Ocean. *Journal of Natural History*, 38: 3273–3314.

Samyn, Y. 2003. Shallow-water Holothuroidea (Echinodermata) from Kenya and Pemba Island, Tanzania. *Std Afrotropical Zoology*, 292, 158 pp.

Samyn, Y. & Van den Berghe, V.E. 2000. Annotated checklist of the echinoderms from the Kiunga Marine National Reserve, Kenya. Part 1: Echinoidea and Holothuroidea. *Journal of East African Natural History* 89: 1–36.

Samyn, Y., Vanden-Spiegel, D. & Massin, C. 2005. Sea cucumbers from the Comoros Archipelago. *SPC Beche-de-Mer Information Bulletin*, 22: 14–18

Samyn, Y., Vanden-Spiegel, D. & Massin, C. 2006. *Taxonomie des holothuries des Comores.* AbcTaxa /Vol 1, 130 pp.

Sella, A. & Sella, M. 1940. L'industria del trepang. *Thalassia*, 4(1): 116 pp.

Semesi, A.K., Mgaya, Y.D., Muruke, M.H.S., Francis, J., Mtolera, M. & Msumi, G. 1998. Coastal resources utilization and conservation issues in Bagamoyo. *Ambio*, 27(8): 635–644.

Shiell, G. 2004a. Questionnaire on the field observations of juvenile sea cucumbers *SPC Bêche-de-Mer Information Bulletin*, 19: 41.

Shiell, G. 2004b. Field observations of juvenile sea cucumbers. *SPC Bêche-de-Mer Information Bulletin*, 20: 6–11.

Shiell, G. 2005. Information on juvenile holothurian communicated by D.B. James. *SPC Bêche-de-Mer Information Bulletin*, 21: 26–27.

Spalding, M. 2006. Illegal fisheries in the Chagos Archipelago. *SPC Beche-de-Mer Information Bulletin*, 23: 32–34.

SPC. 1990 to 2007. *Beche-de-Mer Information Bulletin* No. 1 to 27. (also distributed as pdf files: www.spc.int/coastfish/.CC and data base http://www.spc.int/coastfish/news/search_bdm.asp).

Taddei, D. 2006. Matter and energy transfers in soft-bottom sediments of an anthropogenic reef complex (La Réunion Island, Indian Ocean). Ph.D. Thesis, La Réunion University.

Tehranifard, A., Uryan, S., Vosoghi, G., Fatemy, S. & Nikoyan, A. 2006. Sexual Reproduction Cycle of the *Stichopus herrmanni* a sea cucumber from Kish island Iran (Echinodermata: Holothuroidea). *SPC Beche-de-Mer Information Bulletin*, 24: 22–27.

Terney Pradeep Kumara, P.B., Cumaranathunga, P.R. & Linden, O. 2005. Present status of the sea cucumber fishery in Southern Sri Lanka: a resource-depleted industry. *SPC Beche-de-Mer Information Bulletin* 22: 24–29.

Tewelde, M. & Jeudy de Grissac, A. 2007. Socio-economic survey of Eritrean fisheries for sea cucumber. Unpub. Report ECMIB GEF/UNDP and FAO.

Thandar, A.S. 2007. Additions to the aspidochirotid, molpadid and apodid holothuroids (Echinodermata: Holothuroidea) from the east coast of southern Africa, with descriptions of new species. *Zootaxa*, 1414: 1–62.

Toral-Granda, V. 2006. Fact sheets and identification guide for commercial sea cucumber species. *SPC Beche-de-Mer Information Bulletin*, 24: 49–52.

Toral-Granda, V. 2007a Facts on sea cucumber fisheries worldwide. *SPC Beche-de-Mer Information Bulletin*, 25: 39–41.

Toral-Granda, V. 2007b. Biological and trade status of sea cucumbers in the families Holothuriidae and Stichopodidae. CITES, CoP14 Doc 62 33pp. http://www.cites.org.

Tuwo, A. & Conand, C. 1992. Developments in beche-de-mer production in Indonesia during the last decade. *SPC Beche-de-Mer Information Bulletin*, 4: 2–5.

Uthicke, S. 1997. Seasonality of asexual reproduction in *Holothuria atra, Holothuria edulis* and *Stichopus chloronotus* on the Great Barrier Reef. *Mar. Biol.*, 129: 435–441.

Uthicke, S. 2001. The influence of asexual reproduction on the structure and dynamics of *Holothuria (Halodeima) atra* and *Stichopus chloronotus* populations of the Barrier Reef. *J Mar Freshwater Res.*, 52: 205–215.

Uthicke, S. & Benzie, J.A.H. 2003. Gene flow and population history in high dispersal marine invertebrates: mitochondrial DNA analysis of Holothuria nobilis (Echinodermata: Holothuroidea) populations from the Indo-Pacific. *Molec. Ecology*, 12: 22635–22648.

Uthicke, S. & Conand, C. 2005. Amplified fragment length polymorphism (AFLP) analysis indicates importance of both asexual and sexual reproduction in the fissiparous holothurian *Stichopus chloronotus* (Aspidochirotida) in the Indian and Pacific Ocean. *Coral Reefs*, 24: 103–111.

Uthicke, S., Conand, C. & Benzie, J.A.H. 2001. Population genetics of the fissiparous holothurians and *Stichopus chloronotus* and *Holothuria atra* (Aspidochirotida): a comparison between Torres Strait and La Réunion. *Mar. Biol.*, 139: 257–265.

Uthicke, S., O'Hara, T. & Byrne, M. 2004. Species composition and molecular phylogeny of the Indo-Pacific teatfish (Echinodermata: Holothuroidea) bêche-de-mer fishery. *Mar. Freshw. Res.*, 55: 1–12.

Uthicke, S., Purcell, S. & Blockmans, B. 2005. Natural hybridization does not dissolve species boundaries in commercially important sea cucumbers. *Biol. Journ. Linnean Soc.*, 85: 261–270.

Yamanouchi, T. 1956. The daily activity rhythms of the holothurians in the coral reefs of the Palao Islands. *Publ. Seto Marine Biol. Lab.*, 5: 347–362.

Seychelles: a hotspot of sea cucumber fisheries in Africa and the Indian Ocean region

Riaz Aumeeruddy
Island Conservation Society
Victoria, Seychelles
E-mail: icsscience@seychelles.sc

Chantal Conand
Université de la Réunion
Saint-Denis, France
E-mail: conand@univ-reunion.fr

Aumeeruddy, R.; Conand, C. 2008. Seychelles: a hotspot of sea cucumber fisheries in Africa and the Indian Ocean region. In V. Toral-Granda, A. Lovatelli and M. Vasconcellos (eds). Sea cucumbers. A global review of fisheries and trade. *FAO Fisheries and Aquaculture Technical Paper*. No. 516. Rome, FAO. pp. 195–209.

SUMMARY

The Seychelles Archipelago, comprising 115 islands, is located in the middle of the Western Indian Ocean and has a large Exclusive Economic Zone (EEZ 1.4 millions km^2). Sea cucumbers in Seychelles have been fished for more than a hundred years, but the fishery has recently seen a rapid development. Sea cucumbers are mostly collected by divers using SCUBA gear. They are processed for the export market. The population status is presented for the five main species caught ("pentard", white teatfish, black teatfish, prickly redfish and sandfish) from the estimated stock and the overall density (ind./ha). Catch and effort data have been collected since 1999 by Seychelles Fishing Authority (SFA). The data are provided by the fishers as per the requirements of their fishing license conditions. The catch per unit effort (CPUE), expressed in numbers of sea cucumbers collected per diver per day, shows mostly a downward trend. The sea cucumber fishery in Seychelles was open-access until 1999. As part of a recent FAO-funded project, a management plan for the fishery has been prepared, based on the results of the resource assessment. Management measures were established through the Fisheries (Amendment) Regulations (1999) which provided some control over the fishery through licences for fishing and processing. The total allowable catch (TAC) has been calculated, based on the maximum sustainable yield (MSY) for each species. The trade data show that there are three main export markets, with China Hong Kong Special Administrative Region (SAR) being largely the dominant one. The socio-economic importance to the local fishing community appears from the three operations of harvesting which relates principally to the collection of the sea cucumbers, processing which involves cleaning and drying of the product, and trading which involves sales to different markets. Finally, recommendations for improving fisheries management and conservation through the involvement of all stakeholders are discussed. Evaluation of the pros and cons of listing in the Convention on the International Trade in Endangered Species of Wild Fauna and

Flora (CITES) is presented. This fishery in Seychelles is an example of recent management showing steps toward a sustainable exploitation of the resources.

1. INTRODUCTION

The Seychelles Archipelago is located in the middle of the Western Indian Ocean (Figure 1). Its 115 islands are scattered over a wide area which explains its large Exclusive Economic Zone (EEZ 1.4 millions km^2). The country has a total population estimated at 85 032 people as of June 2007 (Anon., 2007). Most of the population live on the main islands (Mahé and Praslin) and the rest of the islands are sparsely populated or remain uninhabited. This has left them relatively unexploited, and today they still have a rich biodiversity. The coral reefs of Seychelles have been described as being one of the most extensive networks in the Western Indian Ocean (Jennings *et al.*, 2000). Aldabra, the most remote atoll was designated as a World Heritage Site by UNESCO in 1982. The isolation of these islands in the middle of the ocean, the difficulties of accessing many of them and the low levels of exploitation due to a small population makes the Seychelles a safe heaven for many of its natural resources.

Sea cucumbers in Seychelles have been fished for more than a hundred years, with reports of bêche-de-mer exports dating back to the late 1800s. However, quantities harvested were fairly low and it is only in the late nineties that the fishery has seen a rapid development (Aumeeruddy, 2007). Currently, in the Seychelles sea cucumbers are mostly collected by divers using SCUBA gear. The average depth of the Mahé Plateau,

FIGURE 1
Map of the Seychelles archipelago

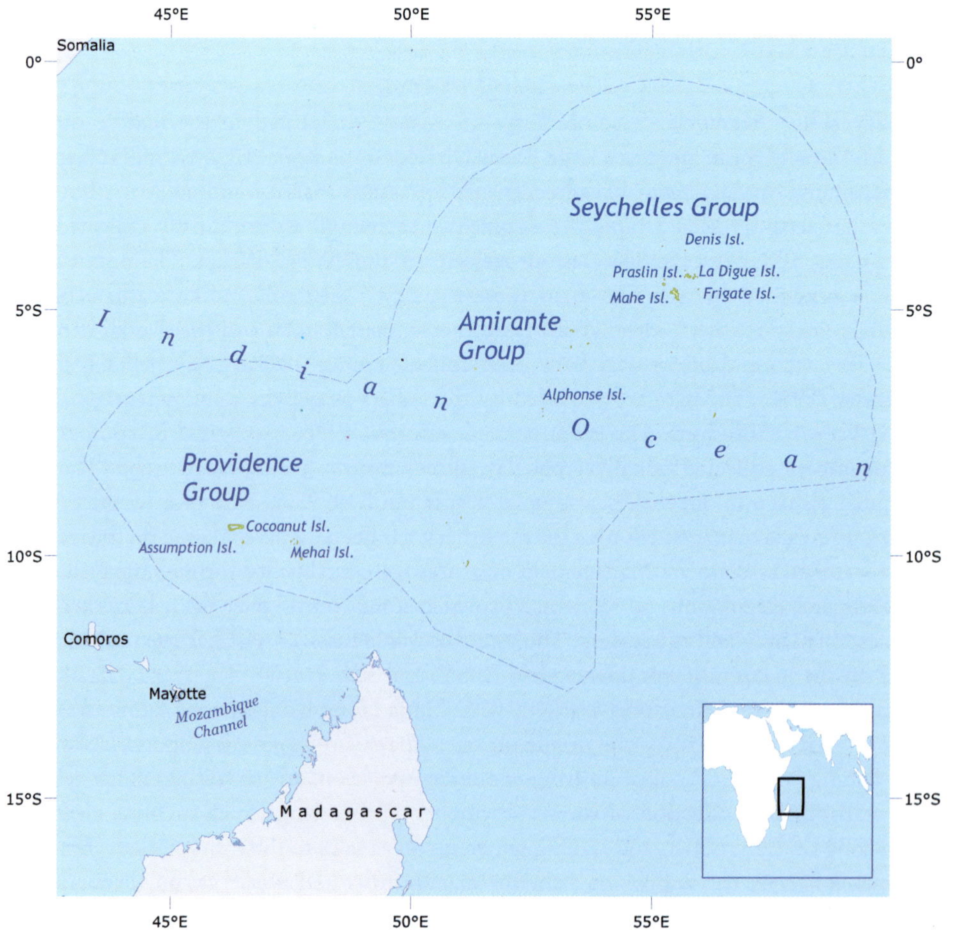

which is one of the main fishing grounds, is around 50 m, and most of the harvesting of sea cucumbers takes place in depths ranging from 10 to 40 m. The hookah system (air supplied to the diver from a compressor at the surface) is not used in the Seychelles.

Sea cucumbers are collected mostly for the export market as there is no local consumption in the Seychelles. There are two categories of fishers: (i) those that use small outboard powered boats and operate mostly in the coastal areas around the three main populated islands (Mahé, Praslin and La Digue) (normally carrying out day trips); and (ii) fishers that use larger decked boats equipped with a diesel inboard engine, which can stay at sea for up to two weeks. These boats employ four divers (maximum authorized under the license conditions) and sea cucumbers are usually pre-processed onboard.

2. BIOLOGICAL AND POPULATION STATUS

2.1 Current species in trade

Very little research had been done on holothurians in the Seychelles. A synthesis by Clark (1984), based on previous publications by Clark and Rowe (1971) listed 151 Echinoderm species, including 35 sea cucumbers, and gives some brief information on the ecology of the most common ones such as *Holothuria atra*, *Stichopus chloronotus*, *Bohadschia marmorata*, *Actinopyga mauritiana* and *Holothuria nobilis*. Clark's work was restricted to the Aldabra atoll and Mahé Island, and out of the 35 species found, seven were observed only at Aldabra but are not endemic. Other observations and reports have been published on Aldabra by Sloan (1982).

A recently completed holothurian population survey has shown that out of the 35 sea cucumber species recorded previously in the Seychelles, there are more than twenty commercial species, some not currently exploited. These are predominantly from the Holothuriidae and Stichopodidae families. However, only about six species are regularly fished locally fetching the highest price on the market (Aumeeruddy and Payet, 2004). These comprise three species of teatfish: the black teatfish (*Holothuria nobilis*), the white teatfish (*H. fuscogilva*) and a teatfish yet to be described and locally named "pentard" (the export name of flower teatfish is used for this species). The two other species found in the catch are the prickly redfish (*Thelenota ananas*) and the yellow surfish (*Actinopyga mauritiana*). The sandfish (*H. scabra*) used to be collected in fairly large numbers, but in the past few years, landings of this species have dropped to very low levels compared to the other five species. Even though this species fetches a high price on the market, the fishers target it less because of its burying behaviour, hence making it more difficult to find. Other species like the lollyfish (*H. atra*) or the elephant trunkfish (*H. fuscopunctata*) are collected in small numbers. The five main species caught (pentard, white teatfish, black teatfish, sandfish and prickly redfish) represented respectively 46.2, 11.3, 4.9, 3.1 and 0.6 percent of the total catch in 2006.

2.2 Population status

The sea cucumber fishery has existed in the Seychelles for over a century, but it was unregulated because of its small size and low catch. The only data collected were exports quantities of dried sea cucumbers by the customs authorities (Marguerite, 2005). No catch and effort data prior to 1999 were collected compared to other artisanal fisheries. The fishery saw a rapid development in the late-1990s, mainly due to the high demand for bêche-de-mer on the Asian markets. This convinced the authorities to collect information on population status and more recently on the biology of the sea cucumbers, if the fishery was to continue on a sustainable level. The Seychelles Fishing Authority (SFA), which is responsible for the management of fisheries in the Seychelles, conducted a stock assessment of holothurians in 2003–2005 with funding obtained from the Food and Agriculture Organization (FAO) under a Technical Cooperation Project (Aumeeruddy *et al.*, 2005).

Due to the large size of the Seychelles' EEZ stock assessment surveys were carried out for the two main sea cucumber fishing grounds, the Mahé Plateau and the Amirantes Plateau, which together represent an area of 48 305 km^2 or 3.45 percent of the total (Aumeeruddy *et al.*, 2005). A total of 246 sites were surveyed throughout the study area in December 2003, March/April and November 2004 for a total of seven weeks. More than 20 species, having a commercial or potentially commercial value, were observed and counted. Table 1 provides the estimated biomass and stock status of sea cucumbers observed during the surveys.

The overall density of commercial species in the study area was 19.78 ind. ha^{-1}, which equated to a standing stock of 95.5 million individuals (± 21.5%; 90% CI) of sea cucumbers (Aumeeruddy *et al.*, 2005). The most abundant species was the lollyfish (*Holothuria atra*) (5.61 ind. ha^{-1}), followed by the black teatfish (*H. nobilis*) (1.98 ind. ha^{-1}), brown sandfish (*Bohadschia marmorata*) (1.70 ind. ha^{-1}) and tigerfish (*B. atra*) (1.62 ind. ha^{-1}). Converted to weight, the highest biomass was that of the tigerfish (*B. atra*) at 13 105 tonnes, followed by the lollyfish (*H. atra*) at 12 722 tonnes, the brown sandfish (*B. marmorata*) at 12 017 tonnes and the black teatfish (*H. nobilis*) at 11 589 tonnes.

TABLE 1
Estimated stock, overall density and status of sea cucumber species. Exploited species are shaded

Species name	Common name	Stock (tonnes)	Overall density (ind./ha)	Status
Holothuria nobilis	Black teatfish	11 588.9	1.98	Underexploited
H. fuscogilva	White teatfish	5 178.1	0.63	Fully exploited
Holothuria sp.	Pentard	3 909.5	0.48	Fully exploited
H. fuscopunctata	Elephant trunkfish	7 876.7	1.03	Underexploited
H. atra	Lollyfish	12 721.7	5.61	Unexploited
H. edulis	Pinkfish	951.4	0.35	Underexploited
H. scabra	Sandfish	ND	ND	Overexploited
H. pervicax	Dark papillae	69.4	0.07	Unexploited
Actinopyga mauritiana	Surfish	233.8	0.11	Overexploited
Actinopyga sp.	White belly	41.1	0.01	Exploited
A. miliaris	Blackfish	4 980.2	1.09	Exploited
A. echinites	Deep water redfish	1 154.4	0.64	Exploited
A. echinites hybrid?	Redfish hybrid	261.5	0.14	Unknown
Bohadschia marmorata	Brown sandfish	12 016.9	1.79	Unexploited
B. atra	Tigerfish	13 105.2	1.62	Unexploited
B. subrubra	Bohadschia white belly	486.4	0.13	Unexploited
Pearsonothuria graeffei	Flowerfish	186.1	0.08	Unexploited
Thelenota ananas	Prickly redfish	6 349.9	0.52	Fully exploited
T. anax	Amberfish	235.6	0.02	Unexploited
Stichopus herrmanni	Curryfish	718.9	0.11	Unexploited
S. chloronotus	Greenfish	203.9	0.21	Unexploited

Source: Adapted from Aumeeruddy *et al.*, 2005.

FIGURE 2
Density of sea cucumbers in shallow (<20 m), intermediate (20–50 m), and deep waters (>50 m)

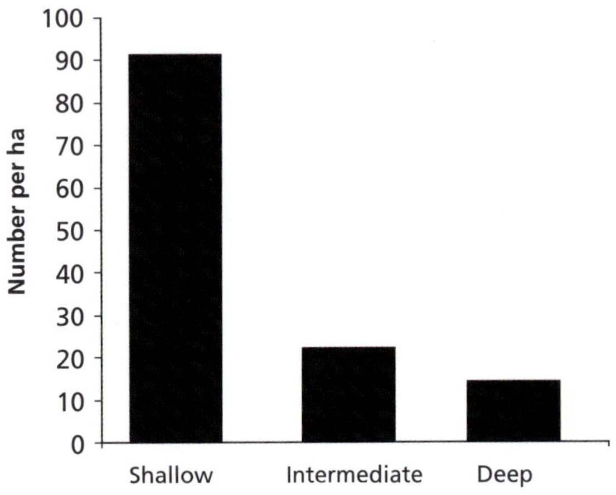

Source: Adapted from Aumeeruddy *et al.*, 2005.

The highest density of sea cucumbers was in the shallow habitats (<20 m) and the lowest in the deep habitats (>50 m) (Figure 2). Habitats in the intermediate depths (20–50 m) had an intermediate density of sea cucumbers, but due to the comparatively large size of that habitat, it had the bulk of the overall standing stock (61.2%). During the surveys size structures and biological data were not collected.

2.3 Catches

Catch and effort data have been collected since 1999 by the SFA. The data are provided by the fishers as per the requirements of their fishing license conditions. The responsibility has been placed on the fishers to provide their catch data, as they normally unload their catch at odd hours and the fishery authority does not have the necessary manpower. Failure to provide catch data may result in the revocation of the fishing licence. Catch data (in numbers) are recorded separately for the five main species. The other species are grouped as "Others". Table 2 gives the total catch per species from 2001 to 2006 and shows a regular increase in the total catch, with the sharpest increase in 2005 (SFA, 2006). This is particularly the case for the "pentard", which is one of the highest valued sea cucumbers on the market making it a prime target for the fishers (Table 2 and Figure 3). The prickly redfish and the "pentard" were not recorded as separate species before year 2001 (grouped with "Others"), but they were considered separately when their importance in the catch increased.

The catch per unit effort (CPUE) expressed in numbers of sea cucumbers collected per diver per day (ind. $diver^{-1}$ day^{-1}) shows mostly a downward trend from 2001 to 2006

TABLE 2
Total number of sea cucumbers harvested in the Seychelles from 2001 to 2006

Year	Black teatfish	Sandfish	White teatfish	Prickly redfish	Pentard	Others	Total
2001	7 794	9 114	17 202	2 517	330	4 978	41 935
2002	7 391	728	38 868	10 987	8 807	43 788	110 569
2003	6 982	22	21 398	12 241	44 185	52 853	137 681
2004	9 897	611	38 162	11 468	57 489	30 582	148 209
2005	11 488	100	44 943	17 142	82 291	98 055	254 019
2006	8 753	1 851	31 899	13 842	130 046	94 970	281 209

FIGURE 3
Total number of sea cucumbers harvested in Seychelles from 2001 to 2006

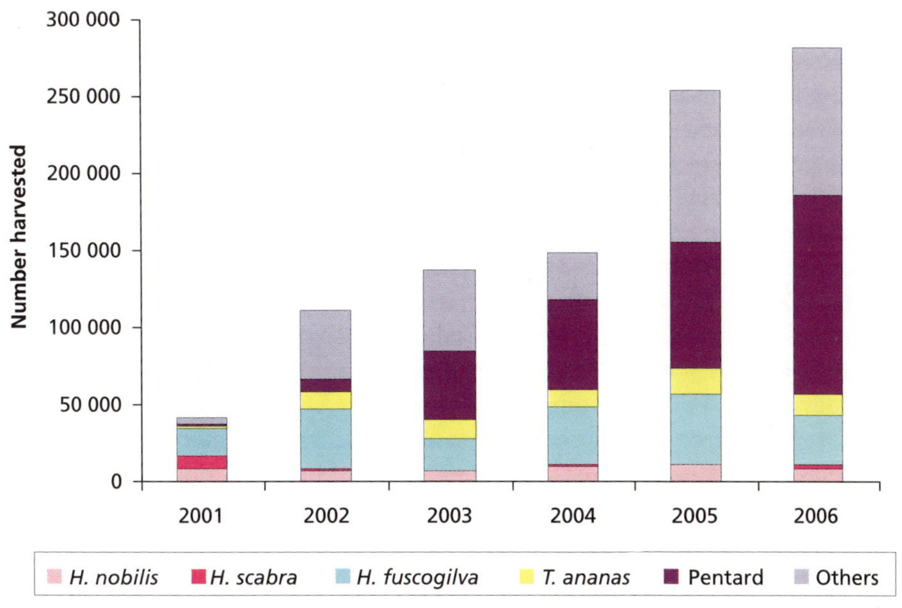

(Figure 4). The only exception is for "pentard" for which the CPUE has kept increasing over the years, except for a drop in 2004 and 2005. The "pentard" has become the main species targeted by the fishers and some of them have reduced their effort on other species to concentrate on this highly demanded species. This was confirmed by fishers during a socio-economic survey conducted in May 2007 (Pinault and Conand, 2007). This fishing strategy could explain why the CPUE has decreased for the other species. Landings of the "pentard" have been above the proposed total allowable catch (TAC) of 71 000 individuals (82 291 were harvested in 2005 and 130 046 in 2006) which is raising some concern for the population status of the species.

FIGURE 4
Catch per unit effort of sea cucumbers by species from 2001 to 2006

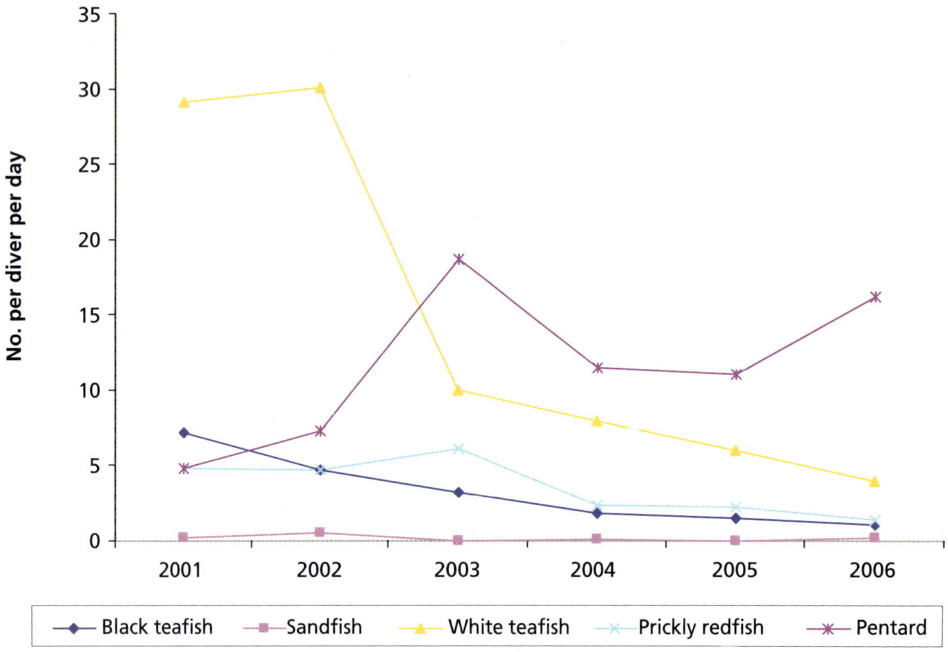

2.4 Management of the fishery

The sea cucumber fishery in Seychelles was open-access until 1999 and therefore no catch and effort data were collected before then. The only information available was exports data (which included dried shark fins) provided by the customs department. Furthermore, no information was available on the number of fishers involved in the fishery or on specific fishing grounds. There were also no quotas or size limits in place. The only no-take zones were the designated marine national parks, where no fishing activity is allowed. Some of the main marine parks (e.g. Sainte Anne, Curieuse, Bay Ternay and Port Launay) are regularly patrolled by rangers from the Marine Parks Authority, which results in some control on the ban of fishing activities. Other smaller parks such as Silhouette are patrolled irregularly due to lack of resources. Enforcement in these parks is poor, and there is the possibility that some fishing activity takes place (Bijoux, J., Seychelles Centre for Marine Research and Technology, Marine Parks Authority, Seychelles, personal communication).

In 1999, with the rapid development of the fishery, there was a need for management measures. Since there was no baseline data on the sea cucumber population, its distribution and abundance, some management measures were taken as a precautionary approach. These precautionary measures were expected to last until a resource assessment of sea cucumbers was done and management measures could then be established based on the population status. Management measures were established through the Fisheries (Amendment) Regulations (1999) which provided some control over the fishery through the following conditions:

- Fishing and processing sea cucumbers require a specific license issued under the licensing law;
- A sea cucumber licence (for fishing or processing) shall be independent of any other licence required by law;
- The holder of a licence shall not fish in protected areas defined by law;
- The holder of a licence shall furnish information related to fishing or processing as required by the SFA;
- A yearly fee of SR 300 (USD 55) is payable in respect of each licence issued under this regulation.

As a precautionary measure, the number of fishing licences was limited to 25 in 2001. A licence is attached to a specific boat, and each boat can have a maximum of four registered divers. The licence holders were also required to complete catch and effort logbooks that need to be submitted to the SFA on a monthly basis.

As part of the FAO-funded project referred to earlier, a management plan for the fishery was also prepared, and is based on the results of the resource assessment. Stakeholders were consulted several times during the preparation of the management plan, and it was completed in August 2005 during a final stakeholders' workshop. This plan is based on a mix of input (limited number of fishing licences) and output controls (TAC for each commercial species) (Payet, 2005). It was agreed to continue with the limit of 25 fishing licences that had been implemented as a precautionary measure in 2001.

The TAC was calculated based on the maximum sustainable yield (MSY) for each species (Aumeeruddy *et al.*, 2005). The total TAC for all species was calculated at 1 707 tonnes landed weight (gutted), from which the high value species (e.g. black teatfish, white teatfish, "pentard", prickly redfish) represent 425 tonnes, which equated to approximately 50 tonnes dry bêche-de-mer using a 14 percent recovery rate; medium value species (e.g. blackfish) made up 121 tonnes and low value species (e.g. lollyfish) made up the remaining 1 161 tonnes. The TAC was calculated for a wide area which is not fully exploited by the fishers. Today fishers say that they have to move further away from the coastal fishing grounds because these areas are fully or overexploited. The Management Plan also made provision for an increase of the licence fees which would help cover some of the management costs. An advisory management committee, composed of representatives of government departments, professionals involved in the sea cucumber industry (boat-owners, divers, processors and exporters) and environmental NGOs was set-up to oversee the implementation of the management plan.

Discussions on the management plan have not been finalized yet, and several other issues are being taken into consideration. These include the possibility of increasing the number of fishing licences, due to the high demand by fishers who are interested in accessing the fishery, as well as introducing a closed season.

One of the main problems in the management of the fishery is the timely provision of catch and effort data by fishers and processing data by processors. While this is a requirement of the license, most do not submit data on time. Sometimes it is only when a license is due for renewal that the licensee provides the required data, as failure to do so could result in the non-renewal of the licence. This makes it difficult to assess the level of catch in a timely manner, and it will be even more difficult to do that assessment when TACs are introduced. Since there are no observers on board, it is also difficult to ascertain whether the fishing location and the amount of catch reported by the fishers are accurate. Fishers also very often land their catch very late in the evening, which makes it difficult for the fishing authority to have field technicians present to record the catch being unloaded. So for the time being management of the fishery relies mostly on the data provided by the fishers regardless whether it is accurate.

It has also proved difficult to reconcile the amount of sea cucumbers sold by fishers to processors and the amount of bêche-de-mer produced by the processors. This would help check whether processors are producing more dried sea cucumbers compared to the amount they bought from licensed fishers, hence laundering illegally caught sea cucumbers. A sales receipt book was introduced where all sales of sea cucumbers from a licensed fisher to a processor are to be recorded, and each receipt needs to be counter-signed by both the seller and the buyer. This has proved difficult to implement, as both the fishers and processors were reluctant to give price information. A new version that requires only quantities sold was introduced in 2006, and this seems to be better accepted by those concerned. Random inspections are carried-out by inspectors from the Monitoring, Control and Surveillance section of SFA to control the stocks of sea cucumbers at the processing factories. Since May 2006, the same inspectors also verify all consignments of bêche-de-mer before export.

The management of sea cucumber fishing licences may sometimes prove problematic. A fishing license is allocated to a specific vessel and is not transferable between vessels. Because of the imposed quota of 25 fishing license, they are all allocated at any one time, with a waiting list that has reached a total of 35. There are times when some of the licensed vessels are not active in the fishery, either because the boat needs repairs or a crew is unavailable and in some cases this can last for several months. Those on the waiting list then complain that the license should be revoked and be allocated to the first on the waiting list. As a temporary measure, the SFA has allowed a boat-owner to use another boat if his boat is unavailable (e.g. boat under repair), with the provision that he uses the same crew.

Having considered all the constraints that hinder the good management of the fishery, the fishing authority has made amendments to the original management plan that was proposed in 2005, and this has now been submitted to the government for endorsement. The main points of the proposed plan are as follows:

- The number of fishing licences will be capped at 25 and be valid for the period of the open fishing season (i.e. 1 October – 31 May), and will be renewable. It is proposed to introduce a close season from 1 June to 30 September, which is usually the period when the sea is too rough for boats and divers to operate safely;
- The management plan will introduce an output quota limit on the amount of catch that can be taken from the fishery. The proposed TAC based on a recent stock assessment is 1 707 tonnes per year. The five most valuable species would represent only 425 tonnes (24.9%) of the TAC. The TAC would be monitored via the catch and effort forms submitted by all fishing licence holders to ensure that they match the sales receipts;
- Real time monitoring will be undertaken at landing sites and SFA will be issuing receipts to fishermen upon verification of their catch which will be done each time a sea cucumber fishing vessel enters into port; vessel monitoring system (VMS) data when available will be used to validate the fishing locations;
- The quota monitoring will also be done through the processors and through export figures. The processors would be required to keep detailed records of their purchases and stocks in a logbook and report on a monthly basis to SFA. The SFA will also do random inspections of sales to processors;
- Boat-owners need to inform SFA of each fishing trip conducted by a vessel licensed to fish sea cucumber (i.e. when the vessel is leaving port and when it is coming back). All licensed sea cucumber fishing vessels with the capacity to have a VMS onboard will have it installed, and this will be a condition of the licence;
- All landings of sea cucumbers must be done within the hours of 6 a.m. to 6 p.m. to facilitate control;

- A licence will only be issued to an individual with a valid local fishing vessel licence and who has the capacity to undertake such an activity. No person can have two sea cucumber fishing licences at the same time. A registered company would be allowed to possess only one sea cucumber fishing licence;
- Any licence which is revoked will be allocated on a first come-first serve basis as per SFA's waiting list. Licences may be revoked if they are inactive for more than four months and also if the catch and effort data are not submitted to SFA on a monthly basis, except for circumstances under *force-majeure;*
- Mothership operations will not be allowed;
- It has been argued that because of the natural closure of the fishery during the South-East Monsoon, there is no need for a specific closed season. However, to manage a quota system it is better to introduce a closed season of four months from 1 June to 30 September. This will allow SFA to analyse the logbooks returned by the fishermen and processors to estimate catch and export levels during the fishing period, and to determine whether quotas have been exceeded. If exceeded, the excess catch will be deducted from the next fishing season's quota;
- Fishing of sea cucumbers inside reef lagoons around the inner islands will be prohibited. Marine parks will continue to be no-take zones;
- All purchases of sea cucumbers must be from licensed sea cucumber fishermen only. Similarly, exporters must purchase only from licensed processors and licensed fishermen. In order to promote value addition, it is proposed that all sea cucumbers must be processed (i.e. dried) for export. Therefore, there shall be no export of fresh sea cucumbers. Prices offered for export of fresh sea cucumbers are very low compared to those offered for the dried product;
- The licence fee, which costs currently RS 300 (USD 55) per year for fishing and processing activities, will be increased. For a fishing vessel it will depend on the size of the vessel. The proposed fee structure per year is as follows:

– Small boats	RS 1 000 (USD 182)
– Medium-size boats	RS 2 000 (USD 364)
– Large-size boats	RS 2 500 (USD 455)

The processing licence fee is proposed at RS 5 000 (USD 909).

Note that all the proposed measures may not be accepted by the government authorities, but most of them are expected to be. The management plan could be implemented in 2008.

3. TRADE

Most of the sea cucumbers harvested are processed to a dried state and exported to the main Asian markets. There is a very small market for local consumption of sea cucumbers, due mostly to the small Chinese community present in the country. Sales for local consumption can be considered as negligible when compared to exports. China Hong Kong Special Administrative Region (SAR), Malaysia and Singapore represent the three main export markets in terms of importance (Table 3), with China Hong Kong SAR, being largely the dominant one. Small quantities may be exported to other countries, but this is on an irregular basis. Dried sea cucumbers are exported mostly by air cargo to the destination markets. One of the local processors also exports small amounts of sea cucumbers in a semi-dried state by sea freight. The export figures when compared to the catch figures given in Table 2 show that there is a discrepancy between them. For example total exports in 2006 amounted to 61 538 kilograms; using an average weight of 167 grams per sea cucumber, this would equate to 368 491 sea cucumbers. In 2006 the reported catch was 281 209 sea cucumbers (see Table 2). The export figures can be considered as accurate, as all shipments of dried sea cucumbers are weighed. This means that reported catch is underestimated by nearly one third, the main reasons for this being that some fishers still do not report or under-report their catch.

TABLE 3
Exports of dried sea cucumbers from the Seychelles (in kg)

	China Hong Kong SAR	Malaysia	Singapore	Others
2001	4 662	2 387	1 729	0
2002	22 805	8 995	2 170	2 625
2003	30 467	13 085	1 075	0
2004	12 555	0	2 605	0
2005	22 858	2 415	6 030	1 725
2006	46 794	8 742	5 762	240

TABLE 4
Export price of bêche-de-mer from the Seychelles (USD/kg)

Species	Export name	Price range (USD/kg)
Holothuria nobilis	Black teatfish	15–25
Holothuria fuscogilva	White teatfish	18–30
Holothuria sp.	Flower teatfish ("pentard")	17–26
Holothuria atra	Lollyfish	5–7
Holothuria fuscopunctata	Elephant trunkfish	3–4
Thelenota ananas	Prickly redfish	6–15
Actinopyga mauritiana	Surf redfish	5–6
Actinopyga echinites	Yellow surfish	4–5

At present there are 25 vessels licensed to fish sea cucumbers, three licensed processors and one licensed exporter. While there is a quota of 25 licenses for fishing, there are no limits on the number of processors or exporters of sea cucumbers. The fishers usually sell their catch to the processors on a unit price basis. The price depends on the species, the size and the grade of the product but it is not easy to obtain the pricing structure from the fishers or the processors. The high valued species such as white teatfish, "pentard" and black teatfish fetch respectively a maximum of USD 17.3, USD 13.8 and USD 8.6 per piece for the best grade and large size individuals. The processors usually export their products themselves or they can also sell to a local exporter. These processors are also involved in the processing of shark fins, and often export both shark fins and bêche-de-mer in the same consignment. Table 4 gives the price range (USD/kg) obtained by the exporters for a number of species, and is based on declarations made on export permits. These are prices for bêche-de-mer exported by air cargo.

In order to reduce the risk of illegal fishing by local fishers, the SFA has introduced a system whereby the processors have to certify that they have bought sea cucumbers only from licensed fishers. For each sale, the seller and the buyer have to complete a form stating the amount of sea cucumbers in the transaction, which is then submitted to the fisheries authority. This information helps the authority check the sale against the catch declaration by the fishers and against the exports declared by the processors. This system was put in place recently because of the potential for processors to buy sea cucumbers collected by unlicensed fishers. Some of the islands in the southern group are isolated and quite far from the main inhabited islands and are thus difficult to control. Some of these islands are uninhabited (e.g. Astove and Cosmoledo atolls), which makes it easy for poachers to fish around these islands. It is known that there are instances where fishers from countries in the region fish illegally around some of the southern islands. In April 2001, a vessel from Madagascar was apprehended while it was fishing illegally near Farquhar Atoll. It had on board 110 fishers and several tons of sea cucumbers (Le Quotidien, La Réunion, 2001).

4. SOCIO-ECONOMIC IMPORTANCE TO LOCAL FISHING COMMUNITY

The sea cucumber fishing and trade business in the Seychelles can be classified into three distinct operations and interrelated groups (Marguerite, 2005). The different operations are namely:

- harvesting, which relates principally to the collection of the sea cucumbers;
- processing, which involves cleaning and drying of the product into the form and quality acceptable to be traded; and
- trading – this involves sales to different markets, both local and overseas.

The groups or operators associated with the harvesting can be further divided into two distinct sub-groups.

- The single boat operators, which fish from small boats between 5–8 m in length, equipped with outboard engines. Fishing trips last usually for one day in the case of undecked boats. Harvesting operations normally take place in the coastal areas around the three most populated islands (Mahé, Praslin and La Digue) of the archipelago.
- The second group concerns those licensed harvesters operating larger vessels (whalers to schooners). Fishing trips last for more than one day and trips usually involve some form of mothership/dory type operation, i.e. where the bigger vessels are accompanied with a smaller dinghy which is used to move around the fishing ground and from which diving normally takes place. These harvesters normally operate on grounds further from the main island of Mahé or in the outer islands. Some form of pre-processing is also done on board which includes eviscerating the sea cucumbers and preserving them in salt.

This fishery has the potential to create employment for a maximum of 170 people, if all licenses are taken and the bigger vessels have an average crew of 6 persons. This would amount to between 9–10 percent of the total employment in the artisanal fisheries sector or about 3 percent of total employment in fisheries and related activities.

Compared to the overall formal employment, the sea cucumber business could be classified as insignificant as it would account for only about 0.4 percent of total formal employment in the country. Fishing and fisheries related employment as estimated by the SFA in 2004 accounted for about 15 percent of total formal employment in the country (SFA, 2004).

Based on the results of the socio-economic survey carried out in 2004, it can be estimated that between 120 to 125 households were to various degrees dependent on the sea cucumber harvesting and processing business (Marguerite, 2005). The potential number of household that can directly or indirectly benefit from this fishery can be estimated at around 180 households. While the revenue from the sea cucumber industry may be insignificant compared to total revenue generated by the fisheries sector, this industry may have a significant importance for those whose livelihood depend mostly or entirely on this business.

The earnings for the fishery vary widely depending on the boat category (small or big), but even within one category the earnings can vary considerably. However the results of the socio-economic survey have shown that in general, it is the smaller boats, doing day trips, which are more profitable.

A recent development in the fishery is the fact that the processors who have access to foreign currency are supplying equipment to some of the fishers, in particular diving equipment. These in turn are then forced to sell their catch to their suppliers at a lower price. There is thus a risk that the whole fishery can end up being controlled by the processors.

5. ADDITIONAL THREATS TO LOCAL SEA CUCUMBER POPULATIONS

Habitat degradation, in particular in the coastal areas would create additional threats to local sea cucumber populations. One of the main causes of habitat degradation is the loss of coral reefs. During the El Niño event of 1998, 50–95 percent of corals died in the coastal areas across the archipelago (Ahamada *et al.*, 2002). Coral recovery has been very slow and most of the carbonate reefs have lost their three dimensional structure, which has thus reduced the habitat for some species that live in these areas. The risk of further coral bleaching and mortality is very much present. There was an episode in 2003 that caused some coral mortality, and in 2007 there are signs of coral bleaching caused by a combination of high water temperatures and strong light penetration in the sea during the months of April and May. It is assumed that the habitat degradation will have an impact on sea cucumber populations also.

Due to the limited amount of flat coastal land available for development purposes on the main islands (Mahé and Praslin), the government has reclaimed land from the sea in the vicinity of the coast. Several small islands were thus created, totalling more than a hundred hectares in area. This was done by building a bund at the perimeter of each island and filling with material pumped from the bottom of the sea. While precautions were taken to limit siltation through the use of filtering cloths (geotextiles) placed inside the bunds, there is now a large area around these newly created islands that have a much heavier load of silt. Even though the total area involved is not that big compared to the total coastal areas, it is still a habitat loss for sea cucumbers (and other invertebrates).

Coastal pollution is not a threat today as the country has very few industries and most of it is light industry. However, the area around the capital Victoria on Mahé Island is where most of the industries are concentrated. These comprise power generating plants using heavy oil, a tuna canning factory which is one of the largest in the world, sewerage treatment plants and other factories. These industries could pose a potential threat if for example raw or partially treated effluents are released either on purpose or by accident in the sea. There have already been cases of fish mortality in that area due to the release of effluents, little information is available on whether other organisms such as sea cucumbers have been also affected. The use of explosives for fishing or bottom trawl is banned by law (Fisheries Act, 1986) and the trade of wild caught fish (both for aquarium and for consumption) is not allowed. This was legislated through a Fisheries Regulations in 2005.

6. RECOMMENDATIONS FOR IMPROVING FISHERIES MANAGEMENT AND CONSERVATION

There are several ways for improving fisheries management and conservation of sea cucumbers. One of the aspects is for the stakeholders' involvement in order to create a sense of ownership of this fishery. Traditionally in the Seychelles, fisheries have been accessible to all people, and there are no community rights to fishing grounds. Except for the lobsters and sea cucumbers fisheries, that require a fishing license and for which there is a restricted number of licenses, there is no limit to the number of fishers who can access the other resources. With this open-access system, fishers have little incentive to conserve the resource, or at least it is more difficult to make them understand the need for conservation.

In the Seychelles, fisheries management has used a top-down approach, where the fisheries authority would take management decisions and impose it on the fishing community, without necessarily consulting the stakeholders. This approach is now changing, and in the case of the sea cucumber fishery, a first step has been taken with the creation of an advisory management committee with representation from the stakeholders. The committee is composed of representatives from the fishery authority, the Department of Environment, the Department of Natural Resources, representatives of sea cucumber boat owners, divers and processors as well as a representative of an environmental NGO. This new system seems to give a higher sense of responsibility to the fishers who better understand the need to fish in a sustainable way.

The recommendation is to control effort on the higher value species and spread effort to the lower value ones while keeping the catch within the recommended TAC. At present, most of the lower value species, which have the highest abundance, are either fished in very low numbers (e.g. *Holothuria atra*) or not fished at all (e.g. all the species from the genus *Bohadschia*). Increased demand for bêche-de-mer products will see a continued increase in the value of this commodity, extending to low value species (Conand, 2004). There is also a need to control fishing effort in areas close to the main islands on the Mahé Plateau for high value species, to alleviate the risk of local depletion. There is at present no minimum size limits for species in the catch. Minimum size limits could be designed to protect individuals until they have spawned once; this will require research on size of maturity. Implementing minimum size limits will not necessarily be an easy task because there are no observers on board the fishing vessels and would require improved monitoring and enforcement capabilities. The fishery authority has begun collecting size information of processed sea cucumber and once sufficient information is available, a minimum processed size could be implemented for some of the species (Aumeeruddy and Conand, 2007). The idea is for fishers to also target the lower valued-species for which a MSY has been determined and a TAC agreed, and not species that were not assessed during the population survey.

There is a need to improve enforcement through controls and patrols, in particular around the outer islands. Most of these islands are far from the populated islands, and some of them are uninhabited. Poaching by both unlicensed local fishers and foreign fishers could cause a lot of damage to sea cucumber populations around these islands, with a threat of local depletion.

7. EVALUATION OF THE PROS AND CONS OF CITES LISTING

There is an ever increasing demand for sea cucumbers on the Asian market and reports from several countries show that sea cucumber populations are suffering from overexploitation and depletion. This has triggered a discussion on whether sea cucumbers from the Holothuridae and Stichopodidae families should be considered for listing by CITES, and the Animals Committee of CITES was asked to review the outcomes of a technical workshop, develop appropriate recommendations, and prepare for consideration at CoP13 a discussion document on the biological and trade status of these sea cucumbers (Anon., 2004). At a CITES meeting in 2005, the Working Group on sea cucumber of the CITES's Animals Committee did not feel that it was appropriate to comment on the value of a CITES listing because there was insufficient information and evaluation at this stage (Pourkazemi, 2005; Bruckner, 2006).

To manage the international trade of holothurians, one alternative would be to include species belonging to these two families in CITES Appendices, and in that case decide if all species should be listed, or only those that are considered threatened; another alternative would be to consider that the conservation of holothurians is the responsibility of national fisheries management, in line with FAO's guidelines for sustainable fisheries.

In the Seychelles, the procedure in place for exports of sea cucumbers is well established. The exporter submits an application for an export permit to the Export Control Division of the Ministry of Finance. The exporter lists the species, their quantities and price, as well as the name and address of the buyer. Before an export certificate is issued, a copy of the application is sent to the SFA whose inspectors physically check that the consignment corresponds to what is declared on the application form. The export permit is issued by the Export Control Division only after SFA has given its approval. CITES permits are issued by the Nature and Conservation Division of the Department of Environment. If sea cucumbers are listed by CITES, then staff from that Division will also have to be involved in checking every consignment for export before the export certificate is issued.

In the case of the Seychelles, a CITES listing in Appendix II may be appealing as most of the sea cucumbers harvested are traded internationally and many populations are under threat. However, considering the small size of the country and the limited human resources, a CITES listing would place increased pressure on these limited resources, without making any major difference in terms of export control on sea cucumbers. There are enough control conditions in place through the management plan to ensure that sea cucumbers are exploited on a sustainable level. The people involved in the sea cucumber industry already consider that there are too many controls in place that hinders their operations. Adding another layer of control through a CITES listing would possibly make them even less willing to cooperate with the authorities. Some may decide to drop out of the business or try to export their products outside the official circuit. Considering the administrative burden that would be placed on local authorities if sea cucumbers were listed either in Appendix II or III, the authors do not see the need for any listing as the populations do not seem under threat. The conservation of sea cucumbers in the Seychelles would be more easily done as part of fishery management in consultation with other national and international expert bodies.

ACKNOWLEDGEMENTS

We would like to thank V. Toral-Granda, A. Lovatelli and M. Vasconcellos for inviting us to contribute to the sea cucumber project. S. Uthicke and D. Herrero-Pérezrul are thanked for their revision of the paper. This review has greatly benefited from data collected through the FAO and WIOMSA sea cucumber projects. Many other colleagues are thanked for their help and contributions.

REFERENCES

Ahamada, S., Bigot, L., Bijoux, J., Maharavo, J., Meunier, S., Moyne-Picard, M. & Naidoo, P. 2002. Status of coral reefs in the South West Indian Ocean Node: Comoros, Madagascar, Mauritius, Réunion and Seychelles. *In: Status of Coral Reefs of the World: 2002*. C. Wilkinson (Ed.). Australian Institute of Marine Science.

Anon. 2007. Population and vital statistics No. 2 of 2007. Statistical Bulletin. National Statistics Bureau. Seychelles. 9 pp.

Anon. 2004. Conservation of and trade in sea cucumbers in the families Holothuridae and Stichopodidae (decisions 12.60 and 12.61). Twentieth meeting of the Animals Committee, Johannesburg (South Africa), 29 March–2 April 2004. CITES. AC20 Doc. 18. 3 pp.

Aumeeruddy, R. 2007. Sea cucumbers in Seychelles. *In*: Conand, C. and N.A. Muthiga (eds) 2007. *Commercial sea cucumbers: a review for the Western Indian Ocean*. WIOMSA Book Series No. 5, 41–52.

Aumeeruddy, R. & Conand, C. 2007. Seychelles' sea cucumber fishery: Data on processed products and other parameters. *SPC Bêche-de-Mer Information Bulletin*, 26: 19–26.

Aumeeruddy, R. & Payet, R. 2004. Management of the Seychelles sea cucumber fishery: status and prospects. *In: Advances in sea cucumber aquaculture and management*. A. Lovatelli, C. Conand, S. Purcell, S. Uthicke, J.-F. Hamel and A. Mercier (eds). *FAO Fisheries Technical Paper*. No. 463. Rome, FAO. 2004. 425p.

Aumeeruddy, R., Skewes, T., Dorizo, J., Carocci, F., Coeur de Lion, F., Harris, A., Henriette, C. & Cedras, M. 2005. Resource assessment and management of the Seychelles sea cucumber fishery. FAO Project Number: TCP/Seychelles/2902(A), November 2005. 49 pp.

Bruckner, A.W. 2006. The Proceedings of the Technical workshop on the conservation of sea cucumbers in the families Holothuridae and Stichopodidae. *NOAA Technical Memorandum*, 44, 239 pp.

Clark, A.M. 1984. Echinodermata of the Seychelles. *In: Biogeography and ecology of the Seychelles Islands*. D.R. Stoddart (ed). W. Junk, The Hague. *Monographiae Biologicae*, 55, 83–102.

Clark, A.M. & Rowe, F.W. 1971. Holothurioidea. *In: Monograph of Shallow Water Indo-West Pacific Echinoderms*. Trustees of the British Museum, London: 171–210.

Conand, C. 2004. Present status of world sea cucumber resources and utilisation: an international overview. *In*: *Advances in sea cucumber aquaculture and management*. A. Lovatelli, C. Conand, S. Purcell, S. Uthicke, J.-F. Hamel and A. Mercier, (eds). *FAO Fisheries Technical Paper*. No. 463. Rome, FAO. 2004. 425p.

Jennings, S., Marshall, S., Cuet, P. & Naim, O. 2000. The Seychelles. *In: Coral Reefs of the Indian Ocean.* McClanahan, T., Sheppard, C. and D. Obura (eds). Oxford University Press, United Kingdom. pp. 383–410.

Le Quotidien (La Réunion). 2001. Illegal beche-de-mer fishing in the Seychelles, Indian Ocean. *SPC Beche-de-Mer Information Bulletin*, 15: 39.

Marguerite, M. 2005. Capacity building in the Resource Assessment and Management of the Seychelles holothurian fishery – Socio-economic analysis. FAO Project No. TCP/Seychelles/2902(A). August 2005. 55 pp.

Payet, R. 2005. Management Plan of the Seychelles Sea cucumber Fishery. Final Report of Project No. FAO/TCP/SEY/2902(A), September 2005. 46 pp.

Pinault, M. & Conand, C. 2007. Contribution to the Seychelles' sea cucumber Masma project. Report from a mission made in May 2007. Unpub. report .19 pp (online www.wiomsa.org).

Pourkazemi, M. 2005. Conservation and trade in sea cucumbers in the families Holothuridae and Stichopodidae. Twenty-first meeting of the Animals Committee Geneva (Switzerland), 20–25 May 2005. CITES. AC21 Doc. 17. 2 pp.

Seychelles Fishing Authority. 2004. Annual Report.

Seychelles Fishing Authority. 2006. Annual Report. (in press).

Sloan, N.A. 1982. Size and structure of Echinoderm populations associated with different coexisting coral species at Aldabra Atoll, Seychelles. *Mar. Biol.*, 66: 67–75.

Population status, fisheries and trade of sea cucumbers in Latin America and the Caribbean

Verónica Toral-Granda

Galapagos Islands: a hotspot of sea cucumber fisheries in Latin America and the Caribbean

Verónica Toral-Granda

Population status, fisheries and trade of sea cucumbers in Latin America and the Caribbean

Verónica Toral-Granda
FAO Consultant
Puerto Ayora, Santa Cruz Island, Galapagos Islands, Ecuador
E-mail: veronica.toral@gmail.com

Toral-Granda,V. 2008. Population status, fisheries and trade of sea cucumbers in Latin America and the Caribbean. In V. Toral-Granda, A. Lovatelli and M. Vasconcellos (eds). Sea cucumbers. A global review of fisheries and trade. *FAO Fisheries and Aquaculture Technical Paper.* No. 516. Rome, FAO. 2008. pp. 213–229.

SUMMARY

The region under study comprises a total of 25 countries where, although there are some sea cucumber fisheries, scant information exists about them. There are eleven species of sea cucumbers currently harvested for commercial use in the region, with legal and illegal fisheries currently occurring in Mexico, Panama, Colombia, Ecuador, Nicaragua, Peru, Venezuela and Chile. In most of the countries where a fishery exists, there is hardly any biological or ecological information as well as little knowledge on the population status and even, in some cases, the taxonomy of the species under commercial exploitation. In most countries with ongoing fisheries, no management measures are in place and new species are normally being incorporated to the fishing activities. Although sea cucumber fishing it is not a traditional activity, some households have become highly dependent on this fishery, with increasing pressure towards decision makers to allow such activity. Despite the total bans on certain countries for this activity, exports are being recorded in China Hong Kong Special Administrative Region (SAR). Furthermore, the available catch and trade statistics reveal that a high level of illegal, unreported and unregulated (IUU) activities are currently taking place. Amongst the major threats to sea cucumber populations in the region, one can mention the development of fisheries with little or no information on the species, its biology, ecology and population status. Additionally, the permanent search of new species to supply the bêche-de-mer markets poses a serious threat to the wellbeing of not only sea cucumber populations but for the ecosystem as well. Sea cucumber fisheries have arrived to the furthermost fishing grounds available. The impact of this activity on the population status and socio-economic dependence by local fishers are noticeable, especially for an area where no knowledge or political will exists so as to avoid an overexploitation spiral that may leave few species in the brink of extinction.

1. THE REGION

The region under study comprises Mexico, Central America and South America (Figure 1). Information from countries in the Caribbean region, including Jamaica, Haiti and Cuba, is also reported.

FIGURE 1
Region under study and countries covered in this review (highlighted in yellow). The region encompasses the FAO Statistical Areas 31 (Western Central Atlantic), 77 (Eastern Central Pacific), 87 (Southeast Pacific) and 41 (Southwest Atlantic)

2. BIOLOGICAL AND POPULATION STATUS

2.1 Key taxonomic groups

Current legal fishing activities involve specimens from the Order Aspidochirotida and Dendrochirotida, with the families Holothuriidae and Stichopodidae as the most common; however, two species from the family Cucumaridae (Dendrochirotida) have also been recorded (Table 1).

TABLE 1
Sea cucumber species under commercial exploitation in Mexico, Central and South America

Order	Family	Species	Commercial exploitation
Aspidochirotida	Holothuriidae	*Actinopyga agassizi*	Panama, Venezuela
		Holothuria mexicana	Nicaragua, Panama, Venezuela
		H. impatiens	Mexico
		H. theelii	Peru
		H. atra	Ecuador
		H. kefersteini	Ecuador
		H. inornata	Costa Rica
		H. arenicola	Nicaragua
	Stichopodidae	*Isotichopus badionotus*	Cuba, Nicaragua, Venezuela
		I. fuscus	Ecuador, Mexico, Panama, Peru
		Stichopus horrens	Ecuador
		Astchopus multifidus	Panama, Caribbean region
Dendrochirotidae	Cucumaridae	*Athyonidium chilensis*	Chile, Peru
		Pattalus mollis	Chile, Peru

On the Pacific coast of Mexico, South and Central America, the current species under legal commercial exploitation are *Isostichopus fuscus* (Ludwig 1875) in the Galapagos Islands (Ecuador) (Toral-Granda and Martínez, 2004) and in Baja California (Mexico) (Castro 1995, 1997; Aguilar-Ibarra and Ramírez-Soberón, 2002), *Holothuria impatiens* and *Parastichopus parvimensis* in Mexico (Castro, 1997), *Holothuria arenicola* in Nicaragua (Palacios and Brenes, 2008), *Holothuria theelii* in Peru (Elliott, González and Ramirez, undated) and *Athyonidium chilensis* and *Pattalus mollis* in Chile and Peru (Guisado, C., Universidad de Valparaiso, Chile, personal communication; Elliott *et al.*, 2000). Illegal activities have started in the Galapagos Islands targeting *Stichopus horrens* (Hearn and Pinillos, 2006; Reyes and Murillo, 2007), *H. atra* (Reyes and Murillo, 2007) and *H. kefersteini* (Toral-Granda *et al.*, 2005).

Guzman and Guevara (2002) reported that the most highly commercial species in the Caribbean region are *Isostichopus badionotus*, *Astichopus multifidus* and *Holothuria mexicana,* although Conand (2006) mentions that there is an incipient fishery for *Actinopyga agassizi* as well. A sea cucumber fishery within Venezuelan waters was registered in mid-1990s, based on *H. mexicana* and *I. badionotus* (Rodríguez-Milliet and Marquez-Pauls, 1998). Unfortunately, recent information on this fishery is not available. In Nicaragua, there are three taxonomically identified species currently exploited, however, there are at least seven more sea cucumber species being targeted. These are known with their local common names.

2.2 Biology and ecology of sea cucumbers

Little is known for all sea cucumber species in the region, with sparse information on few species that could be of commercial interest. Guzman, Guevara and Hernandez (2003) present basic reproductive biology parameters for *I. badionotus* and *H. mexicana* from the Caribbean coast of Panama. Both species present a unimodal population structure with a majority of mature individuals and a 1:1 male to female sex ratio. The size at sexual maturity (SOM) is between 13 and 20 cm total length (TL). These species also showed a continuous reproductive activity throughout the year, with a peak of reproductive activity between July and November for *I. badionotus* and between February and July for *H. mexicana* (Guzman, Guevara and Hernández, 2003). In northwestern Venezuela, the gametogenic cycle of *I. badionotus* is continuous, reaching maximum maturity during July and August (Foglietta *et al.*, 2004); whereas, in Brazil, *I. badionotus* presents a high gonadal index in October and November and in January and February, with a spawning event occurring only in January when the sea temperature is at its highest (Pires-Nogueira *et al.,* 2003). In Bonaire, Netherlands Antilles, spawning of this species was recorded after the full moon in August (Graaf, Geertjes and Videler, 1999). *I. badionotus* is a large species reaching up to 45 cm TL with a wide distribution in the Caribbean; its southern geographic distribution reaching to south of Brazil (Conand, 2006). It is a common shallow water species, inhabiting mud, sand and rocky bottoms. Adults are generally exposed, whilst the juveniles hide under rubble (Conand, 2006). To the author's knowledge, no information on aestivation is available.

H. mexicana is also a large species (reaching up to 50 cm TL), inhabiting offshore reefs between 2 and 10 m depth (Conand, 2006). The gametogenesis and spawning for *H. mexicana* occurs during spring–summer and late summer in southern Florida (Engshorm, 1980; Mosher, 1982), however, individuals with mature gametes can be found throughout the year (Guzman, Guevara and Hernandez, 2003). In Curaçao, 70 percent of the individuals spawned within five days of the full moon between August and October (van Veghel, 1993). Rodríguez-Milliet and Pauls (1998) reported the SOM for *H. mexicana* at 18 cm TL. To the author's knowledge, no information on aestivation is available.

A. multifidus can reach 50 cm TL and it is characteristic of soft bottoms, with muddy or sandy patches, in and around seagrass beds (Conand, 2006). *P. parvimensis* has a geographical distribution from Cedros Island, Baja California to Carmel Bay, California. It occurs subtidally on both rocky and soft habitats down to 60 m depth. *A. agassizi* is also large (35 cm TL) inhabiting rocky areas and seagrass beds. It is a nocturnal species (Conand, 2006).

Out of the species of commercial importance within the region, *I. fuscus* is probably the species most studied. It can be found from Baja California to mainland Ecuador, including Galapagos, Socorro, Cocos and Revillagigedos islands (Deichman, 1958; Maluf, 1991), although Hooker, Solís-Marín and Leellish (2005) include Peru (Islas de Lobos de Afuera) in its geographical distribution. It can be found in the coastal zone from the shallow subtidal to 39 m depth (Deichman, 1958; Maluf, 1991). *I. fuscus* prefers rocky habitats with *Ulva* sp. (Toral-Granda and Martinez, 2007) whilst in Baja California, it is usually found in coral and rocky habitats (Herrero-Pérezrul *et al.,* 1999). In the Galapagos Islands, *I. fuscus* shows reproductive activity all year long irrespective of sea surface temperature (Mercier, Ycaza and Hamel, 2007; Toral-Granda and Martinez, 2007). However, Mercier, Ycaza and Hamel (2007) report that *I. fuscus* spawns every month between one and four days after the new moon. In Baja California the species shows an annual reproductive season influenced by the influx of warm water (Fajardo-León *et al.,* 1995; Herrero-Pérezrul *et al.,* 1999). In both localities, *I. fuscus* presents a unimodal population structure with a majority of mature individuals (Herrero-Pérezrul *et al.,* 1999; Toral-Granda and Martinez, 2007). This species is more active at night (Shepherd, Toral-Granda and Edgar, 2003). A more detailed analysis on the biology and ecology of this species is presented in the hotspot document (Toral-Granda, this volume).

Athyonidium chilensis can be found from Peru to south of Chile. It can be found intertidally down to 7 m in rocky areas with great amount of organic matter. *A. chilensis* presents a continuous reproductive season (Rojas, L. and Guisado, C., Universidad de Valparaiso, Chile, personal communication).

Stichopus horrens is found in the Pacific Ocean from Malaysia to the Society Islands, French Polynesia, and from southern Japan and Hawaii to New Caledonia (Massin *et al.*, 2002) and in the Galapagos Islands. It prefers rocky substrates from 5–20 m depth (Hickman 1998), with abundant population over 30 m in certain islands (Hearn, A. Charles Darwin Foundation [CDF], personal communication). *S. horrens* normally remains hidden during the day to emerge at night to feed (Hearn and Pinillos, 2006). The population showed a unimodal distribution with the absence of smaller animals (Hearn and Pinillos, 2006) perhaps due to different habitat preference. Ongoing research on the reproductive biology of *S. horrens* show that this species reproduces throughout the year (J. Mora, CDF, unpublished information).

No information was available for *Pattalus mollis*, *Holothuria theeli*, *H. arenicola* and *H. impatiens*.

Although sea cucumbers are known to exert an important ecological role, as they recycle nutrients and help in the bioturbation process that enables organic matter to be brought back to the surface for further use by other organisms (Massin, 1982; Birkeland, 1988), no specific studies on a species level were found for the region.

2.3 Population status

In Panama, population surveys showed that *H. mexicana* was the most abundant species (161.8 ind. ha^{-1}) followed by *I. badionotus* (117.4 ind. ha^{-1}) (Guzman and Guevara, 2002), however these values are only for one small region within the country. No information is available on a country basis. In Cuba, *I. badionotus* reaches its highest density (ca. 8 800 ind. ha^{-1}) in the southeastern region and in some bays in the northeastern region (Alfonso *et al.*, 2000; Alfonso, 2006). No population surveys were done in mainland Ecuador before the commencement of the fishery in 1988 and recent surveys showed only one *I. fuscus* in over 3 000 m^2 (Martínez, C., USFQ-ECOLAP, personal communication). No information was available for the remaining countries.

2.4 Sea cucumber fishery

For most of the countries within the region, information on the starting date of sea cucumber fisheries is absent, with some records of illegal activities still in place. Currently, only Cuba, Peru, Chile, Mexico and Ecuador have regulated fishing activities, with Mexico and Ecuador focusing mostly on *Isostichopus fuscus*, and Mexico, to a lesser degree, on *P. parvimensis* (Stichopodiade: Aspidochirotida). Fisheries in Chile target *A. chilensis* and *P. mollis*, (Cucumaridae: Dendrochirotida) (Guisado, C. Universidad de Valparaiso, personal communication), while in Peru the fishery is focused on *P. mollis* (Hooker, Y., Universidad Peruana Cayetano Heredia [UPCH], personal communication). Cuba has an ongoing sea cucumber fishery focusing on *I. badionotus* (Holothuriidae: Aspidochirotida) (Alfonso *et al.*, 2004; Alfonso, 2006).

2.4.1 Panama

Guzman and Guevara (2002) report fishing activities in Panama in 1997 focusing on *I. badionotus*, *Astichopus multifidus* and *H. mexicana*. In 1997, A fishing license was given to one oriental entrepreneur to fish in the Bocas del Toro Archipelago (Caribbean coast), however his fishing permit was revoked after only 41 days as the conditions under which the licence was given were not met (Ministerio de la Presidencia, 2004). Conservative values report a total of 750 000 individuals of all three species caught during the 1997 30-day fishing period, with a total of 25 fishers involved in the activity (Guzman and Guevara, 2002). Currently, this activity is banned for the entire Republic of Panama by means of the Decree 157-2003 (Ministerio de la Presidencia, 2004); nonetheless, there are reports of illegal fishing activities currently going on in Las Perlas and Coiba (Pacific coast) (Pretto, R., Autoridad de los Recursos Acuáticos de Panama, personal communication). Stock assessments of *H. mexicana*, *I. badionotus* and *A. multifidus* in Bocas del Toro indicated that these species have small population sizes and a high risk of collapse in the short term if fishing effort levels were to be maintained (Guzman and Guevara, 2002). Illegal fishing activities also target *I. fuscus* in the Pacific coast of Panama (Serrano, N., personal communication).

2.4.2 Venezuela

Rodríguez-Milliet and Pauls (1998) provide information on a sea cucumber fishery in Venezuela targeting *I. badionotus* and *H. mexicana*. This fishery began in 1991–1992 as an informal activity, and then in 1993 the Venezuelan Fisheries and Aquaculture Office Service of the Ministry of Agriculture issued the first licence (Rodríguez-Milliet and Pauls, 1998). However, this licence was revoked as no technical reports were submitted in order to regulate the fishery. In 1994, four commercial licences were given and two scientific institutions offered to contribute with information so as to help regulate the fishery and the concession of new fishing licences. This fishery was allowed only in the Cubagua and Coche Islands, northeastern Venezuela. The total catch for the 1994 season were 3 285 kg and 1 922 kg (dry weight), respectively. In 1995, an illegal shipment of 930 kg was impounded (Rodríguez-Milliet and Pauls, 1998). Sea cucumber fishing activities are banned in Venezuela since 1996. However, in 1996, 500 kg of *H. mexicana* with a value of USD 150 000 were confiscated in the Archipelago Los Roques National Park[1]. In addition, China Hong Kong SAR, reports approximately 0.5 tonnes of sea cucumbers imported from Venezuela for 2005 (see Table 3). No recent information for Venezuela was available.

2.4.3 Chile

In Chile, sea cucumber fishing activities started in 1992 focusing mainly on *A. chilensis*, although this fishery is not constant and it may be self regulated by specific demand of

[1] http://www.parkswatch.org/parkprofiles/pdf/ronp_spa.pdf.

the product. Currently there is no information on the number of fishers involved in this venture and it is reported not to be of high importance to the local fishers (Guisado, C., Universidad de Valparaiso, personal communication). The fishery is considered artisanal with fishers free diving to collect the individuals. No management measures are in place to safeguard this species (Guisado, C., Universidad de Valparaiso, personal communication). Total catches for 1992 were 237 tonnes, increasing to 1 510 tonnes in 2000 and a showing a drastic decrease to 19 tonnes in 2002. All catches are used for the bêche-de-mer market and the main markets are Taiwan Province of China, China and Singapore (Guisado, C., Universidad de Valparaiso, personal communication). China Hong Kong SAR, reports almost 41 tonnes of sea cucumber imported from Chile between 2000 and 2005 (see Table 3). No information was available for *P. mollis*.

2.4.4 Peru

In Peru the fishery started in the late 1980s originally on *I. fuscus*, but upon depletion of this species the fishing effort moved to other species, nowadays focusing especially on *P. mollis*, although there are records of *I. fuscus*, *H. theelii* and *A. chilensis* also being caught. Historically, *A. chilensis* was traditionally eaten only in the Department of Labayeque. Nonetheless, currently *P. mollis* is readily available on local markets for domestic consumption (Hooker, Y., UPCH, personal communication). This activity is carried out upon request, and sea cucumbers are gathered as a bycatch of other sea food species, and it is regarded as a subsistence and artisanal venture. There are no management plans for this fishery (Hooker, Y., UPCH, personal communication).

2.4.5 Cuba

In Cuba, sea cucumber fishing activities started in 1999 targeting *I. badionotus* (Alfonso, 2006). Other species of sea cucumbers are also present but at much lower densities, and no fishery has been directed towards them (i.e. *A. agassizi*, *H. floridana*, *H. mexicana* and *A. multifidus*).

This activity was deemed viable after a study carried out by the Fishery Research Center in 1997; however, commercial exploitation started only in 1999. This season had a total allowable catch (TAC) of 320 tonnes (dry weight) with one license granted to a Korean company (Alfonso *et al.*, 2004). During the first years of the fishery (1999–2000) over 3 000 000 individuals were caught, with an average Catch per Unit of Effort (CPUE) of 1 153 ind. $boat^{-1}$ day^{-1}. However, both catches and CPUE decreased to less than 500 000 individuals and 350 ind. $boat^{-1}$ day^{-1} for the next two seasons. In 2004, the CPUE showed signs of recovery with over 1 500 ind. $boat^{-1}$ day^{-1} (Alfonso *et al.*, 2004). By 2004, there were a total of 28 fishers and 18 processors operating this fishery (Alfonso, 2006). Fishers venture to sea in speed boats and each fishing outing lasts an average of 20 days, with 10 days resting period. Animals are collected from the sea floor using hookah at depths between 3 and 15 m (Alfonso *et al.*, 2004).

During the first fishing season, various studies were performed in order to provide information on how to manage this activity and to minimize its possible impact. There is a fishing ban over the reproductive season (June–October), minimum landing sizes (22 cm TL for the southwestern region and 24 cm TL for the southeastern region). Additionally, each locality has a TAC which regulates the number of active vessels (Alfonso, I., Fisheries Research Center, personal communication). Moreover, a Technical Operational Procedure (TOP) manual provides measures for accident prevention, information on decompression tables and TAC and bag sizes (Frías *et al.*, 2002) whilst another TOP was developed to improve the quality of the final product and to guarantee better yields (Castelo *et al.*, 2002), which has resulted in an increase of the percentage of Class A product being exported (Alfonso, 2006).

From 1999 to January 2007, approximately 99.4 tonnes (dry weight) *I. badionotus* have been exported from Cuba representing over USD 1.5 million revenue (Alfonso, I.,

Fisheries Research Center, personal communication). With prices per kilo ranging from USD 13.5 in 1999–2001 to USD 22.0 in April 2003 for dry *I. badionotus*; whilst the price for Class B has fluctuated from USD 6.0 to USD 10.0 per kg (Alfonso, 2006). These exports are solely to supply the bêche-de-mer market, whilst there are some studies to evaluate the possible commercial use of gonads and the processing water (the water in which *I. badionotus* has been boiled for producing bêche-de-mer) for bioactive and medicinal extracts, specifically as antifungal medicines (Alfonso, Tacoronte and Mesa, 2007). Cuba does not report any illegal sea cucumber fishery (Alfonso, 2006).

2.4.6 Mexico

In Mexico, the sea cucumber fishery started in Baja California targeting *I. fuscus* in the eastern coast and *P. parvimensis* in the western coast. The fishery for *I. fuscus* started in the 1980s as a small-scale activity that responded to the international demand for bêche-de-mer in oriental countries (Aguilar-Ibarra and Ramírez-Soberón, 2002). The exploitation of *P. parvimensis* started in 1996. However, Castro (1997) state that the fishery for *I. fuscus* started in 1988 and for *P. parvimensis* in 1989, and for *H. impatiens* in 1994. The fishery for *I. fuscus* continued until 1994 when the government imposed a total closure because the species was considered "endangered" (NOM-059-ECOL-1994). This measure was taken because in 1992 and 1993 there was a sudden increase in the fishing effort which led to a considerable decrease of the CPUE and total landings (Castro, 1995). The closure was not obeyed by fishers, who continued fishing and the standing biomass of *I. fuscus* reached ca. 2 percent of the original biomass for the region (Aguilar-Ibarra and Martínez-Soberón, 2002). No significant recovery has been registered despite management measures set in place. The lack of recovery may be a result of continued illegal activities and/or decreased productivity because population densities may be below a minimal threshold level required for successful reproduction (Toral-Granda, 2007). In March 2002, *I. fuscus* was downlisted as a "subject to special protection" which allows the use of this species under restricted conditions. In 2005, an adaptive management plan was developed under a participatory scheme, which included local and national government, scientists and fisher associations. This plan includes the allocation of harvest quotas, catch reports and population monitoring reports to assess continuously the impact of the fishery. This management plan is reviewed on a yearly basis and adapted to the new results obtained (Toral-Granda, 2007).

Currently, the *I. fuscus* fishery is managed under concessions of "*Unidad de Manejo para la Vida Silvestre*" (UMAS) to organized fishers who in return must submit reports on the activity. This information is used to evaluate the impact of the fishery in the population and decide further management measures to ensure its sustainability (Herrero-Perezrul, D., Centro Interdisciplinario de Ciencias Marinas, Instituto Politécnico Nacional [IPN], personal communication). All *I. fuscus* from Mexico is exported to Taiwan Province of China and China Hong Kong SAR, either as chilled or dry products.

For *Parastichopus parvimensis*, the other Mexican species entering international trade, there are separate management provisions (as for any other fishery in Mexico) since this species is not included in the domestic threatened species list. No information was available for this species.

China Hong Kong SAR, reports over 14 tonnes of sea cucumber that enter the country imported from Mexico from 1995 to 2005 (see Table 3), whilst data available at FAO indicates a total of 2 564 tonnes caught from 1998 to 2005, with an average annual catch of 320.5 ± 86.3 tonnes (FAO, 2007). There is no information on catch and export at the species level. Sea cucumber fishery helps the local communities during the closed season of more profitable resources (i.e. red sea urchin (*Strongylocentrotus franciscanus*) and complements income from abalone (*Haliotis fulgens)* and keyhole limpet (*Megathura crenulata*) (Castro, 1997).

2.4.7 Ecuador

Sea cucumber fishing activities arrived to Ecuador in 1988, targeting *I. fuscus*. This fishery was centered mostly on the coast of Guayas and Manabí provinces and was fuelled by Asian entrepreneurs who taught the local fishers how to collect and process the catch (Carranza and Andrade, 1996). There are no official estimates of the total catch of *I. fuscus* in mainland Ecuador, however, Carranza and Andrade (2006) indicate that at least 420 000 individuals were caught monthly. The fishers would normally deplete one area and then move to nearby areas where they would proceed as before (De Paco *et al.*, 1993). This activity lasted about four years in mainland Ecuador (Carranza and Andrade, 1996), when the collapse of the population prompted the migration of this activity to the Galapagos Islands (De Paco *et al.*, 1993; Carranza and Andrade, 1996; De Miras, Andrade and Carranza, 1996). During the beginning of the sea cucumber fishery in the islands, the lack of knowledge on how to process the catch resulted in the waste of great quantities of animals and low prices paid for the final product, resulting in even less benefits to the local fishers of the Galapagos (Carranza and Andrade, 1996). Commercial exploitation of all sea cucumber species is forbidden in mainland Ecuador by the Ministerial Decree #147, RO/26 of 15 September 1992.

The arrival of this activity to Ecuador is also shown in the country reports to FAO (FAO, 2007) which shows exports of three tonnes sea cucumbers in 1988, increasing to 152 in 1992. From 1993 to 1997, there are 12 tonnes exported annually, and from 1998 to 2005, 15 tonnes (FAO, 2007). According to the Department of International Commerce from the Ecuadorian Central Bank over 74 percent of the sea cucumber exports are directed to Taiwan PC, followed by the United States of America and China (Altamirano, M., personal communication). In all, over 90 percent of the total sea cucumber catches are directed to the Asian countries. Only 80 kilograms of dry *I. fuscus* have been exported from an aquaculture venture (L. Ruidiaz, Subsecretaria de Pesca del Ecuador, personal communication), and there is only one sea cucumber aquaculture company legally registered. However, to the author's knowledge there are at least five more companies dedicated to rearing *I. fuscus* in abandoned shrimp ponds.

The Galapagos Islands are the only regulated and legal fishing ground for sea cucumber fisheries, where it focuses on *I. fuscus*, although there are illegal landings of *S. horrens*, *H. atra* and *H. kefesteini*. More detailed information for the Galapagos sea cucumber fishery is presented in the hotspot document (Toral-Granda, this volume).

2.4.8 Costa Rica

There are reports of commercial sea cucumber fishing activities focusing on *H. inornata* and *I. fuscus* in 1993 and 1994, although no official catch records exist (Anonymous, 2006). There were ten licenses, one per fisher, to carry out this activity. The fishers free dived or use hookah to collect the animals from the sea floor. This activity was regulated by means of a two-month fishing season, a minimum landing size of 20 cm, a bag limit of 1 200 individuals per fisher per month, restricted ports for landing, and closed areas (Anon., 2006). Currently, sea cucumber fisheries are banned in Costa Rica (Anon., 2006; Alvarado, J.J., The Nature Conservancy, personal communication). Nonetheless, China Hong Kong SAR reports imports from Costa Rica of approximately 1.3 tonnes from 1999 to 2005 (Hong Kong Census and Statistics Department, personal communication), indicating that there is still a substantial (illegal) fishery in the country.

2.4.9 Nicaragua

Sea cucumber fishing activities in Nicaragua date back to 1994 when the first exports are recorded for this country. In 2005, oriental traders requested a permit to exploit this resource and a one-year permit was granted while scientific research was undertaken in the Pacific coast. No taxonomic information exists on the species harvested during that year. Currently there are three taxonomically identified species under commercial harvest, whilst there are at least seven more unidentified sea cucumbers harvested. In the Pacific coast, the most important species is the "*Pepino rojo con espiculas*" and in the Caribbean *H. mexicana* and *I. badionotus*. There are about 45 fishers that make of this activity their mean of subsistence, with other 150 who alternate this fishery with other. All products are exported to China and Taiwan Province of China as there is no traditional consumption in Nicaragua. From the Caribbean there has been two main shipments with 5 tonnes in 2006 and 182 in 2007; in the Pacific coast, exports started in 2005 with 51 tonnes (dry weight) and increased to 93 tonnes (dry weight) for both 2006 and 2007. Presently, all sea cucumber fishing activities are not regulated, however, there is a plan to start applying total allowable catches and fishing seasons to both Pacific and Caribbean species (Palacios and Brenes, 2008; Brenes, B. Instituto Nicaraguense de la Pesca y Acuicultura, personal communication). In 2006, Nicaraguan fisheries managers also arrived to the Galapagos Islands looking for possible recommendations on how to manage this incipient activity.

2.4.10 Other countries

Currently, sea cucumber fisheries are banned in Venezuela (Rodríguez-Milliet and Pauls, 1998) and Panama (Ministerio de la Presidencia, 2004); whilst there is no fishery in El Salvador (Barraza and Hazbún, 2007), Honduras (Hasbún, C.R., Fundación Zoológica de El Salvador [FUNZEL], personal communication), Jamaica (Aiken, 2006) and Argentina (Tablado, A., Museo Argentino de Ciencias Naturales, personal communication). In Colombia, there is an illegal, unregulated and non-quantified fishery over *I. badionotus* (Navas, G., Instituto Investigaciones Marinas y Costeras [INVEMAR], personal communication). No information was available from Haiti, Brazil, Uruguay and Belize.

Most sea cucumber fishing activities within the region are unregulated or non-existent. Ecuador, Costa Rica, Cuba, Mexico, Panama and Venezuela have enacted regulations to control the harvesting of these species, whilst other countries provide no information on their management strategies. Table 2 summarizes management measures in place for the countries in the region.

TABLE 2
Management measures for the sea cucumber fisheries in Latin America and the Caribbean

Country	Management measures
Argentina	No known fishery
Belize	No information available
Brazil	No information available
Chile	No management measures in place
Colombia	No management measures in place
Costa Rica	Total ban on sea cucumber fishing
Cuba	Minimum landing sizes, temporal closure over the reproductive season, total allowable catches per region
Ecuador (Galapagos Islands)	Fishing season, minimum harvesting size, no take zones, total allowable catch
Ecuador (Mainland)	Total ban on sea cucumber fishing
El Salvador	No known fishery
Haiti	No information available
Honduras	No known fishery
Jamaica	No known fishery
Mexico	Fishery manager inside "*Unidad de Manejo para la Vida Silvestre*" for *I. fuscus*
Panama	Total ban on sea cucumber fishing
Peru	No management measures in place
Uruguay	No information available
Venezuela	Total ban on sea cucumber fishing

3. TRADE AND ILLEGAL, UNREPORTED AND UNREGULATED CATCHES

Although there is no official information on the main export destinations for most of the sea cucumber landings, it can be safely assumed that most of the catches are exported to supply the Asian market demand for bêche-de-mer. In some cases, fishing activities started upon request from oriental entrepreneurs (e.g. Panama and Cuba); whilst for other countries (e.g. Ecuador) most catches are send directly to China.

For many countries within the region, most of the catches seem to originate from illegal, unreported and unregulated (IUU) fishing. For some countries there is clear evidence of illegal trade and catches despite regulations forbidding the capture of the species (e.g. Panama). In other countries (e.g. Colombia) there are fishing activities that are not regulated.

For instance, according to the China Hong Kong SAR Census and Statistics Department there are 14 countries from Latin America that export sea cucumbers to China Hong Kong SAR (Table 3). Although the total percentage of the imports of such countries is generally less than 1 percent of the global imports, it indicates a considerable amount of IUU catches coming from countries in the region.

The greater amounts of sea cucumbers exported from the region originate from Peru (26.1 percent), Ecuador (25.9 percent), Chile (14.1 percent) and Cuba (10.1 percent) which have declared to have fishing activities for sea cucumbers. The remaining percentage (13.7 percent) originates from countries where either this fishery is banned (i.e. Panama and Costa Rica) or has no official record of it (Colombia).

In the FAO catch statistics only Chile, Ecuador, Mexico and Nicaragua report catch figures for sea cucumbers with a total of 6 035 tonnes from 1988 to 2005 (Table 4) (FAO, 2007). Ecuador reported 389 tonnes, with the first figures in 1988, approximate date on which sea cucumber harvesting began in mainland Ecuador. However, there is also substantial catch statistics from periods when the catch of *I. fuscus* was banned (1994–1998). These catches are probably from illegal activities in the Galapagos Marine Reserve, the only source of *I. fuscus* after the depletion of this species in mainland Ecuador. However, the data from Mexico, with 2 564 tonnes, is the second highest value after Chile (3 031 tonnes) and reports sea cucumber catches since 1998, with its peak catch in 2001 (481 tonnes). Chile started reporting in 1992, around the time in which commercial exploitation of *A. chilensis* started. The figures for Chile are quite erratic, with values ranging from 1 tonne in 1997 to 1 510 in 2000. Lastly Nicaragua, reports a total of 51 tonnes only for 2005, which corresponds to the total reported by Nicaragua. However, China Hong Kong SAR reports the first imports from Nicaragua in 2003. As a whole, catches from the region represent 1.46 percent of the total reported sea cucumber catches to FAO between 1988 and 2005 (Table 4).

The data provided by FAO and the China Hong Kong SAR Census and Statistics Department show that there is a substantial trade of sea cucumbers from Mexico, Central and South America; with the FAO figures providing an overall picture of the amount being harvested, whilst the information from China Hong Kong SAR indicates the diversity of countries that export their produce.

4. SOCIO-ECONOMIC IMPORTANCE

Although recent in arrival and importance, sea cucumber fisheries are an important source of income to coastal communities in the region, however, no official quantification of the level of dependence of fishers on the income generated from this activity exists. Nonetheless, it can be assumed that new fisheries keep emerging in the region as a result of previous experiences elsewhere. No quantification of number of fishers, prices, and trade exist, as in most countries, this is an illegal or unregulated activity.

In Ecuador, with the Galapagos Islands as the case study for the region, further information is provided in Toral-Granda (this volume).

TABLE 3
Total volume of sea cucumber imported by China Hong Kong SAR from Latin American and the Caribbean countries from 1996 to 2005 (kg dry weight)

Country	1996	1997	1998	1999	2000	2001	2002	2003	2004	2005	Total
Brazil	1 600	–	–	–	–	444	50	–	–	–	2 094
Chile	–	–	–	–	22 318	7 599	2 906	527	4 485	5 123	42 958
Colombia	–	–	–	–	–	540	–	–	–	1 646	2 186
Costa Rica	–	–	–	108	664	325	–	7	164	–	1 268
Cuba	–	–	–	2920	19 023	13 941	3 800	7 648	5 080	8 641	61 053
Dominican Republic	–	–	–	–	–	–	2 562	45	–	–	2 607
Ecuador	120	–	–	24 567	15 285	991	10 130	3 026	11 322	13 248	78 689
Haiti	–	–	–	–	–	–	1 000	9 680	–	–	10 680
Mexico	405	500	–	–	150	1 818	3 302	1 270	4 294	2 378	14 117
Nicaragua	–	–	–	–	–	–	–	252	–	5 826	6 078
Panama	–	–	–	–	–	–	–	–	281	408	689
Peru	137	2 364	376	4 170	7 331	3 881	1 828	8 354	19 906	31 002	79 349
Puerto Rico	–	–	–	–	–	–	1 300	–	–	256	1 556
Venezuela	–	–	–	–	–	–	–	–	–	456	456
Total - countries from the region[1]	**2 262**	**2 864**	**376**	**31 765**	**64 771**	**29 539**	**26 878**	**30 809**	**45 532**	**68 984**	**303 780**
Total - world wide[2]	**5 021 828**	**4 524 832**	**3 977 324**	**2 924 331**	**4 760 719**	**4 384 273**	**4 419 356**	**4 675 043**	**5 071 839**	**4 480 643**	**44 240 188**
% from Region[3]	0.05	0.06	0.01	1.09	1.36	0.67	0.61	0.66	0.90	1.54	0.69

Source: Hong Kong Census and Statistics Department.

[1] Sum of kg for the countries from the region.

[2] Sum of kg of the total of sea cucumber imported into China Hong Kong SAR from all countries worldwide.

[3] Percentage of the imports from the region in comparison from those worldwide.

TABLE 4
Volume of sea cucumbers reported to FAO by Chile, Ecuador, Mexico and Nicaragua in comparison to the total tonnage reported worldwide (in tonnes)

Year	Mexico	Chile	Nicaragua	Ecuador	Total in the region	World harvest total	Percentage from region
1988	–	–	–	3	3	19 905	0.02
1989	–	–	–	10	10	17 467	0.05
1990	–	–	–	12	12	19 976	0.06
1991	–	–	–	29	29	21 790	0.15
1992	–	237	–	152	389	20 892	1.95
1993	–	13	–	12	25	19 348	0.13
1994	–	4	–	12	16	24 505	0.08
1995	–	106	–	12	118	24 050	0.59
1996	–	115	–	12	127	26 795	0.64
1997	–	1	–	15	16	24 672	0.08
1998	271	30	–	15	316	22 004	1.59
1999	234	108	–	15	357	20 462	1.79
2000	426	1510	–	15	1 951	24 509	9.80
2001	481	107	–	15	603	20 431	3.03
2002	290	106	–	15	411	23 445	2.06
2003	285	307	–	15	607	28 085	3.05
2004	265	234	–	15	514	27 540	2.58
2005	312	153	51	15	531	26 002	2.67
Total	**2 564**	**3 031**	**51**	**389**	**6 035**	**411 878**	**1.46**

Source: FAO Fisheries and Aquaculture Information and Statistics Service 2007.

5. RECENT DEVELOPMENTS

Guisado (2005) reports the interest to develop a mariculture program for *A. chilensis* in central Chile. This initiative will help generate scientific knowledge on the ecology, biology and nutrition of this species, as well as assure the supply of a wanted resource in a known market. Additionally, it will help the artisanal fishers to diversify their activities (Guisado, 2005). However, to the author's knowledge there is no clear evidence that this venture ever took off. Additionally, the Chilean Government, has recently approved a project to start the aquaculture of *Apostichopus japonicus* in northern Chile. This will need the introduction of specimens from China, which will be placed in quarantine and then in ponds. All effluent waters will be dealt according to the local laws. At the moment, this activity is waiting for the approval of the introduction of the specimens in order to start with the mariculture project (Guisado, C., Universidad de Valparaiso, personal communication).

Mercier, Ycaza-Hidalgo and Hamel (2004) report the successful rearing of *I. fuscus* larvae under laboratory conditions and juveniles in abandoned shrimp ponds in mainland Ecuador. *I. fuscus* juveniles can grow up to 8 cm in length in 3.5 months, with a survival rate between 30 and 50 percent. Currently, there is only one legally registered aquaculture firm focusing on *I. fuscus*. However, to the author's knowledge, there are at least five more with aquaculture ventures working with the same species. There is only 80 kilograms of dry *I. fuscus* that has been exported to Taiwan Province of China from these ventures (Ruidiaz, L., Subsecretaria de Pesca, personal communication). Initiatives to develop similar ventures in the Galapagos Islands is analysed thoroughly in the hotspot document (Toral-Granda, this volume).

In Mexico, Gutierrez-García (1999) indicated the potential of starting aquaculture activities over *I. fuscus*, and currently there are studies to start the aquaculture production of *I. fuscus* prompted by the local fishers as an alternative to capture fisheries (Herrero, A., Gobierno del Estado de Baja California Sur, personal communication). No information is available on the status of this project.

6. ADDITIONAL THREATS TO SEA CUCUMBER POPULATIONS

The exploitation of sea cucumber species without the use of the precautionary approach is probably the biggest threat to their populations within the region. For many of the species under legal or illegal fishing there are no or little scientific information available on biology, ecology, population abundance and dynamics. The lack of information to assess the status of stocks, their productivity and resilience to exploitation (e.g. minimum densities required for successful reproduction) precludes the development of management strategies and plans based on sound scientific knowledge.

The constant searches of new species to supply the bêche-de-mer Asian market also pose a threat to the wellbeing of sea cucumbers in the region. Upon depletion of the most valuable commercial species, trade scouts will find new species on which to focus. This procedure is facilitated by the existence of the know-how, the socio-economic dependence on the resource and the constant demand in the international market. This may promote fishing activities, both legal and illegal, in countries that already have it, as well as in others that have not started it yet. It is possible to infer that, without proper management attention, the development of new sea cucumber fisheries will have a similar faith to that observed in other countries and species.

An additional threat is habitat loss and degradation, either due to climate oscillations (e.g. El Niño), environmental disasters (e.g. tropical cyclones) and human-induced causes such as unsustainable fishing practices, coastal pollution and sedimentation. Aiken (2006) suggests that the decrease in the abundance of *H. mexicana* in Jamaica may be due to the effect of two tropical cyclones that damaged the habitat of this species.

Although there are not many ongoing aquaculture ventures, this activity may pose a threat to sea cucumbers populations. In Chile, there are studies to evaluate the possibility of introducing *Apostichopus japonicus* for mariculture. This introduction could become a threat if not enough care is exerted during the procedure and the activity itself. Lubchenco *et al.* (1991) recognised that human-mediated bioinvasions are one of the main threats to the marine environment and, although no information is available on the possible impacts of introduced sea cucumbers, an introduction of an exotic marine species should be evaluated with great care, especially on the eventuality of larval escape hence promoting inter-species competition. In the Galapagos Islands, an interest to explore the feasibility of restocking the sea cucumber *I. fuscus* to enhance depleted populations and as an alternative to the local fishing sector has been proposed. But under the current management policies of the Galapagos Marine Reserve (GMR) it is not advisable, as this may cause the translocation of *I. fuscus* without any knowledge on genetic delineation of stocks, possible destruction of parts of the terrestrial National Park, and potential stress to the individuals under study as no information on larval and juvenile natural history, optimal release conditions, diseases and predation is available (Toral-Granda, 2005). More information on this is presented in the Galapagos as a hotspot (Toral-Granda, this volume).

7. CONCLUSION

Although there are relatively few species under commercial exploitation in Mexico, Central and South America, most of the fishing activities are being carried out at random and without planning. Most of the sea cucumber species exploited lack biology and ecology studies that will help yield management plans that will provide sustainable use of this resource with the consequent economical benefit to local human populations. For the few countries that have management plans in place, there is little enforcement and illegal activities are common.

Sea cucumber fisheries have arrived to the furthermost ground from its place of origin, where it is a non-traditional activity that is changing the fishing behaviour of the local community. The boom-and-bust cycle has already taken its tokens in the region, with few species overexploited and new ones on the brink of overexploitation. This activity does not require qualified workmanship and/or mechanised fishing gear, which means that is a gold-rush fishery, providing high incomes with little work and investment.

The little published information on sea cucumber fisheries in the region should be viewed as an urgent call to researchers on this field to evaluate the extent to which this fishery is happening within their area of expertise, hence helping to draw management plans that could be applied on a regional base or as species specific. As some species have a wide geographical distribution, joint research programmes should be initiated in order to help gather all possible biological and ecological information on which to base decisions about fishing activities.

ACKNOWLEDGEMENTS

My sincere thanks to Alessandro Lovatelli who believed in me and made me part of the team. To Marcelo Vasconcellos and Alex Hearn who provided inputs so as to improve this document. To Annie Mercier and Priscilla Martínez who helped improving this document and gave key inputs so as to make this document more accurate. My sincere thanks to all colleagues within the region who collaborated with information on the sea cucumber fishing activities within their country or area of expertise, especially to Alejandro Tablado, Armando Herrero, Buffy Rebekah-Beth Turner, Camilo Martínez, Carlos Roberto Hasbún, Chita Guisado, Dalva Arosemena, Dinorah Herrero-Perezrul, Eréndira García, Gabriel Navas, Héctor Guzmán, Irma Alfonso, Juan José Alvarado, Lorena Ruidiaz, Luis Morales, Luis Rojas, Nelly Serrano, Nora Palacios, Patricia Fuentes, Richard Pretto and Yuri Hooker. To Tom Poulsom, who provided great cups of tea, support and editorial assistance. This is Charles Darwin Foundation contribution number 1072.

REFERENCES

Aguilar-Ibarra, A. & Ramírez-Soberón, G. 2002. Economic reasons, ecological actions and social consequences in the Mexican sea cucumber fishery. *SPC Beche-de-Mer Information Bulletin,* 17: 33–26.

Aiken, K.A. 2006. National Report – Jamaica, West Indies. *In: Proceedings of the CITES workshop on the conservation of sea cucumbers in the families Holothuriidae and Stichopodidae.* Bruckner, A.W. (ed). NOAA Technical Memorandum NMFSOPR 34, Silver Spring, MD 244 pp.

Alfonso, I. 2006. National Report – Cuba. *In: Proceedings of the CITES workshop on the conservation of sea cucumbers in the families Holothuriidae and Stichopodidae.* Bruckner, A.W. (ed). NOAA Technical Memorandum NMFSOPR 34 , Silver Spring, MD, 244 pp.

Alfonso, I., Frías, M.P., Vilaragut, M., Téllez, I., Morales, J.M. & Reyes, C.A. 2000. Estado actual del recurso Pepino de Mar en la Plataforma sur oriental de Cuba. *Third Congress on Marine Sciences 'MARCUBA 2000'.* Congress Palace, Havana City, Cuba.13 pp.

Alfonso, I., Tacoronte, J.E. & Mesa, J.A. 2007. Isostichotoxin isolated from *Isostichopus badionotus* (Selenka, 1867) sea cucumber processing's byproducts. *SPC Beche-de-Mer Information Bulletin,* 25: 29–31.

Alfonzo, I., Frías, M.P., Aleaga, L. & Alonso, C.R. 2004. Current status of the sea cucumber fishery in the south eastern region of Cuba. *In: Advances in sea cucumber aquaculture and management.* Lovatelli A., Conand, C., Purcell, S., Uthicke, S., J.-F. Hamel & Mercier, A. (eds). FAO, Rome.

Anon. 2006. National Report – Costa Rica. *In: Proceedings of the CITES workshop on the conservation of sea cucumbers in the families Holothuriidae and Stichopodidae.* Bruckner, A.W. (ed.). NOAA Technical Memorandum NMFSOPR 34, Silver Spring, MD 244 pp.

Barraza, J.E. & Hazbún, C.R. 2007. Los equinodermos (Echinodermata) de El Salvador. *Revista de Biologia Tropical,* 53(Suppl 3): 139–146.

Birkeland, C. 1988. The influence of echinoderms on coral-reef communities. *Echinoderm Stud.,* 3: 1–79.

Carranza, C. & Andrade, M. 1996. Retrospectiva de la pesca del pepino de mar a nivel continental. Fundación Charles Darwin para las Islas Galapagos/ORSTOM. Quito. 54 pp.

Castelo, R., Alfonso, I., Frías, M.P. & Aleaga, L. 2002. Procedimiento Operacional Técnico pera el procesamiento del pepino seco. *Fishery Research Centre. XIV Science and Technology Forum,* June 2002. 17 pp.

Castro, L.R.S. 1995. Management options of the commercial dive fisheries for sea cucumbers in Baja California, Mexico. *SPC Beche-de-Mer Information Bulletin* 7: 20.

Castro, L.R.S. 1997. Review of recent developments in the Baja California, Mexico, *Isostichopus fuscus, Holothuria impatiens* and *Parastichopus parvimensis* fisheries. *SPC Beche-de-Mer Information Bulletin,* 9: 26–27.

Conand, C. 2006. Sea cucumber biology: taxonomy; distribution; biology; conservation status. *In: The Proceedings of the CITES workshop on the conservation of sea cucumbers in the families Holothuriidae and Stichopodidae.* Bruckner, A.W. (ed). NOAA Technical Memorandum. 244 pp.

De Miras, C., Andrade, M. & Carranza, C. 1996. Evaluación socioeconómica de la pesca experimental de pepino de mar en Galapagos. Fundación Charles Darwin para las Islas Galapagos/ORSTOM. Quito.

De Paco, C., Hurtado, M., McFarland, C., Martínez, P., Reck, G. & Richmond, R. 1993. Evaluación de la pesquería de pepinos de mar en las islas Galapagos – Ecuador. Informe para la Unión Mundial para la Naturaleza (UICN) como resultado de la Misión realizada a solicitud de la Fundación Charles Darwin para las Islas Galapagos. 25 pp.

Deichmann, E. 1958. The Holothuroidea collected by The Velero II and IV during the years 1932 to 1954, Part II. Aspidochirota. *Allan Hancock Pacific Expeditions,* 11(2): 253–348.

Elliott, R.W., Blas-Flores, N., Gonzáles-Bazalar, R., Ramírez-Quezada, A., Flores-Huamán, O. & Becerra-Olaya, W. 2000. Las pesquerías de recursos pelágicos, demersales, costeros e invertebrados marinos durante el año 2000. Instituto del Mar del Peru, Laboratorio Costero de Huacho. 12 pp.

Elliott, R.W., González, R., & Ramírez, A. undated. Bancos naturales de invertebrados en Hornillos y El Colorado (Huacho Región, Lima). Instituto del Mar del Peru, Laboratorio Costero de Huacho. 6 pp.

Engstrom, N.A. 1980. Reproductive cycles of *Holothuria (Halodeima) floridana, H. (H.) mexicana* and their hybrids (Echinodermata: Holothuroidea) in southern Florida, USA. *Int J Invertebr Reprod,* 2: 237–244

Fajardo-León, M.C., Michel-Guerrero, E., Singh-Cabanillas, J., Vélez-Barajas, J.A. & Massó-Rojas, A. 1995. Estructura poblacional y ciclo reproductor del pepino de mar *Isostichopus fuscus* en Santa Rosalía, B.C.S., Mexico. *Cienc Pesquera,* 11: 45–53.

FAO Fisheries and Aquaculture Information and Statistics Service. 2007. Capture production 1950-2005. FISHSTAT Plus – Universal software for fishery statistical time series [online or CD-ROM]. Food and Agriculture Organization of the United Nations. (http://www.fao.org/fi/statist/FISOFT/FISHPLUS.asp).

Foglietta, L.M., Camejo, M.I., Gallardo, L. & Herrera, F.C. 2004. A maturity index of Holothurians exhibiting asynchronous development of gonad tubules. *J. Exp. Mar. Biol. Ecol.,* 300: 19–30.

Frías, M.P., Alfonso, I., Castelo, R. & Aleaga, L. 2002. Procedimiento Operacional Técnico para la pesca y manipulación a bordo del pepino de mar, en su medio natural. *Fishery Research Centre. XIV Science and Technology Forum,* June 2002. 15 pp.

Graaf, de M., Geertjes, G.J. & Videler, J.J. 1999. Observations on spawning of scleractinian corals and other invertebrates on the reefs of Bonaire (Netherlands Antilles, Caribbean). *Bull Mar Sci,* 64: 189–194.

Gutierrez-García, A. 1999. Potential culture of sea cucumber in Mexico. *SPC Beche-de-Mer Information Bulletin,* 11: 26–29.

Guisado, C. 2005. CONICYT Project: Development of the cultivation technology of the sea cucumber *Athyonidium chilensis*, in the Chilean South Central Region. *SPC Beche-de-Mer Information Bulletin,* 21: 34.

Guzmán, H.M. & Guevara, C.A. 2002. Population structure, distribution and abundance of three commercial species of sea cucumbers (Echinodermata) in Panama. *Caribbean Journal of Science,* 38(3–4): 230–238.

Guzmán, H.M., Guevara, C.A. & Hernández, I.C. 2003. Reproductive cycle of two commercial species of sea cucumber (Echinodermata: Holothuroidea) from Caribbean Panama. *Marine Biology,* 142: 271–279.

Hearn, A. & Pinillos, F. 2006. Baseline information on the warty sea cucumber *Stichopus horrens* in Santa Cruz, Galapagos, prior to the commencement of an illegal fishery. *SPC Beche-de-Mer Information Bulletin,* 24: 3–10.

Herrero-Pérezrul, M.D., Reyes-Bonilla, H., García-Domínguez, F. & Cintra-Buenrostro, C.E. 1999. Reproduction and growth of *Isostichopus fuscus* (Echinodermata: Holothuroidea) in the southern Gulf of California, Mexico. *Mar Biol,* 135: 521–532.

Hickman, C.P. 1998. Guía de campo sobre estrellas de mar y otros equinodermos de Galapagos. Serie vida marina de Galapagos. Sugar Spring Press, Virginia, USA. 83 p.

Hooker, Y., Solís-Marín, F.A, & Lleellish, M. 2005. Equinodermos de las Islas Lobos de Afuera (Lambayeque, Peru). *Rev. Peru. biol.,* 12(1): 77–82.

Lubchenco, J., Olson, A.M., Brubaker, L.B., Carpenter, S.R., Holland, M.M., Hubbell, S.P., Levin, S.A., MacMahon, J.A., Matson, P.A., Melillo, J.M., Mooney, H.M., Peterson, C.H., Pulliam, H.R., Real, L.A., Regal, P.J. & Risser, P.G. 1991. The sustainable biosphere initiative: an ecological research agenda. A report from the Ecological Society of America. *Ecology,* 72: 371–412.

Maluf, L.Y. 1991. Composition and distribution of the central eastern Pacific echinoderms. Nat. Hist. Mus. *Los Angeles Co. Technol. Rep.*, 2: 1-242.

Massin, C. 1982. Effects of feeding on the environment: holothuroidea. In: Jangoux, M., Lawrence, J.M. (eds.). *Echinoderm Nutrition*, pp. 493–497.

Massin, C., Zulfigar Y., Tan Shau Hwai, A. & Rizal Boss, S.Z. 2002. The genus *Stichopus* (Echinodermata: Holothuroidea) from Johore Marine Park (Malaysia) with the description of two new species. *Bulletin de l'Institute Royal des Sciences Naturelles de Belgique,* 72: 73–99.

Mercier, A., Ycaza Hidalgo, R. & Hamel, J.-F. 2004. Aquaculture of the Galapagos sea cucumber, *Isostichopus fuscus. In: Advances in Sea Cucumber Aquaculture and Management.* Lovatelli A., Conand, C., Purcell, S., Uthicke, S., J.-F. Hamel & Mercier, A. (eds.). FAO, Rome.

Mercier, A., Ycaza, R.H. & Hamel, J.-F. 2007. Long-term study of gamete release in a broadcast-spawning holothurian: predictable lunar and diel periodicities. *Marine Ecology Progress Series,* 329: 179–189.

Ministerio de la Presidencia. 2004. Por medio del cual se prohíbe la extracción, posesión y comercialización del organismo marino conocido como pepino de mar, en la Republica de Panama. Decreto Ejecutivo No. 157 (De 31 de diciembre de 2003). *Gaceta Oficial,* 24 963: 3–6.

Mosher, C. 1982. Spawning behaviour of the aspidochirote holothurian *Holothuria mexicana* Ludwig. In: *Proc Int Echinoderm Conf,* Lawrence J.M. (ed). Tampa Bay. Balkema, Rotterdam, pp 467–468.

Palacios, N. and Brenes, B. 2008. Actividad extractiva del pepino de mar en Nicaragua. Internal document.

Pires-Nogueira, R.L., Freire, C.B., Rezende, C.R. & Campos-Creasey, L.S. 2001. Reproductive cycle of the sea cucumber *Isostichopus badionotus* (Echinodermata) from Baía da Ilha Grande, Rio de Janeiro, Brazil (abstract). *In: Abstracts of the 9th international congress of invertebrate reproduction and development*. Rhodes University, Grahamstown, South Africa, 15–20 July 2001.

Reyes, H. & Murillo, J.C. 2007. Esfuerzos de control de pesca ilícita en la Reserva Marina. *In: Informe Galapagos 2006–2007*. FCD-SPNG-INGALA.

Rodríguez-Milliet, E. & Pauls, S.M. 1998. Sea cucumbers fisheries in Venezuela. Proc. 9th Inter. Echinoderm Conf., 513–516.

Shepherd, S.A., Toral-Granda, M.V. & Edgar, G.J. 2003. Estimating the abundance of clustered and cryptic marine macro-invertebrates in the Galapagos with particular reference to sea cucumbers. *Noticias de Galapagos,* 62: 36–39.

Toral-Granda, M.V. 2005. Should stock enhancement and re-stocking be an option for the Galapagos sea cucumber? Charles Darwin Foundation Internal Document. 10 pp.

Toral-Granda, M.V. 2007. Biological and trade status of sea cucumbers in the families Holothuriidae and Stichopodidae. CITES CoP14 Doc. 62, Annex 1.

Toral-Granda, M.V. & Martinez, P.C. 2007. Reproductive biology and population structure of the sea cucumber *Isostichopus fuscus* (Ludwig, 1875) (Holothuroidea) in Caamaño, Galapagos Islands, Ecuador. *Marine Biology,* 151: 2091–2098.

Toral-Granda, M.V., Murillo, J.C., Piu, M., Nicolaides, F., Moreno, J., Reyes, H., Castrejón, M. & Hearn, A. 2005. Evaluación de la pesquería de pepino de mar (*Isostichopus fuscus*) en la Reserva Marina de Galapagos en el año 2005. Fundación Charles Darwin/Parque Nacional Galapagos. Puerto Ayora, Santa Cruz. 41 pp.

van Veghel, M.L.J. 1993. Multiple species spawning on Curacao reefs. *Bull Mar Sci,* 52: 1017–1021.

Galapagos Islands: a hotspot of sea cucumber fisheries in Latin America and the Caribbean

Verónica Toral-Granda
FAO Consultant
Puerto Ayora, Santa Cruz Island, Galapagos Islands, Ecuador
E-mail: veronica.toral@gmail.com

Toral-Granda, V. 2008. Galapagos Islands: a hotspot of sea cucumber fisheries in Central and South America. In V. Toral-Granda, A. Lovatelli and M. Vasconcellos (eds). Sea cucumbers. A global review of fisheries and trade. *FAO Fisheries and Aquaculture Technical Paper*. No. 516. Rome, FAO. pp. 231–253.

SUMMARY

The sea cucumber fishing activities started in the Galapagos Islands, Ecuador, in 1991 after the collapse of this venture in mainland Ecuador. Although there is only one species (*Isostichopus fuscus*) that can be legally harvested in the Galapagos Islands, oriental scouts have promoted the illegal capture of *Stichopus horrens*, *Holothuria kefersteini* and *H. atra*, with no biological or ecological information available for these species. Fishing operations in Galápagos are restricted to the artisanal fishing fleet; however, the current practices fit more with a semi-industrial scale. Fishing for *I. fuscus* is done by means of small wooden ("*pangas*") or fiber glass ("*fibras*") boats that will store the catch in mother boats ("*botes*"); collection of the animals is done using hookah. The catch is boiled on board and then salted and dried in the inhabited ports. All of the catch is sold to middlemen. The fishing for *I. fuscus* used to be the most important economical activity; however, low catches and recent lack of interest have yielded this fishery of less importance. This activity is regulated by means of a season, total allowable catch, minimum landing size, effort control and spatial and temporal closure, along with the implementation of an adaptive and participatory management plan, incorporating the main stakeholders of the Galápagos Marine Reserve. Nonetheless, despite all efforts, the population status of *I. fuscus* has follow a similar faith of other sea cucumbers commercially harvested worldwide, and its population is seriously overexploited.

1. INTRODUCTION

Upon depletion of the traditional fishing grounds for bêche-de-mer in Asia and Pacific Islands Nations and Territories (Conand, 2001), this fishery arrived to mainland Ecuador in the late 1980s and after uncontrolled activities, the resource was depleted, promoting the migration of this activity to the Galapagos Islands, where it settled in 1991 (De Paco *et al.*, 1993; Camhi, 1995; Powell and Gibbs, 1995; Martínez, 2001). This fishing activity in Ecuador and Galapagos was one of the first successful commercial activity in Latin America, although there are informal reports of oriental marketing scouts visiting other countries within the region (Alvarado, J.J., TNC–Costa Rica, personal communication.).

The Galapagos Islands sea cucumber fishery is one of the oldest fisheries of its kind in South America (ca. 16 years), and its management has been a model for many other nations wishing to start such an activity. Although it has been surrounded by conflicts, negotiations and an incipient and inefficient participatory and adaptive management process, the fishery within the Galapagos Marine Reserve (GMR) has not been any different to most other fisheries elsewhere, with the population of *Isostichopus fuscus*, the commercial species, severely depleted (Toral-Granda, 2005b). However, the Galapagos case study represents an interesting scenario from which to evaluate how a non-traditional activity has come to shape the management regime and strategies within the GMR. The sea cucumber fishery also represented a "gold-rush" that modified the socio-economical environment in the Galapagos Islands, creating a wake of migrants aiming to benefit from this activity. Moreover, it provided a high income to local fishers who became used to lucrative alternatives and disregarded their traditional activities.

Additionally, this activity has one of the most comprehensive theoretical management plans ever put in place in the GMR. The Galapagos Islands, although not a highly diverse or commercially important sea cucumber area, was chosen as a hotspot, as a showcase for other Central and South American countries and other non-traditional fishing grounds, wishing to start this activity, to have a similar scenario from which to learn from and avoid mistakes that could lead to the depletion of the resource and the socio-economic distress that follows.

2. BIOLOGICAL AND POPULATION STATUS

2.1 Current species in trade

In the Galapagos Islands there are 38 shallow water species of sea cucumbers (Maluf, 1991). The first and only sea cucumber species to be legally commercially harvested in the Galapagos Islands was the brown sea cucumber *Isostichopus fuscus* (Ludwig, 1875) (Aspidochirotida: Stichopodidae), however due to current market pressure and decrease in abundance of this species within the GMR, catches of *S. horrens* (Hearn and Pinillos, 2006; Reyes and Murillo, 2007), *H. atra* (Reyes and Murillo, 2007) and *H. kefersteini* (Toral-Granda *et al.*, 2005a) have also been recorded. The capture of these species, or any other sea cucumber species, is illegal within the GMR. Informal records from fishers also claim that there are more species under illegal commercial exploitation; however, no formal records or a species list exist.

The sea cucumber *I. fuscus* is one of the most common species found in the Eastern Pacific (Maluf, 1991) and it used to be the most conspicuous invertebrate in the shallow subtidal zone in the Galapagos Islands (Wellington, 1974). This species can be found from Baja California to mainland Ecuador, including Galapagos, Socorro, Cocos and Revillagigedos Islands (Deichman, 1958; Maluf, 1991). Hooker, Solís-Marín and Leellish (2005) include Peru (Islas de Lobos de Afuera) in its geographical distribution. It can be found in the coastal zone from the shallow subtidal to 39 m depth (Deichman, 1958; Maluf, 1991). In Galapagos, *I. fuscus* prefers rocky bottoms where the seaweed *Ulva* sp. is predominant (Toral-Granda and Martínez, 2007) and where it is more active at night (Shepherd, Toral-Granda and Edgar, 2003). It can be found all throughout the archipelago, with higher densities in the Western Islands (Fernandina and Isabela) (Murillo *et al.*, 2002a).

I. fuscus is gonochoric (separate sexes) and does not present sexual dimorphism. It reproduces continuously throughout the year disregarding variation in sea water temperature (Mercier, Ycaza and Hamel 2007; Toral-Granda and Martínez, 2007). However, Mercier, Ycaza and Hamel (2007) establish that *I. fuscus* in captivity has a lunar periodicity in its reproductive cycle, spawning between one and four days after the new moon. Spawning generally occurred at night (Mercier, Ycaza-Hidalgo and Hamel, 2007). It attains sexual maturity (SOM) between 161.0 and 170.9 g drained wet weight (Toral-Granda and Martínez, 2007) and has a planktotrophic larval development (Mercier, Ycaza-Hidalgo and Hamel 2004). The same species in Baja California shows an annual reproductive season influenced by the influx of warm water (Fajardo-León *et al.*, 1995; Herrero-Pérezrul *et al.*, 1999) and prefers coral and rocky habitats (Herrero-Perezrul *et al.*, 1999). In Galapagos, the population has a mean total length (± SD) of 20.8 ± 2.81 cm with sizes ranging from 13 to 31 cm total length (TL), with a population mode of 21 cm TL (Toral-Granda and Martínez, 2007). Only one massive recruitment event has been recorded in the Galapagos Islands, present only in the Canal Bolívar area (Murillo *et al.*, 2002a; Hearn *et al.*, 2005). This pulse was first registered in the "before" fishing survey in 2000 and it ended in the "after" fishing survey in 2002 (Figure 2) (Toral-Granda, 2001). Since 1993, five sites within the Canal Bolívar, have been monitored on a yearly basis (Martinez, P., unpublished information; Toral-Granda and Martínez, 2004), and recruitment indices were very low from 1993 until 1998 (Martínez, 1999). This recruitment event is probably connected to the El Niño in 1997–1998 (Murillo *et al.*, 2002a; Hearn *et al.*, 2005) when increased sea surface temperature aided a successful reproductive event, and a subsequent La Niña event, with a high productivity helped the development and growth of *I. fuscus* juveniles (Murillo *et al.*, 2002a). Studies of larval biology of *I. fuscus* are limited to that of Hamel, Ycaza-Hidalgo and Mercier (2003), who raised larvae in captivity and reported that this species completes its metamorphosis and settles between 22 and 27 days, and attains 35 mm in 72 days under culture treatment (Mercier, Ycaza and Hamel, 2004).

There are no growth rates estimates for *I. fuscus* within the Galapagos Islands, however Herrero-Perezrul *et al.* (1999) gave preliminary growth estimates using FiSAT (Gayanillo and Pauly, 1997) which included estimations of asymptotic length ($L_\infty = 36.118$ cm) and the growth coefficient ($K = 0.180\ y^{-1}$). Reyes-Bonilla and Herrero-Perezrul (2003) obtained similar results ($L_\infty = 29.108$ cm, $K = 0.243\ yr^{-1}$).

I. fuscus is harvested to meet the bêche-de-mer demand in oriental countries where it is sought as a high value species (Clarke, 2002). In China it is favoured for its consistency and flavour, which is similar to that of *A. japonicus*, the preferred species for Chinese consumption (Chen, J., Yellow Sea Fisheries Research Institute, personal communication).

The warty sea cucumber, *S. horrens* is found in the Pacific Ocean from Malaysia to the Society Islands, French Polynesia, and from southern Japan and Hawaii to New Caledonia (Massin *et al.*, 2002) and in the Galapagos Islands, where it is found on rocky substrates from 5–20 m depth (Hickman, 1998), although current evidence show an abundant population over 30 m depth in certain islands (Hearn, A., Charles Darwin Foundation, personal communication). During the day, it is usually found in crevices, cracks and caves where it seeks shelter to emerge at night to feed (Hearn and Pinillos, 2006). The population seems to be comprised of juvenile and adult individuals (9 to 31 cm TL, mode = 20 cm) with the absence of smaller animals (Hearn and Pinillos, 2006) perhaps due to different habitat preference. Ongoing research on the reproductive biology of *S. horrens* show that this species reproduces throughout the year (Mora, J., Charles Darwin Foundation, unpublished information). Hearn and Pinillos (2006) presented a $L_\infty = 37.7$ cm and a $Z/K = 4.95$ with ($r^2 = 0.967$) (Powell-Wetherall analysis).

Stichopus horrens is generally harvested for the medicinal properties of its coelomic fluid or "gamat" in Malaysia (Baine and Choo, 1999) and Madagascar (Rasolofonirina, Mara and Jangoux, 2004) with medicinal purposes and as delicacy in China (Ilias and Ibrahim, 2006). However, in the Galapagos Islands, this species is collected for the production of bêche-de-mer, although fetching a much lower price than *I. fuscus*.

Holothuria atra is a large (20–30 cm TL) and robust sea cucumber that inhabits shallow waters, generally lying exposed on lava substrates or coral rubble. Its geographic distribution includes Mozambique to Hawaii, Clipperton and the Galapagos Islands

(Hickman, 1998). No information is available on the biology or ecology of this species for the Galapagos Islands.

Holothuria kefersteini is also a large (average 20 cm TL) sea cucumber. This species can be found both day and night, although camouflaged with sand. It inhabits both intertidal and subtidal habitats typically exposed on coral sand bottoms. It is often the most common species on white sandy bottoms. It can be found in tropical waters from Mozambique to Hawaii, Clipperton and the Galapagos Islands (Hickman, 1998). No information is available on the biology or ecology of this species for the Galapagos Islands.

2.2 Population status

2.2.1 Fishery independent information

The fishery for *I. fuscus* started in mainland Ecuador in 1989 (Carranza and Andrade, 1996) after the rapid decline of sea cucumber populations in the traditional fishing grounds in Asia and the Pacific Island Nations and Territories (Conand, 2001; Toral-Granda and Martínez, 2004). No stock assessment surveys were carried out in mainland Ecuador before the commencement of the fishery, and it came to a halt upon depletion of the resource. This promoted the migration of the fishers to the Galapagos Islands where extraction began in 1991 (De Paco *et al.*, 1993; Camhi, 1995; Powell and Gibbs, 1995; Martínez, 2001). From the beginning, commercial fishing activities for *I. fuscus* in the Galapagos Islands developed without baseline biological and ecological data and without a monitoring plan (Martínez, 2001) and it was surrounded by political and social problems (Bremner and Perez, 2002). Most fishing activities were centred in Western Isabela and Fernandina, a high population density area (Figure 1).

Although this fishery was suppose to be mainly for local fishers, the arrival of the activity brought many illegal mainland fishers, whose demand for work and access to the resource produced many conflicts that led to the ban of the fishery in 1992 (De Paco *et al.*, 1993; Powell and Gibbs, 1996; Toral-Granda and Martínez, 2004). Due to pressure exerted by the fishers, this activity re-opened in 1994, focusing mainly in the western region of the archipelago; nonetheless, it had to be closed before the end of the season as the original quota of 500 000 individuals was greatly surpassed, amounting between 8 and 12 million individuals (De Miras, Andrade and Carranza, 1996). All fishing activities for *I. fuscus* were illegal between 1994 and 1999.

A pilot stock assessment carried out over 17.3 km of coast of eastern Fernandina in 1993 yielded a mean population density of 6.24 ind. m^{-2} (± 0.20 ind. m^{-2} standard deviation) (Aguilar *et al.*, 1993). Other stock assessments, which included the whole of the Canal Bolívar recorded between 0.8 to 6.2 ind. m^{-2} (Aguilar *et al.*, 1993; Richmond and Martínez, 1993). A stock depletion model ran for the Canal Bolívar area (Hearn *et al.*, 2005) yielded initial densities of adults (≥16 cm TL) which fluctuated between 0.27 and 0.4 ind. m^{-2} from 1994 to 1997, similar to those obtained in 1999 (Toral-Granda and Martínez, 2004). Current density estimates for this region, between 1999 and 2007, range from 0.03 in the post fishery survey in 2005 to 1.03 ind. m^{-2} in the pre fishing survey in 2001 (Figure 2).

Since the re-opening of *I. fuscus* fishing activities in Galapagos in 1999, population surveys were declared mandatory in order to achieve its sustainable management. These surveys were carried out in six of the islands where higher population densities were present and where fishing activities were to take place. These islands are: Fernandina, Isabela, Santa Cruz, San Cristóbal, Floreana and Fernandina (Figure 1). Twice a year, before (between January and April) and after (between June and August) each fishing season, a team, comprised of fishers, managers, naturalist guides and scientists, visit permanent monitoring sites over an average of 10 working days (Toral-Granda and Martínez, 2004). Since 1999, a total of 533 550 m^{-2} have been surveyed in those islands covering a range of depths between 3 and 25 m. In 2006 only one population survey

FIGURE 1
The Galapagos Islands and mainland Ecuador. The shaded section in the Galapagos Islands map is known as the Bolivar Channel

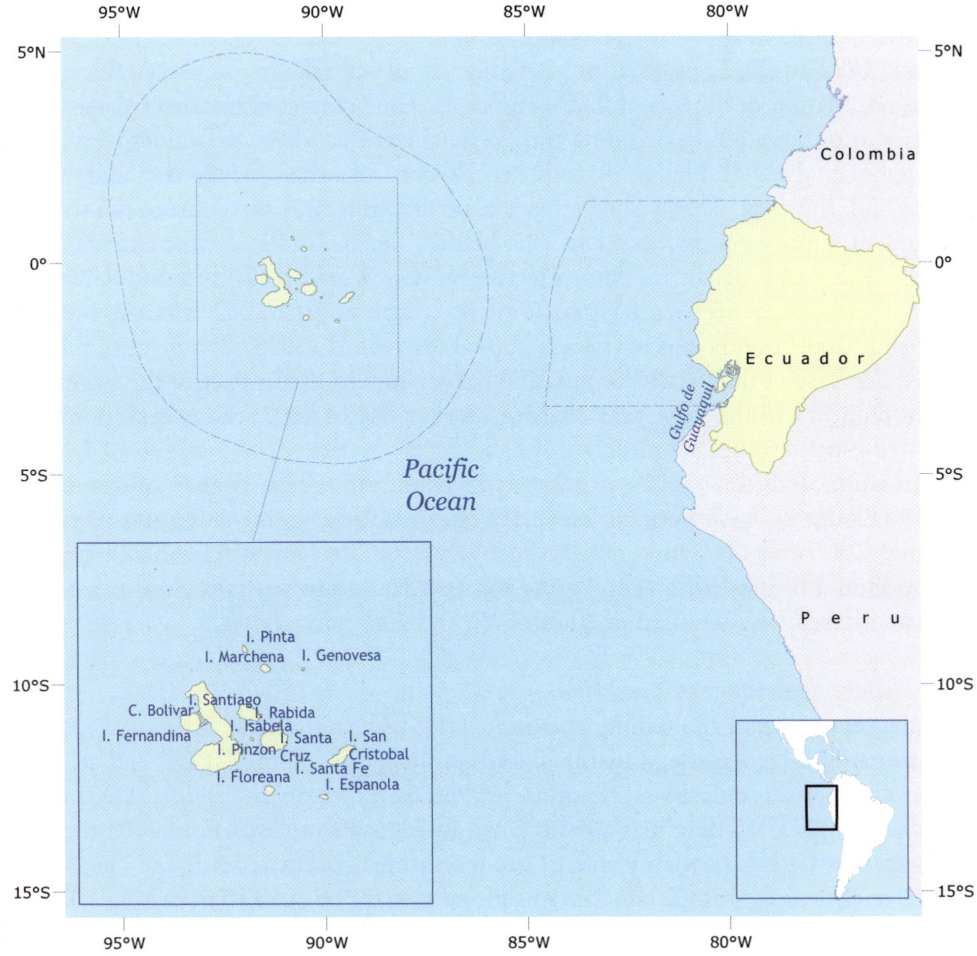

FIGURE 2
General population densities of *I. fuscus* in Española, Fernandina, Floreana, Isabela, Santa Cruz and San Cristóbal before and alter each fishing season, 1999–2007

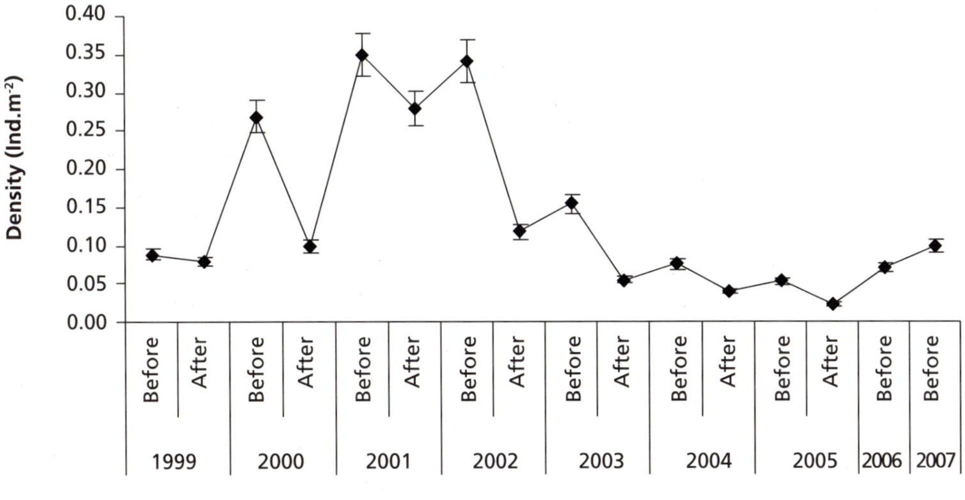

Source: Toral-Granda, 2005a; Avendaño, Llerena and Gravez, 2006; UCOOPEPGAL, 2007.

was carried out as the fishery remained closed, mainly because the fisher sector never exerted enough pressure for its opening, and there was no need to evaluate the status of the population. In 2007, the fishery was opened in five of the six islands (Floreana remained closed due to low population densities with only preliminary results available (Servicio Parque Nacional de Galapagos, 2007).

From 1999 to 2007, population densities in all six islands were variable, with a serious population decline as of 2003 (Figure 2). The greatest variation of total average density was observed in Fernandina and Western Isabela, where the values were almost an order of magnitude higher than those observed in other surveyed islands (Toral-Granda and Martínez, 2004). The high peaks in 2000 and 2001 were due to the strongest recruitment pulse ever recorded in the islands since population monitoring began in 1993. This pulse began in 1999 (Martínez, P.C., World Wildlife Fund–Galapagos, personal communication) and lasted until the beginning of 2002 when it started to diminish (Toral-Granda and Martínez, 2004; Hearn *et al.*, 2005). Every year, there is a marked decrease in the "after" population survey due to the impact of the fishery, and a slight recovery in the next year "before" survey. As of 2003, the overall population presented similar values (Figure 2).

Population densities from April 2007 (0.09 ind. m^{-2}) show similar values to those in 1999 (Figure 2), with Fernandina and Western Isabela with the highest population densities (0.17 and 0.14 ind. m^{-2}, respectively) (UCOOPEPGAL, 2007); however, caution should be used when comparing this last data point with previous years due to possible different sites studied, methodology and surveying depth.

2.2.2 *Fishery dependent information*

With the reopening of the fishing season in 1999 fisheries data such as total catch per unit effort (CPUE) became available. Based on the landings, the CPUE had an overall decrease from 102.6 ind. $diver^{-1}$ $hour^{-1}$ in 1999 to 54.5 ind. $diver^{-1}$ $hour^{-1}$ in 2005 (Toral-Granda *et al.*, 2005). There was a fishing ban in 2006. From 2001 until 2003 there was an increase in CPUE probably due to the recruitment event recorded in the Western Islands; nonetheless, this CPUE contains illegal size (≤19.9 cm TL) individuals. Certain islands, such as San Cristóbal, have presented a continuous decline in CPUE, whilst other, such as Fernandina and Western Isabela show similar values throughout the study (Table 1). Floreana has been kept as a closed zone since 2003 (Toral-Granda *et al.*, 2005).

After a detailed analysis of the fishing sites, it was noticed that areas that were previously characterized by high catches, are now seldom visited by fishers due to low *I. fuscus* density (Toral-Granda *et al.*, 2005; Toral-Granda, 2005b). Although the population density surveys reveal very low numbers of sea cucumbers, the catch

TABLE 1
Catch per unit of effort for *I. fuscus* in the Galapagos Marine Reserve from 1999 until 2005

Fishing zones	CPUE (ind. $diver^{-1}$ $hour^{-1}$)						
	1999	**2000**	**2001**	**2002**	**2003**	**2004**	**2005**
Española	131	62	*	92	*	69	45
Fernandina	178	*	128	138	116	128	86
Floreana	ND	58	89	51	*	*	*
North and East Isabela	ND	108	52	100	67	55	42
Western Isabela	104	*	112	144	115	82	73
Southern Isabela	85	73	64	111	65	58	56
San Cristóbal	75	60	46	*	44	50	39
Santa Cruz	83	76	56	*	83	69	42
Average CPUE	**102.6**	**89.1**	**103.7**	**136.1**	**97.7**	**72.1**	**54.5**

* = Closed to fishing activity for that year.
ND = No data available.
Source: Toral-Granda *et al.*, 2005.

rates were maintained due to the stability in CPUE (sensu Hilborn and Walters, 1992), mainly due to the fact that it is a sedentary resource with patchy distribution. Additionally, there is a change in diver behaviour, who concentrates their effort on the very few places where sea cucumbers remain (Toral-Granda, 2005b). In this case, CPUE may have little value as an indicator of abundance, and perhaps misleading, as often noted for benthic sedentary or strongly aggregating populations (Oresanz, Parma and Hall, 1998).

During the 2007 fishing season a total of 1 200 000 sea cucumbers were collected in 55 days out of which only 12 000 were impounded due to illegal size (Servicio Parque Nacional de Galápagos, 2007), hence denoting the low level of enforcement and political will to follow up the regulations stated in the management plan for the fishery.

After eight years of legal fishing activities for *I. fuscus* within the GMR, the resource is severely depleted (Toral-Granda, 2005a). Little recovery has been observed between fishing seasons (see Figure 2) perhaps due to the constant illegal fishery that takes place in the Islands (Reyes and Murillo, 2007) and the small number of juveniles entering the population (Hearn *et al.*, 2005). Although little information is available for the other sea cucumber species which are being illegally harvested, it is possible to infer that similar fate will await these species, spiralling into overexploitation of sea cucumbers species within the GMR.

3. FISHING ACTIVITY

3.1 Catches

The total catch figures for *I. fuscus* are based on the Fishing Monitoring Programme, a joint venture between the Galapagos National Park Service (GNPS) and the Charles Darwin Foundation (CDF) with the support from the Galapagos fishing cooperatives. No legal catches for *S. horrens*, *H. atra* and *H. kefersteini* are recorded for the Galapagos Islands. From 1999 until 2005, almost 30 million sea cucumbers have been harvested legally within the GMR or the equivalent of over 8 000 tonnes live weight (Table 2). Historically, Western Isabela and Fernandina have been the fishing regions that have yielded the most of the catches (Table 2).

As with CPUE, total catch is highly variable with some islands showing a decrease if compared to initial catches in 1999 (i.e. San Cristóbal, Española and Western Isabela) (Table 2). In 2001 and 2002 catches were higher in most islands if compared to previous

TABLE 2
Total allowable catches (TAC) and total catches in numbers of *I. fuscus* per fishing region and season, 1999–2005

Fishing zone	1999	2000	2001	2002	2003	2004	2005
Fernandina	169 877	*	624 105	758 775	736 006	380 751	72 307
North and East Isabela	282 883	378 418	52 689	1 551 529	267 879	88 514	87 154
Western Isabela	1 641 360	2 615 495	1 735 181	5 395 804	3 054 595	1 187 079	580 417
Southern Isabela	98 724	117 206	73 419	279 913	484 814	462 833	149 326
Española	489 669	256 980	*	79 775	*	117 098	150 852
San Cristóbal	1 163 104	621 405	52 697	*	121 845	327 584	125 030
Floreana	78 980	239 843	47 324	235 652	*	*	*
Santa Cruz	477 060	717 600	86 931	*	340 435	395 043	235 282
Total	**4 401 657**	**4 946 947**	**2 672 346**	**8 301 448**	**5 005 574**	**2 959 091**	**1 400 368**
TAC	4 000 000	4 500 000	4 000 000	-	4 700 000	4 000 000	3 000 000

* = Closed to fishing

Source: Toral-Granda *et al.*, 2005.

years; however, these catches included undersized individuals (≤19.9 cm TL) (Toral-Granda *et al.,* 2005). Additionally, in 2002 there was no TAC and a conflict between the GNPS and the Navy, meant little patrolling and enforcement.

3.2 Type of fishery

The Galapagos sea cucumber is open only to Galapagos artisanal fishers. Currently there are 1 032 registered fishers in the GMR (PR.C.P003.R002, 2007), with the majority of them in San Cristóbal (Murillo, Reyes and Hearn, 2007). Although the sea cucumber fishery is the one with the most fishers involved, during the 2007 fishing season, only 436 were active during the 50-day fishing period (PR.C.P003.R002, 2007). This activity relies on the use of hookah diving and it is normally carried out from small wooden vessels ("*pangas*") or fibreglass fast boats ("*fibras*") that work from a mother boat ("*bote*"). In the 2007 fishing season, there were only 160 vessels (out of 446) carrying out sea cucumber fishing activities (PR.C.P003.R002, 2007). Fishers may dive down to 45 m depth, and although it is believed that there is a positive correlation between population density and depth, this has not been proven scientifically (Toral-Granda *et al.,* 2005).

Although by law, the only fishing activity allowed to be carried out within the GMR is artisanal, the current fishing practices (between 1999 and 2007) coincide more with those of a semi-industrial activity. During the last fishing season (2007), a reduced number of fishers (only 436) were involved in this activity due to the decline of the resource and lack of incentive for the low revenues. All catches are sold to middlemen, generally of oriental origin, and shipped to the Asian markets.

4. MANAGEMENT OF THE FISHERY

4.1 Management tools

Since 1998, with the passing of the Special Law for the Galapagos Islands, the GMR is under a Participatory and Adaptive Management Scheme (PAMS) which consists of a three-pole system with the Participatory Management Board (PMB), the Inter-Institutional Management Authority (IMA) and the GNPS. The PMB members are locally-based and represent the direct users of the GMR; decisions are made based on a consensus. The IMA is on a ministerial level where all decisions taken in the PMB are decided by means of a voting system. The GNPS is the entity in charge of putting into effect all resolutions taken by both the PMB and the IMA (for further information see Altamirano, Toral-Granda and Cruz, 2004; Toral-Granda and Martínez, 2004). All activities within the GMR are regulated by the PMB and the IMA. Since 1999, the only sea cucumber that can be legally fished within the GMR is *I. fuscus.*

The sea cucumber fishing activity has been regulated by means of a fishing season (two months between March and August), a TAC, a minimum landing size (20 cm TL), a minimum dry size (7 cm TL), No-Take Zones (NTZ), which include closure of nursery grounds, specific islands and areas under the coastal zoning scheme (Altamirano, Toral-Granda and Cruz, 2004). Individual Transferable Quotas (ITQ) were used only in 2001 (Altamirano, Toral-Granda and Cruz, 2004). The nursery grounds that have been closed for several seasons, locally known as the Bolivar Channel, is the only area where large amounts of juveniles were found in 2000 and 2001. In this area, small individuals (about 4 cm) were found amongst the adults.

Additionally, all fishers that partake in the fishing activity must have their catch monitored to ensure legal size and to give fishing information (i.e. fishing sites, effective fishing hours, total number of divers) (Table 3). Any illegal catch is impounded. If fresh, it will be returned to an arbitrarily designated nursery ground near-by, disregarding the origin of the individuals to be released; and if processed, will be stored in GNPS containing facility. Information gathered during the fisheries monitoring process will be used to evaluate the fishery and theoretically provide information to proceed with an adaptive management system that seeks sustainable use of the resources.

The two-month fishing season is normally enforced, with specific regulations set in place (i.e. number of days that boats at sea have in order to come back to nearest port, number of days during which all market activities must come to an end). As it does not require a high number of personnel and a large budget it is easier to enforce; however, as the months during which the fishing season take place is variable, total catches and CPUE are hard to compare as oceanographic and weather conditions vary greatly between the months mentioned above. One of the reasons given by fishers to explain low catches is that physical and oceanographic conditions have worsened throughout the years, resulting in lower catches and less income. However, it is unclear to what extent this is a generalized perception among the fishing sector, or a strategy for negotiating larger quotas.

The TAC for each fishing season has normally been given based on political pressure rather than using scientific information. The CDF, as the official scientific advisor of the GNPS and the Ecuadorian Government, provide the best information available on a yearly basis in order to make informed decisions. The different yearly TACs were established on an auction-like procedure where fishers generally proposed very high quantities, with which they expected to meet their economic whims. On the other hand the CDF scientific team proposed scientifically-based quantities, intended to provide a sustainable use of the resource, which were normally low compared to the expectations raised for each fishery. In general, however, the adopted TAC was a quantity that had no scientific basis whatsoever, but rather a figure that satisfied the expectations of the fishing sector.

For 1999, 2000 and 2003, the TAC was exceeded by amounts between 300 000 and 450 000 individuals; on the other hand, for the remaining years, the TAC was not reached, with deficits between 1.0 and 1.6 million individuals. For 2007, the catch was a mere 32 percent of the TAC, just a month prior to the closure of the fishery. For those years in which the TAC was exceeded, this was mainly due to the fact that the most productive fishing grounds are in the far west of the archipelago, where no radio communication is available. Upon reaching the TAC, the fishers have to be notified so that they can stop fishing and return to their port of origin. This caused the difference between the TAC and the real amount landed.

On the other hand, for those years that the TAC was not reached, many reasons were given by the fishers, amongst them: (i) poor fishing due to oceanographic and atmospheric conditions; (ii) changes in fishing method; (iii) poor price paid for the product that did not encourage starting any fishing activities; (iv) fishing grounds too far to reach on day trips and too few mother boats willing to carry the "*fibras*"

TABLE 3
Different management and monitoring measures implemented in the Galapagos sea cucumber fishery from 1999 to 2007

	Fishing season							
Measures	**1999**	**2000**	**2001**	**2002**	**2003**	**2004**	**2005**	**2007**
Two-month season	✓	✓	✓	✓	✓	✓	✓	✓
Total allowable catch (TAC)	✓	✓	✓	–	✓	✓	✓	✓
Individual transferable quotas (ITQ)	–	–	✓	–	–	–	–	–
Fisheries monitoring	✓	✓	✓	✓	✓	✓	✓	✓
Minimum landing size	–	–	✓	✓	✓	✓	✓	✓
Coastal zoning	–	–	✓	✓	✓	✓	✓	✓
Closure of nursery grounds	–	–	✓	✓	✓	✓	✓	✓
Closure of islands (i.e. Floreana)	–	–	–	✓	✓	✓	✓	✓
Population density evaluation	✓	✓	✓	✓	✓	✓	✓	✓

or "*pangas*". Low densities of *I. fuscus* in the fishing grounds were never given as an option to explain the low catches.

Due to the natural morphological plasticity of sea cucumbers, the minimum landing size has not been successful in ensuring that only individuals that have reached sexual maturity are caught. When handled, sea cucumbers may either contract or relax, hence changing their total length. Additionally, the value given as a minimum landing size was not based on science. Toral-Granda (1996) established the SOM for *I. fuscus* at 23.6 cm TL (or 260 g drained weight). The minimum landing size (22 and then 20 cm TL) that is currently used was put in place in order to accommodate the lack of big individuals within the population. Additionally, the minimum landing size in dry state (6 cm) does not correspond to that of a 20 cm animal (estimated at 9 cm TL by Castrejon *et al.*, 2005), and was also set on the PMB in order to help the fishing sector oblige to this regulation.

Since 2000, the coastal perimeter of the Galapagos Islands is under a zoning scheme. This scheme allows the use of 5 percent for inhabited ports (zone 2.4), 78 percent of the coastal zone for fishing activities (zone 2.3), 11 percent for tourism activities (zone 2.2), 6 percent for conservation (zone 2.1) (Calvopiña *et al.*, 2006). Zones 2.1 and 2.2 could be considered as no-take zones as fishing is not legal. Fishing activities are allowed exclusively on zone 2.3, where there is the highest abundance of this species (Edgar *et al.*, 2004). Nonetheless, one of the more common infringements (23 percent of total) by the Galapagos fishing sector is carrying out fishing activities outside the 2.3 zone (Toral-Granda *et al.*, 2005a) hence showing the lack of respect to this management tool.

The area known as the Bolivar Channel (between the islands of Fernandina and Western Isabela) is of high ecological importance, as it comprises the nesting sites of the endemic Galapagos Flightless Cormorant (*Phalacrocorax harrisi*), the Galapagos Penguin (*Spheniscus mendiculus*), as well as the Galapagos sea lion (*Zalophus wollebaeki*) amongst others, and it is one of the most important tourist areas. Additionally, this region showed the highest density of juveniles of *I. fuscus* in 2001 and 2002 and was identify as a nursery ground in a PMB meeting. Due to this importance, this region has been closed to sea cucumber fishing activities since 2000 until 2005 as a way to preserve the most important nursery ground for this species and as the breeding ground for charismatic endemic Galapagos species. Yet again, this area shows high level of illegal fishing and it is where most of the illegal fishing outside the fishing season takes place (Toral-Granda *et al.*, 2005).

Ever since the start of the legal fishing activities in 1999, only six islands were open as fishing grounds; however, due to the adaptive management principle included in the GMR Management Plan, any island may be closed in response to low populations densities, if such are observed in a "before" fishing survey. Because of this, different islands were closed in certain fishing seasons (in 2000 – Fernandina; in 2002 – San Cristóbal and Santa Cruz; in 2003 – Española and Floreana) (Table 3).

In 2004, the "before" fishing survey included six more islands (Genovesa, Pinta, Marchena, Santiago, Pinzón y Rábida; Figure 1). This scouting for new fishing grounds could be regarded as (i) the fishers aimed to identify new fishing grounds that may yield high catches; (ii) the fishers wanted to explore more islands to have a larger fishing area; or (iii) the fishers wanted to use these islands with low densities as a trading item in order to be granted a fishing season in all islands with high density. The interest on those islands was first stated since the fishery reopened in 1999, however, the sea cucumber density were lower than those of the islands commonly fished (Figure 3). Genovesa was the only island to merit some consideration as it had density values similar to Floreana. However, the distance of this island to any of the inhabited ports, made it a non viable option for this fishery. There were no further attempts to open a legal fishery on these islands; nonetheless, although illegal, few catches have been recorded in these islands through the Fisheries Monitoring Programme (CDF, Fisheries Monitoring Programme Database).

Floreana, an island with low productivity, has been closed to sea cucumber fishing activities since 2002 (Table 3). According to a mass-balance model developed by Okey *et al.* (2004) the 1999 and 2000 fishing seasons accounted for 88 percent of the total mortality of *I. fuscus* in Floreana. This mortality greatly exceeded the optimum sustainable capture rate indicating unsustainable fishing pressure (Okey *et al.*, 2004). Monitoring of *I. fuscus* in Floreana showed that population densities have been erratic, with clear evidence of a population collapse between 2002 and 2004, a slight increase in 2005, and values similar to those of 1999 in the pre–fishery survey in 2007 (Figure 4) (UCOOPEPGAL, 2007); although it covered only two out of eight sites that were monitored in previous years. The recent trends in densities indicate a possible recovery of the stock following the closure in 2003. However, figures obtained by UCOOPEPGAL (2007) should be dealt with caution due to difference

FIGURE 3

Population density of *I. fuscus* around 12 islands during the "before" fishing survey in 2004. Columns in red represent the new islands that were prospected for legal fishing activities

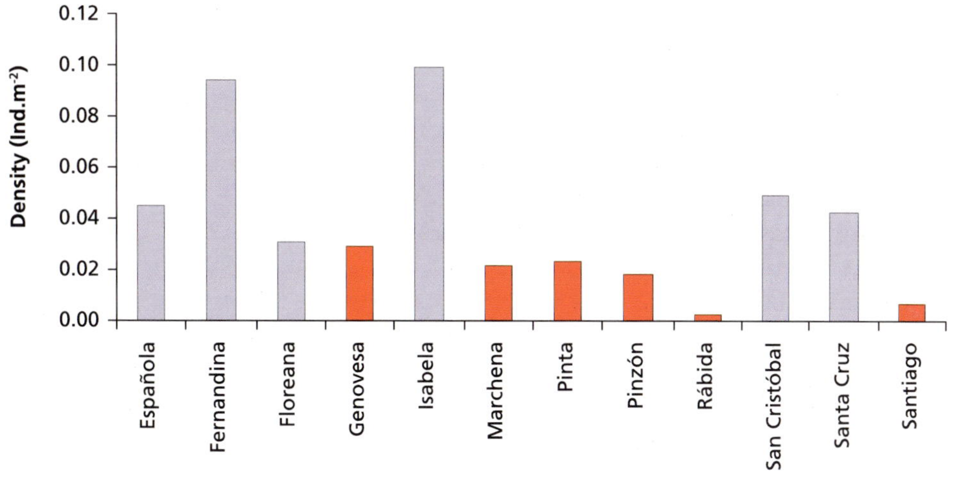

FIGURE 4

Mean population density of *I. fuscus* in Floreana Island declared as a No-Take Zone since 2003

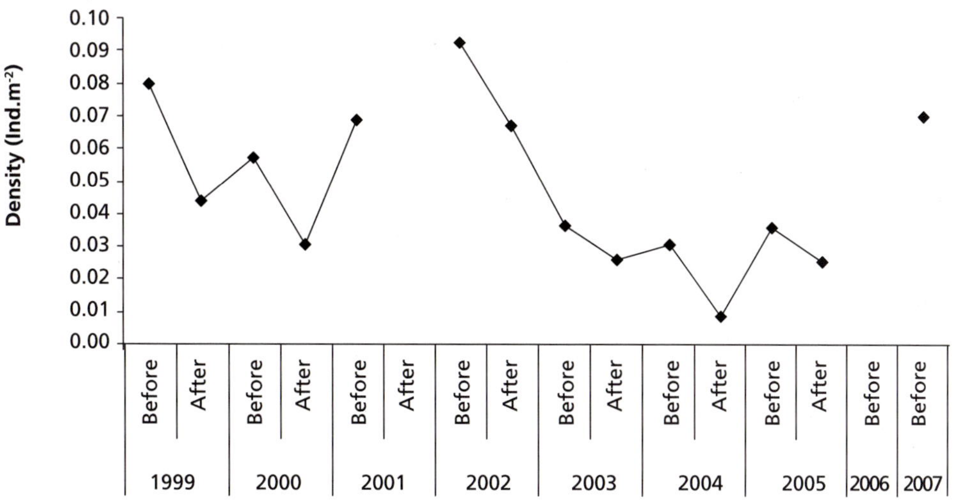

in methodology, sites sampled and depth range surveyed. Also, during the years of the closure of Floreana, the GNPS have registered illegal fishing activities, both during and outside the fishing season (Espinoza, E., Galapagos National Park Service, personal communication).

Individual Transferable Quotas (ITQs) were used only in 2001. This ITQ was obtained by equally dividing the TAC (4 million sea cucumbers) amongst all fishers and fishing vessels, disregarding if they were active or not. This yielded a total of 3 174 sea cucumbers per fisher and per vessel (Murillo *et al.,* 2002b). In the 2001 fishing season, the TAC was not reached and the total number of fishers and vessels were low in comparison to previous years (Murillo *et al.,* 2002b). Most fishers sold their ITQ to other fishers as the total given did not merit starting a fishing operation. Further attempts to implement this form of management have not been successful due to the opposition by the fishing sector that claim that this management tool only seeks to impoverish them more, as they normally work on a cash advance system with the trader. It is claimed that an ITQ will never yield enough to cover these cash advances and provide them an economic benefit. An ITQ system will probably work in the Galapagos if each cooperative cleans-up the list of members, leaving only those that are active fishers rather than opportunistic ones. Some of the registered fishers work as captains or crew on tourist boats, in the construction or public sector and are only active during the sea cucumber fishing season. Murillo *et al.*, (2002b) considered this the most lucrative fishing activity; however, now-a-days it is not considered as such.

The most common management tools used in these eight years of legal fishing activities are: fisheries restricted to two-month season, fisheries monitoring and population density evaluation (see Table 3). In 2002, there was no TAC, as the fishers demanded an open fishery due to the high densities present. As a consequence, a serious population decline was observed in that year (see Figure 2).

4.2 Enforcement

Although the GPNS has the required infrastructure and facilities to carry out patrolling and enforcement within the GMR, currently there are not enough personnel to operate the patrolling vessels or carry out fisheries monitoring activities. Although, the GNPS has three oceanic vessels that can do long-distance and long-term monitoring activities most of them operate only during legal fishing seasons, with little or no control over illegal activities (Reyes and Murillo, 2007).

Additionally, even though the GMR is part of the Ecuadorian National Parks System, any GNPS vessel must carry a Navy officer on board in order to be able to intercept or approach a vessel carrying out any activity within the GMR. The Navy officer must have all expenses paid by the GNPS. During 2004 and 2005, there were serious problems between the GNPS and the Navy, which meant that little patrolling activities were carried out both during and outside the fishing seasons.

Another problem with enforcement is that if the GNPS detects and stops illegal fishers and vessels, they are seldom punished to the full extent of the law. The main reasons being: (i) political pressure exerted to release the vessel and illegal fishers; (ii) the administrative burden within the GNPS that does not allow to process the case in a timely manner, hence releasing the violators as stated in the law; (iii) lack of sufficient number of GNPS lawyers in order to maintain the case going and do follow-up procedures; and (iv) examples from previous similar cases that have set a precedent on not exerting the law on illegal activities.

4.3 The bottom-up management approach for the Galapagos Marine Reserve

The innovative participatory management approach for the GMR was created to ensure the participation of the local stakeholders of the GMR in the decision-making process. It was created in 1998 after a series of major social upheavals mainly related to the sea cucumber crisis that surrounded Galapagos in the 1990s (Stone, 1995; Powell and Gibbs, 1995; Jenkins and Mulliken, 1999; Shepherd *et al.*, 2004). When the Participatory and Adaptive Management System (PAMS) was enacted, it was expected that the local-based decision-making process would promote commitment from the local users favouring management and conservation of natural resources (Viteri and Chavez, 2007). This system incorporates the rights and responsibilities of the local stakeholders in the decision making process, which is a novel approach for the Galapagos Marine Reserve and Ecuador as a whole (Heylings and Bravo, 2007; Viteri and Chavez, 2007).

The GMR is a multi-user marine reserve, and its management system involves the major stakeholders. On the local level, the PMB includes the tourism, fishing, conservation and management sectors. All decisions are approved by consensus. At the national level, there is the IMA, comprised of the Ministers of Defence, Tourism and Fisheries, the Galapagos Chamber of Tourism, the Galapagos fishing sector and the CEDENMA, a conglomerate of environmental groups within Ecuador (Altamirano, Toral-Granda and Cruz, 2004). The GPNS is the secretary of IMA and the CDF is the scientific advisor. Decisions are taken by majority vote, although if any given topic arrives with a consensus from the PMB, it is generally ratified by the IMA (Altamirano, Toral-Granda and Cruz, 2004).

For every fishing season, scientific information has been provided to the PMB and IMA, however, in most cases, it was disregarded and decisions were taken based on the current interests of the members present. Moreover, the fishers accused the scientist of providing false information on the pretence of discouraging the fishing season as a personal attack on the fishing sector. Scientists and conservationists were accused of caring more for the wild than the well-being of the human population that inhabit the Islands. Even though, during each fishing survey and the elaboration of the reports, members of the fishing sector were present, at a later stage, and during PMB meetings, the Fishing sector representative will refuse to acknowledge the members as true fishers, accuse them of having sold out to the conservationists and of going against their own sector. As a consequence, in following years, few fishers would accept to participate in the population surveys or in the making of the final report.

All sites surveyed during the stock assessment trips were chosen with the fishers, so as to ensure that their traditional knowledge was taken into account. Out of these, 17 were chosen as permanent monitoring sites in order to provide comparable data throughout the years. Every year, during the PMB meetings in which the management measures for the fishery are discussed, it is common practice for fishers to disregard the monitoring sites, claiming that those are not traditional fishing grounds and will then refuse to accept the results. This refusal affects the different management measures to be put in place, leading to the auction-like procedure of the TAC and minimum landing size described earlier.

The PMB and the IMA have not proven to be a successful mechanism to manage the sea cucumber fishery in the GMR, although having all possible tools to become a clear example of the benefits of co-management. Political interventions and continue change of the Park Directors (more than 10 in less than two years) debilitate the management process, causing its weakness and allowing the intrusion of the opportunistic politicians in the system. One of the major flaws of the system is that the stakeholders were not ready to be involved in management towards sustainability, but rather took this opportunity to achieve specific goals that would benefit them personally or within their own specific sector. Since then, many of the stakeholders' official representatives have used the PMB as a political platform, and the turn-over rate amongst fishing sector representatives is the highest in comparison to other sectors.

The GMR participative process has been rooted in conflict, argumentation, manipulation and negotiation, with the well being of the natural resources at stake.

Although the shift from conflict to consensus models for environmental management and decision making were at its highest in the 1980s (Peterson, Peterson and Rai Peterson, 2005), it only reached Galapagos in the late 1990s, when it was decided that a consensus will be the most favourable, to both stakeholders and resources, way forward to attain conservation. Peterson, Peterson and Rai Peterson (2005) state that "sustainable development's focus on local conditions, diversity, participation, and locally produced development strengthened this link, particularly as consensus is more readably attainable at smaller, local scales".

Consensus works on a "win-win" strategy which hopes to develop a sense of ownership and responsibility for the resources managed (Peterson, Peterson and Rai Peterson, 2005). However, in most of the PMB meetings that dealt with the management of sea cucumbers, none of the stakeholders left with a clear sensation of "winning", but rather frustrated and disillusioned, as they had had to give up many things in order to achieve something that in turn did not leave them satisfied. This sense of loss is probably the main culprit of the poor level of achievement in the management of the GMR.

In the participatory management of the GMR, the sea cucumber resource has been viewed as the resource that can be sacrificed on behalf of others. The sea cucumber fishery used to be the most lucrative, and by giving in to the political demands favouring the opening of the fishery, the fishing sector will be content and will settle for less-favourable management regimes in other fisheries (e.g. spiny lobster, ban on shark finning, long-lining ban). However, this strategy proved unsuccessful as the spiny lobster fishery is also in bad state (Hearn and Murillo, 2007); there have been considerable seizures of shark fins both in the Galapagos (Murillo, Reyes and Hearn, 2007) and in mainland (Castro, M., Sea Shepherd Conservation Society, personal communication) and the use of long lining is still a constant within the GMR.

The Participatory and Adaptive Management Scheme (PAMS) is currently taking most the blame of the collapse of some resources within the GMR; however, there are some possible explanations that could help elucidate the light a the end of this tunnel: (i) the validity of the participatory management system for *I. fuscus* vs. for the GMR, with a new and incipient system trying to manage a gold-rush fishery, worldwide characterised for boom-and-bust cycles; (ii) the incessant prerogative from fishers not to comprehend the validity of scientific data produced; and, (iii) the perennial decision to overlook scientific evidence in favour of social and economical reasons rather than conservation ones.

Applying the PAMS to the sea cucumber fishery *per se* may not have been the wisest choice. This fishing activity yielded a unprecedented economical gain that lead the fishers to push for its yearly opening, despite the clear scientific evidence provided against the stability of the resource (Toral-Granda and Martínez, 2004; Toral-Granda *et al.*, 2005). The high income generated by this fishery led it to be the target of many politicians, who offered in many cases an open-access fishery in order to gain constituents for the next elections. Perhaps co-management should have started with low income generator fisheries (i.e. white fish fishery) as a pilot study, with inclusion of other fisheries at a later stage.

5. TRADE

In October 2003, *I. fuscus* was included in Appendix III of the Convention on the International Trade in Endangered Species of Wild Fauna and Flora (CITES) becoming the first, and so far the only sea cucumber species included in a CITES Appendix. Appendix III can be a valuable tool by which a CITES Party seeks help from other Parties in the protection of a specific species, especially if such species is under illegal trade to supply to foreign markets (Willock, Burgener and Sancho, 2006). An Appendix III does not include the "introduction from the sea" and "the look-alike" provision.

Appendix III is the easiest one to apply as it can be done unilaterally by a range State in any given time. Catches that are to be imported need to have a CITES Permit stating the origin of the catch and its legality. Other range States wishing to export their catch, must only present a certificate of origin. Importing countries require only such certificate, in order to accept the importation.

Since the 2004 fishing season, the official CITES permits have provided information on the main export destination and trade route for the 2004 and 2005 fishing seasons. In both years, China has been the major importing country with 98.5 percent in 2004 and 95.9 percent in 2005. Minor imports were recorded for Peru and the United States of America (Table 4). Within China the major ports of entry are Keelung in Taiwan Province of China and China Hong Kong Special Administrative Region (SAR).

According to the total catches of *I. fuscus* per year (see Table 2), between 2004 and 2005 there were close to 400 000 sea cucumbers that were caught within the GMR but did not have a CITES permit (Table 4). Since all *I. fuscus* catches are to supply the oriental market, any difference could be explained by misreporting problems originating from Ecuador, and that are maintained through the importing country and a later stage in CITES trade database.

Due to the high interest in *I. fuscus* in the oriental market, the low catches from the Galapagos capture fisheries, and the depletion of this resource in mainland Ecuador, aquaculture ventures started in mainland Ecuador in abandoned shrimp ponds (Mercier, Ycaza and Hamel, 2004). This is the only species under aquaculture production in Ecuador. The listing of *I. fuscus* in Appendix III does not differentiate between wild and farmed individuals, so any exports would require a CITES permit. However, the aquaculture production leaves the country without the CITES permit as it is claimed that it is not *I. fuscus* (Galiano, P., Ministry of Environment, personal communication). Only 80 kilograms of dried *I. fuscus* have been exported from aquaculture production (Ruidiaz, L., Subsecretaria de Pesca del Ecuador, personal communication) although the final destination is not known.

Although China and China Hong Kong SAR are the major importers of sea cucumbers from Ecuador, their trade statistics report that only 91 000 animals that have entered the country (Hong Kong Census and Statistics Department, personal communication) leaving a void of Ecuadorian exports that have not been recorded by China, or China Hong Kong SAR. The Food and Agriculture Organization (FAO) reported a total of 389 tonnes of sea cucumbers from capture fisheries in Ecuador from 1988 to 2005 (FAO, 2007). The highest reported catch (152 tonnes in 1992) coincides with the initial exploitation phase in the Galapagos Islands. From 1993 to 2005 catches are steady at 12 and 15 tonnes per year (FAO, 2007). The recorded caches from 1995

TABLE 4

Number of *I. fuscus* imported per importing countries according to the CITES Appendix III permits and total number of individuals landed in 2004 and 2005

Importing country	Port of entry	2004	2005	Total
China	Hong Kong SAR	760 015	482 931	1 242 946
	Dalian	172 684	0	172 684
	Keelung (Taiwan PC)	1 711 216	343 405	2 054 621
	Taipei (Taiwan PC)	48 790	186 413	235 203
	Qingdao	42 229	41 180	83 408
	Yantai	0	96 085	96 085
China – Total	–	2 734 934	1 150 014	3 884 948
Peru – Total	Lima	36 227	49 494	85 721
USA – Total	Los Angeles	5 832	0	5 832
Total exported		**2 776 993**	**1 199 508**	**3 976 513**
Total catches*		2 959 091	1 400 368	4 359 459
Unaccounted for		182 098	200 860	382 958

Source: GNPS and CITES permit databases; * Table 2 and Toral-Granda *et al.* (2005).

until 1998 may be either from mainland Ecuador as there was a ban on all fishing activities within the Galapagos Archipelago, or could represent illegal harvest from the Galapagos Islands. Ecuador does not report aquaculture production of sea cucumbers to FAO (FAO, 2007).

Between 1996 and 2006, the GNPS has seized over 670 000 individuals of *I. fuscus* (Reyes and Murillo 2007). These impoundments, however, do not show the real amount of catches from Illegal, Unreported and Unregulated (IUU) fishing, and up to date, there is no information on the actual numbers that have been caught. Illegal fishers generally fish at night and deliver their catch to mother boats waiting in the vicinity. Those catches may be landed in mainland Ecuador, where they may be shipped by land to Peru, and eventually exported as Peruvian goods (Willock, Burgener and Sancho, 2006). This is a clear example on how the lack of cooperation among range states of *I. fuscus* is detrimental to the Galapagos populations. Every range State that has not listed this species in Appendix III is a potential channel for illegal caught specimens (Willock, Burgener and Sancho, 2006).

Fishing for *S. horrens* is illegal within the GMR; hence the only figures available are those of impounded catches. Latest official numbers establish a total of over 74 000 individuals raided by GNPS (Reyes and Murillo, 2007) generally from illegal campsites, fishing boats and private houses (Hearn and Pinillos, 2006). Reyes and Murillo (2007) provide one seizure of almost 19 000 *H. atra* within the GMR. No figures exist for *H. kefersteini*.

Sea cucumbers are traded as fresh individuals or per kilogram in brine. The price for either product has greatly varied over time. In 1999, when the fishery started, a single fresh *I. fuscus* had an average price of USD 0.80 which declined continuously until 2002 when it reached the lowest value paid (USD 0.33). With the decline in population, there was a drastic price increase until 2004, when it reached the highest price paid (USD 1.50). For the 2005 fishing season, the price declined to USD 1.20 (Toral-Granda *et al.*, 2005). Prices for 2007 were USD 1.40 (Servicio Parque Nacional de Galápagos, 2007).

6. SOCIO-ECONOMIC IMPORTANCE TO LOCAL COMMUNITIES

In the Galapagos, sea cucumber fishery is a semi-industrial activity, however it has some characteristics of an artisanal fishery, such as the fact that the catch and the processing is done by the fishers themselves or a member of their family unit. The sea cucumber fishery is considered the most lucrative fishing activity within the GMR and has been very profitable for the small number of fishers participating (Bremner and Perez, 2002; Murillo *et al.*, 2002a). No clear values are available of the total income generated by the 1994 fishery, but the total economical gain from this season started the gold-rush fever which lasts up to the present day.

Between 1993 and 2000, during the highest point of sea cucumber activities, the number of registered fishers increased from 392 to 682 (74 percent), probably due the arrival of seasonal migrants from mainland Ecuador and Galapagos native fishers who turned to this highly lucrative fishery (Bremner and Perez, 2002). Currently there are 1 032 fishers operating within the GMR, out of which only 25 percent are active through the year with a greater increase in number of active fishers during the sea cucumber fishing seasons. Of these, 97.7 percent are men with the remaining 2.3 percent women (Murillo, Reyes and Hearn, 2007). During the 2000 fishing season, there was a total of 1 229 active fishers whilst in 2001 (the year with ITQs) there were only 597 (Toral-Granda *et al.*, 2005; Murillo, Reyes and Hearn, 2007). During the early years of this fishery, women and children were known to wade, snorkel and free dive in the shallow coastal areas and to process the catch themselves. Presently, most women are boat owners ("*armadoras*") rather than fishers (Murillo, Reyes and Hearn, 2007).

Full-time fishers rely on the income generated from the sea cucumber and spiny lobster fishery for the whole year. Few of them participate in other fishing activities (i.e. white fish, high seas fishing) although due to the decline in catches in both resources, fishers are starting to do more fishing trips with the aim of catching fish (Murillo, Reyes and Hearn, 2007). Nonetheless, each year, although there is common knowledge that *I. fuscus* populations have declined throughout the marine reserve, fishers will demand the opening of the fishery claiming that the lack of alternatives and the strict management controls and regulations are making them poorer every year.

The easiness of catching, processing and storing bêche-de-mer renders this an activity that could be done all year long even if illegal. Illegal catches can be sold on the black market, hence provide cash flow that will allow every day living and purchase of commodities.

When the resource was still abundant, and prior to the start of a fishing season, the middlemen used to hand out cash advances to fishers in order to help them prepare for the season, and then would buy the catch at a bargain price. The final price paid after the first fishing outing would often not be enough to cover the fishers' basic expenses, hence asking for yet another cash advance, therefore would create a debt cycle that was hard to break. In the 2007 fishing season, no cash advances were given limiting the number of fishers that took part in the fishery (167 active fishers [Servicio Parque Nacional de Galápagos, 2007], a drastic decrease from the 703 fishers in the 2005 season [Toral-Granda *et al.*, 2005]).

7. ADDITIONAL THREATS TO SEA CUCUMBER POPULATIONS

The search for new species to meet the growing international demand for bêche-de-mer is a constant threat for sea cucumber populations in the Galapagos Islands. In the early years, commercial activities focused on *I. fuscus*, as a high value abundant sea cucumber species. However, as the resource diminished and the catches dwindled, new species became the target. Currently there are four species under commercial exploitation (*I. fuscus*, *S. horrens*, *H. atra* and *H. kefersteini*), although only one is legally permitted. Trade and market scouts are always searching for new species, and since the fishing infrastructure and know-how already exist in Galapagos, there is always the chance of new species being harvested, both legally and illegally.

As the *I. fuscus* population is showing signs of severe depletion, and the market demand for it has not declined, there is a current interest to develop mariculture for restocking depleted populations in the Galapagos Islands (Toral-Granda, 2005b). This initiative was prompted by the successful completion of the reproductive cycle in captivity (Hamel, Ycaza-Hidalgo and Mercier, 2003) and rearing of this species in abandoned shrimp ponds in mainland Ecuador (Mercier, Ycaza-Hidalgo and Hamel, 2004). The Galapagos fishing sector is demanding the development of this activity as an alternative form of employment and income.

Nonetheless, Toral-Granda (2005c) suggests a series of studies that should take place before the commencement of this activity, including: a complete genetic delineation of the Galapagos stocks; larval and juvenile natural history; optimal release conditions; as well as studies on diseases and predation. It is also suggested that any mariculture or sea ranching activities be directed to mainland Ecuador where there is the right physical setting. If such initiative would materialize it would certainly reduce fishing pressure within the GMR.

8. RECOMMENDATIONS

8.1 Recommendations for improving sea cucumber fisheries management and conservation

For *I. fuscus* there should be a complete ban on fishing activities in order to allow the stocks to recover. In the mean time, scientific research should be conducted on population genetics, larval biology and ecology. Genetic studies carried out by Lohr

(2003) on populations of *I. fuscus* from the southern islands of San Cristobal, Española and Floreana, indicate historical genetic connection between populations from the Galapagos Islands and mainland Ecuador, suggesting the mainland population as source with higher diversity of alleles and the Galapagos as sink populations. However it is not known when and how this genetic connection is manifested. Such information would certainly have important management implications for the fishery. Nevertheless, further genetic research efforts should concentrate on populations of the Bolivar Channel, western Galapagos, which include the area where the only mass recruitment *I. fuscus* has been recorded. All this could mean that rather than implementing a local management plan, a regional one may need to be developed and implemented with the aid of all countries within the region.

Fished populations should be monitored regularly so as to provide basic information on recovery rates and resilience. As sea cucumbers are an important part of the food web, ecosystem monitoring should be encouraged in order to understand the possible changes caused by their removal. In the Galapagos, *S. horrens* used to be rare (Toral-Granda, M.V., personal observation); at present, their abundance in certain areas is very high (Hearn, A., CDF, personal communication) possibly due to less inter-specific competition and niche release. However, Martínez, P.C. (WWF–Galapagos, personal communication) state that this is not the case, and that this species used to be abundant especially in the early 1990s.

Despite all the scientific information available, the innovative management plans and the commitment of the stakeholders to make the *I. fuscus* fishery a sustainable one, this activity provoked the severe depletion of its natural populations. Lessons should be learned from this fishery, so as not to make the same mistakes and to allow other fisheries to be sustainable over time.

For other sea cucumber species entering the fishery, basic biological and ecological information should be gathered before the start of any fishing activity. Furthermore, an increase communication exchange with the fishing sector of Galapagos is encouraged, if they were to accept any scientific result. For these "new" species, the three "phases" stated in Perry, Walters and Boutillier (1999) should be carried out. For "Phase 0", all information related to the given sea cucumber fishery should be collated, and presented to the stakeholders. For "Phase 1", mistakes should be learned from other sea cucumber fisheries and once key scientific information has been identified, obtain the funds and the commitment to undertake relevant investigations; for "Phase 2", the best management practices or the ones that have yielded satisfactory results on the conservation of the given species, should be kept and evaluated against new ones that could be implemented. This study should also investigate similar problems elsewhere, and seek advice from places where sea cucumber fisheries have not collapsed.

Constant monitoring of NTZ should be undertaken, as these provide shelter for commercial and non-commercial species. Despite the little information available on the benefits of these tools for sea cucumbers, they have a proven record to benefit other species of both demersal and pelagic species. In the GMR, there is 20 percent of coastal perimeter protected as NTZ; specific analysis should be undertaken to understand the benefits, if any, of such a management tool.

A ban of SCUBA diving or hookah can also help conserve deepwater stocks, which would in turn help through reproduction. Otherwise a ban to carry hookah compressors onboard, out of the fishing season could help stop illegal fishing activities.

The current "chain of custody" in place in the GMR for sea cucumbers and spiny lobsters has served a major stepping stone for fishers to understand the need of monitoring and reporting. Thanks to this chain, the GNPS has a clear understanding of the number (and/or kg) that have been extracted, processed and shipped outside the GMR. Additionally, fishers understand the need of the permits and certificates, hence being an active part of the management of the fisheries. Such a programme could be followed in many other places, as it allows control, enforcement and understanding.

As *I. fuscus* is the only sea cucumber species included in CITES, trade monitoring should be improved so as to understand the trade route and identify bottlenecks where IUU catches may be exported to. Better reporting from both the CITES administrative authority (the GNPS in the Galapagos and the Ministry of Environment in mainland Ecuador) should be encouraged, as well as better communication with the CITES Secretariat. Moreover, a thorough evaluation of the effectiveness of the Appendix III listing of *I. fuscus* should be undertaken, which will in turn provide tools for other countries with sea cucumber populations in peril due to international trade.

As a way to ensure accurate and fast identification of *I. fuscus* in the international market, the Ecuadorian Government, through the GNPS, advised by the CDF, have sent samples of dried specimens of *I. fuscus* to China Hong Kong SAR, where they are used for education and training purposes (Kwan, B., Hong Kong Census and Statistics Department, personal communication) and to the US National Oceanic and Atmospheric Administration (NOAA) in Los Angeles, an important re-exporting port for many sea cucumber species (Torres, R., NOAA, personal communication). Similar procedures can be used for other species in the international market.

8.2 Evaluation of the pros and cons of a CITES listing

I. fuscus it is the only sea cucumber species included in the CITES Appendix III. Since its inclusion came into effect (16 October 2003), most of the catches from capture fisheries have been recorded in CITES permits; however, about 400 000 individuals are unaccounted for, perhaps due to early mistakes in recording and registering the permits. The 2007 fishing season will yield more information on how the CITES permits are being handled and will elucidate more on the possible benefits of an Appendix III listing.

Identification of specimens is intrinsic to the proper functioning of a CITES listing, and this may become a handicap for marine species as these can sometimes be traded in processed forms that may complicate visual recognition. Moreover, commodity codes may classify shipments in highly generic terms (Willock, Burgener and Sancho, 2006).

The listing of *I. fuscus* in CITES has advantages and disadvantages. Below a list of some of the advantages and disadvantages:

Advantages:

i) Certainty of the legality of the catch of the exported goods.
ii) Increased awareness of the need to conserve and manage sea cucumber populations.
iii) Possibility of identifying trade bottlenecks where laundering of illegal catches may occur.
iv) Better opportunities for technical assistance, targeted research and capacity building.
v) Creating and putting into place standardized and comprehensive trade reporting codes and data gathering amongst countries.
vi) Catch and export data is centralized in one location allowing faster analysis and interventions.
vii) Understanding of the trade route when the products leave Ecuador.
viii) Understanding that international trade is the major force behind the exploitation of *I. fuscus* in the GMR.
ix) Curtailing international trade by means of an attached CITES permit that ensures the legality of the catch.

Disadvantages:

i) Increased burden to CITES administrative officers (i.e. processing of permits, compilations and submission of annual reports to the CITES Secretariat).

ii) Increased costs to train and educate managers, border patrol and custom officers particularly in species identification.
iii) Slow communication between concerned CITES authorities may delay action responses.
iv) Delay in acquisition of the CITES Secretariat trade reports on CITES species.

Currently, Ecuador is not considering either the inclusion of *I. fuscus* in Appendix II or listing any other sea cucumber species in any CITES Appendix. As a developing country, with different needs and prerogatives, a CITES listing has become more of a burden than a help, and no clear advantages have been shown to managers, fishers or scientists. However, it is well understood that a clear and thorough evaluation of the current listing of *I. fuscus* is a priority, as this would help clarify some possible misconceptions and to aid the government to make the best use of the CITES permits as management tool, which in turn will help in the conservation and sustainable use of sea cucumber populations.

Finally, the inclusion of *I. fuscus* in Appendix III by other range states would possibly help to reduce laundering of this species and ensure a better control of its international trade.

ACKNOWLEDGEMENTS

My sincere thanks to Alessandro Lovatelli who believed in me and made me part of the team. To Marcelo Vasconcellos, Alex Hearn, Annie Mercier and Priscilla Martínez who helped improving the document and gave key inputs so as to make this document more accurate. To Cesar Peñaherrera, who designed the GIS map. To Tom Poulsom, who provided great cups of tea, support and editorial assistance. Data presented in this report was collected thanks to the financial support of the David and Lucille Packard Foundation, the Interamerican Development Bank and WWF–USAID. This is Charles Darwin Foundation contribution number 1073.

REFERENCES

Aguilar, F., Chalen, X., Castro, F., Sonnenholzner, J. & Herrera, M. 1993. Evaluación del recurso pepino de mar Holothuroidea al este de la Isla Fernandina en la Provincia de Galapagos. Guayaquil, Instituto Nacional de Pesca: 20.

Altamirano, M., Toral-Granda, M.V. & Cruz, E. 2004. The application of the adaptive principle to the management and conservation of *Isostichopus fuscus* in the Galapagos Marine Reserve. *In*: *Advances in Sea Cucumber Aquaculture and Management*. Lovatelli, A., Conand, C., Purcell, S., Uthicke, S., Hamel, J.-F. & Mercier, A. (eds). *FAO Fisheries Technical Paper*. No. 463. Rome, FAO. 2004. 425p.

Avendaño, U., Llerena, S. & Gravez, V. 2006. Evaluación poblacional del pepino de mar *Isostichopus fuscus* en la Reserva Marina de Galápagos, Pre-Pesquería 2006. UCOOPEPGAL-COPAHISA-COPESAN-COPES PROMAR-COPROPAG. Informe de Monitoreo. 8 pp.

Baine, M. & Choo, P.S. 1999. Sea cucumber fisheries and trade in Malaysia. pp. 49–63. *In*: *The conservation of sea cucumbers in Malaysia — their taxonomy, ecology and trade: Proceedings of an international conference.* Baine, M. (ed). Department of Agriculture, Kuala Lumpur, Malaysia. February 1999. Heriot-Watt University, Orkney, Scotland.

Bremner, J. & Perez, J. 2002. A case study of human migration and the sea cucumber crisis in the Galapagos Islands. *Ambio*, 31(4): 306–310.

Calvopiña, M., Visaira, R., Cruz, E., Piu, M. & Dumas, J. 2006. Implementación de la zonificación consensuada de la Reserva Marina de Galapagos. *In: Pasos hacia la sustentabilidad de la Reserva Marina de Galápagos.* WWF-USAID, Puerto Ayora, Santa Cruz, Galapagos.

Camhi, M. 1995. Industrial fisheries threaten ecological integrity of the Galapagos Islands. *Conservation Biology*, 9(4): 715–724.

Carranza, C. & Andrade, M. 1996. Retrospectiva de la pesca del pepino de mar a nivel continental. Fundación Charles Darwin para las Islas Galapagos/ORSTOM. Quito. 54 pp.

Clarke, S. 2002. Trade in Asian dried seafood: characterization, estimation and implications for conservation. Wildlife Conservation Society Working Paper No. 22 (15 December 2002).

Conand, C. 2001. Overview of sea cucumbers fisheries over the last decade – What possibilities for a durable management? Echinoderms 2000. Barker (ed.) pp. 339–344.

De Paco, C., Hurtado, M., McFarland, C., Martínez, P., Reck, G. & Richmond, R. 1993. Evaluación de la pesquería de pepinos de mar en las islas Galápagos, Ecuador. Informe para la Unión Mundial para la Naturaleza (UICN) como resultado de la Misión realizada a solicitud de la Fundación Charles Darwin para las Islas Galápagos. Fundación Charles Darwin, Quito, 25 pp.

Deichmann, E. 1958. The Holothuroidea collected by The Velero II and IV during the years 1932 to 1954, Part II. Aspidochirota. *Allan Hancock Pacific Expeditions*, 11(2): 253–348.

De Miras, C., Andrade, M. & Carranza, C. 1996. Evaluación socioeconómica de la pesca experimental de pepino de mar Galápagos. Unpublished report, FCD and ORSTOM: 45 pp.

Edgar, G.J., Bustamante, R.H., Fariña, J.M., Calvopiña, M., Martínez, C. & Toral-Granda, M.V. 2004. Bias in evaluating the effects of marine protected areas: the importance of baseline data for the Galápagos Marine Reserve. *Environmental Conservation*, 31(3): 1–7.

FAO. 2007. Capture production 1950–2005. FISHSTAT Plus – Universal software for fishery statistical time series [online or CD-ROM]. Food and Agriculture Organization of the United Nations.

Fajardo-León, M.C., Michel-Guerrero, E., Singh-Cabanillas, J., Vélez-Barajas, J.A. & Massó-Rojas, A. 1995. Estructura poblacional y ciclo reproductor del pepino de mar *Isostichopus fuscus* en Santa Rosalía, B.C.S., México. *Cienc Pesquera*, 11: 45–53

Gayanilo, F.C. Jr. & Pauly, D. (eds) 1997. FAO-ICLARM stock assessment tools (FiSAT). Reference Manual. FAO Computerized Information Series (Fisheries). No. 8, Rome, FAO. 262 pp.

Hamel, J.-F., Ycaza-Hidalgo, R. & Mercier, A. 2003. Larval development and juvenile growth of the Galapagos sea cucumber *Isostichopus fuscus*. *SPC Beche-de-Mer Information Bulletin*, 18: 3–8.

Hearn, A. & Murillo, J.C. 2007. Life History of the Red Spiny Lobster, *Panulirus penicillatus* (Decapoda: Palinuridae), in the Galapagos Marine Reserve, Ecuador. Pacific Science, 62(2): 191–204.

Hearn, A. & Pinillos, F. 2006. Baseline information on the warty sea cucumber *Stichopus horrens* in Santa Cruz, Galapagos, prior to the commencement of an illegal fishery. *SPC Beche-de-Mer Information Bulletin*, 24: 3–10.

Hearn, A., Martínez, P., Toral-Granda, M.V., Murillo, J.C. & Polovina, J. 2005. Population dynamics of the exploited sea cucumber *Isostichopus fuscus* in the western Galápagos Islands, Ecuador. *Fisheries Oceanography*, 14(3): 1–9.

Herrero-Pérezrul, M.D., Reyes-Bonilla, H., García-Domínguez, F. & Cintra-Buenrostro, C.E. 1999. Reproduction and growth of *Isostichopus fuscus* (Echinodermata: Holothuroidea) in the southern Gulf of California, México. *Mar Biol*, 135: 521–532.

Heylings, P. & Bravo, M. 2007. Evaluating governance: A process for understanding how co-management is functioning, and why, in the Galapagos Marine Reserve. *Ocean & Coastal Management*, 50(2–3): 174–208.

Hickman, C.P. 1998. Guía de campo sobre estrellas de mar y otros equinodermos de Galápagos. Serie vida marina de Galápagos. Sugar Spring Press, Virginia, USA. 83 pp.

Hilborn, R. & Walters, C.J. 1992. Quantitative Fisheries Stock Assessment: Choice, Dynamics and Uncertainty. Chapman and Hall, New York. 570 pp.

Hooker, Y., Solís-Marín, F.A. & Lleellish, M. 2005. Equinodermos de las Islas Lobos de Afuera (Lambayeque, Perú). *Rev. Peru. Biol.*, 12(1): 77–82.

Ilias, Z. & Ibrahim, K. 2006. National Report: Malaysia. *In: Proceedings of the CITES workshop on the conservation of sea cucumbers in the families Holothuriidae and Stichopodidae.* Bruckner, A.W. (ed.). NOAA Technical Memorandum NMFSOPR 34, Silver Spring, MD 244 pp.

Jenkins, M. & Mulliken, T.A. 1999. Evolution of exploitation in the Galapagos Islands: Ecuador's sea cucumber trade. *Traffic Bulletin*, 17: 107–118.

Lohr, H.R. 2003. Genetic variation among geographically isolated populations of the commercially important sea cucumber *Isostichopus fuscus* in the Eastern Pacific. M.S. Thesis, Moss Landing Marine Laboratories, California State University, San Francisco, USA.

Maluf, L.Y. 1991. Composition and distribution of the central eastern Pacific echinoderms. Nat. Hist. Mus. *Los Angeles Co. Technol. Rep.*, 2: 1–242.

Martínez, P.C. 1999. Estado Poblacional y Biología Reproductiva del Pepino de Mar *Stichopus fuscus* en las Islas Galápagos. Informe Técnico. Fundación Charles Darwin, Santa Cruz, Galápagos 13 pp.

Martínez, P.C. 2001. The Galapagos sea cucumber fishery: a risk or an opportunity for conservation? *SPC Beche-de-Mar Information Bulletin*, 14: 22–23.

Massin, C., Zulfigar Y., Tan Shau Hwai, A. & Rizal Boss, S.Z. 2002. The genus *Stichopus* (Echinodermata: Holothuroidea) from Johore Marine Park (Malaysia) with the description of two new species. *Bulletin de l'Institute Royal des Sciences Naturelles de Belgique*, 72: 73–99.

Mercier, A., Ycaza-Hidalgo, R. & Hamel, J.-F. 2004. Aquaculture of the Galapagos sea cucumber, *Isostichopus fuscus. In*: *Advances in Sea Cucumber Aquaculture and Management.* Lovatelli, A., Conand, C., Purcell, S., Uthicke, S., Hamel J.-F. & Mercier, A. (eds). *FAO Fisheries Technical Paper*. No. 463. Rome, FAO. 2004. 425p.

Mercier, A., Ycaza-Hidalgo, R. & Hamel, J.-F. 2007. Long-term study of gamete release in a broadcast spawning holothurian: predictable lunar and diel periodicities. *Marine Ecology Progress Series*, 329: 179–189.

Murillo, J.C., Reyes, H. & Hearn, A. 2007. Aspectos sociales de las pesquerías en Galapagos. *In: Informe Galápagos 2006–2007.* FCD-SPNG-INGALA.

Murillo, J.C., Martínez, P., Toral, M.V. & Hearn, A. 2002a. Pepino de Mar. *In: Reserva Marina de Galápagos, Línea Base de la Biodiversidad.* Danulat, E. & Edgar, G. (eds). Fundación Charles Darwin y Servicio Parque Nacional de Galápagos, Santa Cruz, Galápagos, Ecuador.

Murillo, J.C., Vizcaino, J., Nicolaides, F., Moreno, J., Espinoza, E., Chasiluisa, C., Andrade, R., Born, B., Villalta, M., Yépez, M. & Molina, L. 2002b. Informe técnico final de la pesquería del pepino de mar (*Stichopus fuscus*) en las islas Galápagos, 2001 Análisis comparativo con las pesquerías de 1999 y 2000. Fundación Charles Darwin/ Parque Nacional Galápagos. 29 pp.

Okey, T.A., Banks, S., Born, A.F., Bustamante, R.H., Calvopiña, M., Edgar, G.J., Espinoza, E., Fariña, J.M., Garske, L.E., Reck, G.K., Salazar, S., Shepherd, S., Toral-Granda, V. & Wallem, P. 2004. A trophic model of a Galápagos subtidal rocky reef for evaluating fisheries and conservation strategies. *Ecological Modeling*, 172: 383–401.

Orensanz, J.M., Parma, A.M. & Hall, M.A. 1998. The analysis of concentrating and crowding in shellfish research. *In: Proceedings of the North Pacific Symposium on Invertebrate Stock Assessment and Management.* Jamieson, G.S. and Campbell, A. (eds). *Canadian Special Publication on Fisheries and Aquatic Sciences*, 125: 441–459.

PR.C.P003.R002. 2007. Más de un millón de pepinos de mar se capturaron en la RMG. *Galapagos National Park Press Release*. 20 August 20 2007.

Perry, R.I., Walters, C.J. & Boutillier, J.A. 1999. A framework for providing scientific advice for the management of new and developing invertebrate fisheries. *Reviews in Fish Biology and Fisheries*, 9: 125–150.

Peterson, M.N., Peterson, M.J. & Rai Peterson, T. 2005. Conservation and the myth of consensus. *Conservation Biology*, 19(3): 762–767.

Powell, J.R. & Gibbs, J.P. 1995. A report from Galapagos. *TREE,* 10(9): 351–354.

Reyes-Bonilla, H. & Herrero-Perezrul, M.D. 2003. Population parameters of an exploited population of Isostichopus fuscus (Holothuroidea) in the southern Gulf of California, Mexico. *Fisheries Research,* 59: 423–430.

Reyes, H. & Murillo, J.C. 2007. Esfuerzos de control de pesca ilícita en la Reserva Marina. *In: Informe Galápagos 2006–2007.* FCD-SPNG-INGALA.

Richmond, R.H. & Martínez, P.C. 1993. Sea cucumber fisheries in the Galápagos Islands. Biological aspects, impacts and concerns. World Conservation Union (IUCN) Technical Report: 18 pp.

Servicio Parque Nacional de Galápagos. 2007. Boletín de prensa sobre la pesquería del pepino de mar *Isostichopus fuscus* en la Reserva Marina de Galápagos. 1 pp.

Shepherd, S.A., Toral-Granda, M.V. & Edgar, G.J. 2003. Estimating the abundance of clustered and cryptic marine macro-invertebrates in the Galapagos with particular reference to sea cucumbers. *Noticias de Galápagos,* 62: 36–39.

Shepherd, S.A., Martinez, P., Toral-Granda, M.V. & Edgar, G.J. 2004. The Galapagos sea cucumber fishery: management improves as stocks decline. *Environmental Conservation,* 31(2): 102–110.

Stone, R. 1995. Fishermen threaten Galapagos. SPC *Beche-de-Mer Information Bulletin,* 7: 22.

Toral-Granda, M.V. 1996. Biología reproductiva del pepino de mar *Isostichopus fuscus* en la isla Caamaño, Santa Cruz, Galápagos. Tesis de Licenciatura, Universidad del Azuay Cuenca. 89 pp.

Toral-Granda, M.V. 2001. Monitoreo de las poblaciones de *Isostichopus fuscus* antes y después de la temporada de pesca 2000. Informe interno, Estación Científica Charles Darwin. 25 pp.

Toral-Granda, M.V. 2005a. Estado poblacional del pepino de mar *Isostichopus fuscus* posterior a la temporada de pesca 2005. Informe Técnico. 29 pp.

Toral-Granda, M.V. 2005b. Requiem for the Galapagos sea cucumber fishery? *SPC Beche-de-Mer Information Bulletin,* 21: 5–8.

Toral-Granda, M.V. 2005c. Should stock enhancement and re-stocking be an option for the Galapagos sea cucumber? Charles Darwin Foundation Internal Document. 10 pp.

Toral-Granda, M.V. & Martínez, P.C. 2004. Population density and fishery impacts on the sea cucumber (*Isostichopus fuscus*) in the Galapagos Marine Reserve. *In*: *Advances in sea cucumber aquaculture and management.* Lovatelli, A., Conand, C., Purcell, S., Uthicke, S., Hamel, J.-F. & Mercier, A. (eds). *FAO Fisheries Technical Paper.* No. 463. Rome, FAO. 2004. 425p.

Toral-Granda, M.V. & Martinez, P.C. 2007. Reproductive biology and population structure of the sea cucumber *Isostichopus fuscus* (Ludwig, 1875) (Holothuroidea) in Caamaño, Galapagos Islands, Ecuador. *Marine Biology,* 151: 2091–2098.

Toral-Granda, M.V., Murillo, J.C., Piu, M., Nicolaides, F., Moreno, J., Reyes, H., Castrejón, M. & Hearn, A. 2005. Evaluación de la pesquería de pepino de mar (*Isostichopus fuscus*) en la Reserva Marina de Galápagos en el año 2005. Fundación Charles Darwin/Parque Nacional Galápagos. Puerto Ayora, Santa Cruz. 41 pp.

UCOOPEPGAL. 2007. Evaluación poblacional del pepino de mar *Isostichopus fuscus* en la Reserva Marina de Galápagos. Pre-pesquería 2007. Sector pesquero artesanal de Galápagos (UCOOPEPGAL), Galápagos, Ecuador. 16 p.

Viteri, C. & Chavez, C. 2007. Legitimacy, local participation, and compliance in the Galapagos Marine Reserve. *Ocean & Coastal Management,* 50: 253–274.

Wellington, G.M. 1974. Ambientes marinos de Galápagos. Reporte Técnico para el Departamento de Parques Nacionales y Vida Silvestre. 357 pp.

Willock, A., Burgener, M. & Sancho, A. 2006. First Choice or Fallback? An examination of issues relating to the application of Appendix III of CITES to marine species. *TRAFFIC International.* 57 pp.

Population status, fisheries and trade of sea cucumbers in temperate areas of the Northern Hemisphere

Jean-François Hamel and Annie Mercier

Precautionary management of *Cucumaria frondosa* in Newfoundland and Labrador, Canada

Jean-François Hamel and Annie Mercier

Population status, fisheries and trade of sea cucumbers in temperate areas of the Northern Hemisphere

Jean-François Hamel
Society for the Exploration and Valuing of the Environment
Portugal Cove-St. Philips, Newfoundland & Labrador, Canada
E-mail: jfhamel@seve.cjb.net

Annie Mercier
Ocean Sciences Centre, Memorial University
St. John's, Newfoundland & Labrador, Canada
E-mail: amercier@mun.ca

Hamel, J.-F.; Mercier, A. 2008. Population status, fisheries and trade of sea cucumbers in temperate areas of the Northern Hemisphere. In V. Toral-Granda, A. Lovatelli and M. Vasconcellos (eds). Sea cucumbers. A global review of fisheries and trade. *FAO Fisheries and Aquaculture Technical Paper*. No. 516. Rome, FAO. pp. 257-291.

SUMMARY

Data on sea cucumber fisheries in the temperate northern hemisphere are mainly available for four countries (Canada, United States of America, Russian Federation and Iceland), and commercial harvests are centered on four species (*Parastichopus californicus*, *P. parvimensis*, *Cucumaria frondosa* and *C. japonica*). Both *Parastichopus* species are primarily exploited by divers at a scale that is similar to what occurs elsewhere (e.g. tropical Indo-Pacific regions). However, harvests of the *Cucumaria* species typically involve industrialized processes (i.e. fishing boats, specialized trawls and processing plants). While *Parastichopus* fisheries date back to the early 1970s, most fisheries of *Cucumaria* are fairly new, and most of them are still in the exploratory phase, especially in Canada. The present document outlines the biological and population status, the current catches, the management measures in place, the socio-economic importance of the sea cucumber resources, the current research associated with these fisheries and the threats they may be facing.

1. INTRODUCTION

Compared to the sea cucumber trade in Asia and the Indo-Pacific, which is considered to date back some 1 000 years (Conand, 2001), the commercial harvesting of holothurians in North America is fairly new, having started in the 1970s on the west coast of the United States of America (hereinafter abbreviated to USA) and in the 1980s on the west coast of Canada (Conand and Sloan, 1989; Bruckner, 2005, 2006a,b; Therkildsen and Petersen, 2006). It spread to the east coast of Maine (USA) and to the Atlantic provinces of Canada a little over a decade ago (Therkildsen and Petersen, 2006). However, sea cucumbers have been fished for subsistence by native people for centuries along the west coast of North America (Mathews, Kookesh and Bosworth, 1990) and in Arctic Canada (Wein, Freeman and Markus, 1996).

In response to a growing demand for sea cucumber products and owing to the depletion of traditionally fished stocks, new species have been sought (Conand, 2004), including temperate and polar ones, which are slowly gaining popularity on the market. Hence, despite the abundance of some of these temperate and polar sea cucumbers, they have only recently entered the world trade for bêche-de-mer and other products. The present document covers four different species and four countries for which data are available. Some species/regions have been well studied and are therefore associated with good biological data, landing statistics and management protocols, whereas others are not well documented, leading to an unequal treatment in the following pages.

2. REGION UNDER STUDY

Although most of this chapter is dedicated to North American sea cucumber fisheries, it will also provide data for adjacent countries, which are harvesting sea cucumbers in the temperate-polar waters of the North Atlantic and North Pacific (Figure 1). On the Pacific side, this review will include the Russian Federation, the USA (including Alaska) and Canada. On the Atlantic side, it will focus on Canada, USA, Iceland and the Western Russian Federation coast along the Barents Sea. There are probably small scale fisheries occurring elsewhere, especially in Scandinavia (Norway and Sweden) (Therkildsen and Petersen, 2006), however, no reliable information pertaining to landings has been found and consequently they will not be discussed in the present document. A developmental fishery for the sea cucumber *Stichopus tremulus* has reportedly started in Norway in 2007, however it is still too early to obtain any tangible data on this initiative. Moreover, data for the Russian Federation and Iceland remain scarce.

FIGURE 1

Region under study and countries covered in this review (highlighted in yellow). The map shows the confirmed distribution of the sea cucumber species under study. The region encompasses the FAO Statistical Areas 18 (Artic Sea), 21 (Northwest Atlantic), 27 (Northeast Atlantic), 61 (Northwest Pacific), 67 (Northeast Pacific) and 77 (Pacific Eastern Central)

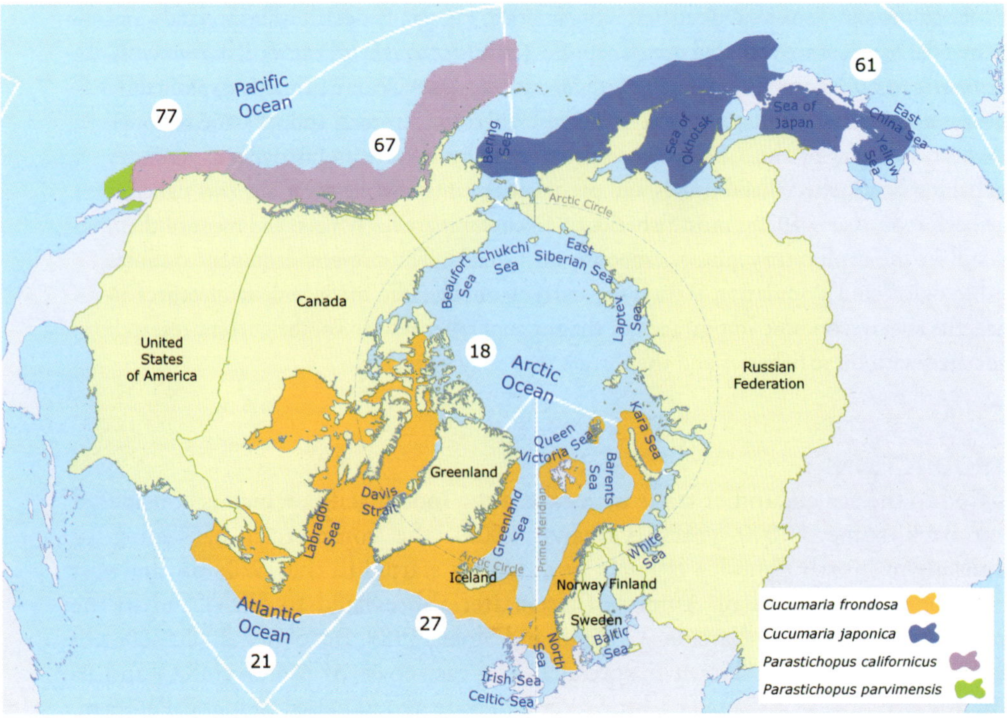

TABLE 1
Commercial species of sea cucumber (Echinodermata: Holothuroidea) found in the North Atlantic and North Pacific

Species	Common name	Order	Family
Cucumaria frondosa (Gunnerus, 1767)	Sea pumpkin; Orange–footed sea cucumber	Dendrochirotida	Cucumariidae
Cucumaria japonica (Semper, 1868)	Japanese sea cucumber; Kinko (in Japanese)	Dendrochirotida	Cucumariidae
Parastichopus californicus (Deichmann, 1937; Stimpson, 1857)	Giant red sea cucumber; Californian sea cucumber	Aspidochirotida	Stichopodidae
Parastichopus parvimensis (Clark, 1913)	Warty sea cucumber	Aspidochirotida	Stichopodidae

3. BIOLOGICAL AND POPULATION STATUS

3.1 Key taxonomic groups

Sea cucumber fisheries in the higher latitudes of the northern hemisphere focus mainly on four species: *Cucumaria frondosa*, *C. japonica*, *Parastichopus californicus* and *P. parvimensis* (Table 1; Figure 1). The first two species are the only two Dendrochirotida species harvested for human consumption in the northern hemisphere; the last two species are included in the order Aspidochirotida (Table 1).

Based on currently available estimates, *C. frondosa* is the most abundant commercial sea cucumber on the globe. It can be found in such high densities that fishers in some areas are able to catch up to 15 tonnes per day, even as bycatch of other sea products (Hamel and Mercier, 1999a) (Figure 2). Bradshaw, Ryan and Cooper (1991) indicated that a tow of five minutes will collect an average of 745 sea cucumbers in Nova Scotia (Canada). In 1988, a small fishery for *C. frondosa* developed on the east coast of the USA and has expended so rapidly that, according to available statistics, the landings of this species alone in 2003 made the USA the world's second largest producer of wild-caught sea cucumber, and Canada the fourth largest producer (Therkildsen and Petersen, 2006). Despite this, the biomass of unexploited *C. frondosa* stocks remains extensive in local areas and most fisheries are still at the exploratory stage.

The landings of *C. japonica*, mostly by Russian Federation fishers in the North West Pacific, remain poorly documented, precluding the evaluation of its importance in the world market (Gudimova, E., Polar Research Institute of Marine Research and Oceanography, Murmansk, Russian Federation, personal communication). Conversely, data for *P. californicus* and *P. parvimensis* are readily available and show that these

FIGURE 2
Harvesting of *Cucumaria frondosa* off Newfoundland and Labrador (Canada)

COURTESY OF L. BARRETT (DFA)

two species contribute little to the world market. Note that some tropical species of commercial value are found along the coast of Florida (USA) however they are not discussed in the present document. Bruckner (2006b) mentioned that there were no known harvests of these species in federal waters of the USA.

3.2 Biology and ecology of sea cucumbers

3.2.1 Cucumaria frondosa

Cucumaria frondosa (Figure 3) is among the most intensively studied species of sea cucumber; it has been the subject of numerous M.Sc. and Ph.D. dissertations, as well as several primary research papers and reports. *C. frondosa* is found in the North Atlantic from the Arctic to Cape Cod and from the Arctic to the northern latitudes of the United Kingdom, in Iceland, in the Barents Sea along the coast of the Russian Federation, in Scandinavia, the North Sea and along the coast of Greenland (Levin and Gudimova, 2000; Bruckner, 2006b). In some locations, such as in the Passamaquoddy Bay, New Brunswick (Canada), *C. frondosa* can be found in tide pools of the lower intertidal zone (Brinkhurst *et al.*, 1975) and is known to commonly occur down to 300 m (Brinkhurst *et al.*, 1975). Small specimens have recently been collected from depths of ca. 800 m along the continental rise of the Grand Banks in Newfoundland. However, *C. frondosa* is typically distributed between 20 and 100 m, with some variation between sites (e.g. Jordan, 1972; Coady, 1973; Hamel and Mercier, 1996b; Singh *et al.*, 2001; Grant, 2006).

This football–shaped sea cucumber is ca. 25–30 cm long (Hamel and Mercier, 1996b). Jordan (1972) mentioned it can reach 50 cm when relaxed. Desrosiers *et al.* (1989) established the average live weight around 250 g in Quebec (Canada). Campagna, Lambert and Archambault (2005) reported an average between 541 and 605 g depending on depth of collection along the coast of Quebec and Ke (1984, 1987) established it around 360 g in Nova Scotia (Canada). The sex ratio of *C. frondosa* is close to 1:1 (Coady, 1973; Hamel and Mercier, 1995, 1996b).

C. frondosa mainly colonizes rocky or pebbly bottoms (Hamel and Mercier, 1996b), though they have been observed occasionally on sandy bottoms in some areas (i.e. St. Pierre Bank) of the Newfoundland Grand Banks (Grant, 2006). The typical colour is light or dark brown, but a small percentage of individuals are a pale shade of orange, or cream, seemingly with little or no pigmentation. Some individuals can be brownish with pale patches scattered over the body. *C. frondosa* uses ten ramified tentacles

FIGURE 3
***Cucumaria frondosa* in its natural habitat (left) and a group of freshly fertilized eggs (right)**

COURTESY OF J.-F. HAMEL AND A. MERCIER (SEVE)

distributed around the mouth to capture particles in suspension in the water column, including phytoplankton and the tiny animalcules of the zooplankton (Jordan, 1972; Hamel and Mercier, 1998b). This species is mainly sessile, although it can display a strong escape response, detaching from the substrate and vigorously contracting its body wall, in the vicinity of one of its predators, such as the sea star *Solaster endeca*. Generally, the latter is very common where large populations of *C. frondosa* are found. Other known predators include the northern wolfish *Anarhichas lupus*. Sheffield *et al.* (2001) also indicated that *C. frondosa* was found to be part of the diet of walruses. Newly settled larvae and juveniles smaller than 1 cm are susceptible to grazing species, especially the green sea urchin *Strongylocentrotus droebachiensis* (Hamel and Mercier, 1996b).

The reproductive cycle of *C. frondosa* is characterized by an annual spawning and a generally highly synchronized gamete release (Hamel and Mercier, 1996b, c). The spawning season varies between populations from different locations (e.g. Jordan, 1972; Coady, 1973; Hamel and Mercier, 1995, 1996b,a,c; Oganesyan and Grigorjev, 1996; Singh *et al.*, 1999; Gudimova and Antsiferova, 2006). Table 2 summarizes the various spawning periods known for *C. frondosa* over its distribution range in North America and Europe.

Gametogenesis in *C. frondosa* is clearly initiated by an increase in day length around January while spawning appears to be triggered by a mix of factors, including tide, high phytoplankton concentrations in the water column and diet components (Hamel and Mercier, 1995, 1996a, 1999b). Males spawn first and the female begin to release their buoyant oocytes when the water column is already filled with spermatozoa (Hamel and Mercier, 1996c). The large reddish oocytes of *C. frondosa* measure around 0.5–0.8 mm in diameter (Figure 3); they are very rich in vitelline reserves, and they develop into lecitotrophic larvae that remain in the plankton for about 35–50 days (Hamel and Mercier, 1996b).

Detailed descriptions of spawning, fertilization and early development in *C. frondosa* are available (Hamel and Mercier, 1995, 1996b, c). Falk-Petersen (1982) reported that the oocytes were 0.65 mm in diameter and that the fecundity varied between 10 000 and 100 000 oocytes. Elena Gudimova (Polar Research Institute of Marine Research and Oceanography, Murmansk, Russian Federation, personal communication) also reported that the oocytes were 0.65 mm in diameter. Gudimova found a mean fecundity between 60 000 and 150 000 oocytes in the Barents Sea (Russian Federation). Settlement occurs preferably on rocks or pebbles, usually in crevices located in the

TABLE 2
Spawning periods of *Cucumaria frondosa* in its distribution range

Location	Spawning	References
St. Lawrence Estuary (Quebec, Canada)	Mid–June (mass spawning)	Hamel and Mercier, 1995, 1996a, b, c
Fundy Bay (New Brunswick, Canada)	April–June	Lacalli, 1981; Singh *et al.*, 2001
Nova Scotia (Canada)	End of March	Sherrylynn Rowe, DFO, Halifax, personal communication
Avalon Peninsula (Newfoundland, Canada)	February until early May	Coady, 1973
St. Pierre Bank (Newfoundland, Canada)	Early spring up to June	Grant, 2006
Avalon Peninsula (Newfoundland, Canada)	End of March up to end of April	Unpublished data
Maine (USA)	Mid–April	Jordan, 1972
New England (USA)	April to June	Medeiros-Bergen *et al.*, 1995
North Sea (Europe)	February and March	Runnström and Runnström, 1919
Northern Norway	April and May	Falk-Petersen, 1982
Barents Sea (Russian Federation)	February–April	Oganesyan and Grigorjev, 1996; Gudimova and Antsiferova, 2006

shaded area under hard substrata (Hamel and Mercier, 1996b). Medeiros–Bergen and Miles (1997) stated that juveniles of *C. frondosa* were also found in mussel beds, or on coralline algae and kelp holdfasts in Maine (USA).

The growth rate of the juveniles is very slow. Data gathered in eastern Quebec (Canada) show that the size at sexual maturity is reached at a size of about 8–10 cm, after approximately 3–4 years of growth and that the individuals grow to 12 cm in 4.5–5.5 years, reaching the commercial size (ca. 25–30 cm) in approximately 10 years (Hamel and Mercier, 1996b). A growth experiment currently under way in Newfoundland (Canada) is hinting at a similar growth rate, if not even slower (So, J., Ocean Sciences Centre, personal communication).

Hamel and Mercier (1996a, 1999b) have used *C. frondosa* to show the role of inter–individual chemical communication in the fine tuning of gametogenesis among congeners, which ensures synchronous development and release of mature gametes. This work confirmed the importance of density and spatial distribution on the reproductive success of benthic marine invertebrates. The detailed chemical composition of *C. frondosa* has been studied in Iceland (Geirsdottir and Stefansdottir, 2004), in Nova Scotia (Ke, Hirtle and Smith-Hall, 1984; Ke *et al.*, 1987) and in Newfoundland (Shahidi, 2006). Data on seasonal variations in the biochemical composition of tissues in *C. frondosa* have also been published (Girard *et al.*, 1990; David and MacDonald, 2002).

3.2.2 Cucumaria japonica

Morphologically, the sea cucumber *Cucumaria japonica* closely resembles *C. frondosa* to the point that their taxonomic relationship is disputed. The two species are commonly mistaken one for the other (Levin and Gudimova, 2000), and many scientists consider *C. japonica* as a sub–species (Saveljeva, 1941; Lambert, 1984) or a variety of *C. frondosa* (Mortensen, 1932; Panning, 1949, 1955). It has been reported by Levin and Gudimova (2000) that the species independence of *C. japonica* has been doubted since Britten (1906). Recently, Levin and Gudimova (2000) demonstrated that previous cataloguing of *C. japonica* as a sub–species of *C. frondosa* is not supported and that they truly are two distinct species. Their external anatomy still makes them hard to differentiate. The geographical distribution of *C. japonica* is also a topic of discussion; it is usually said to encompass the northeastern region of the Yellow Sea, the northeastern coast of the Honshu Island, the coast of Russian Federation in the Sea of Japan, the sea of Okhotsk, the Kuril Islands, the Kamchatka Peninsula and the Bering Sea (Levin and Gudimova, 2000). However, it has recently been argued that *C. japonica* is in fact not present in the Kamchatka and Kuril Islands areas and that another species, *C. okhotensis,* is found in these waters (Gudimova, E., Polar Research Institute of Marine Research and Oceanography, Russian Federation, personal communication). The same source indicated that the Russian Federation is planning to harvest 1 400 tonnes of this species in 2008. Considering the lack of information on *C. okhotensis*, this species will not be further discussed in the present document.

Cucumaria japonica occurs from the intertidal zone to ca. 300 m with a peak density between 30 and 60 m. Juveniles have been found in kelp forests, whereas adults colonize various substrates such as gravel, shell debris, rocks and mud (Levin, 1995; Levin and Gudimova, 2000). Like *C. frondosa*, *C. japonica* is a plankton feeder, extending its ten bucal tentacles to capture food items from the water column (Levin, 1995; Levin and Gudimova, 2000). The body of this sea cucumber is dense, cylindrical, barrel–shaped, slightly curved dorsally, especially in live animals, with a rounded or slightly stretched posterior end. The tube feet are mainly localized on the ventral surface which lies in contact with the substrate. The tube feet from the dorsal side are smaller. The maximum size recorded is 40 cm in length with an average around 20 cm (Levin and Gudimova, 2000). Large animals can reach 1.5–2.0 kg but the average is around 500 g. The colour of the body wall can be dark brown, dark purple, brown, greyish, and yellowish with the dorsal side clearly lighter. Completely white animals are also known (Levin and Gudimova, 2000). Levin (1995) stated that *C. japonica* is usually sessile and that, according to surveys, aggregations of this species move to shallow areas where the sea becomes warmer.

Kinosita and Sibuya (1941) indicated that the breeding season is a long period between early March and mid-November, and that ripe gametes are present in the gonad throughout the year. The structure of the male gonad and its maturation during the reproductive cycle was examined by Reunov *et al.* (1994) in the Popov Island (Japan Sea, Russian Federation). In winter, the male gonad is at its minimum state of development. In spring and summer, the quantity of spermatids and spermatozoa increase to reach a peak around August. The females synthesise around 300 000 oocytes of ca. 0.5 mm in diameter (Levin, 1995). Naidenko and Levin (1983) indicated that males spawn first, followed by females 2–5 hours later. According to the same investigators, the spawning season spreads from mid-April to mid-October; the green oocytes are very buoyant and reach the surface after being broadcasted. Levin (1995) pointed out that spawning occurs twice a year between April–June and September-October.

The larval development of *C. japonica* follows the pattern typical of many holothurians. The development is characterized by lecithotrophic larvae (Mokretsova and Koshkaryova, 1983; Naidenko and Levin, 1983), which are not feeding from the plankton but rather depend on endogenous nutrients (Mashanov and Dolmatov, 2000). Naidenko and Levin (1983) described the developmental biology and early growth of juveniles: the embryos hatch as pear-shaped larvae (ca. 0.5 mm) that develop into pentactulae after 6 to 10 days. Mashanov and Dolmatov (2000) wrote that the pentactula stage with complete digestive system is reached after 6 days and settlement occurs after 7 or 8 days of development. After 2 months, the juveniles were 9 mm long and the respiratory tree appeared to be fully developed. For details on the development of the pentactulae and early juveniles, see Dolmatov and Mokretsova (1995).

3.2.3 Parastichopus californicus

This species is found from the Gulf of Alaska (USA) to Cedros Island, Baja California (Mexico) between the low intertidal zone and 250 m (Cameron and Fankboner, 1986, 1989; Lambert, 1997). This is the largest holothurian species on the west coast of North America, reaching a maximum of 50 cm in length (Lambert, 1997; DFO, 2002). The average wet weight of specimens from Alaska is 225–250 g; however some can reach 500 g (Meredith, B., Alaska Department of Fish & Game, personal communication).

FIGURE 4

Parastichopus californicus in its natural habitat. The insert shows a male spawning

Source: Alaska Department of Fish & Game, Commercial Fisheries.

COURTESY OF BO MEREDITH

Parastichopus californicus is cylindrical with slightly tapered ends (Figure 4). The dorsal side of the body wall is covered by ca. 40 large papillae and many papillae of smaller size. The adults are usually mottled brown, the juveniles usually a more solid brown or red; some white specimens were also observed. The ventral surface of the body wall is covered by numerous rows of tube feet (Lambert, 1997). *P. parvimensis*, which lives south of Monterey Bay (see section below), is very similar and can grow to nearly the same size, but the body is chestnut brown dorsally and much paler underneath; furthermore, the tips of its papillae are tipped with black instead of red. Zhou and Shirley (1996), Woodby, Smiley and Larson (2000) and Bruckner (2006b) mentioned that *P. californicus* can be found on a wide variety of substrates and under various current regimes but are most abundant in areas of moderate current on cobbles, boulders, crevassed bedrock, shells, sand and algae. The species seems to avoid mud bottoms and areas of freshwater runoff. However, a preference for harder substrates has been observed (Woodby, Smiley and Larson, 2000).

Parastichopus californicus is a slow moving deposit feeder which collects organic matter and associated micro-organisms using its 20 peltate feeding tentacles and a copious amount of mucus to trap particles such as bacteria and fungi. Maximum densities are therefore found in areas of accumulation in highly organic sediments (Cameron and Fankboner, 1989). This species ceases feeding and becomes dormant between September and early March (Lambert, 1997). Adults undergo fluctuations in body mass, body wall thickness, and muscle weight during their annual cycle as they reabsorb and regenerate their internal organs (Fankboner and Cameron, 1985; DFO, 2002). Sea stars of the genus *Pycnopodia* and *Solaster* are known predators of this sea cucumber (Cameron and Fankboner, 1989; Rogers-Bennett and Ono, 2001). A few anecdotal reports suggest that sea otters may also prey on these sea cucumbers. Muse (1998) proposed that *P. californicus* can swim to some extent to avoid predation by sea stars.

Parastichopus californicus is either male or female, but does not display any external sexual dimorphism; the sex ratio is about 1:1 (Cameron and Fankboner, 1986). This sea cucumber exhibits an annual reproductive cycle. Initiation of gametogenesis is visible by the increase in gonadal index in January or February; the maximum gonad size was recorded in June and July in British Columbia, Canada (Cameron and Fankboner, 1986). Spawning occurs in June, July and August, although the species can sometime spawn a bit earlier (Muse, 1998). Gametes are broadcasted in the water column after males and females adopt the typical posture observed in many tropical species (i.e. anterior body raised) (Cameron and Fankboner, 1986; McEuen, 1988). Spawning seems to be correlated with bright sunshine days and high phytoplankton productivity, although epidemic spawning has not been observed (Cameron and Fankboner, 1986; Muse, 1998). After spawning, the gonad begins to decrease in size (i.e. degeneration phase of the fecund tubules) to reach its smallest size at the end of the year (Fankboner and Cameron, 1985; Cameron and Fankboner, 1986). A more precise seasonal cycle of the various classes of gonadal tubules has been published by Smiley (1988). This species is apparently able to undertake seasonal migrations to different depths (DFO, 2002). The shallower populations (less than 16 m deep) seem to be the most susceptible to be involved in reproduction (Rogers-Bennett and Ono, 2001; Bruckner, 2006b). In fact, Courtney (1927), Cameron and Fankboner (1986) and Smiley *et al.* (1991) mentioned that migration to shallower waters for the purposed of spawning is observed from late April to August. Compared to plankton feeders like *C. frondosa* and *C. japonica* which are mostly sessile, *P. californicus* moves randomly, covering distances of up to 3.9 m per day (Cameron and Fankboner, 1989). Muse (1998) indicated that *P. californicus* can move as much as 100 m in 24 hours.

The light orange mature oocytes of *P. californicus* measure ca. 200 µm in diameter (Lambert, 1997). The larvae feed on plankton and remain in the water column for a period of 35 to 52 days (Lambert, 1997). However, Cameron and Fankboner (1989) observed that the pelagic phase could be as long as 65 to 125 days just to reach the auricularia stage and another 2 days were necessary for the pentactula to develop. This suggests that the spawning stock may be quite far from where the larvae settle and grow (Muse, 1998). Settlement was observed in dense mats of filamentous red algae, algae holdfast, under rocks or in crevices (Cameron and Fankboner, 1989; DFO, 2002). Cameron and Fankboner (1989) added that they can also be found on tubes of polychaete worms. The recruitment is sporadic and a high natural mortality rate is suspected (Rogers-Bennett and Ono, 2001; Bruckner, 2006b). Juvenile growth is slow, reaching 0.5 to 2 cm after a year and 4 to 10 cm after 2 years (DFO, 2002). Sexual maturity is reached after 4 years (Cameron and Fankboner, 1989; Lambert, 1997). Based on an analysis of length frequency data for the first three years of growth, commercial-sized specimens were estimated to be at least 4 years old; the oldest animals could be around 12 years of age (Cameron and Fankboner, 1989).

3.2.4 Parastichopus parvimensis

This species is common from Baja California (Mexico) to Monterey Bay (California, USA), although scattered individuals were reported to occur north, up to Point Conception (California, USA). *Parastichopus parvimensis* is found mainly in low energy environments from the intertidal down to 30 m and can reach a maximum length of 30–40 cm (Bruckner, 2006b). Yingst (1982) found that *P. parvimensis* are most abundant where the organic content of the substrate is highest, the sea cucumber being able to ingest the first two centimetres irrespectively of the size of particles. Population of small individuals from Santa Catalina Island were observed to feed on fine particulate material collected from rock rubble under the kelp canopy. Larger individuals avoid rocks; they dwell and feed on granular sediments. According to Muscat (1982, 1983) *P. parvimensis* feeds both during the day and at night, and densities of sea cucumbers increase on hard substrates, being up to ten times higher than on soft bottoms. Nonetheless, the largest specimens are found on sand and an intermediate size class on the surfaces of rocks. Muscat (1982) also mentioned that different movement patterns on soft and hard substrate are noticed, sea cucumbers on sand moving significantly more than those found on rocks.

Yingst (1982) observed an annual evisceration cycle, which affected 60 percent of individuals during October and November, causing them to cease feeding for about four weeks until minimum gut regeneration was accomplished. Muscat (1982) brought evidence of an annual pattern of intestinal growth and development, with sea cucumbers undergoing spontaneous evisceration in September and October. They can regenerate lost parts in one or two months, reaching a maximum overall body weight during the winter, probably due to resumed feeding and accumulation of gonad materials. There are dramatic seasonal fluctuations in population densities, with nearly complete disappearance from shallow water from August to November in Southern California (Muscat, 1982). There appears to be a migration downslope from the warmer, shallower areas, which occurs on both hard rock and sand substrates.

Parastichopus parvimensis does not become sexually mature until it reaches ca. 40 g in total body weight (Muscat, 1983). Muscat (1982) mentioned that this species expresses a definite reproductive cycle with spawning in May and June possibly in response to increasing water temperatures. Furthermore, according to the same author, the gonad is completely reabsorbed during September and October and the gonadal growing phase starts around January. The vitellogenic oocyte reach an average of 180 µm in diameter (Muscat, 1983). Eckert *et al.* (2001) indicated that *P. parvimensis* larvae reached the doliolaria stage in 27 days; on day 28, larvae began to settle on under sides of rocks and kelp holdfast. Juveniles reached ca. 3.5 cm long after a year of growth. Juveniles between 2 and 6 cm long are found under rocks only, whereas individuals between 8 and 12 cm dwell both on and under rocks (Muscat, 1983). This different distribution between adult and juveniles could be due to predatory pressures since juveniles are susceptible to be preyed upon by fish (Muscat, 1982). Large individuals are at least five years old (Muscat, 1983).

3.3 Background of sea cucumber fisheries

Harvesting of temperate sea cucumbers in the northern hemisphere, even when taking into account the extended geographical range covered in the present document, is focused on relatively few species: one species (*C. frondosa*) in the North Atlantic and three species (*C. japonica*, *P. californicus* and *P. parvimensis*) on the Pacific coast. For this reason, the origin and status of the fisheries will often take in account the main regional variations, when applicable.

The commercial exploitation of sea cucumbers in the northern sectors of the Pacific and Atlantic oceans has only started a few decades ago (Conand and Sloan, 1989). Harvesting of *P. californicus* began in Washington State (USA) in the early 1970s, spreading to California, Alaska and Oregon over the late 1970s and early 1980s (Bruckner, 2005). In British Columbia (Canada), the first official landings of *P. californicus* date back to 1980 (Sloan, 1986), although sea cucumbers were already harvested a decade earlier. During the same period, a small fishery for *C. frondosa* developed in Maine on the east coast of the USA, expanding into the Atlantic provinces of Canada during the 1990s (Bruckner, 2005; Therkildsen and Petersen, 2006). In Iceland, the sea cucumber fishery opened in the early 2000s (Olafsson, K., Reykofninn-Grundarfirdi ehf, personal communication). Unfortunately, we were not able to obtain historical data for the Russian Federation.

3.3.1 United States of America

West Coast

The commercial sea cucumber fishery in the USA began in the state of Washington and focused on a single species: *P. californicus*. Between 1971 and 1987, sea cucumbers were harvested by divers without any seasonal or spatial restrictions. In 1987, when depletion of sea cucumber populations became noticeable, the state established harvest districts and a restricted harvest season (ADFG, 1990). Today, most of the harvesting is done by hand using scuba diving or hookah, with only limited trawl collections under experimental fishery permits. Collection by divers for personal use/consumption is subject to a daily limit of 10 specimens (Bruckner, 2005).

In the state of California, the fishery started in 1978 and focused on two species: *P. californicus* and *P. parvimensis*. Both diving and targeted trawling are used for the capture, however trawls are restricted mainly to the south of the state and their use has declined over the past few years (Bruckner, 2006b). In 1997, divers with permits for sea urchins and abalones were allowed to collect sea cucumbers as well (Rogers-Bennett and Ono, 2001; Bruckner, 2006b).

In the state of Alaska, harvesting of *P. californicus* began in 1981 as an experimental fishery; however the first true commercial landing was recorded in 1983 in Ketchikan (Southern Alaska). Sea cucumbers are collected by SCUBA divers. The fishery was initially not subject to any regulation, and the number of divers increased rapidly when, in 1990, the Alaska Department of Fish and Game closed down the fishery to develop a management plan. In 1995–1996, 424 divers were issued permits in southeast Alaska and this number decreased to 235 in 2001–2002 (Hebert and Pritchett, 2002) and 174 in 2006–2007 (Table 3). There is currently a fishery in Kodiak and the Aleutian Islands areas (Ruccio and Jackson, 2002). Sea cucumbers were traditionally used as a food on the Northwest Coast culture area, of which southeast Alaska communities are a part. The earliest reference comes from the 1804 voyage of Urcy Lisiansky into southeast Alaskan waters (Mathews, Kookesh and Bosworth, 1990).

In the state of Oregon, the fishery began in 1993 targeting *P. californicus* exclusively (Bruckner, 2006b). Collection is by SCUBA divers and by trawl with experimental gear permits (McCrae, 1994; Bruckner, 2006b). The fishery of *P. californicus* is under the Developmental Fisheries Program. Only nine divers requested a permit for the

TABLE 3
Historical perspective for sea cucumber harvests in southeast Alaska

Season (years)	Guideline harvest levels (tonnes)	Average wet weight (grams)	Ex-vessel price (USD kg^{-1})	Number of divers
1990–91	319.6	–	1.25	143
1991–92	380.6	–	1.61	187
1992–93	499.2	–	1.94	240
1993–94	362.5	–	2.27	320
1994–95	612.8	213.3	3.45	261
1995–96	525.0	227.0	2.82	424
1996–97	426.1	203.9	2.82	294
1997–98	404.8	204.3	3.65	226
1998–99	465.5	217.9	3.41	219
1999–00	716.7	213.3	4.27	200
2000–01	509.2	213.8	4.91	220
2001–02	646.5	198.8	3.85	235
2002–03	742.7	214.6	2.77	201
2003–04	742.8	231.8	3.23	195
2004–05	626.5	224.8	4.66	194
2005–06	658.5	221.9	4.75	198
2006–07	725.2	–	4.42	174

Source: Bo Meredith, Alaska Department of Fish & Game, Commercial Fisheries.

fishery in 1993 even though 44 permits had been made available, whereas 22 divers were recorded in 1994 (Bruckner, 2006b). The number of harvesters in Oregon after 1994 remained low: five in 1997, two in 1999, one in 2000, two in 2001 and two in 2003 (Bruckner 2005).

East Coast

The fishery of *C. frondosa* on the east coast of the USA reportedly dates back to the 1970s (Sutterlin and Waddy, 1975; Seatech Investigation Services Ltd., 1981; Bradshaw, Ryan and Cooper, 1991). True commercial harvests began in 1988, expanding significantly in 1994 when the Asian markets opened up to this new resource (Chenoweth and McGowan, 1997). Scallop chain sweeps or light urchin drags were used as fishing gear. Fishing activity was concentrated in Washington and Hancock Counties with catches landed in Winter Harbor, Jonesport, Beals Island and in Eastport. The fishery has recently experienced great increases in landings, corresponding to expanding export markets (Feindel, 2002). It began in Maine with one operator and has expanded, especially since 1994. In the mid-1990 this sea cucumber fishery employed 15–20 harvesters and between 75 and 100 workers to process the product (Feindel, 2002; Bruckner, 2005).

3.3.2 Canada

West Coast

Sloan (1986) indicated that the first official landing of *P. californicus* in British Columbia (considered the first commercial harvest of sea cucumber in Canada) was recorded in 1980, although exploitation may have started as early as 1971 (DFO, 2002). The fishery itself was first regulated under commercial licence on an experimental basis in 1980 (DFO, 2002; Campagna and Hand, 2004). *P. californicus* is the only species fished on the west coast of Canada, the fishery being centered on Vancouver Island in the Georgia and Johnstone Straits and in the Puget Sound. The fishery was an open access through 1990. The number of licenses rose from 40 in 1985 to 215 in 1990. The number of vessels went from 21 in 1985 to 126 in 1990. In 1988, 124 divers where involved and 163 in 1990 (Muse, 1998). However, Muse (1998) mentioned that British Columbia's Natives harvested sea cucumbers as a traditional food in much earlier times.

The fishery in British Columbia expanded rapidly after 1980 with annual landings exceeding 1 900 tonnes in 1988. Fishing was initially permitted in the South Coast areas only and the majority of landings were taken in the Strait of Georgia until 1987. The north coast was opened in 1986 with a total annual recommended catch of 500 tonnes, although fishing did not occur there until 1987. To date, landings of sea cucumbers have been recorded from all Canadian Pacific fisheries management areas, with the exception of the north and west Queen Charlotte Islands. The central and north coasts currently support about 80 percent of the fishery (DFO, 2002).

East Coast

Although there had been sporadic attempts to initiate a fishery for *C. frondosa* in Atlantic Canada since 1980, the state of Maine (USA) was the first to develop a substantial fishery for this species in the 1980s (Therkildsen and Petersen, 2006). After the onset of the Maine fishery, provincial and federal governments in some Canadian provinces began assessing the feasibility of starting a fishery for *C. frondosa*, which developed timidly in several locations (Therkildsen and Petersen, 2006). In 1989, a small Nova Scotia fishing company located in St. Mary's Bay identified a potential market for *C. frondosa*. Today, the fishery in the Scotia–Fundy Fisheries Management Sector, which extends from the northern tip of Cape Breton to the New Brunswick–Maine border, accounts for the highest landings of sea cucumber in Canada. The fishery began with harvesters using urchin licenses, but it was rapidly restricted to holders of experimental sea cucumber licenses (DFO, 2005b). Churchill (1996) mentioned that Newfoundland's interest in *C. frondosa* began in late 1995 when large quantities of sea cucumbers were noted in the results of seismic surveys carried out on the Grand Banks. During the same period, representatives of the Department of Fisheries and Aquaculture (DFA) attended the first International Chinese Food Industry Exposition in Toronto, Canada, to promote non-traditionally harvested products, including sea cucumber, from Newfoundland and Labrador (Dooley, McDonald and Rumboldt, 1995). In Atlantic Canada, the bulk of the sea cucumber fisheries were initially carried out using modified trawls and/or scallop gear. Now, at least for Newfoundland and Labrador, a specifically designed sea cucumber drag (a government approved, standardized science tool used by the Department of Fisheries and Oceans [DFO]) is used for science surveys and directed commercial activity. The sea cucumber gear appears to be highly efficient at capture and is designed, in part, to address bycatch and bottom impact concerns that are associated more so with other harvesting gear such as scallop dredges and/or otter trawls. *C. frondosa* is also commonly harvested as a result of bycatch from scallop fisheries.

In 1997, the Canada-Newfoundland Cooperation Agreement for Fishing Industry Development carried out an exploratory trial for sea cucumber in the province and concluded that a future fishery for this species was possible. In 2002, the Department of Fisheries and Aquaculture, under the Fisheries Diversification program, surveyed potential sea cucumber fishing grounds and alternative harvesting methods. In Newfoundland and Labrador, the government issued eight exploratory licenses for the St. Pierre Bank area and licence holders have an exploratory annual quota of ca. 612 tonnes in NAFO subdivision 3Ps; in addition, two exploratory licenses have been issued in NAFO area 4R with an annual quota of ca. 90 tonnes (Barrett, L., DFA, personal communication). Also, initial surveys suggested that diving might be a possible alternative harvesting method and exploratory trial dives yielded up to 2 700 kilograms per day in some areas. However, experimentation with this harvest method in the Scotia–Fundy region suggests that it will not be financially viable (DFO, 2005a). In New Brunswick, the fishery began in 2000 with two permits emitted by the DFO in Moncton (Coulombe and Campagna, 2006).

Sea cucumber fisheries are not yet developed in Quebec (Coulombe and Fillion, 2002), although GERMA (GERMA, 1989), Ondine Inc. (Ondine, 1991), Hamel and

Mercier (1998a) and Campagna, Lambert and Archambault (2005) all surmised that *C. frondosa* could represent an interesting commercial potential. On June 11, 2002, Les Biotechnologies Atrium Inc. of Quebec and Cusimer (1991) Inc. in Mont–Louis (Gaspesian Peninsula) received funding to develop an experimental project to valorize nutraceutical products from this sea cucumber. Coulombe and Campagna (2006) stated that the interest in this resource increased in Quebec following the visit of potential buyers from the USA around 2002. Despite the interest and initial efforts, there is still no official fishery of *C. frondosa* in the St. Lawrence Estuary and Gulf areas at the moment. However, a sea cucumber processor in New Brunswick was exploring the establishment of a sea cucumber processing plant on the Gaspésie Peninsula in 2007; it is therefore anticipated that commercial sea cucumber harvests may begin in the near future in the St. Lawrence area.

3.3.3 Iceland

To the authors' knowledge, Iceland is the only European country to harvest sea cucumbers (if western Russian Federation is excluded). *C. frondosa* is common all around Iceland and is often discarded as bycatch during bottom trawls for other species. The first small landing of *C. frondosa* in Iceland is reported in 1995 (Valtysson, 2001). Therkildsen and Petersen (2006) mentioned that the future of this fishery is uncertain because the stocks seem to be lower than expected. Nonetheless, Kari Olafsson (Reykofninn-Grundarfirdi ehf, personal communication) indicated that the fishery is small and still under development but that it would continue in 2007. The same source mentioned that the company bears all costs of development with a minimal assistance from the government. There has only been one licence issued to date.

3.3.4 Russian Federation

In the Barents Sea, fishers recently began harvesting *C. frondosa* (Gudimova, 1998; Organesyan and Grigorjev, 1998). According to Therkildsen and Petersen (2006) there is no directed fishery for this species with all landings coming from bycatch of the scallop fishery. Like in Canada and the USA, bycatch was originally discarded; however, since 2000, harvesters have been able to sell this resource for processing (Gudimova, Gudimov and Collin, 2004; Therkildsen and Petersen, 2006). Landing has not exceeded 200–500 tonnes per year (Gudimova, Gudimov and Collin, 2004) and no regulation is governing this resource at the moment (Therkildsen and Petersen, 2006; Gudimova, E., Polar Research Institute of marine Research and Oceanography, Russian Federation, personal communication). The fishery of *C. japonica* on the eastern coast of the Russian Federation in the Pacific Ocean remains obscure and we were not able to find any tangible information. However, *C. japonica* is a common component of the fauna of the Sea of Japan and is also one of the most important commercial species among the echinoderms of the Far East (Mashanov and Dolmatov, 2000).

3.4 Species in trade

3.4.1 Food products

The main products derived from *C. frondosa* in Maine are: (i) the muscle bands, which are vacuum packed and flash frozen; and, (ii) the boiled and dried body wall (Feindel, 2002; Bruckner, 2006b; Therkildsen and Petersen, 2006). Ke *et al.* (1984, 1987) mentioned that the body walls were boiled for 60 min. in 3 percent NaCl utilizing a steam cooker. They were subsequently dried in a vacuum drying oven at room temperature and finally at 60 °C. In New Brunswick (Canada), the animals are first relaxed at 40 °C, then cut longitudinally and eviscerated; the muscles are removed and packed separately, the body wall is boiled and dried mechanically (Coulombe and Campagna, 2006). Bruckner (2005) indicated that the processing of sea cucumbers involved removing the anterior end with the tentacles, slitting the body lengthwise to remove the viscera, and scraping the muscles off the body wall. The body and muscles are typically boiled, dried and salted before export, while lesser quantities are marketed as a frozen, pickled or live product.

In eastern Canada, especially in Newfoundland, several new products from *C. frondosa* were tested including: (i) the cooked, salted and dried skin with the meat attached; (ii) muscles and skin (fresh frozen together or separately); (iii) aquapharyngeal bulbs (called flowers), gut, gonad and respiratory tree (fresh frozen and packed separately); and, (iv) fresh frozen muscles and skin separated (Barrett, L., DFA, personal communication). The industry tried these products over the last five years in an effort to adapt to relatively unfamiliar markets in Asia with a non–traditional Newfoundland product.

In Canada and other parts of North America, the processing of sea cucumbers is generally carried out in a semi-industrialized environment (Figure 5). An industry in New Brunswick is employing as many as 120 people during the fishing season; they are processing up to 35–40 tonnes of sea cucumber daily. This processing plant, as well as the one on Fogo Island, Newfoundland and Labrador, is currently experimenting with the prototype of an eviscerating/skinning machine devised by engineers from DFA and Memorial University (Figure 6). According to the plant owners in both provinces, the results are promising and although they would prefer to retain the services of workers for most of the processing, they can envision using the mechanical gutter/skinner for part of the process during leaner times to reduce labour related production costs.

In Iceland, *C. frondosa* is processed into a dried product (skin with muscles) and an intermediate product consisting of gutted, boiled and individually quick frozen (IQF) skin with muscles (Olafsson, K., Reykofninn-Grundarfirdi ehf, personal communication).

The main product derived from *C. japonica* captured in Far East Russian Federation is bêche-de-mer according to suppliers that are currently advertising on the Internet. However, an older report (Levin, 1995) stated that *C. japonica* was rarely used as dry product, as the Japanese consumed them raw. All harvested sea cucumbers were reportedly boiled, sliced in small pieces and sold in local shops as a salad (Levin, 1995). They also produced medicines for domestic animals and an additive for toothpastes, creams, etc. (Levin, 1995). Elena Gudimova (Polar Research Institute of Marine Research and Oceanography, Russian Federation, personal communication) indicated that the product from the Barents Sea is transformed into bêche-de-mer, though some is canned or used for medicinal purposes (see below). It would appear that 100 percent of *C. frondosa* and 80 percent of *C. japonica* from the Russian Federation are used solely for domestic consumption.

In USA, both species of *Parastichopus* are boiled, dried and salted (bêche-de-mer), some are marketed frozen, and as live product (Bruckner, 2006b). Mathews, Kookesh and Bosworth (1990) and Bo Meredith (Alaska Department of Fish and Game, personal communication) indicated that the muscle bands are also processed (frozen) in Alaska. The Alaskan industry has created its own market for the "meat strips" that line the inside of the body wall (Ess, 2007). We assume they are referring to the muscle bands.

In British Columbia, *P. californicus* are cut open longitudinally and viscera and internal fluids are removed in a process called splitting. The animals are processed into two products; frozen muscle strips and dried skin (DFO, 2002).

Traditional sea cucumber products can also be found marginally, for instance within native communities in Alaska (Mathews, Kookesh and Bosworth, 1990). In the past, sea cucumbers were often eaten fresh from daily catches. Drying and smoking were also described as traditional methods of preserving sea cucumbers for storage. With the advent of home freezers in the late 1940s and early 1950s, drying was abandoned in favour of freezing. Today consumers of sea cucumbers preserve them by canning or

FIGURE 5

Sea cucumber processing in New Brunswick (Canada) – fresh product is trucked to the processing plant where workers proceed with evisceration (top left); freshly cooked skin on drying racks (top right); fresh muscles bands, or meat (left centre); packaged meat (right centre); dried skin and dried aquapharyngial bulbs, or "flowers" (bottom left); dried skin being packaged (bottom right)

COURTESY OF L. BARRETT (DFA)

FIGURE 6

Prototype of an eviscerating/skinning machine

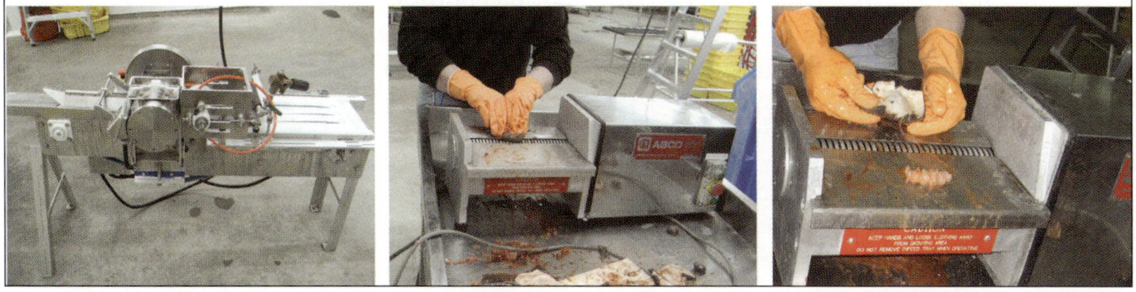

COURTESY OF L. BARRETT (DFA)

FIGURE 7
Processing of *Parastichopus californicus* in Alaska – body wall (left) and meat strips (right)

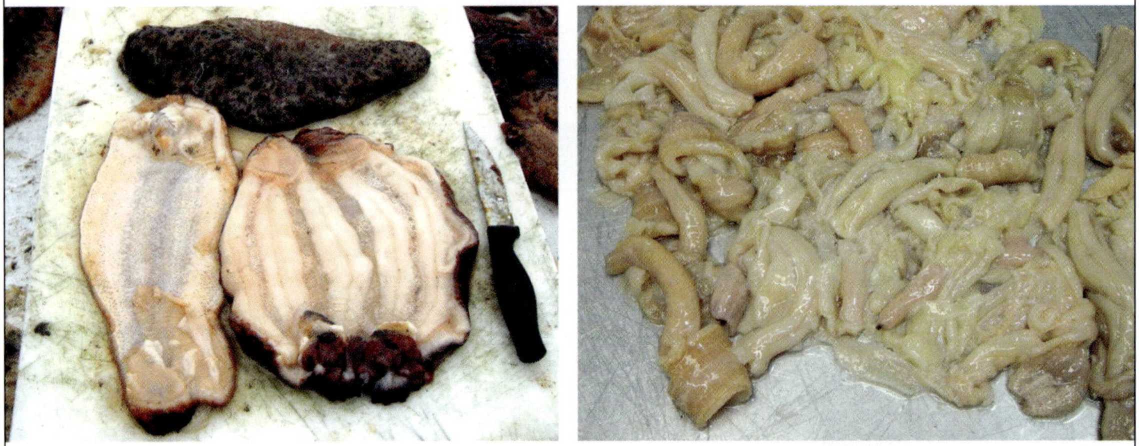

Source: Alaska Department of Fish & Game, Commercial Fisheries.

freezing. The canning process consists of par-boiling the meat, sterilizing the jars and lids, inserting meats, and sealing, whereas freezing involves boiling the sea cucumber meat, draining and storing it into plastic freezer bags.

3.4.2 Other derivatives

Over the years, several scientists and companies have developed ways to valorize the processing by-products of sea cucumbers by extracting nutraceuticals from them. Red Oil produced from *C. frondosa* intestines and body–wall were used in an adjuvant arthritis model in rats to study their anti-inflammatory properties (Glenn, 1966; Winter and Nuss, 1996). Bruckner (2006b) mentioned that the by-product of processing from *C. frondosa* is marketed in the USA as nutritional supplement providing chondroitin (NutriSea®), and is also sold as a treatment for arthritis in humans (ArthriSea® and SeaCuMAX®) and pets (Sea Jerky®) on the USA and Canadian markets (Coastside Bio Resources, Stonington, Maine, USA). The latter also offer a full line of nutritional formulas for humans, dogs, cats, horses and pigs[1]. The by-product is also marketed as compost in Maine (Bruckner, 2006b).

It is believed that *C. frondosa* contains chemicals that inhibit harmful prostaglandins involved in causing pain and arthritis. They are also rich in nutrients needed by cartilage including chondroitin and mucopolysaccharides, and several vitamins and minerals. One product supplies a unique jerky-type treat made of sea cucumber which dogs find quite palatable. Each jerky treat provides 1 200 mg of chondroitin. These treats are perfect for dogs and/or cats that are difficult to medicate. The jerky treats can also be used in conjunction with other similar pill supplements, as it is unlikely to overdose a pet on glucosamine or chondroitin[2]. Several supplements made from *C. frondosa* extracts are sold for human consumption as well[3].

Extracts from the sea cucumber *C. japonica* may be useful in the therapeutic management of the Epstein–Barr viral infection (Spainhour, 2005). Mulcrone (2006) indicated that, in eastern Russian Federation, an increasing demand on *C. japonica* has led to concern for this species for both food and cosmetic products.

---

1 *Source*: http://www.coastsidebio.co.uk/.

2 *Source*: http://www.glucosamine-arthritis.org/arthritis-pets/holistic-approaches4.html.

3 *See for example the following sources*: http://www.vuru.com/supplements/product?productId=1389&lastCategoryId=313; http://www.ihealthtree.com/buried-treasure-joint-ease-complete-16oz.html.

3.5 Population status

Within the vast territory covered by this review, which encompasses extremely long coast lines and countless islands and bays, estimations of biomass, density and catch per unit effort (CPUE) are limited and sometime incomplete. Furthermore, biomass and density are not always established in the most suitable location for the global understanding of stock levels and management needs. Notwithstanding these limitations, this report has tried to compile the most complete and up to date sets of data available for the territory and species under study.

3.5.1 Cucumaria frondosa

Therkildsen and Petersen (2006) indicated that most fisheries for *C. frondosa* are still at the early exploratory phase, and it is difficult to predict whether the available stocks can sustain long term exploitation. Indeed, the large *C. frondosa* landings in eastern Canada and USA may not be sustainable in view of the slow growth of this species (Hamel and Mercier, 1996b).

Campagna, Lambert and Archambault (2005) demonstrated that *C. frondosa* was found in relatively high abundance in Quebec along the north coast of the Gaspesian Peninsula. In general, the species was more abundant at 10 m and became less abundant with increasing depth. The lowest values were found between 50 and 80 m. The highest average CPUE was 2 138 kg m^{-1} h^{-1} with an average density of 1.6 ind. m^{-2} at 10 m depth. The highest values of biomass recorded were 3 798 tonnes, 5 677 tonnes and 5 010 tonnes. The total biomass for this area of 813 km^2 would be around 61 000 tonnes.

Bradshaw, Ryan and Cooper (1991) described CPUE based on capture using ring scallop bags of 182 sea cucumbers per minute in Nova Scotia in 1991. They indicated that with each tow lasting 10 minutes, it would take about 3 tows to fill a boat that can hold 2 300 kilograms (Bradshaw, Ryan and Cooper, 1991). Coulombe and Campagna (2006) mentioned that a single boat in the Bay of Fundy, New Brunswick (Canada) could harvest 9–10 tonnes per day.

DFA (2004) reported that the CPUE for *C. frondosa* in the Strait of Belle Isle, Newfoundland, was equivalent to 553 kilograms over a 15 minutes tow. Dive surveys in Pistolet Bay yielded up to 2.2 ind. m^{-2}. An earlier report (DFA, 2003) measured catches of 275–325 kilograms in 10-minute tows in the Strait of Belle Isle, northern Newfoundland. The Marine Institute (MI, 2006) indicated that densities of *C. frondosa* on the St. Pierre Bank (Grand Banks, Newfoundland) were 0.22 ind m^{-2} on rocky substrata, 0.34 to 0.44 ind. m^{-2} on sand and shell bottoms, and 0.72 ind. m^{-2} on gravel-cobble beds. The Marine Institute (MI, 2004) noted sea cucumber densities between 0.16 to 0.58 ind. m^{-2} on the St. Pierre Bank. The total catch on St. Pierre Bank during one 24-hour exploratory fishing trip using 8 tows of 10–30 minute duration was 1 190 kilograms of sea cucumber. From some exploratory dive fishing trips, a single diver reportedly collected 907 to 1 360 kilograms per day in Notre-Dame Bay, Newfoundland (DFA, 2002b). Specifically, a 2 to 3.5 hours dive in Bonavista Bay will yield a landing of 1 450 kilograms, whereas diving in Notre-Dame Bay will yield 149 kilograms from the lesser productive areas.

Bruckner (2005) stated that there is concern associated with the Maine sea cucumber fishery in that most of the fishing effort is concentrated within three locations in eastern Maine, and there are anecdotal reports of some sites having been fished out. Nonetheless, Bruckner (2006b) indicated that the population densities can reach 5 ind. m^{-2} and populations can comprise up to 50 percent of the benthic biomass. Dive surveys using transects showed the patchy nature of the species, with abundances ranging from 0.01 to 7.45 ind. m^{-2}, with substantial differences in size and weight of animals between sites (Bruckner, 2006b).

In Iceland, the natural density of *C. frondosa* is not known; the best 10-minute tow yielded 1 200 kilograms, but the usual harvest is commonly between 1 and 500 kilograms per 20-minute tow (Olafsson, K., Reykofninn-Grundarfirdi ehf, personal communication). Although densities were not stated, captures of *C. frondosa* in the Barents Sea are reportedly stable (Gudimova, E., Polar Research Institute of Marine Research and Oceanography, Russian Federation, personal communication).

3.5.2 Cucumaria japonica

Data on this species are scarce. Dulepov, Scherbatyuk and Jiltsova (2003) indicated that densities fetched 7.6 ind. m^{-2} for the Great Peter Bay in Far East Russian Federation based on data gathered by a semi autonomous underwater vehicle. Because this value was established using image analysis, it might be overestimated. In Aniva Bay (Russian Federation) *C. japonica* has been observed to display a patchy distribution with reported maximum densities of 0.3–0.9 ind. m^{-2} [4]. Levin (1995) reported that the highest registered density was around 40 ind. m^{-2}. More precisely, possible catches were estimated at 2 300 tonnes in the Sea of Japan, 2 000 tonnes in Kuril Islands, and 11 800 tonnes in the Sea of Ochotsk.

3.5.3 Parastichopus californicus

The DFO (2002) mentioned that early southern British Columbia harvesters targeted populations of *P. californicus* in areas close to the harbour, where diving was easy and the resource abundant. Researchers and managers felt that only a small proportion of the stock was being harvested, and that many sea cucumbers were left untouched in these areas. The fishery therefore expanded to more remote northern areas, but remained targeted on very accessible locations. Still today, large areas of the coast have not been visited by the commercial sea cucumber fleet. Surveys conducted in various areas of the coast indicate that sea cucumber population densities vary considerably with habitat type. Density estimates from almost all surveys are significantly higher than the conservative estimate of 5.08 sea cucumbers per meter of shoreline. *P. californicus* populations extend below the safe diving depth of 20 m where extensive harvesting cannot be conducted (DFO, 2002). In British Columbia, the densities are estimated at less than 0.25 ind. m^{-2} (Bruckner, 2006b). The CPUE values supplied by Muse (1998) for *P. californicus* in British Columbia are in kg per diver per hour: 372 in 1983, 347 in 1987 and 617 in 1992.

Woodby, Smiley and Larson (2000) indicated that the depth distribution of *P. californicus* was investigated in the vicinity of Sitka Sound, Alaska. Sea cucumber densities were greatest in shallower waters with ca. 70 percent of the sea cucumbers observed above 15 m, and 70 percent above 20 m. Average densities were 0.03 m^{-2} in deeper water and 0.3 m^{-2} at SCUBA diving depths. The deepest sea cucumber was observed at 87 m (Woodby, Smiley and Larson, 2000). In Southeast Alaska, Zhou and Shirley (1996) used a submersible to measure densities which reportedly varied from 0 to 267 ind. ha^{-1}. The highest density recorded was 0.23 ind. m^{-2}.

Bruckner (2006b) reported that harvested sites in California showed densities that were 50–80 percent lower than in the non-fished areas. For instance, at an established reserve in northern California, densities averaged 2 200 ind. ha^{-1}. Another set of data from a newly established reserve (Punta Gorda Ecological Reserve) ranged from 250 to 790 ind. ha^{-1}, taking into account the large size classes of sea cucumbers exclusively (Rogers-Bennett and Ono, 2001; Bruckner, 2006b). Until 1996, an average of 75 percent of the annual catch came from the trawl fishery in southern California. Between 1997 and 1999, the dive fishery accounted for 80 percent of the take. Recent surveys showed a 50–60 percent decline in abundance between 1994 and 1998, but no correlation

[4] *Source*: http://www.sakhalin2.ru/en/documents/doc_lender_eia_12.pdf.

was noted between decline in abundance and data on landings. The only increase in abundance (39 percent) was noted at two no-take reserves (Rogers-Bennett and Ono, 2001; Schroeter *et al.*, 2001).

In Washington State, CPUE from 1983 show that the minimum in those days was around 75 kg diver h^{-1} and the maximum around 130 kg diver h^{-1} (Bradbury, 1994). Data recorded in this state at the time clearly demonstrates the impact of the fishery on the stock of sea cucumbers with a decline of up to 70 percent: sea cucumber densities were around 0.35 ind. m^{-2} before the fishery opened and dropped to 0.1 ind. m^{-2} just after (Bradbury, 1994).

In Oregon, *P. californicus* densities are between 0.1 to 0.22 ind. m^{-2} at depths of 80 to 130 m (Bruckner, 2006b). In Washington, the CPUE between 1995 and 1998 (Bradbury, 1999) varied from 56 to a maximum of 80 kg diver h^{-1}.

3.5.4 Parastichopus parvimensis

Schroeter *et al.* (2001) provided density values for *P. parvimensis* in California before the onset of the fishery, varying from 0.2 to 21.1 individuals per 10 m^2. The onset of the fishery in the same site marked a drastic decline (by 33–83 percent) in population densities. The CPUE expressed in kg per boat per day varied from a site to another between 1993 and 1999. For instance, it fluctuated around 453 kg in San Nicolas Island, around 340 kg in Santa Barbara Island, and around 181 kilograms in San Miguel (Schroeter *et al.*, 2001).

3.5.5 Accuracy of stock assessments

Accurate estimates of available stocks are instrumental to the sustainable management of any fishery. At present, the assessment of commercial stocks is largely based on fishery-dependent data such as catch per unit effort (CPUE), which can generate biased indicators of stock abundance (Schroeter *et al.*, 2001). It has been stated that marine reserves could play a valuable role in providing more reliable information on stock assessment (Schroeter *et al.*, 2001). A number of studies on fish have revealed higher abundances and larger sizes in no-take reserves when compared to the outside; others have reported higher abundance and larger fish in a given site after its designation as a no-take zone (Rowley, 1992). However, such comparisons between harvested and control areas are not appropriate in the case of most environmental impact studies; the best assessment is obtained by comparing population changes in affected and pristine areas before and after the impact occurs (Underwood, 1993; Stewart-Oaten and Bence, 2001). Regrettably, the so-called BACI "before-after, control-impact" design is generally inapplicable to established fisheries because data on stocks before fishing and/or data from a suitable control site are not available. The situation may be different with emerging fisheries, for which no–take reserves that use proper monitoring programs offer a powerful means of assessing the status of harvested stocks and their likelihood for sustainability (Schroeter *et al.*, 2001).

3.6 Catches

This section presents traditional and commercial activities as well as landing figures for three of the four species covered by this review, and for all countries except the Russian Federation where records were not found (Gudimova, E., Polar Research Institute of Marine Research and Oceanography, Murmansk, Russian Federation, personal communication). When possible, the landing data are further detailed by state or province (for USA and Canada), otherwise it is provided for the whole country, such as Iceland.

Various methods have been used to fish sea cucumber in North America: both *P. californicus* and *P. parvimensis* are mainly harvested by divers (Bruckner, 2006a). In Alaska, harvesting is restricted to hand picking. Divers use SCUBA tanks or surface supplied air and gather the animals in mesh bags for transport to the surface (Woodby, Smiley and Larson, 2000). In Washington, Oregon and California, both dive and trawl fisheries are being used; in British Columbia divers use converted salmon boats between 7 and 13 m. The typical operation involves three to four people, including two divers and a tender (Muse, 1998). All Canadian and USA fisheries on the east coast use dredging gear similar to the one used in Maine and these activities can therefore be considered more "industrialized". Scallop gears were initially used, but they were eventually replaced by the modified urchin drags that appear to yield less bycatch an have a lighter impact on the seabed (DFA, 2002a). In Iceland, harvesters use the same gear as in eastern Canada (Olafsson, K., Reykofninn-Grundarfirdi ehf, personal communication), whereas in the Russian Federation, the scallop dredging gear is used for sea cucumber harvesting (Gudimova, Gudimov and Collin, 2004).

Harvest methods for subsistence by aboriginal communities in southeast Alaska, while incorporating new technology, do not differ very much from traditional methods. Mathews, Kookesh and Bosworth (1990) indicated that harvests were confined to low tide using poles or bare hands for collection. Most harvesters were found to use a wooden pole of 2.5–5 m. Fish hooks of various types and lengths were often tied to end of the pole to lift the sea cucumbers closer to shore. Another type of pole used consisted of a bamboo pole 2.5–3 m long with sixteen penny nails driven through, and used to rake the cucumbers in. Other harvesters used cockle rakes, small dipnets, and brailers (Mathews, Kookesh and Bosworth, 1990). Another harvest method is incidental catch by commercial seining in which aboriginal communities are involved. Seine nets used in shallow waters sometimes become fouled with sea cucumbers, which then have to be removed manually from the nets. Individuals working on the seine boats may retain the cucumbers for family use (Mathews, Kookesh and Bosworth, 1990).

Sea cucumber landings in the USA have been inconstant but generally rising since 1990 (Figure 8). Catches on the Atlantic coast (*C. frondosa*) are much more important and therefore account for the general trend with a peak around 9 000 tonnes in 2003 and a subsequent decrease to values around 4 000 tonnes in 2004 and 2005 (Figure 8). Landings for *P. californicus* on the Pacific coast are an order of magnitude lower with maxima of 600–700 tonnes in 1994–1995 and values of ca. 200–400 tonnes in 2003–2005 according to FAO Fishery Statistics (Figure 8). However, a recent report (Ess, 2007) states that guideline harvest levels throughout southeast Alaska have stayed healthy since 2000 at around 600–700 tonnes (Table 3).

FIGURE 8
Commercial landings for the United States of America

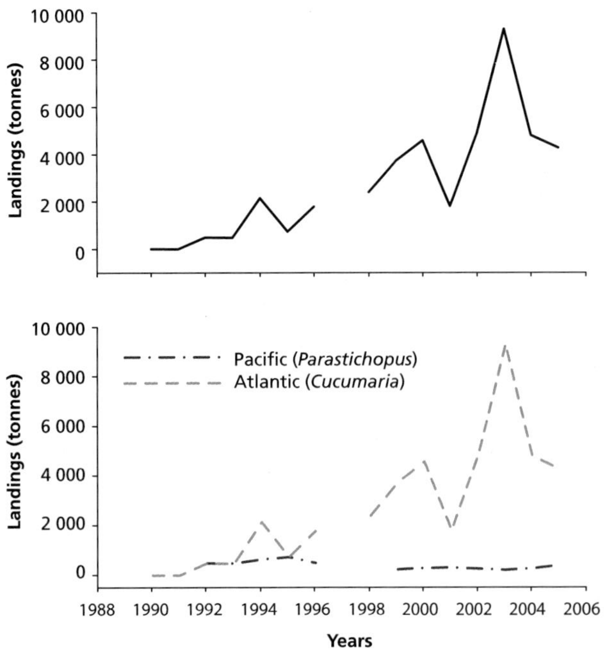

Source: FAO Fishery Statistics 1950–2005.

Official landings for Iceland are very fragmentary, with reported values of 2 tonnes in 1995, 27 tonnes in 2003 and 208 tonnes in 2005 (FAO Fishery Statistics 1950–2005). However, there is information that catches roughly totalized 50 tonnes in 2003, 200 tonnes in 2004, 300 tonnes in 2005 and 150 tonnes in 2006 (Olafsson, K., Reykofninn-Grundarfirdi ehf, personal communication).

Unlike the USA and Iceland, Canada does not keep separate records of landings for sea cucumbers but rather chooses to include them within the larger category of benthic invertebrates (e.g. FAO Fishery Statistics). Nevertheless, data were obtained from provincial representatives of the various DFO offices to illustrate the trend of this growing activity in Canada. Landings for *P. californicus* in British Columbia are shown in Figure 9, whereas landings of *C. frondosa* in New Brunswick, Newfoundland, Nova Scotia and Scotia-Fundy regions are detailed in Figure 10.

Unfortunately, data from the Russian Federation remain unconfirmed despite numerous attempts to contact authorities involved in fisheries regulation and management.

3.7 Management measures in place

There have been some problems associated with sea cucumber fisheries, including the potential for overexploitation, habitat damage, removal of bycatch species, illegal fishing, and conflicts with other resource usages. In many locations, sea cucumber landings increased rapidly following the exploratory phase, but some fisheries have also experienced decreases in the number of sea cucumbers landed per dive or per trawl, and overexploitation has been reported in some areas (Bruckner, 2005).

FIGURE 9
Commercial landings of *Parastichopus californicus* in British Columbia (Canada)

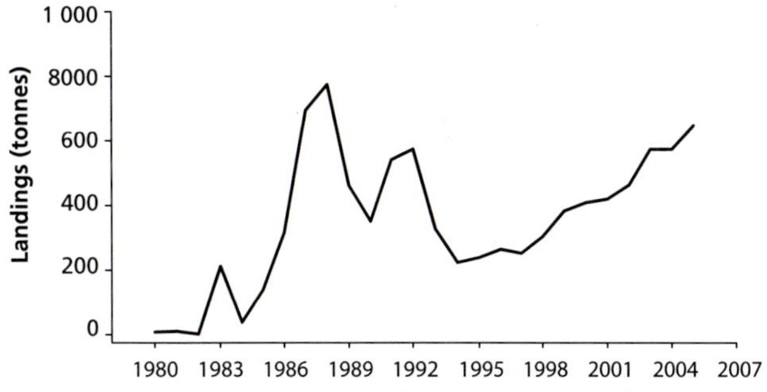

FIGURE 10
Landings of *Cucumaria frondosa* in the Atlantic provinces of Canada based on data provided by DFO and DFA

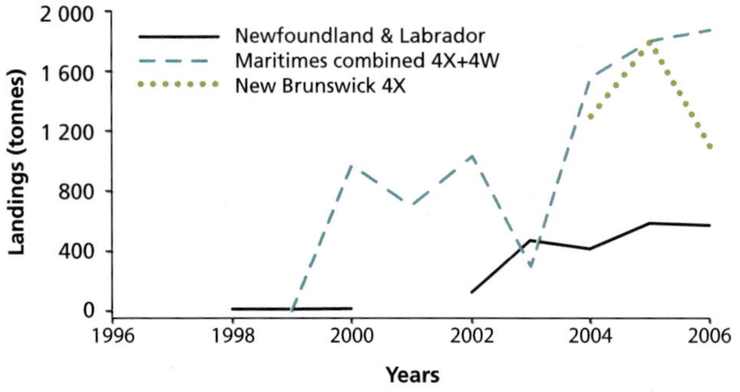

3.7.1 United States of America

In the USA, sea cucumber fisheries can be separated into those occurring in state waters, each being managed by individual states; and those going on in the 200-mile zone off the coast, which are managed by NOAA Fisheries in coordination with Regional Fishery Management Councils (Bruckner, 2006b). In the present document, we will discuss only cases where state management is involved.

Current sea cucumber management practices in Alaska have provided sustainable harvests and consistent quality. Divers rotate their effort between 16 harvest areas, some of which are divided into more than 20 sub-areas in an effort to maintain sustainability throughout the fishing grounds (Ess, 2007).

In southeast Alaska, each fishing area is run on a three year rotation and harvested at a rate of 6 percent a year (Meredith, B., Alaska Department of Fish & Game, personal communication). Thus, in any given area that is open for harvest, approximately 18 percent of the surveyed biomass is removed; this area will remain closed for the next two years. Before the sea cucumber fishing season opens on the first Monday of October, dive surveys are conducted in each of the harvest areas to be targeted that year. Woodby, Smiley and Larson (2000) added two conservative measures into the development of the harvest rate managed by the Alaska Department of Fish and Game: (1) a 50 percent reduction to account for the possibility that the model assumption is incorrect, and (2) an approximate reduction of 30 percent to account for sampling error in the assessment survey. A third safety measure consists of counting only sea cucumbers occurring at depths above 15 m in the population size estimates. Ruccio and Jackson (2002) also mention that the Alaskan fishery of *P. californicus* closes down during the spawning season. Finally, harvest divers can only obtain permits for sea urchins or for sea cucumbers but not both (Bruckner, 2006b).

In Washington, the *P. californicus* fishery is a year-round dive operation. From 1971 to 1986, the fishery was opened in all areas. However, following signs of overfishing, the Washington State Department of Fisheries implemented a rotational harvest from 1987 to 1992 (Bradbury, 1994). Now, the experimental trawl fishery is closed during soft–shell Dungeness crab period and in shrimp areas. An experimental trawl fishery occurs in specific locations using beam trawl gear with a maximum beam width or otter trawl with a minimum mesh size (Bruckner, 2005). For each area, the quota is determined using surplus production models and estimates of biomass from catch-effort data, video surveys, and dive surveys. Bruckner (2005) stated that the current management in Washington includes spatial closures, licensing of collectors and an annual quota. Seven area closures for the dive fishery have been established in the current management plan and trawling is prohibited in shrimp areas. Other regulations for trawlers include no fishing in waters less that 20 m deep. Divers in Washington need to submit their logbooks every month with data on date, depth, location and amount of captures (Bruckner, 2006b).

In Oregon, the fishery of *P. californicus* is conducted by divers. Harvest by trawl required an experimental gear permit until 2003. The target species has been placed under category B of the Development Fisheries List (species in category B are underutilized and are not under another state or federal management plan, and have not shown the potential to be a viable fishery), which includes species with less potential for viable fisheries; a permit is therefore no longer required (McCrae, 1994).

In California, both *P. californicus* and *P. parvimensis* are harvested. A special permit was required for sea cucumber harvest in 1992–1993. Separate permits for each gear type and a limit on the total number of permit were implemented in 1997. There are no restrictions on catch (Rogers-Bennett and Ono, 2001; Schroeter *et al.*, 2001). Bruckner (2005) adds that a maximum of 111 dive permits and 36 trawl permits were issued in 1997, and this declined to 95 divers and 24 trawl permits in 2004. There are no restrictions on catch but trawling is prohibited in some conservation areas and along the shore of most islands (Bruckner, 2005).

In Maine, for the fishery of *C. frondosa*, the "urchin drag" gear used is limited to 167 cm in width and 670 cm in length, and a head bail constructed of <3.8 cm round steel stock (Feindel, 2002). Regulations were implemented in March 2000 under the 1999 Sustainable Development of Emerging Fisheries Act (Bruckner, 2005). The act included restrictions to limit the fishing season with closure between 1 July and 31 September. It also defined gear size, and established a maximum number of endorsements, with licences given only to individuals that had sold 100 000 kilograms during the previous year. Licensed fishermen are required to submit logbooks that provide information on catch, time at sea, area fished and catch value (Feindel, 2002). No incidental takes are allowed, only take through targeted, licensed fishery (Bruckner, 2006b).

3.7.2 Canada

In British Columbia, the annual *P. californicus* fishery lasts for about three weeks in October, when muscle weight is greatest and the animals have reabsorbed their internal organs (DFO, 2002). The commercial fishery is a small limited–entry (i.e. 85 licences) dive fishery that is managed by individual quota (DFO, 2002). Of the 85 licences delivered in 2003 in British Columbia, 15 belonged to native peoples (James, 2003). Quotas are calculated by multiplying estimates of shoreline length, sea cucumber density, individual weight and harvest rate (DFO, 2002). The total allowable catch can be increased in surveyed areas when the measured density estimates are calculated to be higher than the precautionary baseline density estimate. Only 25 percent of the coast is open to the commercial fishery. Up to an additional 25 percent may be used to conduct research, and the remaining 50 percent is closed to harvesting until biologically-based management is possible. Abundance surveys and experimental fisheries are being conducted to estimate biomass and evaluate exploitation rate options (DFO, 2002). All landings are monitored by an independent industry-funded firm, dockside landings are only at designated ports, and licence holders pay a fee (Bruckner, 2006b).

In general, the fisheries for *C. frondosa* are still at the early exploratory stage, and it is difficult to predict whether the available stocks can sustain long term exploitation (Therkildsen and Petersen, 2006). Sea cucumbers tend to be highly vulnerable to overfishing, and their harvest throughout the world are generally characterized by overexploitation (Conand, 2004; Uthicke and Conand, 2005). The relatively low value of *C. frondosa* on the market and the high cost of labour in North America require harvesting to be significant in order to be cost effective (Therkildsen and Petersen, 2006) generating a real concern over the sustainability of this resource.

The Scotia-Fundy sector of Nova Scotia and New Brunswick has issued five exploratory licenses, and a number of regulations are in place, including a closed season from April to December, and a series of closed areas (Therkildsen and Petersen, 2006). In addition, no night time fishing and simple gear restrictions are in place as well as a minimum sea cucumber size of 10 cm in length for landings (DFO, 2005b).

In New Brunswick the fishing season is from 1 January to 31 March, in order to protect the breeding animals and the juveniles (Coulombe and Campagna, 2006). Unfortunately, there is no quota or size restriction. An observer is present on the fishing vessels 20 percent of the time to validate the logbook and bring samples back. The two fishermen involved collected about 500 tonnes each for about 9–10 tonnes a day (Coulombe and Campagna, 2006).

In Newfoundland, although the *C. frondosa* fishery is still in an exploratory phase, the season is set by industry with processor(s) and fishers agreeing to an opening and closing date (Barrett, L., DFA, personal communication). Closure is usually after all fishers have harvested their respective 76 500 kilograms quota (for a total of 612 000 kilograms, wet weight, free water included). In 2006, the fishery opened around the first of June and closed in the second week of September. In previous years, harvesting went on longer into the season. The processor usually tries to work it such that employees get the maximum hours of labour without too much overlap with other species. The minimum length requirement for harvesting is 10 cm. This is measured at dockside while the animal still retains at least part of its free water content. Undersized catches are usually estimated by length but with some level of uncertainty considering the plastic nature of the animals. In Newfoundland and Labrador, the government has issued 10 exploratory licenses (i.e. 8 in NAFO subdivision 3Ps and 2 in NAFO area 4R) in 2006 and reports an exploratory annual quota of ca. 612 tonnes in 3Ps and ca. 90 tonnes in 4R (Barrett, L., DFA, personal communication).

3.7.3 Iceland

No real management plans are in place in Iceland. Kari Olafsson (Reykofninn-Grundarfirdi ehf, personal communication) indicated that their fishing boats have cameras onboard recording as they enter the fishing area and as they leave the area but, unfortunately, they have no means of comparing these images adequately.

3.7.4 Russian Federation

There is apparently a basic management plan in the Russian Federation for both *C. frondosa* (Barents Sea) and *C. japonica* (North Pacific) according to Elena Gudimova (Polar Research Institute of Marine Research and Oceanography, Russian Federation, personal communication). *C. frondosa* from the Barents Sea is subject to a fixed fishing season from August to December. For *C. japonica*, a fishing season is established with no-take-zones and gear restriction. Unfortunately, we could not obtain more information.

4. TRADE

Commercial markets for *P. californicus* exist locally in cities such as New York and San Francisco (USA), and Vancouver (Canada), as well as in Asia (Mathews, Kookesh and Bosworth, 1990). The main exports of *P. californicus* from British Columbia are toward China Hong Kong SAR, Taiwan Province of China, mainland China and Republic of Korea, and toward some domestic markets in Canada and in the USA (DFO, 2002; Lambert, 2006). In Alaska, up to the early 2000s, the meat strips were send to China Hong Kong SAR, and the skins to Taiwan Province of China (Ess, 2007). A few years ago, markets shifted and everything now goes to China. The muscles are routed to southern China via China Hong Kong SAR while skins go to markets in northern China (Ess, 2007).

While guideline harvest levels throughout southeast Alaska have stabilized at around 600–700 tonnes since 2001, ex-vessel prices (i.e. prices received by a captain for the catch) have been varying (Table 3). The 2000–2001 season saw ex-vessel offers climb to values around USD 4.91 kg^{-1}, watched them slip to a low of USD 2.77 kg^{-1} during the 2002–2003 season and enjoyed a rebound to values around USD 4.50 kg^{-1}since 2004 (Ess, 2007) (Table 3).

Bruckner (2005) indicated that sea cucumbers harvested in the Pacific USA are exported to China Hong Kong SAR, Taiwan Province of China, mainland China and Republic of Korea. Exports of *Parastichopus* (both species) were worth USD 0.07 kg^{-1} in early the 1980s, a value that increased in 2005 to USD 0.82 kg^{-1}; processed sea cucumbers sold for up to USD 9 kg^{-1} (Bruckner, 2005). Barsky (1997) reported that ex–vessel prices in California averaged USD 0.32 kg^{-1} in 1995.

Ex–vessel prices per tonne in British Columbia were USD 195 in 1984, USD 854 in 1989 and USD 1 821 in 1994 (Muse, 1998). Native people in British Columbia earned CAD 200 000 (ca. USD 199 500) in 2003 from the fishery of *P. californicus* (James, 2003).

Most *C. frondosa* products are sold on Asian markets. In Maine, fishermen are paid USD 0.05–0.06 per unprocessed sea cucumber. Internal muscle bands and dried body wall are the primary exports, and are currently worth about USD 1.59 kg^{-1}. These sea cucumbers are processed in Maine or shipped to Seattle for processing. Subsequently they are shipped to mainland China, China Hong Kong SAR, Republic of Korea, Singapore, Taiwan Province of China and Japan. Chinese markets in San Francisco, New York and other cities purchase a portion of the catch (Bruckner, 2005). Although *C. frondosa* is smaller than most tropical commercial species and has a thinner body wall, a market acceptance for this species developed over the past decade. However, *C. frondosa* remains a low grade species that yields much lower prices than most commercial species (Therkildsen and Petersen, 2006). Generally, fishermen in Maine are paid USD 0.10 kg^{-1} and the dried product yields only about USD 6–10 kg^{-1} (Feindel, 2002).

In Canada, similar prices are paid per kilo to the fishermen; however precise figures are not always openly divulged by local processors. In 2006, the price paid to the harvester in Newfoundland was USD 0.48 kg^{-1} for wet weight and was based on a company dockside sampling for "free water" retention (Barrett, L., DFA, personal communication). In New Brunswick, the 2006–2007 prices were around USD 0.27 kg^{-1} for gross weight including water content, whereas in Nova Scotia, the reported price for 2006 was ca. USD 0.32 kg^{-1} for what comes out of the vessels (Barrett, L., DFA, personal communication). The same source indicated that the New Brunswick and Newfoundland sea cucumber catches are offloaded at dockside, monitored by the DFO, and then trucked to the plant for processing. Plant workers make between USD 7 and 23 h^{-1}; on average, they can extract ca. 23 kilograms of muscle/meat per day and receive wages of ca. USD 3.78 kg^{-1} (Barrett, L., DFA, personal communication). Another source reported that fishers in New Brunswick received between USD 0.13–0.18 kg^{-1} for non-drained weight of fresh product and the main buyers were in China Hong Kong SAR (Coulombe and Campagna, 2006). There is only one trader in Newfoundland with its main office in Toronto and a possible transit via British Columbia before exportation toward Asia. Products in decreasing order of value are: the dry skin (USD 13–15 kg^{-1}), the muscle bands or "meat" (USD 9–10 kg^{-1}) and the dried aquapharyngial bulbs or "flowers" (USD 3.75 kg^{-1}) (Barrett, L., DFA, personal communication).

In Iceland, the fishermen are paid a fixed monthly salary for the duration of the fishery. The products are shipped by Reykofninn-Grundarfirdi ehf to buyers in Asia (China, Korea, Singapore and Malaysia) either directly or through private sales companies (Icelandic or foreign) (Olafsson, K., Reykofninn-Grundarfirdi ehf, personal communication). Prices paid to fishers and exporters in Iceland were not divulged. No data are available for Russian Federation.

5. SOCIO–ECONOMIC IMPORTANCE

5.1 Commercial fisheries

North American fisheries are threatened by the same factors that are affecting global fisheries, including collapsing fish and invertebrate stocks due to overexploitation and to environmental degradation resulting from pollution and habitat loss. With the closure of some fisheries and the decline of others, fishers in North America are avidly looking toward new resources to increase or maintain their incomes. In some remote locations such as isolated villages on the eastern and western coasts of Canada, a number of social issues arise from employment shortages. As resources from the sea become depleted, new generations of young workers are leaving their home town to seek out new opportunities.

For thousand of communities along the coast of North America, fishing and fish processing are still important components of the local economy, although far fewer now depend exclusively on fishing than in the past. Generally speaking, communities close to the sea, and not just those that rely on the fishery, maintain strong ties to their region and demonstrate fierce independence. And even though fishing activities and processing operations are highly modernized, these occupations have always been seen as a way of life.

Mason (2002) mentioned that with the collapse of the Newfoundland cod stocks the local fishers of various regions were the group most closely identified with the tragic social and economic fallout from the closure of the fishery, as many families and communities were almost totally dependent on the fishery for their livelihoods.

Many fishing communities have developed other activities alongside fishing, like peat harvesting and processing, agriculture, mining and forestry. Tourism, which has been a natural addition to the economy of coastal regions, offers some interesting potential in the long term. There are, however, a number of maritime areas that have never been able to create an alternative to their fishing industry, and they continue to benefit from only a small percentage of the value of their fisheries resource (Beaudin, 1997).

There is no real dependence on sea cucumber harvests anywhere in North America, these fisheries either being experimental or too small to provide substantial income, at least when compared to full-fledged fishing activities. Sea cucumber fisheries are seldom considered essential but one interesting complement is the "recycling" of sea cucumber bycatch to complete net income. Sea cucumbers are a valuable resource that, when harvested in a sustainable fashion provides or could provide important income to many North American fishers and their families. While sea cucumber landings are relatively small in comparison to that of other sea products, they have the potential to become a significant source of income. Most industries related to sea cucumbers in Canada represent just a small portion of total provincial gross domestic product, although they contribute to job creation and the overall economy. The value of sea cucumber harvests may appear negligible in the context of the entire North American fishing industry but it is valuable on a small scale to those individuals who depend on it at times when economic opportunities are limited especially in remote, coastal outposts.

5.2 Traditional harvests

Sea cucumbers are of continuing importance to aboriginal people of the First Nations in Canada, who harvest them for food, as well as for social and ceremonial use. The level of sea cucumber harvest by First Nations in British Columbia is unspecified at this time. A small recreational fishery occurs for sea cucumbers, however landings for this activity remain unknown (DFO, 2002). The Inuit of Belcher Island (Nunavut, northern Canada) are traditionally a marine-oriented community and their diet includes the sea cucumber *C. frondosa* (Wein, Freeman and Markus, 1996). Ninety-three of 102 families poled confirmed that they consumed *C. frondosa*, making sea cucumber a dominant component of their diet (Wein, Freeman and Markus, 1996).

In the USA, more precisely in Alaska, the sea cucumber *P. californicus* was found to be one of the main species harvested for subsistence in native communities. Sea cucumbers were traditionally used as a food in the Northwest Coast culture area (Mathews, Kookesh and Bosworth, 1990). Not surprisingly, commercial harvests account for the largest quantity of sea cucumbers removed from the field, although subsistence harvest by autochthones can be quantified. Mathews, Kookesh and Bosworth (1990) provide data expressed as the total number of 25-litre buckets of gutted sea cucumbers harvested per southeast Alaska community, including estimated total weight harvested in kg. These data are interesting in that they allow comparison between what is harvested through the recently opened fisheries and through the long lasting traditional activities. For instance, in the village of Klawock, 239 buckets of gutted sea cucumbers were harvested in 1987, corresponding to ca. 4 336 kg. In Wrangell, the harvest was estimated at 3 773 kg; in Thorne Bay, 2 213 kg; in Sitka, 2 068 kg; in Craig, 980 kg; in Hollis, 925 kg; in Metlakatla, 707 kg; and in Edna Bay, 526 kg. According to Mathews, Kookesh and Bosworth (1990) some community members placed sea cucumbers high on their list of important subsistence resources and harvested them in substantial numbers. Others harvested sea cucumbers only if they were readily available and if other more desired subsistence species were not available or became limited. In 1990, yearly harvests ranged from 150 to 700 sea cucumbers per household in the studied communities (Mathews, Kookesh and Bosworth, 1990). In 1987, over half of the households in Hollis reported gathering sea cucumbers, and nearly 58 percent of the households reported making use of sea cucumbers. Furthermore, 50 percent of all households in Edna Bay reported harvesting sea cucumbers in 1987, in amounts averaging 1.4 buckets or 25 kilograms of gutted sea cucumbers per household. In Klawock, in 1987, 18.3 percent of the households harvested sea cucumber, and 34.1 percent reported using sea cucumbers for food (Mathews, Kookesh and Bosworth, 1990).

According to Mathews, Kookesh and Bosworth (1990) sea cucumber management in the Alaskan region illustrates a typical issue having to do with the effects of commercialization of a resource which has a history of being used solely for subsistence. One aspect of this problem relates to the likelihood of overfishing the resource to the point of rendering it unusable for subsistence activities. In the specific case of sea cucumbers, it is likely that the resource might be removed from subtidal areas, through a commercial dive fishery, to a degree that is commercially sustainable but with the insidious result of decreasing the availability of sea cucumbers in locations where they are gathered as subsistence food. In this respect, identifying "refuge" or "community use" areas, not accessible to commercial fishing, can become a desirable management tool to protect important subsistence harvest areas. Of course, this approach would require a better definition of subsistence harvest areas. Another management option would be to set harvest quotas in specific areas that would not remove enough of the available sea cucumber biomass to disrupt subsistence activities. This would require awareness of the size of the sea cucumber stock in a given area, prior to any harvesting. These and other management approaches are being considered as a part of the sea cucumber management planning effort in Alaska (Mathews, Kookesh and Bosworth, 1990).

Another key aspect of the commercialization issue relates the involvement of the community in the commercial development of the resource, in order to provide local economic benefits and some degree of local management control (Mathews, Kookesh and Bosworth, 1990). A governance policy that would favour local participation in the fishery might include weekly fishery closures, or requirements for regional, shore-based processing. This would entail a significant shift in perception as commercial sea cucumber exploitation is largely viewed in many communities as a phenomenon whereby large, well-financed, out-of-state vessels and crews would compete with local harvesters (subsistence and commercial), for a limited resource. In certain areas, informal local agreements have been used to avoid such conflicts among user groups (Mathews, Kookesh and Bosworth, 1990).

6. CURRENT PROJECTS AND STUDIES

According to Bruckner (2006b) there is no sea cucumber farming in the USA for either *P. californicus* or *P. parvimensis*. However, preliminary investigations of co-culture scenarios have been undertaken: Ahlgren (1998) examined the assimilation of fouling debris from salmon net pens by *P. californicus*, and Paltzat *et al.* (2006) studied the co-culture of the Pacific oyster *Crassostrea gigas* and *P. californicus*.

Therkildsen and Petersen (2006) recently mentioned that an attempt was being made at starting sea cucumber farming in Norway. In western Canada, aquaculture of *P. californicus* was reportedly attempted some years ago by using settled juveniles on oyster culture lines and relocating them elsewhere for growing (Sutherland, 1996); however, the results of this initiative could not be found. In eastern Canada, to our knowledge, no land-based rearing has been attempted so far, except for the reproduction and growth of *C. frondosa* within the boundaries of academic research, such as the M.Sc. project of Justin So under the supervision of Annie Mercier (Ocean Sciences Centre, Memorial University of Newfoundland). The latter project is also looking at various biological parameters, including gene flow of *C. frondosa* over its geographical range, health status of populations found on various substrates, and impact of predators on the natural stocks. This study is partly being funded by government agencies and the local industry wishing to acquire the biological data required by the novel Emerging Fisheries Policy before opening a new commercial fishery. In this context, various other projects related to *C. frondosa* are being developed in Newfoundland. Stock assessments are being made by the DFO and DFA in collaboration with various fishers on the St. Pierre Bank (Newfoundland Grand Banks).

The Fisheries Technical University in Vladivostok in the Russian Federation has a sea cucumber breeding programme at Slavanka, which rears holothurians that reproduce in mariculture facilities much more rapidly than wild counterparts. The sea cucumbers are exported to the Republic of Korea and China (Johnson, 2004). Kari Olafsson (Reykofninn-Grundarfirdi ehf, personal communication) indicated that there was no aquaculture programme involving *C. frondosa* in Iceland.

7. ADDITIONAL THREATS TO SEA CUCUMBER POPULATIONS

Because of intense naval traffic, the coasts of the Atlantic and Pacific oceans in the northern hemisphere are susceptible to oil spills which can impact the ecosystem, including sea cucumber populations, especially within the shallow subtidal or intertidal zones where sea cucumbers commonly dwell.

In eastern Canada, it is common to setup micro–electric power plants on rivers, thereby modifying the hydrodynamics and indirectly unbalancing the primary productivity, which can consequently interfere with the spawning, development and recruitment of some coastal benthic populations. Deforestation, which can increase runoff of terrigenous material, could also have an impact on already established populations. In some areas like the Hudson Bay (northern Canada), an increase in freshwater runoff generates decreases in salinity in some coastal locations, which could affect some populations of sea cucumbers (*C. frondosa*). These bursts of freshwater are generally the results of hydroelectric power dams producing more or less electricity, thus alternately retaining and releasing large amounts of freshwater. Populations from Hudson Bay, the Nunavut Hudson Bay Inter-Agency Working Group, or Nunavuummi Tasiujarjuamiuguqatigiit Katujiqatigiingit, presented a body of evidence from scientific and traditional knowledge, suggesting that the decreases in salinity, increases in ice cover and changes in the bays' currents could affect sea cucumber populations among other resources (Nunatsiaq-News, 2006).

Industrial countries like USA, Canada and Russian Federation are releasing significant amounts of industrial and domestic wastes in coastal waters, which can affect the sustainability of the resource by reducing their reproductive capability.

Overall, the biggest threat remains overfishing despite the regulations or management programmes that are implemented. Illegal fishing has been reported to occur in California, whereas some regions like southeast Alaska, are concerned about potential conflicts between subsistence and commercial harvesters (Bruckner, 2005).

Furthermore, the use of dredges can destroy the benthic ecosystem and directly or indirectly affect the sea cucumber populations and other resources as well including non target species. The DFO (DFO, 2002) mentioned that conservative management was appropriate until the impact of localized removal of *P. californicus*, a detritus feeder, on seabed ecosystems could be ascertained.

All North American commercial species, to our knowledge, have long embryonic and larval periods and present very slow growth rates, therefore minimizing their potential for aquaculture and stock enhancement programmes, unless novel more cost-effective applications can be discovered and developed to make the investment in aquaculture worthwhile (e.g. pharmaceuticals, nutraceuticals).

ACKNOWLEDGEMENTS

We would especially like to acknowledge the contribution of Lewis Barrett (DFA, Government of Newfoundland & Labrador), and to thank Verónica Toral-Granda, Marcelo Vasconcellos, Alessandro Lovatelli, Chantal Conand and Poh Sze Choo for their valuable support and comments. This review has greatly benefited from input and assistance graciously offered by numerous people all over the world: Jérome Laurent (Centre spécialisé des pêches, Grande–Rivière, Quebec), Bruno Lévesque (DFO, Quebec), Sherrylinn Rowe (DFO, Halifax, Nova Scotia), Valerie Bradshaw (DFO, Halifax, Nova Scotia), Don Stansbury (DFO, St. John's, Newfoundland & Labrador), Ken Budden (Fogo Island Co–op, Newfoundland & Labrador), Guy Parsons (DFA, Government of Newfoundland & Labrador), Justin So (Ocean Sciences Centre, Memorial University, Newfoundland & Labrador), Eric Way (DFO, Newfoundland & Labrador), Bo Meredith (Alaska Department of Fish & Game, USA), Kari P. Olafsson (Reykofninn-Grundarfirdi ehf, Iceland), Elena Gudimova (Polar Research Institute of Marine Research and Oceanography, Murmansk, Russian Federation).

LITERATURE CITED

ADFG. 1990. Community profile database, Division of subsistence. Alaska Department of Fish and Game, Juneau, Alaska, USA.

Ahlgren, M.O. 1998. Consumption and assimilation of salmon net pen fouling debris by the red sea cucumber *Parastichopus californicus*: implication for polyculture. *J. World Aquacult. Soc.,* 29: 133–139.

Barsky, K. 1997. Fishery in California in 1995. *SPC Beche-de-Mer Inf. Bull.,* 9: 12.

Beaudin, M. 1997. L'adaptation économique des régions maritimes de pêche: Le cas des communautés du golfe du Saint-Laurent (Economic adjustment of maritime fishery regions: communities in the Gulf of St. Lawrence). Ph.D. Thesis, Université de Nantes.

Bradbury, A. 1994. Sea cucumber dive fishery in Washington State: an update. *SPC Beche-de-Mer Inf. Bull.,* 6: 15–16.

Bradbury, A. 1999. Holothurian fishery in Washington (USA). *SPC Beche-de-Mer Inf. Bull.,* 12: 25.

Bradshaw, V., Ryan, C. & Cooper, C. 1991. Experimental harvesting and processing of sea cucumber (*Cucumaria frondosa*) in Saint Mary's Bay, Nova Scotia, June 1990. Fisheries Development and Fishermen's Service Division, Scotia-Fundy Region, DFO, Project report 161, 14 pages.

Brinkhurst, R.O., Linkletter, L.E., Lord, E.I., Conners, S.A. & Dadswell, M.J. 1975. A preliminary guide to the littoral and sublittoral marine invertebrates of Passamaquoddy Bay. Special Publication, Huntsman Marine Science Centre, New Brunswick (Canada).

Britten, M. 1906. Holothurien aus dem Japanischen und Ochotskischen Meere. *Bull. l'Acad. Imper. Sci. St Petersburg,* 5: 123–157.

Bruckner, A.W. 2005. The recent status of sea cucumber fisheries in the continental United States of America. *SPC Beche-de-Mer Inf. Bull.,* 22: 39–46.

Bruckner, A.W. 2006a. Management. *In*: *Proceedings of the CITES workshop on the conservation and conservation strategies and practices for sea cucumbers.* A.W. Bruckner (Ed.), NOAA Technical Memorandum NMFS-OPR-34, USA, pp. 74–103.

Bruckner, A.W. 2006b. Sea cucumber population status, fisheries and trade in the United States. *In*: *Proceedings of the CITES workshop on the conservation of sea cucumbers in the families Holothuriidae and Stichopodidae.* A.W. Bruckner (Ed.), NOAA Technical Memorandum NMFS-OPR-34, USA, pp. 192–202.

Cameron, J.L. & Fankboner, P.V. 1986. Reproductive biology of the commercial sea cucumber *Parastichopus californicus* (Echinodermata: Holothuroidea). I. Reproductive periodicity and spawning behavior. *Can. J. Zool.,* 64: 168–175.

Cameron, J.L. & Fankboner, P.V. 1989. Reproductive biology of the commercial sea cucumber *Parastichopus californicus* (Stimpson) (Echinodermata: Holothuroidea). II. Observations on the ecology of development, recruitment, and the juvenile life stage. *J. Exp. Mar. Biol. Ecol.,* 127: 43–67.

Campagna, S. & Hand, C. 2004. Baseline density estimates from sea cucumber (*Parastichopus californicus*) surveys conducted in British Columbia, Canada.

Campagna, S., Lambert, J. & Archambault, P. 2005. Abondance et distribution du concombre de mer (*Cucumaria frondosa*) et prises accidentelles obtenues par dragage entre Matane et Cap-Gaspé (Québec) en 2004, 70.

Chenoweth, S. & McGowan, J. 1997. Sea cucumbers in Maine: Fishery and Biology. Maine Department of marine resources. (www.state.me.us\dmr\publications\seacucumbers.html).

Churchill, G.L. 1996. Harvesting, processing and marketing sea cucumbers. *CAFID project #2148*: 6 pages.

Coady, l.W. 1973. Aspects of the reproductive biology of *Cucumaria frondosa* (Gunnerus, 1770) and *Psolus fabricii* (Düben and Koren, 1846) (Echinodermata: Holothuroidea) in shallow waters of the Avalon Peninsula, Newfoundland. M.Sc. Thesis, Memorial Univeristy of Newfoundland.

Conand, C. 2001. Overview of sea cucumbers fisheries over the last decade – What possibilities for durable management? *In*: *Echinoderms 2000*, pp. 339–344.

Conand, C. 2004. Present status of world sea cucumber resources and utilisation: An international overview. *In*: *Advances in sea cucumber aquaculture and management*. A. Lovatelli, C. Conand, S. Purcell, S. Uthicke, J.-F. Hamel and A. Mercier (eds), FAO Fisheries Technical Paper. No. 463, Rome, pp. 13–23.

Conand, C. & Sloan, N.A. 1989. World fisheries for echinoderms. *In*: *Marine Invertebrate fisheries: their assessment and management*. J.F. Caddy (Ed.), Wiley, Chichester, pp. 647–663.

Coulombe, F. & Fillion, N. 2002. Portrait de l'industrie maritime des espèces émergentes au Québec en 1999. MAPAQ-DIT, No 144, 60 page.

Coulombe, F. & Campagna, S. 2006. Mission d'observation sur la pêche au concombre de mer au Nouveau-Brunswick du 29 au 31 août 2004. Les publications de la direction de l'innovation et des technologie. MAPAQ DIT, No 27, 4 pages.

Courtney, W.D. 1927. Fertilization in *Stichopus californicus*. *Publ. Puget Sound Stn. Univ. Wash.*, 5: 257–260.

David, V.M.M. & MacDonald, B.A. 2002. Seasonal biochemical composition of tissues from *Cucumaria frondosa* collected in the Bay of Fundy, Canada: feeding activity and reproduction. *J. Mar. Biol. Assoc. U.K.*, 82: 141–147.

Desrosiers, A., Lavallée, J., Michaud, J.-C. & Wilson, J.R. 1989. Évaluation du potentiel commercial du concombre de mer (Cucumaria frondosa), Rapport 32. Programme de développement de l'Est du Québec, Ministère des Pêches et Océans région du Québec, division du développement, 116 p.

DFA. 2002a. Sea cucumber dive harvest – gear design and testing. Department of Fisheries and Aquaculture, Goverment of Newfoundland and Labrador, Project FDP 424–3.

DFA. 2002b. Sea cucumber – Fall, 2002, Dive Harvest – Bonavista and Notre Dame Bays; Towed gear design and testing – Dantzic Point and St. Pierre Bank. Department of Fisheries and Aquaculture, Government of Newfoundland and Labrador, Fisheries diversification program, Emerging Fisheries Development, Project report FDP 424-3.

DFA. 2003. Sea cucumber survey. Strait of Belle Isle, September to November 2003. Department of Fisheries and Aquaculture, Government of Newfoundland and Labrador.

DFA. 2004. Sea cucumber survey Strait of Belle Isle August – September 2004. Department of Fisheries and Aquaculture, Government of Newfoundland and Labrador, 31 p.

DFO. 2002. Giant red sea cucumber. DFO Can. Sci. Advis. Sec. Stock Status Rep. C6-10, 5 pages.

DFO. 2005a. Integrated fisheries management plan: sea cucumber by dive. (www-ops2.pac.dfo-mpo.gc.ca).

DFO. 2005b. Conservation harvesting plan: Scotian Shelf sea cucumber *Cucumaria frondosa*, DFO New Brunswick, 8 pages.

Dolmatov, I. & Mokretsova, N.D. 1995. Morphology of different-age pentactula larvae of the holothuroid, Cucumaria japonica (Dendrochirota, Holothuroidea). *Hydrobiol. J.; Zool. Zh*, 31(1): 83–91.

Dooley, T., McDonald, P. & Rumboldt, M. 1995. Report on Chinese Canadian International food expedition.

Dulepov, V., Scherbatyuk, A. & Jiltsova, L. 2003. Investigation of bottom habitant in Great Peter Bay using semi AUV TSL. Institute for Marine Technology, Russia, p. 182–187.

Eckert, G.L., Pinkard, D.R., Reed, D.C. & Schroeter, S.C. 2001. Larval development, settlement, and juvenile growth of the sea cucumber, *Parastichopus parvimensis*. *Gulf Mex. Sci.*, 19: 187.

Ess, C. 2007. North Pacific Sea Cucumber, stable, high-quality product, demand bode well for sea cucumber divers. *Market reports. (www.nationalfisherman.com/month-content.asp?ItemID=1662&pcid=343&cid=347&archive=yes).*

Falk-Petersen, I.B. 1982. Breeding season and egg morphology of echinoderms in Balsfjorden, Northern Norway. *Sarsia,* 67: 215–221.

Fankboner, P.V. & Cameron, J.L. 1985. Seasonal atrophy of the visceral organs in a sea cucumber. *Can. J. Zool.* 63: 2888–2892.

Feindel, S. 2002. Status of the Maine sea cucumber (*Cucumaria frondosa*) fishery. Report to the standing legislative committee on marine resources, Maine Department of Marine Resources, USA, 35 pages.

Geirsdottir, M. & Stefansdottir, K.A. 2004. Icelandic sea cucumbers, chemical and functional properties, 1–23.

GERMA. 1989. Évaluation du potentiel commercial du concombre de mer (*Cucumaria frondosa*). Programme de développement de l'Est du Québec. Ministère des Pêches et Océans, Rapport # 32.

Girard, M., Belanger, J., ApSimon, J.W., Garneau, F.X., Harvey, C. & Brisson, J.R. 1990. Frondoside A. A novel triterpene glycoside from the holothurian *Cucumaria frondosa. Can. J. Chem./J. Can. Chim,* 68: 11–18.

Glenn, E.M. 1966. Adjuvant-induced arthritis: effects of certain drugs on incidence, clinical severity and biochemical changes. *Am. J. Vet. Res.,* 27: 339–352.

Grant, S.M. 2006. Biological resource assessment of the orange footed sea cucumber (*Cucumaria frondosa*) occurring in the Strait of Belle Isle. Centre for Sustainable Aquatic Resources Fisheries and Marine Institute, Memorial University of Newfoundland, 33 p.

Gudimova, E.N. 1998. Prospects of sea cucumber fishing in the Barents Sea. *Rybn Khoz,* 3: 45–46.

Gudimova, E. & Antsiferova, A. 2006. Gonadal morphology and oogenic stages of Cucumaria frondosa from the Barents Sea: comparative aspect. 12th International Echinoderm Conference, New Hampshire, USA, 7–11 August (Book of Abstracts).

Gudimova, E.N., Gudimov, A. & Collin, P. 2004. A study of the biology for fishery in two populations of *Cucumaria frondosa*: In the Barents Sea (Russia) and in the Gulf of Maine (USA). *In*: *Echinoderms Munchen: Proceedings of the 11th Internation Echinoderm Conference* T. Heinzeller and J.H. Nebelsick (eds), A. A. Balkema Publishers, Leiden, pp. 269–275.

Hamel, J.-F. & Mercier, A. 1995. Spawning of the sea cucumber *Cucumaria frondosa* in the St Lawrence Estuary, eastern Canada. *SPC Beche-de-Mer Inf. Bull.,* 7: 12–18.

Hamel, J.-F. & Mercier, A. 1996a. Evidence of chemical communication during the gametogenesis of holothuroids. *Ecology,* 77: 1600–1616.

Hamel, J.-F. & Mercier, A. 1996b. Early development, settlement, growth, and spatial distribution of the sea cucumber *Cucumaria frondosa* (Echinodermata: Holothuroidea). *Can. J. Fish. Aquat. Sci.,* 53: 253–271.

Hamel, J.-F. & Mercier, A. 1996c. Gamete dispersion and fertilisation success of the sea cucumber *Cucumaria frondosa. SPC Beche-de-Mer Inf. Bull.,* 8: 34–40.

Hamel, J.-F. & Mercier, A. 1998a. Recent developments in the commercialisation of the northern sea cucumber Cucumaria frondosa. *SPC Beche-de-Mer Inf. Bull.,* 11: 21–22.

Hamel, J.-F. & Mercier, A. 1998b. Diet and feeding behaviour of the sea cucumber *Cucumaria frondosa* in the St Lawrence estuary, eastern Canada. *Can. J. Zool.,* 76: 1194–1198.

Hamel, J.-F. & Mercier, A. 1999a. Recent developments in the commercialization of the northern sea cucumber *Cucumaria frondosa. SPC Beche-de-Mer Inf. Bull.,* 11: 21–22.

Hamel, J.-F. & Mercier, A. 1999b. Mucus as a mediator of gametogenic synchrony in the sea cucumber *Cucumaria frondosa* (Holothuroidea: Echinodermata). *J. Mar. Biol. Assoc. U.K.,* 79: 121–129.

Hebert, K. & Pritchett, M. 2002. Report to the board of fisheries, miscellaneous dive fisheries. Regional Information Report No IJ02-45. Alaska Department of Fish and Game, Division of commercial fisheries, Junea, Alaska (USA), 10.1–10.30.

James, M. 2003. Native participation in British Columbia commercial fisheries – 2003. Ministry of Agriculture, Food & Fisheries, British Columbia, Canada, 29 p.

Johnson, T.L. 2004. Russian Far East Fisheries. Alaska Sea Grant program, University of Alaska Fairbanks. 25 p.

Jordan, A.J. 1972. On the ecology and behavior of *Cucumaria frondosa* (Echinodermata: Holothuroidea) at Lamoine Beach, Maine. Ph.D. dissertation, University of Maine at Orono.

Ke, P.J., Hirtle, R.W. & Smith-Hall, B. 1984. Technical studies on resource utilization: Atlantic sea cucumber (*Cucumaria frondosa*), Technical Studies Interim Manuscript, No 504, October, 51 p.

Ke, P.J., Smith-Lall, B., Hirtle, R.W. & Kramer, D.E. 1987. Technical studies on resource utilization of Atlantic sea cucumber (*Cucumaria frondosa*). *Can. Inst. Food Sci. Technol. J.*, 20: 4–8.

Kinosita, T. & Sibuya, S. 1941. On the spawning season of the commercial Hokkaido sea cucumber *Cucumaria japonica. Bulletin of the Japanese Society of Scientific Fisheries,* 10: 166–170.

Lacalli, T. 1981. Annual spawning cycles and planktonic larvae of benthic invertebrates from Passamaquoddy Bay, New Brunswick. *Can. J. Zool.,* 59: 433–440.

Lambert, P. 1984. British Columbia marine faunistic survey report Holothurians from the Northeast Pacific. *Canadian Tech. Rep. Fish. Aquat. Sci., N.* 1234: 1–30.

Lambert, P. 1997. Sea cucumbers of British Columbia, Southeast Alaska and Puget Sound. UBC Press, Vancouver, Canada, 166 pp.

Lambert, P. 2006. National report – Canada. Proceedings of the CITES workshop on the conservation of sea cucumbers in the families Holothuriidae and Stichopodidae. *In: NOAA Technical Memorandum NMFS-OPR-34, USA, 203-204.* A.W. Buckner (ed.).

Levin, V.S. 1995. Japanese sea cucumber *Cucumaria japonica* in the eastern seas of Russia. *SPC Beche-de-Mer Inf. Bull.,* 17: 18–19.

Levin, V.S. & Gudimova, E.N. 2000. Taxonomic interrelations of holothurians *Cucumaria frondosa* and *C. japonica* (Dendochirotida, Cucumariidae). *SPC Beche-de-Mer Inf. Bull.,* 13: 22–29.

Mashanov, V.S. & Dolmatov, I.Y. 2000. Developmental morphology of a holothurian, *Cucumaria japonica* (Dendrochirota, Holothuroidea), a species with accelerated metamorphosis. *Invertebr. Reprod. Dev.,* 37: 137–146.

Mason, F. 2002. The Newfoundland cod stock collapse: a review and analysis of social factors. Electronic Green Journal 17.

Mathews, V., Kookesh, M. & Bosworth, R. 1990. Subsistence harvest and use of sea cucumber in southeast Alaska, Division of subsistence, Alaska Department of Fish and Game. Technical Paper No. 190, 1–43.

McCrae, J. 1994. Oregon development species. Sea cucumber *Parastichopus* sp. *Oregon Department of Fish and Wildlife.*

McEuen, F.S. 1988. Spawning behaviors of northeast Pacific sea cucumbers (Holothuroidea: Echinodermata). *Mar. Biol.,* 98: 565–585.

Medeiros-Bergen, D.E. & Miles, E. 1997. Recruitment in the holothurian *Cucumaria frondosa* in the Gulf of Maine. *Invertebr. Reprod. Dev.,* 31: 123–133.

Medeiros-Bergen, D.E., Olson, R.R., Conroy, J.A. & Kocher, T.D. 1995. Distribution of holothurian larvae determined with species-specific genetic probes. *Limnol. Oceanogr.,* 40: 1225–1235.

MI. 2004. Habitat utilization and density of the sea cucumber (*Cucumaria frondosa*) on St. Pierre Bank, Newfoundland: preliminary observations using a towed camera sled. Marine Institute, Memorial University of Newfoundland, Canada, 14 p.

MI. 2006. Habitat utilization and density of the sea cucumber (*Cucumaria frondosa*) on St. Pierre Bank, Newfoundland: observations using a towed camera sled in 2004 and 2005. Marine Institute, Memorial University of Newfoundland, Canada, 16 p.

Mokretsova, N.D. & Koshkaryova, L.N. 1983. Some data about the distribution, ecology and reproduction biology of *Cucumaria japonica* (Semper). *In*: *Mariculture in the Far East*, Vladivostok, pp. 46–51.

Mortensen, T. 1932. The Godthaab expedition 1928. *In*: *Echinoderms*, pp. 46, Fig. 47.

Mulcrone, R.S. 2006. Animal diversity, class Holothuroidea (sea cucumbers). (www.animaldiversity.ummz.edu\site\accounts\information\Holothuroidea.html).

Muscat, A.M. 1982. Aspects of the biology of the sea cucumber, *Parastichopus parvimensis*: a developing commercial fishery (R-RD-14). The Planning and Management of California's Coastal Resources. USC Sea Grant Institutional Program 1981-82, Trainee Report, University of Southern California, pp. 25–27.

Muscat, A.M. 1983. Population dynamics and the effect on the infauna of the deposit-feeding holothurian *Parastichopus parvimensis* (Clark). Ph.D. Thesis, University of Southern California, Los Angeles, California.

Muse, B. 1998. Management of the British Columbia sea cucumber fishery, Alaska commercial fisheries entry commission, Juneau, Alaska, and US Department of the Interior.

Naidenko, V.P. & Levin, V.S. 1983. Rearing of the holothurian *Cucumaria japonica* in the laboratory. *Biol. Morya (Vladivost.)*, 61–65.

Nunatsiaq-News. 2006. Hydro scheme will change environment, Inuit tell Hydro-Québec. *Nunatsiaq News, 12 May 2006.*

Oganesyan, S.A. & Grigorjev, G.V. 1996. Reproductive cycle of the holothuroid *Cucumaria frondosa* in the Barents Sea. Echinoderms: San Francisco. Proceedings of the Ninth International Echinoderm Conference, San Francisco, California, August 6–10.

Ondine. 1991. Prospection des marches pour le concombre de mer (*Cucumaria frondosa*). Rapport Ministère des Pêches et des Océans dans le cadre du programme essais et expérimentations. Les Entreprises Ondine Inc., Sainte-Anne-des-Monts (Québec), 37 p.

Organesyan, S.A. & Grigorjev, G.V. 1998. Reproductive cycle of the holothuroid *Cucumaria frondosa* in the Barents Sea. *In*: *Proceedings of the 9th International Echinoderm Conference, San Franscisco (USA)*. R. Mooi and M. Telford (eds), A.A. Balkema Publishers, pp. 489–492.

Paltzat, D., Pearse, C., Barnes, P. & McKinley, S. 2006. Biological feasibility of suspended Pacific oyster and California sea-cucumber polyculture. *Aquaculture update, April, 6 pages.*

Panning, A. 1949. Versuch einer Neuordnung der Familie Cucumariidae (Holothuroidae, Dendrochirota). *Zool. Jarb. Abt System. Okol. Geogr. Tiere.*, 53: 404–470.

Panning, A. 1955. Bemerkungen uber die Holothurien – Familie Cucumariidae (ordnung Dendrochirotida). *Mitt. Hamburg Zool. Mus. Inst.*, 53: 33–47.

Reunov, A.A., Bodrova, O.V. & Eliseikina, M.G. 1994. Structure of the testis and changes shown during the annual reproductive cycle in *Cucumaria japonica* (Echinodermata: Holothuroidea). *Invertebr. Reprod. Dev.*, 25: 83–86.

Rogers-Bennett & Ono, D.S. 2001. Sea cucumbers. California living marine resources: a status report, California Department of Fish and Game, 131–134.

Rowley, R.J. 1992. Impacts of marine reserves on fisheries: a report and review of the literature. *Science and Research Series. Vol. 51, Department of Conservation, Wellington, New Zealand.*

Ruccio, M.P. & Jackson, D.R. 2002. Red sea cucumber and green sea urchin commercial fisheries management plans for the registration area J (westward region), 2002–03. Regional Information Report No. 4K02-44. Alaska Department of Fish and Game Division of Commercial Fisheries. 14 p.

Runnström, J. & Runnström, S. 1919. Uber die Entwicklung von *Cucumaria frondosa* Gunnerus und *Psolus phantapus* Strussenfelt. Bergens Museums Aarbok 1918-1919. *Naturvidenskabelig raekkenr.*, 5: 1–99.

Saveljeva, T.S. 1941. On the fauna of holothurians of the Far East seas. *Invest. Of the USSR seas,* 1: 73–103.

Schroeter, S.C., Reed, D.C., Kushner, D.J., Estes, J. & Ono, D.S. 2001. The use of marine reserves in evaluating the dive fishery for the warty sea cucumber (*Parastichopus parvimensis*) in California, USA. *Canadian Journal of Fisheries & Aquatic Sciences,* 58: 1773.

Seatech Investigation Services Ltd. 1981. *Studies of handling, processing and quality assurance techniques for underutilized species: sea cucumbers.*

Shahidi, F. 2006. Chemical composition of fresh and processed Newfoundland sea cucumber (*Cucumaria frondosa*), Report for the Department of Fisheries and Aquaculture, Government of Newfoundland and Labrador, 1–19.

Sheffield, G., Fay, F.H., Feder, H. & Kelly, B.P. 2001. Laboratory digestion of preys and interpretation of walrus stomach contents. *Mar. Mamm. Sci.,* 17: 310–330.

Singh, R., MacDonald, B.A., Thomas, M.L.H. & Lawton, P. 1999. Patterns of seasonal and tidal feeding activity in the dendrochirote sea cucumber *Cucumaria frondosa* (Echinodermata: Holothuroidea) in the Bay of Fundy, Canada. *Mar. Ecol. Prog. Ser.,* 187: 133–145.

Singh, R., MacDonald, B.A., Lawton, P. & Thomas, M.L.H. 2001. The reproductive biology of the dendrochirote sea cucumber *Cucumaria frondosa* (Echinodermata: Holothuroidea) using new quantitative methods. *Invertebr. Reprod. Dev.,* 40: 125–141.

Sloan, N.A. 1986. Les concombres de mer. Le monde sous-marin, Pêches et Océans Canada.

Smiley, S. 1988. The dynamics of oogenesis and the annual ovarian cycle of *Stichopus californicus. Biol. Bull.,* 175: 79–93.

Smiley, S., McEuen, F.S., Chaffee, C. & Krishnan, S. 1991. Echinodermata: Holothuroidea. *In*: *Reproduction of marine invertebrates, Vol. VI.* A.C. Giese, J.S. Pearse and V.B. Pearse (eds), The Boxwood Press, Pacific Grove, California, pp. 663–750.

Spainhour, C.B. 2005. Natural products. Drug Discovery Handbook. S.C. Gad (ed.), John Wiley and Sons, Inc., pp. 12–63.

Stewart-Oaten, A. & Bence, J.R. 2001. Temporal and spatial variation in environmental impact assessment. *Ecol. Monogr.,* 71: 305–337.

Sutherland, I. 1996. Sea cucumber culture developments on the west coast of Canada. *SPC Beche-de-Mer Inf. Bull.,* 8: 42–43.

Sutterlin, A.M. & Waddy, S. 1975. Harvesting, processing, and composition of sea cucumber meal as a possible substitution for herring meal in trout diets. *Fish. Mar. Serv. Res. Dev. Tech. Rep.,* 531: 1–8.

Therkildsen, N.O. & Petersen, C.W. 2006. A review of the emerging fishery for the sea cucumber *Cucumaria frondosa*: Biology, policy, and future prospects. *SPC Beche-de-Mer Inf. Bull.,* 23: 16–25.

Underwood, A.J. 1993. On beyond BACI: sampling designs that might reliably detect environmental disturbances. *Ecol. Appl.,* 4: 3–15.

Uthicke, S. & Conand, C. 2005. Local examples of beche-de-mer overfishing: an initial summary and request for information. *SPC Beche-de-Mer Inf. Bull.,* 21: 9–14.

Valtysson, H.P. 2001. The sea around Icelanders: catch history and discards in Icelandic waters. Part Two: North-eastern North Atlantic. The Marine Research Institute, University of Akureyri, Iceland, 36 p.

Wein, E.E., Freeman, M.M.R. & Markus, J.C. 1996. Use of and preference fro traditional food among the Belcher Island Inuit. *Arctic,* 49: 256–264.

Winter, C.A. & Nuss, G.W. 1996. Treatment of adjuvant arthritis in rats with anti-inflammatory drugs. *Arthritis. Rheum.,* 9: 394–404.

Woodby, D., Smiley, S. & Larson, R. 2000. Depth and distribution of *Parastichopus californicus* near Sitka, Alaska. *Alaska Fishery Research Bulletin,* 7: 22–32.

Yingst, J.Y. 1982. Factors influencing rates of sediment ingestion by *Parastichopus parvimensis* (Clark), an epibenthic deposit-feeding holothurian. *Estuarine, Coastal and Shelf Science,* 14: 119–134.

Zhou, S. & Shirley, T.C. 1996. Habitat and depth distribution of the red sea cucumber *Parastichopus californicus* in a southeast Alaska bay. *Alaska Fishery Research Bulletin,* 3: 123–131.

Precautionary management of *Cucumaria frondosa* in Newfoundland and Labrador, Canada

Jean-François Hamel
Society for the Exploration and Valuing of the Environment
Portugal Cove-St. Philips, Newfoundland & Labrador, Canada
E-mail: jfhamel@seve.cjb.net

Annie Mercier
Ocean Sciences Centre, Memorial University
St. John's, Newfoundland & Labrador, Canada
E-mail: amercier@mun.ca

Hamel, J.-F.; Mercier, A. 2008. Precautionary management of *Cucumaria frondosa* in Newfoundland and Labrador, Canada. In V. Toral-Granda, A. Lovatelli and M. Vasconcellos (eds). Sea cucumbers. A global review of fisheries and trade. *FAO Fisheries and Aquaculture Technical Paper*. No. 516. Rome, FAO. pp. 293–306.

SUMMARY

The sea cucumber *Cucumaria frondosa* has been reported almost circumglobally in the high latitudes of the northern hemisphere. The earliest fisheries emerged in the late 1980s in Maine (United States of America), and exploitation has only started in the past decade or so on the eastern coast of Canada. This chapter highlights the unprecedented efforts made by the provincial government of Newfoundland and Labrador to implement ecosystem-based guidelines at the onset of this emerging fishery, under the Canadian New Emerging Fisheries Policy. Mandatory information required before the commercial exploitation can be sanctioned include scientific data on the standing stocks (e.g. density, abundance), and on the biology of the species (e.g. reproductive cycle, size at sexual maturity, growth rate, predatory pressure). The present document outlines the rationale and objectives of the New Emerging Fisheries Policy, and the contributions made to date by the main stakeholders wishing to develop a sustainable fishery for *C. frondosa* in Newfoundland and Labrador.

1. INTRODUCTION AND HISTORICAL PERSPECTIVE

This chapter is unique compared to other hotspots covered in the present document. Instead of identifying and discussing an area where sea cucumber fisheries are particularly important, threatened or otherwise affected, this section will present the case study of the province of Newfoundland and Labrador (Canada), where an exemplary management policy is being implemented before opening a commercial fishery for the sea cucumber *Cucumaria frondosa*.

The area under study (Figure 1) is a historically well known fishing ground that has suffered from mismanagement at the regional and international levels in the recent past (Steele, Andersen and Green, 1992; Finlayson, 1994; Hannesson, 1996; Hutchings, 1996; Sinclair, 1996; Walters and Maguire, 1996). While the Grand Banks that lie south of the Canadian island of Newfoundland have enjoyed a great popularity ever since the colonization era (it is alleged that when John Cabot sailed by in 1497, cods were so abundant they could be caught with a basket) the past half century has taken its toll on the seemingly bottomless resources.

In the aftermath of World War II, the famed fishing area saw harvesters from more than two dozen nations regularly join with Canadians and Americans in search of various fish species. This intense pressure and the use of increasingly sophisticated fishing technology soon threatened to decimate most of the once abundant fish stocks off Canada's east coast. Despite early conservation efforts, which included the creation of the International Convention for the Northwest Atlantic Fisheries (ICNAF) in 1949, Canada's claim to a 200-mile economic zone in 1977 and the ensuing reduction in foreign fishing, resources continued to suffer. Even as the yields reached a historic high on the east coast in the early 1980s, Canadian scientists were advising the implementation of a total ban on fishing of certain species (i.e. capelin and cod) on the Grand Banks. The advice was not heeded. Moreover, members of the Northwest Atlantic Fisheries Organization (NAFO), formed in 1979 to replace ICNAF, were

FIGURE 1

Map of eastern Canada showing the area under study and the NAFO subdivisions

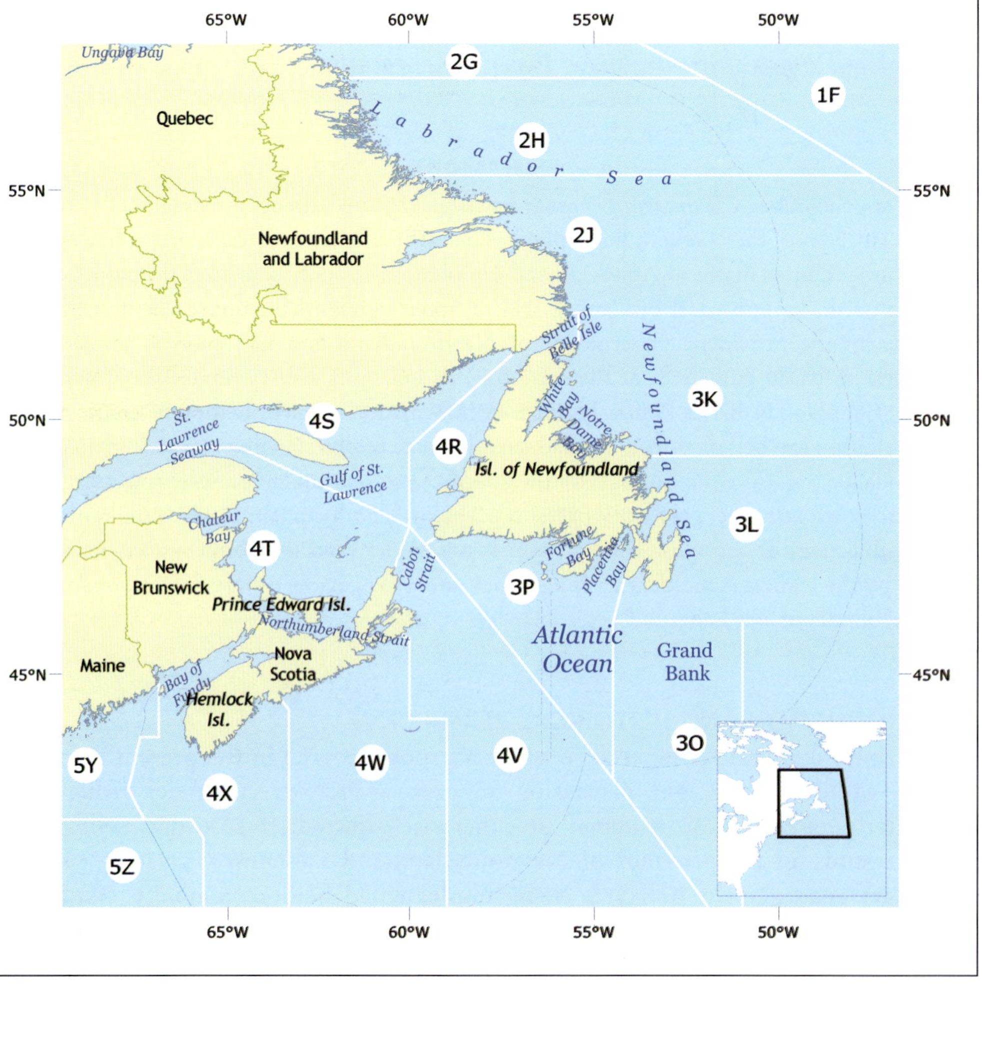

dissatisfied with their quotas inside the 200-mile zone and therefore turned to the Nose and Tail of the Grand Banks outside the limit, registering their vessels in non-NAFO countries to bypass international rules. Between 1986 and 1991, non-NAFO vessels allegedly caught more than 200 000 tonnes of fish in areas outside Canada's jurisdiction, which were important nursery grounds for a variety of species. Furthermore, from 1986 to 1992, the European Union set its own quotas and reportedly caught five times the NAFO quota.

While Canada feared and criticized overfishing by foreign fleets, it also ignored warnings from its own scientists and an independent assessment of the northern cod stocks which stated that the resource was threatened and fishing should be reduced. Canada and other NAFO members still maintained their quotas from fear of generating unemployment that would have eventually shut down the industry. However, by 1992, with the collapse of several fish stocks becoming imminent, the Canadian Government implemented major quota reductions that effectively closed down much of the fishing industry in eastern Canada. In 1994, the introduction of the Coastal Fisheries Protection Act empowered the Department of Fisheries and Oceans (DFO) officers to board and arrest foreign vessels in violation of conservation measures. NAFO eventually agreed to a partial ban on some stocks, but to this day many fishing nations are still uncommitted to the principle of sustainable development. Meanwhile, strong of lessons hard learned, Canada is steering toward ecosystem-based precautionary approaches to improve fisheries management. This new vision is crucial to Newfoundland and Labrador, where the marine resources and the fisheries that they sustain are considered a dynamic and significant component of life, affecting communities and rural areas throughout the province. This vital sector is evolving to include a broader base of species and activities, including aquaculture, sport fishing and value-added processing, all aiming to create a vibrant, self-sufficient and sustainable industry in coastal and rural communities.

One of the main lessons learned by Canada over the past decades is that insufficient biological knowledge of a species can seriously impede its sustainable management (Walters and Maguire, 1996). A key objective is therefore to implement a precautionary approach and support the collection of sound scientific data before authorizing any new fisheries at the commercial level. This novel approach is described under the "New Emerging Fisheries Policy". While it presents appealing goals and expectations on paper, actually enforcing this new policy requires an exceptional and unusual collaboration at various levels of the hierarchy to include harvesters, scientists, the private sector and government authorities. For this reason, it has taken several years for anything concrete to unravel. Among the unexploited or underutilized marine species that thrive off the Canadian east coast, and specifically around Newfoundland and Labrador, sea cucumbers *C. frondosa* have been identified as a target resource and were therefore recently listed under the Emerging Fisheries Policy.

The next sections will present information on how Newfoundland and Labrador is preparing its new sea cucumber fishery. Detailed data on *C. frondosa* (i.e. population status, landings and management, trade and socio-economic importance) have already been provided in the chapter "Population status, fisheries and trade of sea cucumbers in temperate areas of the Northern Hemisphere" and consequently only a few relevant data will be summarized here for the sake of clarity. This will be followed by a description of the "New Emerging Fisheries Policy". The whole initiative and the preliminary results compiled at various levels (i.e. government, academic, private sector and fishing industry) will be evaluated and discussed. Finally the chapter will close by reflecting on the policy and whether or not CITES listing could assist in the development of good management practices to avoid overfishing and promote conservation of sea cucumber populations in eastern Canada.

2. THE SPECIES

The sea cucumber *C. frondosa* is the only species of temperate-polar holothurian with commercial interest that can be harvested along the eastern coast of North America. Although the harvest of *C. frondosa* is still at the exploratory phase in most areas, a commercial fishery with recorded landings began in 1988 in the state of Maine (United States of America). Despite local management efforts, signs of overfishing are already being reported along the coast (Bruckner, 2005; Hamel and Mercier, this volume). The rapid development of the sea cucumber fishery in Maine has raised a question similar to the one addressed by the new emerging fisheries policy in Canada: "How should we be harvesting this resource to ensure a continuous supply in the future?" The answer will be difficult to ascertain. Just as the sea cucumber has been ignored as a commercial product in the past, it has been ignored as a subject of study by fishery scientists. Our knowledge of the biology of most sea cucumbers and how their populations will respond to fishing is rudimentary at best.

The main difference this time in Newfoundland and Labrador is that the authorities will have the tools and means to evaluate this fishery before its onset, an exemplary mode of management based on good science and respectful harvesting practices. The challenge is to gather everyone around the same table in order to design a plan of action, to find the funds, to achieve the objectives, and finally to translate the results into management practices.

3. MANAGEMENT SCHEME

3.1 New Emerging Fisheries Policy

In Canada, this policy applies to all new fisheries undertaken in marine or freshwater areas where the DFO manages the fishery. Aboriginal communities retain their constitutional right, second only to conservation, to fish for food, social and ceremonial purposes. Box 1 summarizes the key contents of the "New Emerging Fisheries Policy" published in 2001 to replace the "Emerging Fisheries Policy" established 1997[1].

3.2 The case of *Cucumaria frondosa*

The significant decline experienced by traditional fisheries in Canada has created opportunities for the study of several new and emerging species. So far, very few species have been identified under the New Emerging Fisheries Policy. In the Maritimes, common whelk fisheries are being developed using some level of precautionary measures. Other fisheries (e.g. toad crab, rock crab, shrimp beam trawl) have also been managed pursuant to the policy as it came into effect in 1998, though they did not benefit from the thorough science that is now being applied. The case of *C. frondosa* in Newfoundland and Labrador appears to be unique, with only the Atlantic hagfish being developed under comparable guidelines.

As soon as the commercial potential of *C. frondosa* was identified by the Department of Fisheries and Aquaculture (DFA) of the Government of Newfoundland and Labrador, an exemplary collaboration was established at all levels in order to work efficiently toward the success of the emerging sea cucumber fishery. The ultimate goal is to acquire data deemed necessary by the DFO to make a guided decision concerning this resource. Sustainability for future generations is at the heart of current preoccupations. To achieve this goal, the DFO, the DFA, the Marine Institute (Memorial University), the Ocean Sciences Centre (Memorial University), the Canadian Center for Fisheries Innovation (CCFI), the processing industry (fish plants) and numerous harvesters are joining hands in an unprecedented effort to promote a new way of managing the fishing industry in Canada, specifically in Newfoundland and Labrador. Below is a summary

---

[1] http://www.dfo-mpo.gc.ca/communic/fish_man/nefp_e.htm.

of the work conducted so far and the studies that still need to be completed before the official opening of a commercial sea cucumber fishery in this Canadian province.

For the sea cucumber *C. frondosa* everything began in 1988 as a small fishery emerged on the east coast of the United States of America. This fishery developed rapidly because of demand from foreign markets. Sea cucumbers are harvested using a bottom trawl gear similar to the one used in the scallop fishery; vessels are typically between 13 and 30 m in length. Landings for Maine increased from 2 900 kilograms in 1990, to over 4.3 million kilograms in 2000, and then dropped to 1.5 million kilograms in 2001. The highest value recorded was USD 614 937 in 2000. In Atlantic Canada, *C. frondosa* is mainly harvested as a result of bycatch from scallop fisheries between May and November. In 1992, the Resource Development Associates (RDA) prepared a proposal following discussions with the DFO concerning a significant biomass of sea cucumbers which had been identified in nearshore Newfoundland waters. This first approach defined a broad objective to undertake a preliminary assessment of market potential for *C. frondosa* in selected target markets (RDA, 1992).

In 1997, the Canada/Newfoundland Cooperation Agreement for Fishing Industry Development (CAFID) carried out an exploratory sea trial for sea cucumbers in the province and concluded that this species had potential based on biomass and accessibility of the resource. In 2002, the DFA, under the Fisheries Diversification programme, surveyed prospective sea cucumber fishing grounds and alternative harvesting methods. This study estimated that certain areas of the province could maintain a viable industry and recommended that alternative methods of harvest be used, such as a modified beam trawl, green sea urchin drag and modified Labrador scallop buckets.

Between 1997 and the early 2000s, the DFO and the DFA surveyed several areas along the coast of Newfoundland and Labrador for sea cucumbers, tested different harvesting methods and investigated ways of minimizing bycatch (Figure 2).

Since then, there have been a number of companies who have carried out preliminary processing and marketing studies of *C. frondosa*. In 2001, new funding was approved through the Fisheries Diversification Programme to conduct sea cucumber surveys in various areas of the province: dive harvests in Notre Dame Bay; scallop bucket harvests in St. Pierre Bank, Placentia Bay and Labrador (DFA, 2001). During 2002, Fogo Island Co-op processed upwards of 160 000 kilograms from the dive harvest fishery in the Notre Dame Bay and Bonavista Bay areas and from a towed gear fishery along the Burin Peninsula. Efforts toward the development of a sea cucumber fishery continued in 2002 with three projects completed by DFA Fisheries Development Officers. These projects included two dive harvest projects and a tow gear design and testing project. The dive harvest studies were carried out in Bonavista Bay and Notre Dame Bay while the tow study was conducted at Dantzic Point and near Fortune and St. Pierre

FIGURE 2
Scientific surveys carried out jointly by the Department of Fisheries and Oceans and the Memorial University on St. Pierre Bank, Newfoundland and Labrador, Canada

COURTESY OF J.-F. HAMEL AND A. MERCIER

BOX 1
The Canadian New Emerging Fisheries Policy

The New Emerging Fisheries Policy was developed to clearly lay out the requirements that must be met and the procedures to follow before a new fishery can be initiated. A cornerstone of the new policy is provision for the establishment of a scientific base with which stock responses to new fishing pressures can be assessed. Not only does the Emerging Fisheries Policy provide applicants with a transparent process to follow, it also gives DFO managers a procedure that can be applied fairly and consistently. This policy is also precautionary in its approach to the development of new fisheries. The objective is to diversify fisheries and increase economic returns while ensuring conservation of the stocks and realizing the sustainable use of fisheries resources. These new guidelines are not restricted to sea cucumbers or to Newfoundland and Labrador; they apply to all prospective species throughout Canada.

The New Emerging Fisheries Policy is guided by the following principles: (i) New fisheries must provide for a reasonable scientific basis for their management. The process by which new fisheries will be managed must include the requirement for stock assessment information in the early stages; (ii) New fisheries should contribute positively to the economical viability of a fishery enterprise on an ongoing basis; (iii) Under the proposed policy all requests from applicants must include proposals that outline research, management and conservation approaches as well as cost of these approaches; (iv) Conservation will not be compromised, a precautionary approach will guide decision making. Information on the abundance, distribution, and productivity of the target species is identified as the key scientific requirement for development of precautionary management strategies; (v) The potential impact or interaction of any new fishery or its gear on associated or dependent species, fishing or gear type and on habitat will be assessed; (vi) Based on biological and environmental information, including input from industry, provinces/territories and the public, DFO will establish conservation standards, set conditions for harvest, and monitor their application; (vii) Users are accountable for compliance with conservation standards for meeting harvesting conditions; and, (viii) In allocating DFO's personnel and financial resources, priority will be given to the research, management and enforcement of established fisheries.

As a general rule, new fisheries involve three stages:

Stage I: The preliminary feasibility stage. (Scientific Licences) The objective of this stage is to determine if harvestable quantities of the species/stock known to be present in a particular fishing area exist, if the species/stock can be captured by a particular gear type, identify multi-species and habitat impacts, if markets exist and, the best approach for proceeding further, e.g. to Stage II.

Stage II: The commercial and stock assessment stage. (Exploratory Licences) The objective of this stage is to determine whether a species/stock can sustain a commercially viable operation and to collect biological data in order to build a preliminary database on stock abundance and distribution.

Stage III: The commercial fishery stage. (Commercial Licences) This stage is reached once it has been determined that a species/stock can sustain (commercially and biologically) commercial fishing operation. A formal Integrated Fisheries Management Plan is introduced.

The preliminary assessment (Stage I)

Applicants will submit proposals/applications that: (i) identify the target species/stock, fishing area and fishing method for which a licence is requested; (ii) summarise current knowledge about the target species, and provide an indication of how other species and/ or the environment might be affected by the proposed activity; (iii) provide a detailed plan outlining proposed fishing activities, e.g. applicable inspection requirements, harvest level by management area, harvesting method to be used, vessels to participate, start-up time, duration of harvesting activity, interactions with other fishing activities; (iv) provide information of product use, e.g. product forms, on-board product forms, onshore production if any, likely market distribution; (v) provide proof of public notification/consultations which allows for industry/public review and input; (vi) successful applicants must, in consultation with DFO, prepare a catch and effort record system; and (vii) identify sources of funding.

NOTA: 1. Applications to access new fisheries for which existing licences have been issued and for which no new licences are being issued because they are fully subscribed or where overcapacity will not be considered.

NOTA: 2. All new fisheries for which Stage I activity has been completed and for which applications/proposals are being sought for additional licences for Stage II activities should be communicated publicly by DFO in a regional press release.

The commercial and stock assessment (Stage II)

An exploratory harvesting strategy for the new fishery, including number of licensees, access criteria, including, where applicable, regional/provincial distribution, catch monitoring and reporting strategy, bycatch limits, seasons, etc., will be developed by DFO or a Committee.

1) In consultation with DFO Science, prepare a protocol to be used for the stock assessment component of the new fishery broken down as follows: data collection, data analysis, data recording and report preparation. The following additional information will be required for applications involving new fisheries at the commercial and stock assessment stage (Stage II).
2) The proposed processing and marketing strategies, including product forms, plants to be used and market destinations.

Upon the receipt of applications for new fisheries, the DFO is responsible for initiating a review of all applications for new fisheries. If the number of qualified applicants exceeds the number of licenses, a selection process will take place.

The licensing process (Stage III)

Once a decision on licensees for a new fishery is made, the Department will initiates new fisheries as follows:

1) Licence conditions for the new fishery are established, including, fishing areas, season, gear restrictions, licensing period, gear-up deadline, by-catch limits, etc.
2) The responsibilities of the licensee with regard to scientific, enforcement and/or management responsibilities and associated costs, as outlined in the exploratory harvesting plan, are included in a Memorandum of Agreement (MOA) or where appropriate as conditions of licence.
3) Successful applicants are notified of their selection and advised that issuance of licences is subject to receipt of a signed MOA by the Department.
4) Once signed MOAs are received by the Department, licences are issued to participants as follows: scientific or experimental licences for Stage I new fisheries and, exploratory licences for Stage II new fisheries.
5) Participation requirements will be introduced as a condition of exploratory licence issuance.
6) DFO will be responsible for analysis of information received from Stage I and II in a timely fashion so as to provide information base for assessing progression to further stages.
7) Scientific licence holders (Stage I) will be given priority for exploratory licences (Stage II).
8) Exploratory licence holders (Stage II) will be given priority for regular licences (Stage III).
9) Scientific/experimental or exploratory licences are not reissuable between individuals.
10) The names of successful applicants are released.

Bank (DFA, 2002). At this point it was mentioned that an environmentally friendly and acceptable method of harvesting should be used to establish the extent of the sea cucumber resource. DFA (2002) reported that dive harvests could be around 900–1 400 kg day^{-1}, this method being much less destructive than the results obtained with the towed gear (scallop gear) used to catch the sea cucumber, which also captured a wide variety of non-target species such as crabs, sea stars, sea urchins, molluscs and fishes.

DFA also looked at the Strait of Belle-Isle in 2003 (DFA, 2003) and again in 2004 (DFA, 2004a). The gear tested allowed to catch up to 375 kilograms of sea cucumber in 15–20 minutes tows. Also in 2004, the fishers and the DFA explored other areas including the southern coast of Labrador with surveys to determine if sufficient quantities of sea cucumbers could be found to sustain a commercial fishery (DFA, 2004b). The results obtained were not satisfying enough to warrant further studies of this species inside the survey areas.

The harvesting of sea cucumber began in Newfoundland and Labrador in 2002 with most of the effort concentrating on Newfoundland's Northeast Coast (DFA, 2004a). The fishery has been slow to develop mainly due to the lack of knowledge relative to resource abundance. Initially, dive technology was used to identify areas of concentration of this species. In 2002, a drag gear technology was transferred to Newfoundland and Labrador from Maine (United States of America) and proved to be very effective with improved catch rates particularly on the St. Pierre Bank.

In 2003, the project refined its target with a sea cucumber resource survey in NAFO subdivision 3Ps (Figure 1) in waters deeper than 20 m, as defined in the Emerging Fisheries Exploratory Sea Cucumber licence condition issued to the participating fishers.

The DFO, the fishers and the Fish, Food and Allied Workers have agreed to jointly undertake scientific survey work supporting the development of a commercial sea cucumber fishery in the NAFO area subdivision 3Ps. This Joint Project Agreement (JPA) is a five year project to be completed by 31 March 2009. The objective from the participants perspective is to: (i) collect reliable scientific information to assist science in carrying out a stock assessment and enable resource managers to make informed decisions for the effective management of an economically and biologically sustainable commercial sea cucumber fishery; and (ii) help the industry acquire knowledge of sea cucumber harvesting, processing and marketing to enlighten the decision making process required within a commercial fishery. The objective from a resource management perspective is to ensure that the industry actively participates in the development and management of any emerging sea cucumber fishery. Through this agreement, activities will be conducted to assist in the provision of scientific advice that can be used in the development of management strategies and potential implementation of a limited-entry fishery under the auspices of the Emerging Fisheries Policy.

The objective of DFO's science branch is to gain from the surveys reliable information on sea cucumber biomass and distribution in NAFO subdivision 3Ps. This objective will be achieved by acquiring detailed data on locations, catches, bycatches and catch rates, through collecting and processing biological samples, analysing data, and providing scientific advice on an on-going basis.

With the information gathered and compiled, it will then be possible to determine which management tools are appropriate to develop a sustainable and stable fishery for sea cucumber in NAFO subdivision 3Ps.

The paragraphs below summarize the respective roles and goals of the parties involved as defined before the onset of the joint collaboration (after JPA, 2004):

DFO Fisheries Management Branch:

- Provide liaison between DFO Branches and participants in the implementation of the various goal;

- Provide Fisheries Management direction and liaise with DFO Branches and the participants for the duration of the various stages;
- Arrange a public draw, if necessary, for selection of vessels;
- Conduct performance review to ensure project objectives are met;
- Conduct consultation meetings for the development of the Annual Project Work Plans and Management Strategies for this fisheries;
- Monitor fishing practices;
- Prepare post-season analysis and annual report on the project results;
- Meet with industry to discuss study results and future plans.

DFO Science Branch:

- Work collaboratively with the participants toward the determination of essential biological parameters on which to supply scientific advice for management strategies for the sea cucumber fishery and monitor the development and performance of the fishery;
- Assist in the development of project design;
- Review and assist, as necessary, work completed by DFA, the Marine Institute (MI) and others to develop and apply measurements to determine relationships between live weight, drained weight and dry weight;
- Conduct analysis of biological data;
- Analyse fishing logs to chronicle the fishery, including spatial distribution of fishing effort and the changes in catch per unit of effort;
- Monitor fishing practices;
- Prepare annual report on the project scientific results;
- Exchange information and establish dialogue between all parties by meeting with the participants and Fisheries Management;
- Participate in consultation meetings for the development of the Annual Project Work Plans and Management strategies for the fisheries.

The fishers and the Fish, Food and Allied Workers (FFAW):

- Install sea cucumber drags and appropriate gear and equipment necessary to carry out sea cucumber fishery on vessels, as defined by DFO;
- Conduct fishing activities and record data in prescribed areas as outlined by DFO;
- Assist at collecting sea cucumber data in sea cucumber logbooks;
- Adhere to a fishing schedule as agreed upon by the fishers, the FFAW and DFO;
- Maintain minimum and maximum weekly sea cucumber landings as agreed upon by the Fishers, the FFAW and DFO;
- Land at designated ports agreed upon by Fishers, the FFAW and DFO;
- Arrange for dockside monitoring and sampling of sea cucumber to determine live weight at port;
- Exchange information between the parties by meeting with DFO to discuss the survey and project results.

The actual role and objectives of the participants has been modified as other partners joined in the venture and they will continue to evolve before the completion of the proposed three stages.

To our knowledge, the sea cucumber project in Newfoundland and Labrador really began in 2003 as an initiative or JPA between the DFO, the fishers, the FFAW and the Fogo Island Cooperative Society Limited (JPA, 2004). These partners, along with considerable input/contribution from DFA and CCFI, will be leading the project entitled "Sea cucumber resource survey in NAFO subdivision 3Ps" in waters greater than 20 m until 2009 (Figure 3).

FIGURE 3
Harvesters are licensed by Department of Fisheries and Oceans to assist in carrying out the biomass surveys

COURTESY OF L. BARRETT

The core partners enlisted the help of the Marine Institute of Memorial University (Newfoundland and Labrador) which published a first report in 2004 (Winger, Keats and Grant, 2004). The research found that sea cucumbers were easily distinguishable by camera in the field (through a camera placed on a sledge allowing scientists to see how the net works and which species are captured and in what amount). Investigators also gathered information on substrates occupied by *C. frondosa* in 3Ps: sand, gravel and cobble. They estimated the density between 0.16 to 0.58 ind. m^{-2}.

In May 2004, the Marine Institute of Memorial University, in partnership with DFA, published a report on the quality assessment of the product as per market stipulations (MI, 2004). They also published a report that described a sorting system to allow minimal size categorization and easy exclusion of incidental or bycatch along with the removal of gravel, pebbles and debris (MI, 2005a). There was also an assessment of water-loss in *C. frondosa* upon harvest (MI, 2005b). Yet another study was conducted by DFA and MI on a handling system designed to quickly transfer sea cucumbers from the holding tank to the fish hold, minimizing the effort of the crew in handling the product and helping grade out undersized animals and bycatch (MI, 2006a). Finally, the Marine Institute published results on habitat utilization and density of sea cucumber (Winger, Keats and Grant, 2004; MI, 2006b). The results indicate that *C. frondosa* is capable of colonizing different types of habitat.

A couple of studies were conducted in view of building databases for stock assessment purposes in the St. Pierre Bank and in the Strait of Belle Isle (Grant, 2006; Grant, Squire and Keats, 2006). The reports summarize a large amount of data on the morphology, sex ratio, size at sexual maturity, spawning season and constitutes the first set of biological data available on *C. frondosa* from the St. Pierre Bank and the Strait of Belle Isle since Newfoundland and Labrador initiated their investigation of this new emerging species.

In 2004 the DFA asked Dr Fereidoon Shahidi (Memorial University) to perform a detailed analysis of the chemical composition of fresh and processed sea cucumber from Newfoundland and Labrador (Shahidi, 2006).

In 2005, in his protocol for the new emerging fisheries, DFO scientist Dr Don Stansbury from St. John's (Newfoundland and Labrador) indicated that the review process of the sea cucumber fishery should include a large science component. It was noted that data relevant to decision making within this new fishery was still sparse, and that fisheries science was important in the development of new fisheries. Information identified as missing included: spatial distribution and abundance (some data gathered during previous surveys were still being compiled), reproduction, development, recruitment, growth, predator pressure on the resources and gene flow.

One of the most recent steps in the process of gathering precise biological data on *C. frondosa*, has been undertaken in 2006 by Justin So (MSc candidate, Memorial University) as part of a project funded by DFA, CCFI and the industry (harvesters/processors). The objective of this new study is to provide answers to some of the as yet unaddressed questions identified by DFO relative to a potential commercial sea cucumber fishery in Newfoundland and Labrador: the growth rate from fertilization to commercial size, the physiological condition of individuals from different habitats (sand, pebbles, bedrock), the pattern of gene flow throughout the distribution range of the species (to determine the impact of larval dispersion and assess connectivity between distant populations), and the influence of predatory pressure, mainly from the sea star *Solaster endeca* on the resource.

3.3 Newfoundland and Labrador compared to other Atlantic Provinces

At this time, Newfoundland and Labrador (NL) has taken the lead in terms of the New Emerging Fisheries Policy where it concerns *C. frondosa*. The province is enforcing the policy in a much stricter fashion than other Canadian provinces currently wishing to develop the sea cucumber fishery. It is however worth mentioning that this may not necessarily be the case for other species with commercial potential which are presently being studied in Canada under the New Emerging Fisheries Policy.

Nova Scotia (NS) and New Brunswick (NB) are both interested in exploiting sea cucumber populations in the North Atlantic, and they have indeed begun accessing this new resource over the past few years. The province of New Brunswick has adopted a much more liberal approach to the Canadian policy relative to emerging resources. In that province, only two sea cucumber harvesters are working under the scheme to provide survey data and they are working over an area of about 5 km^2, from which they extract a total of approximately 1 310 tonnes annually. In contrast, eight harvesters from NL are participating in the study, which is conducted over an area of 5 200 km^2 (i.e. 1 000 times larger than in NB), and the overall quota they share is currently set at ca. 650 tonnes (i.e., roughly half the NB quota).

Because the New Emerging Fisheries Policy is a national initiative, there should not be such inconsistency in the application of the protocol. Canada is therefore looking at standardizing the approach within all its provinces in this domain. But this will certainly prove to be a huge challenge, especially in light of the different views expressed by harvesters and policy makers where fisheries are concerned. It is always difficult to convince fishers to participate in a scientific study during which they are not making any profits, notwithstanding the fact that they will eventually be the ones benefiting from the exploitation of a new resource in a sustainable manner. Nevertheless, Canada has adopted a precautionary approach and policy makers are urged to enforce the new measures as rigorously as possible. This confrontation will be playing out every time the sea cucumber fishery re-opens until April 2009, when the preliminary scientific work will be completed.

4. RECOMMENDATIONS

4.1 General

A new and expanding fishery requires some new structures and processes in management. DFO has a mandate to manage the fishery which should be concerned with maximizing the benefits to the people of Canada. This aspect has been lost in the ensuing hardships of a decline in fish stocks. Allocation issues have been at the forefront of fisheries management and will continue to be. They can only be resolved with a large amount of goodwill among all parties to the discussion. When these last objectives are completed, and as more data on the biomass becomes available, no later that in 2009, all prerequisites required by the DFO will have been fulfilled. The latest estimates of biomass on St. Pierre Bank (NAFO 3Ps) by the DFO evoke very interesting prospects. With an average CPUE of ca. 300 kg tow^{-1} and an available biomass of about 300 000 tonnes (unpublished DFO statistics, St. John's, NL) the sea cucumber fishery may face a promising future if it is well managed and if the fishers respect the regulations established by the DFO based on their data acquisition.

A success story would certainly set a great example for other species (e.g. box crab, tanner crab, Pacific milky Venus clam, varnish clam, purple sea urchin, sea mussel, neon flying squid, Pacific sardine) which are currently being considered for development in experimental fisheries by the Canadian Government.

4.2 CITES considerations

Is there any advantage in listing *C. frondosa* under the Convention on International Trade in Endangered Species (CITES)? If the development of the fishery follows strict protocols, perhaps not. However, fisheries rarely develop according to plan and the history of sea cucumber harvesting worldwide highlights the likelihood of rapid stock depletion. Thus, despite presumably high biomasses, the perspective of *C. frondosa* becoming a threatened species cannot be discarded completely if a commercial fishery is to be opened. CITES is an established organisation with a sound basis in international law, and experience at monitoring international trade at the species level. It is argued that the lack of solid biological and fisheries data currently constrains conservation and management efforts with sea cucumbers elsewhere. CITES listing could potentially provide better quantitative data on sea cucumbers in trade: species, products, volumes. It is also argued that it could provide improved controls for protecting threatened populations from potential overexploitation. On the other hand, a CITES listing would not reduce bycatch, incidental capture, fisheries targeting national consumption (i.e. non-traded resources), nor the fishing impacts on other species on which sea cucumbers may be dependent.

In the specific case of *C. frondosa* and its exploitation in eastern Canada, little thought has been given to the potential benefits associated with CITES listing. According to DFA Fisheries Development Officer Lewis Barrett, steps could be initiated with relevant government departments and agencies to explore the process for having *C. frondosa* listed in CITES Appendix II, should this precaution become necessary in the near future. He believes that with adequate background documentation, the DFA Planning Branch would probably support such an initiative in keeping with the recent focus and collaborative efforts between DFA, DFO (i.e. involving Canada's *Oceans Act*, Marine Protected Areas [MPAs], etc.), and Environment Canada. Although such a project would surely fall under federal jurisdiction, provincial departments can still review it on merit and decide whether they would be prepared to support it. Especially since it will likely complement the guiding principles under DFO's New Emerging Fisheries Policy in laying the ground work for a strategy that future generations will embrace as an overall "futuristic, positive thinking approach" to reverse the global environmental slide on wild fish stocks. It might well be to the industry's advantage to support and endorse a CITES listing as more and more companies select to abide by new consumer rules and requirements that are identified with sustainable fishing practices. An industry led initiative in this respect would be good and governments would naturally follow suit (Barrett, L., DFA, personal communication).

ACKNOWLEDGEMENTS

We would like to acknowledge the valuable contribution of Lewis Barrett (DFA, Government of Newfoundland and Labrador), and to thank Verónica Toral-Granda, Marcelo Vasconcellos, Alessandro Lovatelli, Chantal Conand and Poh Sze Choo for their valuable support and comments. This document has greatly benefited from input and assistance graciously offered by numerous people: Sherrylinn Rowe (DFO, Halifax, Nova Scotia), Valerie Bradshaw (DFO, Halifax, Nova Scotia), Don Stansbury (DFO, St. John's, Newfoundland & Labrador), Ken Budden (Fogo Island Co-op, Newfoundland & Labrador), Guy Parsons (DFA, Government of Newfoundland & Labrador), Justin So (Ocean Sciences Centre, Memorial University, Newfoundland & Labrador), Eric Way (DFO, Newfoundland & Labrador).

REFERENCES

Bruckner, A.W. 2005. The recent status of sea cucumber fisheries in the continental United States of America. *SPC Beche-de-Mer Inf. Bull.*, 22: 39–46.

DFA. 2001. Sea cucumber survey, dive harvest – Notre Dame Bay; Labrador scallop bucket – Placentia Bay and St. Pierre Bank, 2001. Department of Fisheries and Aquaculture (Government of Newfoundland and Labrador), Fisheries diversification program, Emerging Fisheries Development, Project report FDP 358–4.

DFA. 2002. Sea cucumber – Fall, 2002, Dive Harvest – Bonavista and Notre Dame Bays; Towed gear design and testing – Dantzic Point and St. Pierre Bank. Department of Fisheries and Aquaculture, Government of Newfoundland and Labrador, Fisheries diversification program, Emerging Fisheries Development, Project report FDP 424–3.

DFA. 2003. Sea cucumber survey. Strait of Belle Isle, September to November 2003. Department of Fisheries and Aquaculture, Government of Newfoundland and Labrador.

DFA. 2004a. Sea cucumber survey Strait of Belle Isle August – September 2004. Department of Fisheries and Aquaculture, Government of Newfoundland and Labrador, 31 p.

DFA. 2004b. Sea cucumber exploratory survey-south Labrador coast 2004. Department of Fisheries and Aquaculture, Fisheries Development Branch Labrador Region (Government of Newfoundland and Labrador), 43 pp.

Finlayson, A.C. 1994. Fishing for truth: A sociological analysis of northern cod stock assessments from 1977–1990. Memorial University of Newfoundland, Institute of Social and Economic Research, St. John's, Newfoundland, Canada.

Grant, S.M. 2006. Biological resource assessment of the orange footed sea cucumber (*Cucumaria frondosa*) occurring in the Strait of Belle Isle. Centre for Sustainable Aquatic Resources Fisheries and Marine Institute, Memorial University of Newfoundland, 33 p.

Grant, S.M., Squire, L. & Keats, C. 2006. Biological resource assessment of the orange footed sea cucumber (*Cucumaria frondosa*) occurring on the St. Pierre Bank. Centre for Sustainable Aquatic Resources Fisheries and Marine Institute, Memorial University of Newfoundland, 75 pp.

Hannesson, R. 1996. Fisheries mismanagement: the case of the north Atlantic cod. Oxford, England: Fishing News Books.

Hutchings, J.A. 1996. Spatial and temporal variation in the density of northern cod and a review of hypotheses for the stock's collapse. *Can. J. Fish. Aquat. Sci.*, 53: 943–962.

JPA. 2004. Sea cucumber resource survey in NAFO Subdivision 3Ps. Joint project agreement, Schedules A, B, C, D, and E, Newfoundland and Labrador (Canada) 13 pp.

MI. 2004. Sea cucumber quality assessment. Marine Institute, Memorial University (Newfoundland and Labrador), Canada, 27 pp.

MI. 2005a. Evaluation of sea cucumber sorting grid system. Marine Institute, Memorial University of Newfoundland, Prepared for the DFA (Government of Newfoundland), project P 1471.

MI. 2005b. Determining water-loss in sea cucumbers. Marine Institute, Memorial University of Newfoundland, Prepared for the DFA (Government of Newfoundland), project P 1453, 53 pp.

MI. 2006a. Development of on-board handling system for sea cucumbers. Centre for Aquaculture and Seafood Development, Marine Institute, Memorial University, Newfoundland and Labrador, Canada, 18 pp.

MI. 2006b. Habitat utilization and density of the sea cucumber (*Cucumaria frondosa*) on St. Pierre Bank, Newfoundland: observations using a towed camera sled in 2004 and 2005. Marine Institute, Memorial University of Newfoundland, Canada, 16 pp.

RDA. 1992. A preliminary examination of market potential for sea cucumber in selected Asian countries. Resource Development Associates (Newfoundland and Labrador, Canada), 27 pp.

Shahidi, F. 2006. Chemical composition of fresh and processed Newfoundland sea cucumber (*Cucumaria frondosa*), Report for the Department of Fisheries and Aquaculture, Government of Newfoundland and Labrador, 1–19.

Sinclair, P.R. 1996. Sustainable development in fisheries dependant regions? Reflections on Newfoundland cod fisheries. *Sociologia Ruralis,* 36: 225–235.

Steele, D.H., Andersen, R. & Green, J.M. 1992. The managed commercial annihilation of northern cod. *Newfoundland Studies,* 8: 34–68.

Walters, C. & Maguire, J.-J. 1996. Lessons for stock assessment from the northern cod collapse. *Rev. Fish Biol. Fish.,* 6: 125–137.

Winger, P.D., Keats, C. & Grant, S. 2004. Habitat utilization and density of sea cucumber (*Cucumaria frondosa*) on St. Pierre Bank, Newfoundland: preliminary observations using a towed camera sled. A report submitted to the Department of Fisheries and Aquaculture, Canadian Centre for Fisheries Innovation, and Fisheries and Oceans Canada, 14 pp.

Annex 1 – Workshop agenda

International Workshop on the Sustainable Use and Management of Sea Cucumber Fisheries

Puerto Ayora, Galapagos Islands, Ecuador, 19–23 November 2007

WORKSHOP AGENDA

Monday, 19 November 2007	
Morning	
08h30	Arrival and registration
09h00	Welcome note by: Graham Watkins – Executive Director of the Charles Darwin Foundation Raquel Molina – Galápagos National Park Director Marcelo Vasconcellos – Fishery Resources Officer, FAO Introduction of participants
09h30	Scope of the Workshop
10h00	Adoption of Agenda
11h00	*Coffee break*
11h20	Presentation of RR: Asia (HTP Philippines)[1] – Poh Sze CHOO
11h40	Presentation of RR: Western Central Pcific (HTP Papua New Guinea) – Jeff KINCH
12h00	Presentation of RR: North America (HTP *C. frondosa* fishery in Newfoundland, Canada), Jean-François HAMEL/Annie MERCIER
12h20	Presentation of RR: Latin America and the Caribbean (HTP Galápagos Islands) – Verónica TORAL-GRANDA
12h40	*Lunch break*
Afternoon	
14h00	Presentation of RR: Africa and Indian Ocean (HTP Seychelles) – Chantal CONAND
14h20	Plenary discussions: Brainstorming on Table of Contents of the Technical Guidelines and Setting up of working groups
16h00	*Coffee break*
16h30	Working groups discussion and drafting
18h00	*Adjourn for the day*

[18h00-18h45 Internal meeting between the Scientific Committee and FAO]

[1] **RR**: Regional Review; **HTP**: Hotspot.

Tuesday, 20 November 2007

Morning

08h30	Working groups discussions and drafting
12h30	*Lunch break*

Afternoon

14h00	Plenary presentations and discussions
18h00	*Adjourn for the day*

[18h00-18h45 Internal meeting between the Scientific Committee and FAO]

Wednesday, 21 November 2007

Morning

08h30	Working groups discussions and drafting
12h30	Bus trip to Rancho Aquelarre for lunch
13h00	*Lunch breack*

Afternoon

15h00	Hike to "El Chato" (Galapagos giant tortoise reserve), Gemelos Craters, Scalesia Forest, Lava tunnels
17h00	Informal meeting with the fishing sector of Santa Cruz (Charles Darwin Station)
18h00	*Adjourn for the day*

[18h00-18h45 Internal meeting between the Scientific Committee and FAO]

Thursday, 22 November 2007

Morning

08h30	Working groups discussions and drafting
12h30	*Lunch break*

Afternoon

14h00	Plenary presentations and discussions
18h15	*Adjourn for the day*

[18h00-18h45 Internal meeting between the Scientific Committee and FAO]

Friday, 23 November 2007

Morning

08h30	Working groups discussions and drafting
12h30	*Lunch break*

Afternoon

14h00	Plenary discussions on draft Technical Guidelines
16h00	Identification of future steps
17h30	Closing remarks
18h00	*End of Workshop*

[18h00-18h45 Internal meeting between the Scientific Committee and FAO]

20h00	*Closing dinner at Red Mangrove Inn Hotel*

Annex 2 – List of participants

EXPERTS

Jun AKAMINE
Associate Professor
School of Humanities and Social Sciences
Nagoya City University
Yamanohata 1, Mizuho-cho, Mizuho-ku
Nagoya City, Aichi
Japan, 4678501
Tel.: +81-52-8725812
Fax: +81-52-8725812
E-mail: akamine@hum.nagoya-cu.ac.jp

Poh Sze CHOO (Ms)
Science and Policy Specialist
WorldFish Center
PO Box 500 GPO
10670 Penang
Malaysia
Tel.: +60-4-6202219
Fax: +60-4-6265690
E-mail: p.choo@cgiar.org

Chantal CONAND (Ms)
Emeritus Professor
Laboratoire ecologie marine
Université de la Réunion
F-97715 Saint-Denis
France
Tel.: +33-(0)47-6450137 (Private)
Lab: +262-938130 (Laboratory)
Fax: +262-938685
E-mail: conand@univ-reunion.fr

Eduardo ESPINOZA
Acting Head of Marine Resources Management
Coordinator of Marine Research Program
Galapagos National Park Service
Puerto Ayora, Galapagos Islands
Ecuador
Tel.: +593-5-2527410
Fax: +593-5-2527410 ext. 146
E-mail: eespinoza@spng.org.ec

Kim FRIEDMAN
Senior Reef Fisheries Scientist
Secretariat of the Pacific Community (SPC)
BP D5
98848 Nouméa
New Caledonia
Tel.: +687-265474
Fax: +687-263818
E-mail: kimf@spc.int

Ruth GAMBOA (Ms)
Associate Professor
Department of Biology
Collage of Science and Mathematics
University of the Philippines Mindanao
Davao City
8022 Philippines
Tel.: +63-82-2930302
Fax: +63-82-2930302
E-mail: ruthupmin@yahoo.com

Jean-François HAMEL
Research Scientist
Society for the Exploration and Valuing
of the Environment (SEVE)
21 Phils Hill Road
Portugal Cove-St. Philips
Newfoundland & Labrador
Canada, A1M 2B7
Tel.: +1-709-8953257
E-mail: jfhamel@seve.cjb.net

Alex HEARN
Coordinator of Fisheries Research
Charles Darwin Foundation
Puerto Ayora
Santa Cruz, Galapagos Islands
Ecuador
Tel.: +593-5-2526 146/7 ext. 123
Fax: +593-5-2526 146/7 ext. 102
E-mail: alex.hearn@fcdarwin.org.ec

María Dinorah HERRERO-PÉREZRUL (Ms)
Assistant Researcher
Ave. IPN s/n Col. Playa Palo de Santa Rita
La Paz, Baja California Sur
CP 23096, Mexico
Tel.: +52-612-1225344 ext. 2425
Fax: +52-612-1225322
E-mail: dherrero@ipn.mx
dainoper@hotmail.com

Jeff KINCH
Coastal Management Advisor
PO Box 829
Alotau
Milne Bay Province 211
Papua New Guinea
Tel.: +685-21929
Fax: +685-20231
E-mail: jeffreyk@sprep.org
kinch.jeff@gmail.com

Priscilla C. MARTÍNEZ (Ms)
Marine Officer
World Wildlife Fund
Galapagos Eco-Region
Ave. Charles Darwin y Seymour
Santa Cruz, Galapagos Islands
Ecuador
Tel: +593-(0)5-2526053
Fax: +593-(0)5-2526053
E-mail: priscilla.martinez@wwfgalapagos.org

Annie MERCIER (Ms)
Assistant Professor
Ocean Sciences Centre
Memorial University
St. John's Newfoundland and Labrador
Canada A1C 5S7
Tel.: +1-709-7372011
Fax: +1-709-7373220
E-mail: amercier@mun.ca

Steven PURCELL
Ecologist
WorldFish Center
BP D5
Nouméa Cedex 98848
New Caledonia
Tel.: +687-262000
Fax: +687-263818
E-mail: s.purcell@cgiar.org
steven.w.purcell@gmail.com

Verónica TORAL-GRANDA (Ms)

FAO Consultant
Casilla Postal 20-10-10
Puerto Ayora, Galapagos
Ecuador
Tel.: +593-(0)5-2527425
Fax: +593-(0)5-2527425 ext. 102
E-mail: veronica.toral@gmail.com

Sven UTHICKE
Research Scientist
Australian Institute of Marine Science
PMB No. 3, Townsville MC
Queensland 4810
Australia
Tel.: +61-7-47534483
E-mail: suthicke@aims.gov.au

Matthias WOLFF
Head of Marine Research and Conservation
Charles Darwin Foundation
Puerto Ayora
Santa Cruz, Galapagos Islands
Ecuador
Tel.: +593-(0)5-2526146/7 ext. 123
Fax: +593-(0)5-2526146/7 ext. 102
E-mail: matthias.wolff@fcdarwin.org.ec

FAO FISHERY EXPERTS

Alessandro LOVATELLI
Fishery Resources Officer (Aquaculture)
Food and Agriculture Organization
Fisheries and Aquaculture Department
Aquaculture Management and Conservation Service
Viale delle Terme di Caracalla
00153 Rome, Italy
Tel.: +39-06-57056448
Fax: +39-06-57053020
E-mail: alessandro.lovatelli@fao.org

Marcelo VASCONCELLOS
Fishery Resources Officer
Food and Agriculture Organization
Fisheries and Aquaculture Department
Fisheries Management and Conservation Service
Viale delle Terme di Caracalla
00153 Rome, Italy
Tel.: +39-06-57056469
Fax: +30-06-57053020
E-mail: marcelo.vasconcellos@fao.org

Annex 3 – Participant profiles

AKAMINE, Jun – Dr Akamine currently teaches Southeast Asian Studies and Maritime Ethnology at the School of Humanities and Social Sciences, Nagoya City University. His academic career started as a descriptive linguist and anthropologist in the Philippines. One of his major fieldwork activity was conducted in southern Palawan and Sulu Archipelago where the majority of the people have been dependent on sea cucumber fisheries. This is the reason for which he has become interested in the "sea cucumber food system", which covers the historical development of fishery, trade and consumption in the world. Resource management is also of great concern to him. His research has extended to the fisheries communities, both in eastern Indonesia and Japan. He is interested in collaborative resource management between traders and fishermen. He attended several international workshops on holothurian resource management: Advances in Sea Cucumber Aquaculture and Management (ASCAM) in Dalian, China (2003); Convention on International Trade in Endangered Species of Wild Fauna and Flora (CITES) Expert meeting in Kuala Lumpur, Malaysia (2004); and CITES Conference of the Parties (CoP) 14, the Hague, The Netherlands (2007). Dr Akamine has published over 50 research papers on both peer-reviewed and non peer-reviewed journals and books.

CHOO, Poh Sze – Before joining The WorldFish Center as a Science and Policy Specialist in 2002, Ms Choo worked for more than 30 years in several capacities in the Fisheries Research Institute, Penang, Malaysia. Her last appointment being Head of the Aquatic Ecology Branch, a position she held from 1991–2002. In the early 1990s she collaborated with scientists from the Heriot-Watt University on a sea cucumber research project, "The Taxonomy, Life History and Conservation of Malaysian Holothurians" supported by the Darwin Initiative for the Survival of Species, United Kingdom. Her interest in sea cucumber research is mainly in fishery and trade. In her new position at the WorldFish Center, she maintains an active interest in sea cucumber research and conservation issues. Ms Choo holds a MSc in nutritional sciences from the University of Guelph, Canada and a BSc (Hons) in Ecology from the University of Malaya, Malaysia. Her professional interest is in aquaculture, marine conservation research, and gender research in the fisheries sector.

CONAND, Chantal – Since 1964, Prof Conand has been a lecturer and subsequently professor in several French Universities (Marseille, Brest, Réunion) and Senegal (Dakar) and detached (4 years) in the research agency IRD (New Caledonia). Her research was initially on fish biology, followed by Echinoderms, mostly sea cucumbers, in 1978,. From 1994–2003 she was Director of the Marine Ecology Laboratory in Réunion. Her very long and constant investment in sea cucumber fisheries has made her an international expert (with reports and meeting for SPC, FAO, CITES, WIOMSA). She is the scientific editor of the bêche-de-mer Information Bulletin published by SPC. Prof Conand is presently co-responsible of a three years Western Indian Ocean (WIO) programme on sea cucumber fisheries granted by WIOMSA involving five countries. She holds a first French thesis on the reproduction of a commercial migratory fish from Senegal and a French State Doctorate on the fishery biology of sea cucumbers from New Caledonia (partly translated in English by SPC). She has directed several PhDs in reef ecology and has many scientific publications, many of them on echinoderms mostly sea cucumbers.

ESPINOZA, Eduardo – Mr Espinoza is the Acting Head of Marine Resources Management and Conservation Officer to the Galapagos Marine Reserve, Galapagos National Park Service. Mr Espinoza has been involved in both research and management of the Galapagos sea cucumber *Isostichopus fuscus*, first as a fisheries biologist at the Charles Darwin Foundation for five years and then as Marine Conservation Officer in the Galapagos National Park Service. He has a thorough knowledge of the different managerial, technical and law processes that have taken place in Galapagos over the past years, which are what, have forged the current status of the GMR and its sea cucumber fishery. Mr Espinoza holds an MSc from the Fisheries University in Tokyo, Japan. His professional interest are fisheries management and the development of alternatives to the local fishers, patrolling and enforcement and the sound management of the GMR based on technical and scientific information.

FRIEDMAN, Kim – For the last five years Dr Friedman has been responsible for invertebrate resource assessments within SPC's Reef Fisheries Observatory. In this role, he and a small team of researchers have surveyed both commercial and subsistence fisheries within 17 Pacific Island countries and territories. These inshore assessments will provide comparative information on the state of coastal fisheries for a major inshore fisheries study that is due to be completed by late 2008. Dr Friedman was born in South Africa, but completed most of his tertiary education, including his Masters degree (Fisheries and Aquaculture) in the United Kingdom. He holds a doctorate awarded from James Cook University in Australia. Dr Friedman's working past includes salmon farming in Scotland, and a short spell with the New South Wales Fisheries Department in Sydney. He also led pearl oyster research for WorldFish Center in the Western Pacific for over four years (Solomon Islands and Fiji) before joining the Western Australian Fisheries Department in Perth, where he was the senior research scientist for abalone, pearl oyster and sea cucumber fisheries. These fishery assessments spanned the state, from temperate abalone waters near South Australia up to the pearl grounds and sea cucumber fisheries near the tropical Northern Territory.

GAMBOA, Ruth – Dr Gamboa has been with the University of Philippines (UP) at Mindanao for the past 11 years. After finishing her PhD at Hokkaido University, Japan, she went on a 1-year training in hatchery production of sea cucumbers. Together with the sea cucumber team of Dr Annette Menez at the Bolinao Marine Laboratory, UP Marine Science Institute, she pioneered the successful hatchery production of sandfish in the Philippines in 2002. At present, this team is significantly involved in drawing up a national research and development programme for sustainable fisheries of sea cucumbers. She has established good working relations with relevant national government agencies as well as state universities in the Davao Region and also with civil society groups such as NGOs and People's Organizations involved in coastal resource management. Dr Gamboa has experience in writing technical reports and publications and has close contacts with researchers overseas.

HAMEL, Jean-François – Dr Hamel has been studying holothurians since 1986. He has conducted his BSc, MSc and PhD work on the boreal species *Cucumaria frondosa* and *Psolus fabricii* (Laval University, UQAR, Canada). His expertise later sent him travelling around the globe (Marshall Islands, Solomon Islands, Ecuador) to collaborate with academic institutions and private industries as part of internationally and locally funded projects. He has developed expert knowledge of many tropical species (i.e. *Holothuria scabra*, *H. atra*, *H. leucospilota*, *Bohadschia argus*, *B. marmorata*, *Isostichopus fuscus*, *Thelenota ananas*, *T. anax*). Since 2000, he has been overseeing a study of *I. fuscus* which has enabled the onset of an aquaculture industry in Ecuador. While at the College of the Marshall Islands in 2001, he co-designed and helped build the Marshall Islands Science

Station (MISS). In 2003, he was session chair and editor at the FAO sea cucumber workshop (Dalian, China). Over the years, he has been a member of various scientific and managing committees, and has published dozens of peer-reviewed papers under the affiliation of the SEVE (which he co-founded in 1995). He has also co-authored five books and over 400 articles in popular magazines worldwide. Now back in Canada, Dr Hamel remains active internationally and continues to study holothurians and other echinoderms. His research focuses on the control of gametogenesis/spawning, the chemical ecology and the general biology of coastal and deep-sea invertebrates. He is currently involved in projects related to the emerging fishery of *Cucumaria frondosa* in eastern Canada.

HEARN, Alex – Dr Hearn has worked at the Charles Darwin Foundation for five years as coordinator of fisheries research. His team of researchers have studied aspects of fisheries in the Galapagos Marine Reserve (GMR) with a view to providing technical advice on sustainable management practices to the stakeholders and authorities in the GMR. They have also studied aspects of population dynamics of key exploited species, such as the slipper lobster, spiny lobster, and two species of sea cucumber: *Isostichopus fuscus* and *Stichopus horrens*. Dr Hearn and his team are currently studying the spatial dynamics of spiny lobsters and hammerhead sharks. They are also applying ecological risk assessment techniques to fisheries, and are leading the development of a Fisheries Management Plan for the GMR. He has published several articles and reports on sea cucumber population dynamics and fisheries, including the use of stock depletion models to predict changes in populations under different fishing scenarios. He currently advises the National Park Service and Ecuadorian Government on issues related to sea cucumbers and other fisheries in the GMR. He holds a PhD in Fisheries Biology (2001) from Heriot-Watt University in the Orkney Islands, United Kingdom. His areas of interest include population dynamics, spatial dynamics, small-scale fisheries management and management of marine protected areas (MPAs).

HERRERO-PÉREZRUL, María Dinorah – Since 2002, Ms Herrero-Pérezrul has been working as an Assistant Researcher at the Centro Interdisciplinario de Ciencias Marinas (Interdisciplinary Center of Marine Sciences). She is a fisheries biologist working on marine invertebrates (holothurians, echinoids, bivalves, gastropods) under current exploitation. Her work involves reproductive biology, population dynamics, fisheries management and echinoderm ecology. She is a scientific advisor for the Natural Protected Areas in the monitoring plan for invertebrates. She is also technical advisor to the Mexican Government and has helped establish the first Technical Committee for *Isostichopus fuscus* management in Baja California Sur, where she acts as spokeswoman. In her free time, she is the scientific advisor for fishery groups (stakeholders) who exploit echinoderms. She has published some important papers on the reproductive biology and population dynamics of *I. fuscus* and addressed the problem of variability in size measurement. She attended the CITES International Technical Workshop for the Conservation in the Families Holothuridae and Stichopodidae. Ms Herrero-Pérezrul holds a PhD in Marine Sciences (2004) from the Instituto Politécnico Nacional (National Polytechnic Institute). Her professional interests are conservation biology, participatory management in fisheries (including biological and population dynamics aspects), and monitoring of reef invertebrates.

KINCH, Jeff – Since completing his Anthropology PhD fieldwork in the Milne Bay Province, Papua New Guinea in 1999 on marine resource management regimes and livelihood strategies, Mr Kinch has worked on numerous marine resource and fisheries conservation, development and management programmes and projects throughout Melanesia for governments, NGOs and regional organizations. Mr Kinch was formerly the Coastal Fisheries Advisor at the University of Papua New Guinea, and currently works as a private consultant for the FAO on this sea cucumber project, and also served as an expert on the "CITES Workshop on the Conservation of Sea Cucumbers". In the last ten years he has lead several abundance surveys, socio-economic studies and management reviews of the bêche-de-mer fisheries in PNG and the Solomon Islands. Mr Kinch's professional interests include artisanal fisheries development and management, coral reef conservation, integrated coastal zone management and sustainable livelihoods.

MARTÍNEZ, Priscilla – Ms Martínez has been involved with marine biology research in the Galapagos Islands since 1978. Her studies with sea cucumbers began in Guam, Micronesia in 1990, where she worked as a research assistant at the Guam Marine Laboratory on reproductive and larval biology of *Holothuria nobilis*, *Thelenota ananas* and *Actinopyga mauritiana*. In 1993, she formed part of a team of consultants for IUCN in order to conduct the first biological assessment on the impact of sea cucumber fisheries in Galapagos. From 1994–2000, she joined the Marine Biology Department of the Charles Darwin Foundation (CDF), where she started studies of population and reproductive biology of *Isostichopus fuscus*. During this period, she also worked as representative of the scientific sector on the revisions of the Management Plan for the Galapagos Marine Reserve (GMR) and as a representative of the Participatory Management Board for the GMR. Ms Martínez is presently a PhD student from the University of Melbourne, Australia. Her dissertation focuses on *I. fuscus* growth and reproductive biology; preferences of habitat for juveniles and adults, and changes in the density and population structure of the species through the years of exploitation. She has participated as a field leader of the sea cucumber population surveys of 1999, 2001 and 2002 fishing season. Ms Priscilla Martínez and Manfred Altamirano proposed the inclusion of the sea cucumber *I. fuscus* to Appendix III of CITES in 2002. In 2006, she joined Cathy Hair and her team of researchers from the Aquaculture and Stock Enhancement Facility at the Northern Fisheries Centre, Cairns, Queensland (Australia), to collaborate on the culture and restocking project of *Holothuria scabra*.

MERCIER, Annie – Since completing her PhD in oceanography in 1997 (UQAR, Canada), Dr Mercier has worked on several projects related to sea cucumbers in Canada and abroad. She was a research associate at WorldFish Center for three years, studying the ecology of *Holothuria scabra* in the Solomon Islands through a CIDA grant. She was later hired at the College of the Marshall Islands where she led two USDA-funded projects and co-designed and supervised the construction of the marine laboratories at the Marshall Islands Science Station (MISS). Since 2000, she has been a consultant for several sea cucumber aquaculture firms, including one in Ecuador, where she helped develop techniques to reproduce and grow the Galapagos sea cucumber *Isostichopus fuscus*. She was a guest speaker and session editor during the sea cucumber workshop organized by FAO in 2003. For the past three years, she has been a professor at Memorial University (Canada), where she is leading a multi-disciplinary integrative research programme focusing on the reproduction and chemical ecology of coastal and deep-sea invertebrates. Her current projects include the study of the local commercial sea cucumber, *Cucumaria frondosa*. Over the past 15 years, Dr Mercier has published dozens of primary research articles on sea cucumbers and other echinoderms in peer-reviewed journals. She also co-authored five books and over 400 popular articles on wildlife and environmental issues, including the biology and exploitation of sea cucumbers.

PURCELL, Steven – Dr Purcell is an Australian working as an Ecologist with the WorldFish Center in New Caledonia. He is a trained Marine Biologist, with a PhD in quantitative marine ecology from James Cook University (1998). Dr Purcell has worked with WorldFish Center and the SPC since 2001, leading research projects on sea cucumber restocking and aquaculture. His previous work dealt with *Trochus* fisheries enhancement and management, and reef ecosystem dynamics. His published studies on sea cucumbers have spanned topics such as restocking, survival and growth, marking methods, movement, genetic stock structure, pond culture and disease. He currently leads a sea cucumber population assessment and management project in New Caledonia, and is technical specialist for a Philippines-based project on sandfish sea ranching and restocking. Dr Purcell was involved in an ACIAR workshop and preparation of a guidebook "Sea Cucumber Fisheries Management: A Manager's Toolbox", and is the SPC's nominated specialist for contact about sea cucumbers aquaculture.

TORAL-GRANDA, Verónica – Ms Toral-Granda worked at the Charles Darwin Foundation for over 10 years as a fisheries biologist on marine species (i.e. sea cucumbers, spiny lobsters, finfish) subjected to artisanal fisheries. Her research focused on the reproductive biology, natural history, population assessment, and impact evaluation of such fisheries at the single species level. During this time, Ms Toral-Granda also served as a technical advisor to the Galapagos National Park Service and the Ecuadorian Government on marine conservation matters, fisheries management and development of new fishing activities within the Galapagos Marine Reserve. She has been involved in different aspects of the listing of *Isostichopus fuscus* in Appendix III of CITES, the only sea cucumber species to have been included in a CITES appendix, and in 2005 and 2006, she authored the discussion document on sea cucumbers which served as a base in the twenty-second Animal Committee of CITES and in the fourteenth Conference of the Parties respectively to further the analysis of the inclusion of sea cucumbers taxa on a CITES Appendix. She attended the CITES CoP 14 in The Hague as the Ecuadorian marine species specialist with special focus on the sea cucumber item of the agenda. Ms Toral-Granda has published research papers on both peer-reviewed and non peer-reviewed journals on sea cucumber biology and fisheries. Currently she is working as a consultant for FAO on this sea cucumber project, and also served as an expert on the FAO expert consultation "Implementation Issues Related to Listing Commercially-Exploited Aquatic Species on CITES Appendices". Ms Toral-Granda holds a MSc in Conservation Biology (1998) from the University of Cape Town. Her areas of professional interest are marine conservation, participatory fisheries management, specifically small-scale fisheries management and development, and coastal zone management.

UTHICKE, Sven – For the last 14 years Dr Uthicke has worked on the ecology and population genetics of commercially used holothurians. He has completed an honours thesis on the population dynamics of two species and a PhD thesis investigating the importance of these animals in the ecosystem of coral reefs. Since then, he has conducted numerous studies on the population genetics of holothurians, testing how widely populations are interconnected, i.e. whether local or regional management strategies are more important. Other studies were concerned with the recovery of holothurian stocks after over fishing and using genetic methods to clarify the taxonomy of several important holothurian species. In addition to a wide research experience, Dr Uthicke has worked in environmental management and also as a private consultant. He worked for approximately one year each at the German Environmental Protection agency on general water quality related issues, and at the Great Barrier Reef Marine Park Authority (Australia) as project manager Research and Monitoring. In addition to consultancy work for several state agencies and the fisheries industry, Dr Uthicke has also worked as a consultant for the FAO since the Advances in Sea Cucumber Aquaculture and Management workshop in 2003. Dr Uthicke currently works at the Australian Institute of Marine Science on biofilms as indicators for water quality on coral reefs, however, he also keeps a keen interest in sea cucumber ecology, genetics and management.

WOLFF, Matthias – Dr Wolff graduated at Kiel University, Germany in 1981, where he also completed his PhD in 1985, on population dynamics of the Peruvian scallop under the impact of El Niño 1982/83. In 1986 he moved to Chile to engage as a lecturer in Fisheries Science at the University of Coquimbo sponsored by the German Agency for Academic Interchange (DAAD). In 1991 he returned to Germany to work at the Centre for Tropical Marine Ecology (ZMT), Bremen, as a researcher. In 2000 he was appointed Director of research cooperation in marine sciences between the Centre for Tropical Marine Ecology (Bremen) and several Cuban research institutions (Instituto de Oceanología, Universidad de Habana, Centro de Investigaciones Pesqueras) and leader of a first joint cooperative research project on the sustainable management of the Batabanó Gulf, southern Cuba. Since 2004, he has been leader of a working group on data integration and modelling under an EU funded project (CENSOR) on assessing climate variability impacts on the resources and resource users of the Humboldt Current System. In 2007 he has been engaged as Head of the Department of Marine Research and Conservation at the Charles Darwin Research Station in Puerto Ayora, Galapagos. Dr Wolff has eight years of working experience in Latin America (mainly in Peru and Chile). He has several years experience as a referee for several international journals of marine research and is a permanent member of editorial board of the Revista de Biologia Tropical. He has published numerous scientific papers.

LOVATELLI, Alessandro – Mr Lovatelli, a trained marine biologist and aquaculturist, obtained his Bachelor (BSc) and Master of Science (MSc) degrees at the universities of Southampton and Plymouth (United Kingdom), respectively. His first experience with FAO dates back to 1987 working as the bivalve expert attached to an FAO/UNDP regional seafarming development project. His next FAO assignment took him to Mexico working on a regional aquaculture development project which was funded by the Italian Government. From 1993 to 1997 Mr Lovatelli worked in Viet Nam, Somalia and then again in Southeast Asia. In Viet Nam he headed the aquaculture and fisheries component of a large EU project developing, among other activities, 10 regional aquaculture demonstration, training and extension centres. Following this he moved to Somalia working as the lead aquaculture and fisheries consultant for the European Commission. During the 1-year consultancy he provided the Commission technical advice towards assisting the fisheries sector through the EC Rehabilitation Programme for Somalia. Following an additional year in Viet Nam as one of the Team Leaders under the Danish-funded Fisheries Master Plan Project, he was recruited by FAO as the Aquaculture Advisor attached to the FAO-EASTFISH project based in Denmark. In October 2001 Mr Lovatelli once again joined the FAO and took up the post of Fishery Resources Officer (Aquaculture). The main activities currently focused on are marine aquaculture development, transfer of farming technologies, coastal resources management and development of human resources. In October 2003 he organized an international FAO Workshop on the Advances in Sea Cucumber Aquaculture and Management which produced recommendations on sea cucumber fisheries and management as well as aquaculture.

VASCONCELLOS, Marcelo – Dr Vasconcellos is a Fishery Resources Officer at the Food and Agriculture Organization of the United Nations (FAO). Dr Vasconcellos is trained in biological oceanography and holds a PhD in resource management and environmental studies from the University of British Columbia, Canada. Before working for FAO, he worked as a Research Associate at the Department of Oceanography, University of Rio Grande, Brazil, and at the Fisheries Centre, University of British Columbia. One of the central themes of his research has been the development of ecosystem approaches to fisheries assessment and management. At FAO, he is currently responsible for the implementation of trust fund projects related to "Capacity building for an ecosystem approach to fisheries" and to "CITES and commercially-exploited aquatic species".

Annex 4 – **Experts group photograph**

Standing (left to right):
Sven Uthicke, Alex Hearn, Kim Friedman, Ruth Gamboa, Choo Poh-Sze, Jeff Kinch, Matthias Wolff, Annie Mercier, María Dinorah Herrero-Pérezrul, Steven Purcell, Chantal Conand, Marcelo Vasconcellos, Priscilla Martínez.

Seated (left to right):
Jean-François Hamel, Alessandro Lovatelli, Verónica Toral-Granda, Jun Akamine, Eduardo Espinoza.